Handbook of Flowering Plants of Nepal

About the authors

Dr. Krishna Kumar Shrestha, Professor of Botany, is the Former Head of the Central Department of Botany, Tribhuvan University. He had been involved in teaching and research in Plant Systematics, Biodiversity, and Ethnobotany since he obtained his M. Sc. (Botany) degree in 1978 from Tribhuvan University. He has obtained his Ph.D. degree in Plant Systematics from the Komarov Botanical Institute, St. Petersburg, Russia in 1993, and deputed as the Post Doctoral Darwin Fellow at the Natural History Museum, London during 1997-1999, especially to work on the 'Repatriation of Plant Information for Flora of Nepal'. He has coordinated more than 25 research projects on plant biodiversity and ethnomedicinal plants, published seven books, 52 research papers in peer-reviewed international journals, 33 papers in national journals, and over 30 popular articles in the issues of biodiversity and ethnobotany. Prof. Shrestha was involved as a Biodiversity Expert in the Kangchenjunga Landscape Conservation and Development Initiative Project, at the Research Centre for Applied Science and Technology (RECAST), TU. Prof. Shrestha is one of the Editors of Flora of Nepal, an international initiative of the institutions of Nepal, Japan, and U.K. He is also contributing as the Advisory Board Member of Flora of Pan Himalaya, lead by Institute of Botany, Chinese Academy of Sciences, Beijing, China.

Dr. Shandesh Bhattarai obtained M.Sc. in Botany (Plant Systematics and Biogeography) in 2002, and awarded Ph.D. degree in Botany from the Tribhuvan University in 2010. Specialized in the Himalayan Ethnobotany, Dr. Bhattarai was awarded by the Science and Technology Youth Award 2010 for his contribution on "scientific study of the traditional knowledge of Himalayan herbs". Dr. Bhattarai has been working as a Senior Scientific Officer in the Nepal Academy of Science and Technology and involved in the Nepal Flora and Darwin Initiative project for the last seven years. He has received training on flora writing and its database management from the Royal Botanic Garden Edinburgh-UK in 2013 and completed some revisionary and Flora writing. In his credit, he has over two dozen papers published in the national and international journals. He worked in the international and national collaborating projects such as NUFU-Norway, VolksWagen Foundation-Germany, DFID-UK, IUCN Bangkok, etc.

Prabin Bhandari obtained M.Sc. Degree in Botany (specialization in Plant Biodiversity and Systematics), from the Tribhuvan University, and secured first position, in 2015. For his M.Sc. dissertation, he had worked on 'Flora of Panchase Protected Forest, Kaski District, Central Nepal', and published three research papers in the international journals and one in the national journal. He is currently working as a Biodiversity Researcher in Kangchenjunga Landscape and Conservation & Development Initiative (KLCDI) since 2015. Similarly, he has been awarded a research grant by Rufford Foundation to work on Orchids in Panchase Protected Forest, Central Nepal. Mr. Bhandari's fields of interest are Biodiversity, Plant Taxonomy, Ethnobotany, and Conservation Biology.

Handbook of Flowering Plants of Nepal

Volume 1. Gymnosperms and Angiosperms: Cycadaceae - Betulaceae

Krishna Kumar Shrestha
Shandesh Bhattarai
Prabin Bhandari

SCIENTIFIC PUBLISHERS

Published by:
Scientific Publishers
5A, New Pali Road, P.O. Box 91
Jodhpur 342 001, India
E-mail: info@scientificpub.com
Website: www.scientificpub.com

Cover page: *Mahonia napaulensis* DC. (Berberidaceae)
Photo credit: Neera Joshi Pradhan
Photos inside back cover page: *Alnus nepalensis* (Betulaceae), *Butea monosperma* (Fabaceae), *Caltha palustris* (Ranunculaceae), *Coelogyne nitida* (Orchidaceae), *Dactylicapnos macrocapnos* (Papaveraceae), *Ficus hispida* (Moraceae), *Fritillaria cirrhosa* (Liliaceae), *Magnolia kisopa* (Magnoliaceae), *Miscanthus nepalensis* (Poaceae), *Nelumbo nucifera* (Nelumbonaceae), *Pinus wallichiana* (Pinaceae), and *Saxifraga parnassifolia* (Saxifragaceae)

Photo credit: Krishna K. Shrestha
Inside cover page: Map of Nepal
Map credit: Dibas Shrestha

ISBN: 978-93-87307-10-0 (Hardbound)
ISBN: 978-93-87991-73-6 (E-book)

Citation: Shrestha KK, Bhattarai S, Bhandari P (2018) Handbook of Flowering Plants of Nepal (Volume 1. Gymnosperms and Angiosperms: Cycadaceae – Betulaceae), Scientific Publishers, Jodhpur, India

Visit the Scientific Publishers (India) website at
http://www.scientificpub.com
Printed in India

FOREWORD

Nepal has a tremendous diversity of plants and the Nepalese people have a long history of using them to support their everyday livelihoods. Plants provide for our basic needs of food, shelter and heat, but also spices, oils, fibres, medicines, and much more. Plants enrich our lives with their beauty, and a great many have an indispensible role in religious ceremonies in Nepal. These plants, and the habitats in which they grow, also provide ecosystem services to a wide area of Asia, impacting on millions of lives in surrounding countries and beyond. Nepal is a major component of the Himalayan biodiversity hotspot, and so the ca. 7000 species of higher plants thought to grow within its borders are important locally, nationally and globally.

With growing pressures from habitat destruction, climate change, invasive plants, and over exploitation, it is more important that ever to conserve the Nepalese flora, use plants sustainably, and ensure that they will be available for future generations. However, if you don't know about the plants you have, what their names are, what are their characteristics, and where are they found, then it is difficult to conserve them effectively. Such knowledge is encapsulated within a Flora - a comprehensive documentation of plants in an area, together with tools to aid identification (usually keys, but also illustrations and photographs). Nepal is one of the last countries in the northern hemisphere yet to produce a completed Flora, although work on this is well underway with the *Flora of Nepal* (www.floraofnepal.org).

Producing a Flora is a long term, nation-building project. It is not an easy task and requires significant resources. It takes many years to generate the comprehensive knowledge base, which taxonomic botanists need to conduct their research. Building the in-country capacity to undertake these studies is also vitally important, as is developing the international networks of partners to share knowledge and experience. No one country has expertise in all plant groups, and so collaborations need to be made with experts working in other countries to produce the most up to date, authoritative accounts of these plants.

Botanists began collecting scientific samples (herbarium specimens) in Nepal in 1802, but the most extensive fieldwork has been since the early 1950s, when Nepal opened its doors to foreign visitors. These collections enabled the first florisitic works to be published - checklists of plants known to occur in Nepal, with basic information on distribution. Botanical exploration continues and such publications go out of date almost as soon as they are printed. Species new to science are regularly discovered, new records are found for plants previously known from other countries, major extensions are made to distribution and altitude ranges, and new taxonomic revisions adjust the species limits and generic placements of previously known species. These advances in knowledge sometimes result in changes of scientific names, well-known names being placed into synonymy and new species names being proposed.

This new *Handbook of Flowering Plants of Nepal* updates previous checklists with these kinds of new information, and incorporates the latest classifications put forward by the international taxonomic community. It also extends our knowledge on Nepalese plants by including data on vernacular names, uses and life forms. This will be very helpful for those needing to use an updated classification and nomenclature of Nepalese plants, and for applied researchers looking for information on how plants are used. Students of biodiversity will find a great deal of useful information on Nepal, its plants and the history of botanical exploration in the detailed Introduction. Taxonomy students are particularly well served by the incorporation of the APG definitions of plant families and their relationships based on molecular studies of DNA. This book represents a major stepping-stone on the pathway to completing the *Flora of Nepal*, and is an indispensible resource for anyone working on Nepalese plants.

September 13, 2017

Dr Mark F. Watson
Editor-in-Chief, Flora of Nepal
Royal Botanic Garden Edinburgh, UK

PREFACE

Globally, the number of species belonging to Plant Kingdom, Animal Kingdom (except invertebrates), and Kingdom Mycota (Fungi) is fairly catalogued, whereas information on the Kingdom Monera (Bacteria) and Protoctista (Algae, Diatoms, and Protozoa) are largely unknown, except Green Algae. In general, documentation of biodiversity resources is either incomplete or data deficient, with many doubtful or uncertain species. Nepal is considered as one of the biodiversity rich countries in the world on the basis of number of species versus area of the country. It has been realized that integrated and updated information on plants, animals and microoorganisms of Nepal should be available to the wide range of stakeholders including students, researchers, policy makers, and planners, especially for the sustainable use of the biological resources and conservation of rare, threatened and endemic species.

The 'Flora of Nepal' is a dream project for Nepal, which has been initiated by the Department of Plant Resources (then Department of Medicinal Plants), Government of Nepal since the early 1960s. '*Prodromus Florae Nepalensis*' (Don 1825) was the first publication on Flora of Nepal, based on the plant collections by Buchanan Hamilton (1802-1803), and Nathalien Wallich (1820-1821). With the collaboration of the Natural History Museum, London and University of Tokyo, Japan published 'An Enumeration of the Flowering Plants of Nepal: Vols. 1-3 (Hara *et al* 1978, Hara and Williams 1979, and Hara *et al* 1982), followed by the publication of 'Name List of Flowering Plants of Nepal' (Koba *et al* 1994), and 'Annotated Checklist of the Flowering Plants of Nepal' (Press *et al* 2000). The updated version of the Annotated checklist was published on-line (www.efloras.org) in 2004 with the support of Missouri Botanical Garden, USA.

The Department of Plant Resources published 'Catalogue of Nepalese Vascular plants (1976), which has later been updated (Rajbhandari and Baral 2010, Rajbhandari *et al* 2011, 2012, 2015). Since 2000, many new species and new records to Nepal Flora, as well as significant nomenclatural changes have been noticed. Moreover, the classification system adopted by the Enumeration (Bentham and Hooker system, 1867-1892), and Flora of Nepal Vol. 3 (Engler and Diels 1911) has been revised and replaced in recent Flora and taxonomic treatments by the Angiosperm Phylogeny Group (APG III and IV).

The sequences of the orders and families in the present book has been arranged according to APG IV system (2016). In addition, the new species and new records reported from Nepal, as well as recent nomenclatural changes have been adopted. Moreover, we have incorporated valid synonyms, Local Nepali names (in Devanagari), English/Common names, life form (plant habit), elevation ranges, and use value of majority of plants in the Volume 1. Gymnosperms and Angiosperms: Cycadaceae – Betulaceae, and upcoming volume II. Angiosperms: Coriariaceae – Apiaceae which is in progress, will be published soon.

We are grateful to Neera Joshi Pradhan (Studio Petals, Khumaltar, Lalitpur) and Priyanka Malla (London, UK) for their contribution in colour drawing of cover page and cover page design accordingly. Similarly, Dibas Shrestha (Central Department of Hydrology & Meteorology, TU) is acknowledged for preparing the map of Nepal, and Buda Dangol (Manchester, UK) for editing the language. The authors sincerely acknowledge the support of our colleagues Narendra Nath Tiwari (Ayurveda Campus, Institute of Medicine, TU), Mohan Siwakoti, Sangeeta Rajbhandary and Suresh Kumar Ghimire (Central Department of Botany, TU), Ila Shrestha (Patan Multiple Campus, Lalitpur), Jyoti Prasad Gajurel and Bishnu Rijal (former students, Central Department of Botany, TU), and Keshab Raj Rajbhandari, Sanjeev Kumar Rai, Ramesh Basnet and Subhash Khatri (Department of Plant Resources, MoFSC/GoN) for their cooperation in preparing the book.

The first author greatly acknowledges the contribution of Steve Blackmore, JR Press and DA Sutton (The Natural History Museum, London, UK), Peter Raven (Missouri Botanical Garden, St.Louis, USA) and Mike Gilbert (Royal Botanic Gardens, Kew, UK) for their extensive support during the publication of previous version of the checklist 'Annotated Checklist of the Flowering Plants of Nepal', and online version of the checklist (www.efloras.org) respectively. These resources are considered as the fundamental source of information while publishing this book.

Our special thanks goes to Mark F. Watson, Colin A. Pendry, Bhaskar Adhikari (Royal Botanic Garden Edinburgh, UK), and Ram Prasad Chaudhary (Research Centre for Applied Science & Technology, TU) for their critical comments, valuable suggestions and feedback to update the book. The authors also extend their sincere gratitude to the authorities of Tribhuvan University (TU), Nepal Academy of Science and Technology (NAST), Department of Plant Resources (DPR), Royal Botanic Garden Edinburgh, UK and University of Tokyo, Japan for their support and cooperation. Finally, we are thankful to the Scientific Publishers, Jodhpur, India for publishing the book.

The Authors

CONTENTS

Flowering plants of Nepal: Representative species

1. *Pinus wallichiana*

2. *Magnolia kisopa*

3. *Fritillaria cirrhosa*

4. *Coelogyne nitida*

5. *Miscanthus nepalensis*

6. *Dactylicapnos macrocapnos*

Flowering plants of Nepal: Representative species

7. *Caltha palustris*

8. *Nelumbo nucifera*

9. *Saxifraga parnassifolia*

10. *Butea monosperma*

11. *Ficus hispida*

12. *Alnus nepalensis*

FORMAT

The previous version of Enumeration (Hara *et al* 1978, Hara and Williams 1979, and Hara *et al* 1982) and Checklist (Press *et al* 2000) have been updated and modified based on new addition of species to Nepal Flora, nomenclatural changes, and additional information on Local names, English/Common names, Use Values, IUCN Red List Categories, and CITES Appendices. The Latin (Scientific) names in bold letters are accepted names, following Flora of Nepal (2011), Flora of China (1994-2013) and Flora of Bhutan (1983-2002). Online databases such as TROPICOS (www.tropicos.org), Global Biodiversity Information Facility (www.gbif.org), The Plant List (www.theplantlist.org), as well as Catalogue of Life (www.catalogueoflife.org), and International Plant Name Index (www.ipni.org) are also consulted, wherever necessary.

Distribution for species within Nepal is based on herbarium specimens and Flora books, while Flora of Nepal Vol. 3 (Watson *et al* 2011) is followed for distribution format of taxa. Elevation ranges are based on the herbarium specimens and Flora books. The Use and parts used are based on published literature; and IUCN Red List Categories, and CITES Appendices are based on the respective websites.

Sample 1. Description of flowering plants

CUPRESSALES [Order]

TAXACEAE [Family]

Taxus L. [**Genus**: Generic name with authority]

Taxus wallichiana Zucc. [**Species**: Latin name with authority) [4, 5, 15, 19]

Taxus baccata subsp. *wallichiana* (Zucc.) Pilg. (1, 71); *Taxus yunnanensis* W.C.Cheng & L.K.Fu [**Synonyms**]

Literature citation [1 = Enumeration Vol. 1; 4 = Checklist, 2000; 5= Catalogue, 2009, etc…..]

लौठ सल्ला, बर्मे सल्ला, ढेंग्रे सल्ला (Lauth sallaa, Barme sallaa, Dhengre sallaa); East Himalayan Yew [**Nepali/ English name**]

Medium-sized evergreen trees. C, E Nepal: 1900-3300(-3500) m. Nepal, E Himalaya, Tibetan Plateau, Assam-Burma and E Asia. [**Life form/Distribution in Nepal/Elevation ranges/General distribution**]

Use(s)/Parts used: Timber, Medicine (bark, leaves), Food (fruit), Fodder.

IUCN Category: **Endangered** A2acd (ver 3.1)

CITES Appendix: II.

Sample 2. Description of Endemic plants

FABACEAE (LEGUMINOSAE)

Oxytropis DC.

Oxytropis nepalensis Vass., *Nov. Syst. Pl. Vasc.* **14**: 175. 1977. [2, 4, 8, 11, 12, 27, 29]

[**Species name/ Protologue** (original publication citation)]

Perennial stemless herbs. Nepal: 3500-4500 m. (**Endemic to Nepal**). [**Life form/ Elevation/ Distribution**]

Central: *Mustang*, Tukucha, Kali Gandaki 13500 ft. (4110 m); JDA Stainton, WR Sykes and LHJ Williams 1135 (Holotype: BM). [**Type specimen**]

■■■■

Distribution Format

Distribution for species and infraspecific taxa occurring outside Nepal is indicated by a list of geographical regions, which are defined according to political borders, with countries or provinces grouped to form regions underlying biogeographic basis. The names used for the regions are intended to be descriptive and non-political (Watson *et al* 2011). The present book has also adopted the distribution pattern as follows:

W Himalaya	NW India (Jammu & Kashmir [including Siachen Glacier, regions disputed with Pakistan], Himachal Pradesh, Uttarakhand), N Pakistan (Khyber Pakhtunkhwa [previously known as North West Frontier Province], Gilgit-Baltistan [also known as Northern Areas, region disputed with India], Azad Kashmir [region disputed with India]).
E Himalaya	N India (Sikkim & Darjeeling), Bhutan, NE India (Arunachal Pradesh [disputed border with S Tibet, China]).
Tibetan Plateau	China (Xizang [including Aksai Chin and Shaksgam Valley regions disputed with India], Qinghai).
Assam-Burma	NE India (Assam, Nagaland & Manipur), Myanmar.
S Asia	E Pakistan (Punjab, Sind, Islamabad), Peninsular India, Sri Lanka, Bangladesh, Maldives.
E Asia	China (excluding Xizang, Xinjiang, Qinghai), N & S Korea, Japan, Taiwan.
SE Asia	Thailand, Laos, Cambodia, Vietnam, Malaysia, Indonesia, Philippines, New Guinea.
N Asia	China (Xinjiang), Russia, Mongolia.
C Asia	Kazakhstan, Uzbekistan, Turkmenistan, Tajikistan, Kyrgyzstan.
SW Asia	Afghanistan, W Pakistan (Baluchistan, Federally Administered Tribal Areas), Iran, Middle East, Arabian Peninsula, Turkey, Azerbaijan, Armenia, Georgia.

Asia	Collective term for all above areas of Asia.
Europe	includes Ukraine, Belarus, Baltic republics.
Africa	includes Madagascar.
N America	includes C America south to Panama.
S America	south of Panama.
Australasia	Australia, New Zealand, Pacific Islands.
Cosmopolitan	collective term for a generally worldwide distribution.

Source: Watson *et al* 2011 Flora of Nepal Vol. 3, Royal Botanic Garden Edinburgh, UK

■■■■

Abbreviations and Acronyms

AD	Anno Domini
APG	Angiosperm Phylogeny Group
BM	Herbarium, Natural History Museum (British Museum), London
BPP	Biodiversity Profiles Project
C	Central
ca.	circa (about, approximately)
CAL	Central National Herbarium, Calcutta
Cat. no.	Catalogue number
CBD	Convention on Biological Diversity
CDBTU	Central Department of Botany, Tribhuvan University
CE	Critically Endangered
CITES	Convention on International Trade in Endangered Species of wild fauna and flora
DD	Herbarium, Forest Research Institute, Dehradun
DD	Data Deficient
DPR	Department of Plant Resources
E	Herbarium, Royal Botanic Garden Edinburgh
E	East, Eastern
ed.	Edited
EN	Endangered
et al.	et alia – and others
EW	Extinct in the Wild
Fig.	Figure
ft.	Feet

G	Herbarium, Conservatoire et Jardin Botaniques de la Ville de Geneve
G-DC	Geneva – De Candolle's Herbarium
GEF	Global Environment Facility
GH	Gray Herbarium, Harvard University Herbarium, Cambridge, USA
GRM	Muséum d'Histoire Naturelle de Grenoble
ICIMOD	International Centre for Integrated Mountain Development
IUCN	International Union for Conservation of Nature
K	Herbarium, Royal Botanical Gardens Kew
KATH	National Herbarium and Plant Laboratories (NHPL)
KUN	Herbarium, Kunming Institute of Botany, Chinese Academy of Sciences, Kunming, China
K-W	Herbarium, Royal Botanical Gardens Kew-Wallich's collections
KYO	Herbarium, Kyoto University
LC	Least Concern
LE	Herbarium, Komarov Botanical Institute, Saint Petersburg, Russia
LINN	Herbarium of Linnaean Society of London, UK
m	Meter
MENRIS	Mountain Environment and Natural Resources Information System
MoFSC	Ministry of Forest and Soil Conservation
N	North, Northern
NA	Not Available (Data Deficient)
NAST	Nepal Academy of Science and Technology
NT	Near Threatened
P	Herbarium, Museum National d'Histoire Naturelle Paris
PE	Herbarium, Institute of Botany, Chinese Academy of Sciences, Beijing, China
RBG Kew	Royal Botanic Gardens Kew
Rep. ed.	Reprinted edition
SE	South East, South-Eastern
s.n.	sine nomine (without collection number)

SW	South West, South-Western
TI	Herbarium, University Museum, University of Tokyo, Tokyo, Japan
TNS	National Museum of Nature and Science, Tsukuba, Japan
TUCH	Tribhuvan University Central Herbarium
UK	United Kingdom
US	Herbarium, Smithsonian Institution Washington DC, United States
USA	United States of America
Vol.	Volume
VU	Vulnerable
W	West, Western
WCMC	World Conservation Monitoring Centre
?	Question
%	Percentage
*	Asterisk
~	Approximately
<	less than
>	greater than
0	Degree

CHAPTER 1

INTRODUCTION

INTRODUCTION

Origin of the earth, evolution of life on earth, diversity of living organisms, biodiversity richness, and documentation of flora and fauna of the countries are some key concern issues in understanding the global biodiversity. Biological diversity or Biodiversity refers to the variety of life such as plants, animals and microorganisms on the planet (*species diversity*), genetic differences within the species (*genetic diversity*), and the variety of ecosystems where they exist (*ecosystem diversity*). The Earth's biological resources are linked with human civilization, which has been realized as one of the great assets for present and future generations, and vital to the economic and social development of the world. Estimating how many biological species occurs, and what is the status of species, are essential steps in setting conservation priorities for rare and threatened species.

1.1 Origin and diversity of living organisms

It is believed that the earth is nearly 4.6 billion years old, and origin of life in the earth started during 3.5-4 billion years ago. The period of Earth's formation is known as the Pre-Cambrian Period. Earth's initial life forms were bacteria, which could survive in the highly toxic atmosphere. According to the existing fossil records, it is evident that by the end of the Archean Period and beginning of Proterozoic Period, blue-green algae were believed to be the earliest forms of plants, which originated 2.5 billion years ago (www.ecology.com). Moreover, it is believed that evolution of all forms of life including animals and human beings have originated from primitive plants. The Cambrian Period began about 525 million years ago, during this period almost all of the major groups of plants and animals developed.

Tracing the history of living organisms, the Paleozoic era (300-550 million years ago) is important due to the existence of plants, animals, microorganisms, and occurrence of extensive forests throughout the world (Groombridge and Jenkins 2002). About 2.2 million years ago, the first human species *Homo habilis* emerged (www.ecology.com), whereas the modern human being (*Homo sapiens*) evolved in Africa about 200,000 years ago (www.humanorigins.si.edu).

The Earth may comprises 5 to 30 million species, although some estimate range as low as 3 million to 100 million. Despite significant contributions of scientists in exploring and documenting the Earth's living resources since last 300 years, only 1.78 million species or 12 percent of the biodiversity have been reported (Govaerts 2001, Prance 2001, Groombridge and Jenkins 2002, Mora *et al* 2011). So far nearly 5,000 species belonging to the *Kingdom Monera*, the prokaryotic living organisms (Bacteria and Archaea) have been described; however, the estimates of prokaryotes may range from 50,000 species to hundreds of millions. Within the microorganisms, 4000 species of viruses, 10,000 species of bacteria are listed so far. In the *Kingdom Protoctista* or *Protista* (algae, diatoms and protozoa) 80,000 species have been described, and the estimate may range up to 250,000 species. Similarly, the Kingdom *Mycota* (Fungi) comprises 72,000 described species, with estimate of up to 1,500,000 species (Ziegler 2007).

Carl Linnaeus (1707-1778), a Swedish Botanist, had estimated that there were nearly 10,000 species of plants in the world, of which he had described nearly 6,000 species. In the tenth edition of *Systema Naturae*, published in 1758, Linnaeus listed around 9,000 species of plants and animals (Ziegler 2007). At present, about 13,000-18,000 new species of living organisms are being described each year (Jeffries 2005, Connor 2009). About 75 percent of the new species described in 2007 were invertebrates, 11 percent vascular plants and nearly 7 percent were vertebrates (Connor 2009). Number of vascular plants reported as new to science in 2015, as indexed in International Plant Names Index, were 2,034 (RBG Kew 2016).

Nevertheless, thousands of species might have disappeared or in the process of extinction, before understanding their existence in the earth. It is believed that nearly 2 million species have already become extinct (Ziegler 2007), and probably over 5 billion species (nearly 99.9 percent), that ever existed on the Earth became extinct due to rapid environmental changes (Kunin and Lawton 1996).

It is estimated that *ca* 325,000 species of green plants in the *Plant Kingdom* (*Plantae*), including 250,000 species of flowering plants, have been discovered so far, and the estimated number may go up to 500,000 species. The *Animal Kingdom* is the largest kingdom of living organisms, so far 1.32 million species of animals, ranging from microscopic creatures to insects, amphibians, reptiles, birds and mammals have been documented, of which nearly 90 percent of species belongs to insects (Frodin 2001, Zeigler 2007, Chapman 2009, Mora *et al* 2011). Furthermore, recently it has been estimated that 390,900 species of vascular plants are known to science, of which number of flowering plants known to science are 369,400 (RBG Kew 2016).

It is evident that most of higher plants and animals are well documented, for example; only 10 percent of flowering plants and 10 percent of vertebrates remain to be described. On the other hand, it has been believed that 95 percent of fungi and 95 percent of protozoa are yet to be discovered (Groombridge and

Jenkins 2002, Ziegler 2007). Scientists reckon that the number of species existing in the earth is far more than what has been discovered so far. For example, there may be 50,000-1 million species of viruses, 10,000-3 million species of bacteria, 600,000-1.2 million species of Protoctista (algae, diatoms and protozoa), 200,000-2.7 million fungi, 300,000-500,000 species of green plants (flowering and non-flowering plants), and up to 10.6 million species of animals (Groombridge and Jenkins 2002, Jeffries 2005, Zeigler 2007).

1.2 Plant Diversity

Regarding the diversity of plants 268,600 species of flowering plants, 1,021 species of gymnosperms, 12,000 species of ferns and allies, 16,236 species of bryophytes, 98,998 species of fungi, 17,000 species of lichens, 1,899,587 species of microorganisms and related groups have been identified (Table 1). It is estimated that 10 percent of flowering plants, whereas up to 95 percent of the protozoa and more than 95 percent of fungi are yet to be identified (Groombridge and Jenkins 2002, Ziegler 2007).

Table 1. Described and estimated number of species in Fungi and Plants (Chapman 2009)

Category	Group(s)	Subgroup	Described species	Estimated species
Plants			310,129	~390,800
	Flowering plants		~268,600	352,000
	Gymnosperms		1021	~1,050
	Pteridophytes (ferns and allies)		~12,000	~15,000
	Bryophytes		~16,236	~22,750
		Liverworts	~5,000	~7,500
		Mosses	~11,000	~15,000
		Hornworts	236	~250
Fungi			98,998	1,500,00
		Ascomycota	64,056	
		Basidiomycota	31,503	
		Chytridiomycota	706	
		Microsporidia	(>1,300)	
		Zygomycota	1,065	
Lichens			17,000	~25,000
Others			1,899,587	~11,327,630
	Prokaryota (bacteria)		7,643	400,000–1,000,000

Category	Group(s)	Subgroup	Described species	Estimated species
	Cyanophyta (cyanobacteria)		2,664	NA
	Chromista (chromistan 'algae', chromistan 'fungi')		~25,044	~200,500
	Viruses		2,085	400,000
	Protoctista (mainly protozoa — others included under fungi, algae, chromista and diatoms)		~28,871	>1,000,000

1.3 Diversity of Flowering plants

Estimates for the number of described species of flowering plants in the world vary from 223,300 (Scotland and Wortley 2003), 258,650 (Thorne 2002, IUCN 2004), about 270,000 (Groombridge and Jenkins 2002, Chapman 2009), 315,903 (Kier *et al* 2009) and up to 369,400 species (RBG Kew 2016).

The number of families in angiosperms are 416 (APG IV 2016), whereas APG I (1998) enlisted 462 families in the flowering plants. According to latest literature (Byng 2014), the top ten largest families of flowering plants are Orchidaceae (25,000 species), Asteraceae (24,000 species), Fabaceae (19,500 species), Rubiaceae (13,000 species), Poaceae (11,500 species), Lamiaceae (7,530 species), Euphorbiaceae (5,750 species), Cyperaceae (5,400 species), Apocynaceae (5,100 species), and Melastomataceae (5,000 species).

Similarly, the top 10 largest genera of flowering plants in the world (Byng 2014) are: *Astragalus* (Fabaceae: 2,400-3,330 species), *Bulbophyllum* (Orchidaceae: 2,200 species), *Carex* (Cyperaceae: 2,000 species), *Euphorbia* (Euphorbiaceae: 1,930 species), *Begonia* (Begoniaceae: 1,590 species), *Dendrobium* and *Epidendrum* (Orchidaceae: 1,500 species each), *Psychotria* (Rubiaceae: 1,400 species), *Solanum* (Solanaceae: 1,300-1,400 species), and *Croton* (Euphorbiaceae: 1,185 species).

1.4 Species Richness and Flora of the World

Distribution of living organisms is not uniform globally. Due to warm climate and high primary productivity, the tropical regions harbour more species and thus rich in biodiversity, in comparison to temperate regions of the world, which comprises comparatively less number of species. Furthermore, the Antarctica hosts limited number of living organisms, due to dry habitat and permanent snow cover. It is noteworthy that although tropical forest ecosystem comprises less than 10 percent of the earth's surface, 90 percent of world's species occur here.

In terms of species richness, the South America is the richest continent having 85,000 species of flowering plants (nearly one third species of the world), followed by Asia (50,000 species), and Africa (40,000-45,000 species). Similarly, North America comprises 20,000 species, Australia with 18,000 species, and Europe with 12,500 species. The seventh continent Antarctica, which is dry and covered with permanent snow, includes 300-400 species of algae, 250 species of lichens, 125 species of bryophytes, 20 species of macrofungi, and only two species of flowering plants (British Antarctic Survey, www.bas.ac.uk, Shrestha 2016).

The top ten countries in the world with highest number of vascular plants are as follows: Brazil (56,215 species), Colombia (51,220 species), China (32,200 species), Indonesia (29,375 species), Mexico (26,071 species), South Africa (23,420 species), Venezuela (21,073 species), USA (19,473 species), Ecuador (19,362 species) and India (18,664 species). Similarly, top ten plant diversity rich countries in different continents are as follows (Groombridge and Jenkins 2002; Shrestha 2016):

South America: Brazil (56,215 species), Colombia (51,220 species), Venezuela (21,073 species), Ecuador (19,362 species), Bolivia (17,367 species), Peru (17,144 species), Argentina (9,372 species), Paraguay (7,851 species), Guyana (6,409 species), and Chile (5,284 species).

North America and Central America: Mexico (26,071 species), United States (19,473 species), Costa Rica (12,119 species), Panama (9,915 species), Guatemala (8,681 species), Nicaragua (7,590 species), Cuba (6,522 species), Hondurus (5,680 species), Dominican Republic (5,657 species), Haiti (5,242 species), and Canada (3,270 species).

Asia: China (32,200 species), Indonesia (29,375 species), India (18,664 species), Malaysia (15,500 species), Thailand (11,625 species), Vietnam (10,500 species), Philippines (8,931 species), Iran (8,000 species), Myanmar (7,000 species), and Nepal (6,973 species).

Africa and Madagascar: South Africa (23,420 species), Democratic Republic of Congo (11,007 species), Tanzania (10,008), Madagascar (9,505 species), Cameroon (8,260 species), Gabon (6,651 species), Ethiopia (6,603 species), Kenya (6,506 species), Mozambique (5,692 species), Angola (5,185 species), and Uganda (4,900 species).

Europe and Eurasia: Russia (11,400 species), Italy (5,599 species), Ukraine (5,100 species), Portugal (5,050 species), Spain (5,050 species), Greece (4,992 species), France (4,630 species), Croatia (4,288 species), Austria (3,100 species), and Switzerland (3,030 species).

Oceania (Australasia): Australia (15,638 species), Papua New Guinea (11,544 species), New Caledonia (3,250), Solomon island (3,172), New Zealand (2,382 species), Fiji (1,518), Vanuatu (870 species), W. Samoa (737 species), Tonga (463 species), and Guam (330 species).

In South Asia, India has higher number of species (18,664 species), followed by Nepal (6,973 species), Bhutan (5,468 species), Bangladesh (5,000 species), Pakistan (4,950 species), Afghanistan (4,000 species), Sri Lanka (3,314 species), and Maldives (260 species).

1.5 Physiography and Biodiversity in Nepal

Nepal occupies only 0.1 percent of the earth surface area; however, it harbours 3.2 percent of world's flora (MoFSC 2014a). Nepal is one of the biodiversity-rich countries, stands 27th position in the world and 10th position in Asia and second position in South Asia (Shrestha 2016). One of the reasons for being biodiversity rich countries is due to the extreme altitudinal gradient of Nepal's topography, which allows the occurrence of a broad range of climatic condition from tropical in the lowland Tarai to nival zone in High Himalaya. Using floristic and phytogeographical data, Stearn (1960) divided Nepal into three regions (Western, Central and Eastern), which is equivalent to Karnali, Gandaki, and Koshi sections of Nepal.

Western Nepal: from Kumaon frontier to 83ºE (Karnali basin)

Central Nepal: from 83ºE to 86º30 E (Gandaki basin), and

Eastern Nepal: from 86º30 E to Sikkim frontier (Koshi basin)

For the convenience, most of the publications on Flora of Nepal adopted Stearn's classification such as 'Enumeration of flowering plants of Nepal (1978-1982)', 'Annotated Checklist of the Flowering Plants of Nepal (2000)', 'Catalogue of Nepalese Flowering Plants (2010-2012, 2015)', and the present book 'Handbook of Flowering Plants of Nepal Volume 1 (2018)'.

1.6 Major Botanical explorations in Nepal

Explorations, documentations and publications on biological diversity of Nepal were not initiated till the 18th century. The first initiation in botanical exploration was made by the British Naturalist and a Scottish Physician Sir Francis Buchanan-Hamilton (1756-1829), who visited Nepal from the southern part of the country via Bhimphedi, Chitlang to Kathmandu during March 1802. He stayed in Kathmandu Valley for nearly 14 months and explored the plants of Kathmandu Valley (Press and Shrestha 1999). He had collected nearly 1100 species of plants, of which 800 species are believed to be new to science. Later, Nathaniel Wallich (1786-1854), a Danish Physician, also visited Kathmandu Valley during 1820-1821 following the same route, further explored the plants of Kathmandu Valley and its vicinities, and documented the plants from northern parts of the country, up to Gosainkunda region. As a result, he had discovered hundreds of plant species new to the world. Detailed account of the botanical explorations in Nepal is given in the History of Botanical explorations in Nepal: 1802-2015 (Rajbhandari 2016).

With the special permission from the Government of Nepal, Sir Joseph Dalton Hooker, one of greatest British botanists and explorers of 19th century, travelled to Olangchung Gola, Taplejung, via Nepal-India border in the east, during 1848. He had explored the plants of easternmost parts of Nepal and discovered several species new to science, especially the Rhododendrons. Exploration in the western parts of Nepal was initiated by John F. Duthie, who visited Darchula and adjoining regions of Far-west Nepal during 1884. Botanical exploration of Central and Western Nepal was set up by Captain Lal Dhwoj and Khdanada Sharma (KN Sharma), two amateur Nepali collectors during 1928-1934, primarily for the British naturalists. Although limited botanical explorations were carried out till 1950, especially by the British naturalists, the activities were accelerated since the 1950's, due to the beginning of a new era with the dawn of democracy in Nepal. Since then additional British explorers, especially JDA Stainton, Oleg Polunin, LHJ Williams, WR Sykes, visited Nepal from 1949 until 1975 and extensively collected flowering plants of Nepal from east to west and north to southern parts of Nepal. The herbarium specimens, estimated to be nearly 100,000 specimens, are deposited in the British Herbaria namely, Natural History Museum London (BM), Royal Botanical Garden Kew (K, K-W), Royal Botanic Garden Edinburgh (E), Herbarium of Linnaean Society of London (LINN), and other small herbaria of the United Kingdom.

During early 1950's to 1970's Japanese botanists explored the plant resources of Nepal, and collected thousands of plant specimens from Central Nepal (Kathmandu–Trishuli–Dhading–Gorkha–Manang), and later to Eastern Nepal (from Kathmandu to Taplejung). Their collections are deposited in the Herbaria of University Museum, University of Tokyo (TI), Kyoto University (KYO) and several other herbaria of Japan. Several botanists from USA, France, Switzerland, India, Russia, and many other countries had visited and collected thousands of herbarium specimens from different physiographic zones of Nepal and deposited in reputed international herbaria (Rajbhandari 2016).

The botanical explorations made during 1802-1950 were not in large scale, only a few hundreds of specimens were collected and deposited in the British herbaria. Later extensive collection of plant specimens from west to east Nepal, and from south to north part of Nepal was initiated by professional naturalists and botanists, including Mohan Lal Banerji from India (1948-1967), especially the plants of East Nepal; Oleg Polunin (1949-1950), JDA Stainton (1954-1969), Hiroshi Hara (1962-1972), Hiroo Kanai (1963-1983), Dan Nicolson (1966-1967), Jean Francois Dobremez (1969-1975), Christopher Grey-Wilson (1973-1981), Mitsuo Suzuki (1983-2010) and Mark Watson (2003-2014). Their collections are deposited in the world leading Herbaria of the United Kingdom (BM, E, K), Japan (TI, KYO, TNS), USA (GH, US), France (P, GRM), Switzerland (G, G-DC), India (CAL, DD), Russia (LE) and China (PE, KUN).

CONTRIBUTION OF NEPALI BOTANISTS

The contribution of Nepali botanists, in particular by the staff of Department of Plant Resources (Ministry of Forest and Soil Conservation) are significant in exploring and documenting plants of Nepal. Some of the notable plant collectors are Tirtha B Shrestha, Puspa R Shakya, Samar B Malla, Narayan P Manandhar, Damodar P Joshi, Prabha Pradhan, Vidya L Gurung, Madhusudan Bista, Lokendra R Sharma, Keshab R Rajbhandari, Mahesh K Adhikari, Mahendra N Subedi and Kuber J Malla. Their collections are deposited in the National

Herbarium and Plant Laboratories (KATH), and several international herbaria. Further information is available in Catalogue of Flowering Plants of Nepal Part I-III, and Supplement 1 (Rajbhandari and Baral 2010; Rajbhandari *et al* 2011, 2012, 2015).

Several botanists of Tribhuvan University have also contributed in exploring the Flora of Nepal. Some of the notable contributors are Brahmi D Pandey, Dibya D Bhatt, Dayananda Bajracharya, Ram P Chaudhary, Krishna K Shrestha, Mohan Siwakoti, Lokesh R Shakya, Devendra M Bajracharya, Sashi N Jha, Nirmala Pradhan, Dharma R Dangol, Suresh K Ghimire, Sangeeta Rajbhandary, Abishkar Subedi, Bhakta B Raskoti, Prabin Bhandari, etc. Since the establishment of Tribhuvan University Central Herbarium (TUCH) in mid 1990's at the Central Department of Botany, Tribhuvan University, nearly 20,000 herbarium specimens, collected by the students, researchers and faculty members, have been deposited. A recent publication 'Catalogue of Herbarium specimens deposited in TUCH' highlights the collections in TUCH (Rajbhandari *et al* 2016).

Based on the plant specimens collected from Nepal, especially by Hamilton, and Wallich, David Don published the first account of Flora of Nepal 'Prodromus Florae Nepalensis' (Don 1825), with the description of 766 species of flowering plants, and described several species new to science. Later Nathalien Wallich published 'Tentamen Florae Nepalensis' (Wallich 1824-26), 'A Numerical List of dried specimens of plants in the East India Company and Museum' (Wallich 1829-32), and Plantae Asiaticae Rariores (1832), with the catalogue, description and illustrations of Plants of Nepal. In addition, publications of JD Hooker such as 'Himalayan Journals' (Hooker 1854), and 'Flora of British India' in seven volumes (Hooker 1872-1897) highlighted the flowering plant diversity in Nepal and adjoining areas.

Based on plant collections in Nepal by the foreign and Nepali botanists, series of publications on Flora of Nepal are being published. Some of the relevant publications are: Fauna and Flora of Nepal Himalaya (Kihara 1955), Flora of Eastern Himalaya Part 1 (Hara 1966), Part 2 (Hara 1971), and Part 3 (Ohashi 1975), Forests of Nepal (Stainton 1972), Phytogeography and vegetation map of Nepal (Dobremez 1972), Flora of Langtang (Malla *et al* 1976), Enumeration of the Flowering Plants of Nepal in 3 volumes (Hara *et al* 1978, Hara and Williams 1979, and Hara *et al* 1982), Flowers of the Himalaya (Polunin and Stainton 1984), Flowers of the Himalaya – Supplement (Stainton 1988), Flora of Kathmandu Valley (Malla *et al* 1986), Flora of Ganesh Himal (Ohba and Ikeda 1999), Flora of Hinku and Hunku Valley, East Nepal (Ohba and Ikeda 2000), Annotated Checklist of Flowering Plants of Nepal (Press *et al* 2000), Catalogue of Type Specimens from Nepal (Shrestha and Press 2000), Flora of Mustang (Ohba *et al* 2008), Endemic Flowering Plants of Nepal Part I-III (Rajbhandari and Adhikari 2009, Rajbhandari and Dhungana 2010, 2011; Rajbhandari *et al* 2016), Catalogue of Nepalese Flowering Plants Part I-III (Rajbhandari and Baral 2010, Rajbhandari *et al* 2011, 2012, 2015) and Flora of Nepal Volume 3 (Watson *et al* 2011).

1.7. Vegetation, Forest types, and Floristic composition

Nepal is considered as a transitional zone between two major floristic kingdoms, the Holarctic kingdom in the north and the Palaeotropic kingdom in the south. Nepal acts as a corridor-linkage between several floristic regions, the Sino-Japanese, the Irano-Turanian, the Central Asian, the Indian and the Indo-Chinese. The East Himalaya shared the temperate and subalpine mountain flora with the Sino-Japanese floristic region, while the Western Himalayan flora is influenced by Irano-Turanian Floristic regions and the Central Asiatic floristic region has influenced the alpine flora of Nepal showing affinities with the Central Tibetan province. Indo-Gangetic elements are found in lowland Terai and are linked to Indo-Chinese and Indo-Malayan floristic region (Miehe *et al* 2015).

There are 12 global ecoregions found in Nepal namely, Eastern Himalayan Broad leaves Forests, Eastern Himalayan Conifer Forests, Eastern Himalayan Alpine Shrub and Meadows, Himalayan Subtropical Broadleaves Forests, Himalayan Subtropical Broadleaves Forests, Western Himalayan Broadleaves Forests, Western Himalayan Alpine Shrub and Meadows, Upper Gangetic Plains Moist Deciduous Forests, Lower Gangetic Plains Moist Deciduous Forests, Tarai-Duar Savannas and Grasslands and Rock and Ice (Olson *et al* 2001, Wikamanayake *et al* 2002).

Dobremez (1970) identified 136 ecosystems which were later reduced to 118 by ICIMOD/MENRIS in 1995 (BPP 1995). It includes 112 forest ecosystems, four agricultural ecosystems, one wetland ecosystem, and one glacier/ rock ecosystem. The mid hill (1000-3000 m) comprises 53 ecosystems, followed by 38 ecosystems in high hills (above 3000 m), 14 ecosystems in lowland Tarai (below 500 m), and 12 ecosystems in Siwaliks (500-1000 m); whereas the wetland ecosystem occurs in all physiographic zones.

a. Vegetation Types in Nepal (modified after Dobremez 1972, BPP 1995)

The vegetation of Nepal has been variously presented based on altitudinal belts, climatic zones, humidity types, plant life forms and human impacts.

The vegetation type was reduced to 59 and later to following 36 types as follows:

Zone 1 (3 types): Upper Alpine Meadow, Dry Alpine Scrub, and Moist Alpine Scrub.

Zone 2 (10 types): Trans Himalayan Steppe, Trans Himalayan Lower *Caragana* Steppe, Trans Himalayan High Alpine vegetation, Fir-Blue Pine (*Abies spectabilis-Pinus wallichiana*) forest, Birch–Rhododendron (*Betula utilis-Rhododendron*) forest, Fir (*Abies spectabilis*) forest, Larch (*Larix griiffithiana*) forest, Fir-Oak-Rhododendron (*Abies-Quercus-Rhododendron*) forest, Fir-Hemlock-Oak (*Abies-Tsuga-Quercus*) forest, and Sub Alpine Mountain Oak forest.

Zone 3 (17 types): Upper Temperate Blue Pine (*Pinus wallichiana*) forest, Temperate Juniper (*Juniperus* species) forest, Spruce (*Picea smithiana*) forest, West Himalayan Fir-Hemlock-Oak (*Abies-Tsuga-Quercus*) forest, Temperate Mountain Oak (*Quercus* species) forest, *Lithocarpus* forest, *Rhododendron* forest, Mountain Oak-Rhododendron (*Quercus-Rhododendron*) forest, Deciduous Maple-Magnolia-Sorbus (*Acer-Magnolia-Sorbus*) forest, Mixed Rhododendron-Maple (*Rhododendron-Acer*) forest, Cedar (*Cedrus deodara*) forest, Cypress (*Cupressus torulosa*) forest, Mixed Blue Pine-Oak (*Pinus wallichiana-Quercus*) forest, Lower Temperate Oak (*Quercus*) forest, Deciduous Walnut-Maple-Alder (*Juglans-Acer-Alnus*) forest, East Himalayan Oak-Laurel forest, and *Olea* forest.

Zone 4 (4 types): Chir Pine (*Pinus roxburghii*) forest, Chir Pine and Broad leaved forest, *Schima-Castanopsis* forest, and *Eugenia-Ostodes* forest.

Zone 5 (2 types): Hill Sal (*Shorea robusta*) forest, and Lower Tropical Sal and Mixed Broad leaved forest.

b. Forest Types

Stainton (1972) recognized 35 different forest types under four major categories *viz*. Tropical and Subtropical, Temperate and alpine broadleaved, Temperate and alpine conifer and Minor temperate and alpine associations (Stainton 1972).

1. **Tropical and Subtropical forests (10 types):** Sal (*Shorea robusta*) forest, Tropical deciduous riverain forest, Tropical evergreen forest, Subtropical evergreen forest, *Terminalia* forest, *Dalbergia sissoo-Acacia catechu* forest, Subtropical deciduous hill forest, *Schima-Castanopsis* forest, Subtropical semi-evergreen hill forest, and *Pinus roxburghii* forest.
2. **Temperate and Alpine Broad-leaved forest (11 types):** *Quercus incana-Quercus lanuginosa* forest, *Quercus dilatata* forest, *Quercus semecarpifolia* forest, *Castanopsis tribuloides-Castanopsis hystrix* forest, *Quercus lamellosa* forest, *Lithocarpus pachyphyllus* forest, *Aesculus-Juglans-Acer* forest, Lower temperate mixed broad-leaved forest, Upper temperate mixed broad-leaved forest, *Rhododendron* forest, and *Betula utilis* forest.
3. **Temperate and Alpine Conifer (8 types):** *Abies spectabilis* forest, *Tsuga dumosa* forest, *Pinus excelsa* (*P. wallichiana*) forest, *Picea smithiana* forest, *Abies pindrow* forest, *Cedrus deodara* forest, *Cupressus torulosa* forest, and *Larix* forest.
4. **Minor Temperate and Alpine Associations (6 types):** Alnus (*Alnus nepalensis*) woods, *Populus ciliata* woods, *Hippophae* scrub, Moist alpine scrub, Dry alpine scrub and *Juniperus wallichiana* forest.

Recently, Miehe *et al* (2015) have classified the forest types into three major categories (vegetation types of the southern slope of the Himalayas, vegetation types of the inner valleys and vegetation types of Arid Zone) which include fifty-five primary forest types. The floristic composition in different vegetation zones is summarized below, based on the life form such as trees, shrubs/climbers, and herbs.

1. Southern slopes of Himalayas comprises of four vegetation belts namely Tropical belt (5 forest types), Subtropical belt (7 forest types), Cloud forest belt (15 forest types), and Alpine belt (3 forest types).
   a. Tropical belt: *Shorea robusta* forest, *Dalbergia sissoo-Acacia catechu* riverine forest, *Bombax* riverine forest, *Terminalia* and *Anogeissus* forest, and Riverine grassland.
   b. Subtropical belt: *Schima-Castanopsis* forest, *Pinus roxburghii* forest, *Alnus nepalensis* riverine forest, *Toona ciliata-Albizia julibrissin* riverine forest, *Quercus lanata* forest, *Euphorbia royleana* grasslands, Thickets and pastures.
   c. Cloud forest belt: *Quercus semecarpifolia* forest, *Tsuga dumosa* forest, *Rhododendron arboreum* forest, *Abies spectabilis* forest, *Juniperus recurva* forest, *Quercus lamellosa* forest, *Lithocarpus pachyphyllus* forest, *Quercus floribunda* forest, *Rhododendron hodgsonii* forest, *Abies densa* forest, *Juniperus recurva* thickets, *Rhododendron* thickets, Bamboo thickets, Tall forb communities, and Upper treeline.
   d. Alpine belt: *Rhododendron* dwarf thickets, *Kobresia nepalensis* mats, and crustose lichen covers of rock walls.
2. Vegetation types in the Inner Valleys comprises three vegetation belts namely Subtropical belt (one forest type), Temperate belt (13 forest types), and Alpine belt (5 forest types).
   a. Subtropical belt: *Olea europea* subsp. *cuspidata* woodlands
   b. Temperate belt: *Cedrus deodara* forest, *Aesculus-Acer* forest, *Pinus wallichiana* forest, *Picea smithiana* forest, *Abies pindrow* forest, *Betula utilis* forest, *Cupressus torulosa* forest, *Juniperus indica* forest, *Larix* forest, *Hippophae* riverine woodlands, *Caragana sukiensis* thickets, *Rhododendron lepidotum* shrublands, and *Rosa-Berberis-Cotoneaster* shrublands.
   c. Alpine belt: *Juniperus squamata* dwarf shrublands, *Kobresia pygmaea* dwarf mats, High alpine cushion communities, *Hippophae tibetana* riverine dwarf thickets, and Pioneer plant successions in glacial forelands.
3. Vegetation types in Arid Zone comprises six forest types as follows:

   Drought line of forests, *Caragana gerardiana* open dwarf shrublands, *Caragana versicolor* open dwarf shrublands, Alpine steppe, Salt meadows, and Plant communities of wastelands.

c. Species diversity

The diversity of non-flowering plants in Nepal comprises 1,001 species of Algae (Prasad 2013), 2,182 species of Fungi (Kost and Adhikari 2015), 850 species of Lichens (Miehe *et al* 2015), 1,213 species of Bryophytes (Pradhan 2016) and 550 species of Pteridophytes (Fraser-Jenkins *et al* 2015). Regarding the flowering plants in Nepal, nearly 6,500 species occur in Nepal (Hara *et al* 1978, Hara and Williams 1979, Hara *et al* 1982, Press *et al* 2000, Frodin 2001, WCMC 1992). However, it has been documented that Nepal comprises 6,973 species of flowering plants (Groombridge and Jenkins 2002, MoFSC 2014b). The synopsis of reported species from Nepal is given below (Table 2).

Table 2. Plant diversity in Nepal

Organisms (*Kingdom*)	Species in Nepal	References	Species in World	Estimated species in World	References	Nepal (%)
Lichens	465*	Sharma (1995)				
	850	Miehe *et al* (2015)	17,000	25,000	Chapman (2009)	5
Bryophytes	1,150*	Pradhan and Joshi (2008)				
	1,213	Pradhan (2016)				
	1,138	Miehe *et al* (2015)	16,236	22,750	Chapman (2009)	7.01
Pteridophytes	534*	Thapa (2002)				
	550	Fraser-Jenkins *et al* (2015)	12,000	15,000	Chapman (2009)	4.58
Gymnosperms	26*	Bista (2006)	1,021	1,050	Chapman (2009)	2.55
Angiosperms (Flowering plants)			250,000		Ziegler (2007), Byng (2014)	
	6653	Press *et al* (2000), Kunwar *et al* (2010)	268,600	352,000	Chapman (2009)	2.48
	6,973*	Groombridge and Jenkins (2002)	369,400	500,000	RBG Kew (2016)	

Source: Modified version (Chaudhary 1998, *MoFSC 2014b)

Endemic Plants

The endemic plants are most unique plants in the world, which occur in particular or specific area of the country, and nowhere else. In terms of endemism, Nepal comprises 315 species of endemic flowering plants (Groombridge and Jenkins 2002), which is comparatively low, comparing with other countries (WCMC 1992), for example, China has 18,000 species of endemic plants (55 percent of its flora), Indonesia with 15,000 species (66.7 percent), Australia 12,000 species (80 percent), Venezuela 8,000 species (38 percent), Madagascar 7,000 species (68.4 percent), India 5,000 species (32 percent), Philippines 3,500 (39.3 percent), Cuba 3,229 (49.6 percent), Chile 2,698 (51.1 percent), and New Zealand has 1,942 endemic species (81.9 percent). The United Kingdom has only 16 endemic species, which is just one percent of its flora.

Shrestha and Joshi (1996) reported 246 species of endemic flowering plants of Nepal. Bhuju *et al* (2007) documented 399 taxa including species, subspecies and variety, endemic to Nepal. A series of work published by the Department of Plant Resources (DPR) reported 282 flowering plant species endemic to Nepal (Rajbhandari and Adhikari 2009; Rajbhandari and Dhungana 2010, 2011). Recently, Rajbhandari *et al* (2016) updated the list and revealed occurrence of 324 species of endemic flowering plants, belonging to 132 genera and 45 families. A synopsis of the endemic plants listed in Handbook of Flowering Plants of Nepal (Volume 1) is summarized in Appendix 2.

Regarding the endemic species in non-flowering plants, 30 species of bryophytes are reported as endemic to Nepal (Pradhan 2016). Similarly, it has been estimated that 48 species of endemic lichens (Bhuju *et al* 2007), and 12 species of endemic algae (Baral 1995) occur in Nepal. Furthermore, estimation of 142 species of fungi are endemic to Nepal (Kost and Adhikari 2015). Much of the endemic species shows its distribution in the sub-alpine zone (3000-4000 m), particularly in Central Nepal (Vetaas and Grytnes 2002).

d. Floristic diversity

Nepal comprises a wide range of flowering plants (trees, shrubs, climbers, and herbs) occurring in various ecological zones. Some common species of flowering plants are mentioned below:

Tropical zone (60 m to1000 m)

Trees: *Acacia catechu* (khayar), *Adina cordifolia* (karma), *Bombax ceiba* (simal), *Butea monosperma* (palans), *Cassia fistula* (raj brikcchya), *Dalbergia latifolia* (satisal), *Dalbergia sissoo* (sisau), *Lagerstroemia parviflora* (bot dhainyaro), *Mallotus philippensis* (sindure), *Shorea robusta* (sal), *Trewia nudiflora* (bhellor), *Terminalia tomentosa* (asna/Saaj).

Shrubs and climbers: *Bauhinia vahlii* (bhorla), *Caesalpinia decapetala* (arile kanda), *Cissus javana* (jogi lahara), *Dioscorea deltoidea* (bhyakur), *Gnetum montanum* (bhote lahara), *Leea crispa* (galeni), *Murraya koenigii* (kadi patta), *Mussaenda macrophylla* (dhobini), *Thespesia lampas* (ban kapas), *Zizyphus mauritiana* (bayer).

Herbs: *Abelmoschus manihot* (ban kapas), *Acorus calamus* (bojho), *Achyranthes aspera* (datiwan), *Cyperus esculentus* (mothe), *Eclipta prostrata* (bhringraj), *Evolvulus alsinoides* (khunkhune jhaar), *Ipomoea aquatica* (kalmi saag), *Ludwigia adscendens* (poke jhaar), *Mimosa pudica* (lajjawati), *Nelumbo nucifera* (kamal), *Piper longum* (pipla), *Saccharum spontaneum* (kaans).

Subtropical zone (>1000 m to <2000 m)

Trees: *Alnus nepalensis* (utis), *Castanopsis indica* (dhalne katush), *Castanopsis tribuloides* (musure katush), *Cinnamomum tamala* (tejpat), *Cyclobalanopsis glauca* (phalant), *Lyonia ovalifolia* (angeri), *Myrica esculenta* (kaphal), *Pinus roxburghii* (khote salla), *Rhododendron arboreum* (lali gurans), *Schima wallichii* (chilaune).

Shrubs and climbers: *Berberis asiatica* (chutro), *Cissampelos pareira* (batul pate), *Coriaria nepalensis* (macchaino), *Cyathula tomentosa* (bhede kuro), *Maesa chisia* (bilaune), *Princepia utilis* (dhatelo), *Rosa brunonii* (bhainsi kaanda), *Tetrastigma serrulatum* (charchare lahara), *Rubus ellipticus* (ainselu), *Viburnum mullaha* (molo).

Herbs: *Ageratina adenophora* (ban mara), *Anaphalis contorta* (buki phul), *Artemisia dubia* (tite pati), *Barleria cristata* (bhende kuro), *Begonia picta* (magar kanche), *Boenninghausenia albiflora* (uduse jhaar), *Cautleya spicata* (paani saro), *Centella asiatica* (ghod tapre), *Cirsium wallichii* (sungure kaanda), *Dendrobium densiflorum* (sungabha), *Floscopa scandens* (simkane ghans), *Remusatia vivipara* (jaluko).

Temperate zone (>2000 m to <3000 m)

Trees: *Aesculus indica* (lekh pangro), *Betula alnoides* (saur), *Cedrus deodara* (devdar), *Pinus wallichiana* (gobre salla), *Lithocarpus pachyphyllus* (arkhaulo), *Magnolia doltsopa* (raani chanp), *Quercus semecarpifolia* (khasru), *Rhododendron barbatum* (chimal), *Sorbus cuspidata* (chyuli), *Taxus wallichiana* (lauth salla).

Shrubs and climbers: *Actinidia callosa* (theki phal), *Astilbe rivularis* (budho okhati), *Clematis buchananiana* (pinase laharo), *Codonopsis viridis* (ban mula), *Daphne papyracea* (lokta), *Gaultheria fragrantissima* (dhasingare), *Hedera nepalensis* (pipal pate), *Holboelia latifolia* (gufala), *Piptanthus nepalensis* (suga phul), *Schisandra grandiflora* (singato).

Herbs: *Allium wallichii* (ban lasun), *Arisaema erubescens* (baanko), *Chlorophytum nepalense* (ban pyaj), *Habenaria intermedia* (thunma), *Lilium nepalense* (ban lasun), *Lotus corniculatus* (nakhar simbi), *Paris polyphylla* (satuwa), *Parochetus communis* (chengi phul), *Stellaria monosperma* (jethi madhu), *Swertia chirayita* (chiraito), *Valeriana hardwickii* (nakali jatamansii).

Subalpine zone (>3000 m to <4000 m)

Trees: *Abies spectabilis* (talispatra), *Acer acuminatum* (firfire), *Betula utilis* (bhojpatra), *Cinnamomum glanduliferum* (sugandha kokila), *Corylus jacquemontii* (daante okhar), *Cupressus torulosa* (raj salla), *Juniperus indica* (dhupi), *Larix griffithiana* (himali sallo), *Lithocarpus pachyphyllus* (baante), *Rhododendron campanulatum* (chimal).

Shrubs and climbers: *Berberis mucrifolia*, *Clematis acuminata*, *Codonopsis convolvulacea*, *Cotoneaster acuminatus* (dhalke phool), *Daphne bholua* (lokta), *Ephedra gerardiana* (somlata), *Lonicera myrtillus*, *Rhododendron lepidotum*, *Ribes himalense* (khamar), *Rosa sericea* (daarim paate).

Herbs: *Aconitum ferox* (bikh), *Allium hypsistum* (jimbu), *Cynanthus hookeri*, *Dactylorhiza hatagirea* (paanchaule), *Epilobium angustifolium*, *Gentiana depressa*, *Kobresia laxa*, *Morina longifolia* (kaanda masi), *Rheum australe* (padamchal), *Swertia petiolata.*

Alpine zone (>4000 m to <5000 m): *Anemone obtusifolia*, *Juniperus recurva* (dhupi), *Cyperipedium himalaicum*, *Meconopsis dhwojii*, *Primula sikkimensis*, *Rhododendron anthopogon* (sunpati), *Rhododendron lepidotum*, *Salix daltoniana*, *Stellaria congestiflora*, *Swertia multicaulis* (sharma guru).

Nival zone (>5000 m): *Allium hypsistum* (jimbu), *Aconitum nepalense*, *Arenaria bryophylla*, *Corydalis latiflora*, *Delphinium brunonianum*, *Ermania himalayensis*, *Gentiana pentasticta*, *Saxifraga andersonii*, *Sibbaldia tetrandra*, *Stellaria decumbens*.

Threatened plants

The threatened plants are the species which are vulnerable to environment and at the risk of extinction in future. According to the IUCN Redlist threat categories (www.iucnredlist.org), threatened species are divided into following categories: Extinct in the wild (EW), Critically Endangered (CE), Endangered (EN), Vulnerable (VU), and Near Threatened (NT). It has been estimated that nearly 21 percent flowering plants of the world are threatened (RBG Kew 2016). Two additional categories are often used, when information are either lacking or incomplete, such as Least Concern (LC) and Data Deficient (DD).

Initial assessment of threatened species in Nepal revealed two endangered species, five vulnerable species, and two nearly threatened species (MoFSC 2014b). Similarly, 15 species of plants in Nepal are listed under the Convention on International Trade in Endangered Species of wild fauna and flora (CITES). One species of fern in CITES Appendix 1, four species of gymnosperms in CITES Appendix II and III, and eight species of flowering plants as well as all species of orchids (Orchidaceae) are placed under CITES Appendix II and III (MoFSC 2014b).

Habitat loss in the recent years has pushed many plants to extinction and some of the species have already been extinct. Some of the common threatened plants of Nepal include *Butea monosperma* (palans), *Cycas pectinata* (thaakal), *Dalbergia latifolia* (satisal), *Dioscorea deltoidea* (bhyakur), *Flickingeria macraei* (jiwanti), *Oroxylum indicum* (tatelo), *Pterocarpus marsupium* (bijaya sal), *Rauvolfia serpentina* (sarpagandha), *Taxus wallichiana* (lauth salla), and *Ulmus wallichiana* (dhamina).

1.8 Useful Plant Resources and Livelihood

It is evident that our ancestors existed in the earth since about six million years, and the modern human being evolved about 200,000 years ago (Howell 2015). The world population has increased from 1.6 billion to 6.1 billion in the 20th century, and currently the human population is 7.57 billion (as of October 2017). Nearly 100 million new people to the human population are added every year (Lovejoy 1997). The United States, China, India, Pakistan and Bangladesh are among the nine countries contributing half of the world's population. If the current trend of population growth continues, the population will be 9.2 billion in 2050, and up to 11 billion by the end of 2100 AD.

Plants and their parts, and virtually all genetic diversity, useful for human beings can be considered as biological resources. The ecosystem services support the life of human beings by providing a range of services such as food, herbal medicine and hydroelectricity (provisioning services), clean air and water, control of diseases and pests, and protection of flooding (regulating services), and regulates nutrient cycling and seed dispersal (supporting services).

It is estimated that some 70,000 species of plants have been used for medicinal purpose, of which the Chinese Pharmacopoeia lists over 5,700 traditional medicine, mostly of plant origin, whereas in Ayurveda (traditional Indian medicine) about 2,000 species are considered to have medicinal value (Chevallier 1996).

About 31,128 species of flowering plants have been documented as useful plant resources of the world, of which 17,810 species have been documented as medicinal plants, 5,538 species as human food, and 3,649 species as fodder and forage plants (RBG Kew 2016). Nepal comprises 1,950 species of medicinal plants, including 1,614 native species (Shrestha *et al* 2001, Ghimire 2008), of which 143 species are of trade value. Harvesting of medicinal plants is the primary source of income for the mountain people, contributing from 3-44 percent (with an average of 12 percent) of the annual household income, thus constitute an integral part of the livelihood strategies.

It is estimated that nearly 7,000 species of plants are collected for consumption worldwide, of which 200 species have been domesticated for food. Globally, more than 20 percent of the food supply is obtained from only 20 species of plants, and 12 crop plants including rice, wheat, maize, potato, tea, sugar cane and beans, and provide 75 percent of world's calorie intake (Gaston and Spicer 1998). Nearly 600 species of food crops are consumed in Nepal, of which 374 species are wild food crops (Shrestha and Shrestha 1999). A large number of fodder and forage crops, which are the main source of food for the domestic livestock and wild animals, are derived from over 1,000 species of plants.

Useful Plant Resources: On the basis of their use value, useful plant resources are divided into different categories. Some of these important categories and potential plant resources in Nepal are as follows:

Agricultural/Household tools: *Bambusa balcooa* (dhanu bans), *Betula utilis* (bhoj patra), *Pouzolzia rugulosa* (daar), *Calamus acanthospathus* (bet), *Lagerstroemia parviflora* (bot dhaiyaro), *Pyracantha crenulata* (ghangaru), *Schima wallichii* (chilaune), *Terminalia alata* (saaj), *Thysanolaena latifolia* (amriso), *Shorea robusta* (saal).

Beverages: *Aegle marmelos* (bel), *Bombax ceiba* (simal), *Citrus limon* (kaagati), *Cocus nucifera* (nariwal), *Coffea arabica* (coffee), *Eleusine coracana* (Kodo), *Hippophae salicifolia* (daale chuk), *Rhododendron arboreum* (lali gurans), *Thea sinensis* (chiyapate).

Detergents/Shampoo: *Acorus calamus* (bojho), *Albizia lucidior* (padke siris), *Antidesma acidum* (archale), *Diplocyclos palmatus* (ban kankri), *Garuga pinnata* (dabdabe), *Machilus gamblei* (kaathe kaulo), *Oroxylum indicum* (tatelo), *Pogostemon benghalensis* (rudilo), *Sapindus mukorossii* (rittha), *Senegalia rugata* (sikakai).

Dye-yielding plants: *Alnus nepalensis* (utis), *Berberis asiatica* (chutro), *Impatiens balsamina* (tiuri), *Lawsonia inermis* (mehandi), *Maharanga emodi* (maharangi), *Magnolia champaca* (chaanp), *Mallotus philippensis* (sindure), *Rheum australe* (padhamchal), *Rubia manjith* (majitho), *Shorea robusta* (saal).

Edible plants (fruits, vegetables, condiments and oil seeds): *Berberis asiatica* (chutro), *Choerospondias axillaris* (lapsi), *Diploknema butyracea* (chyuri), *Juglans regia* (okhar), *Myrica esculenta* (kaafal), *Phyllanthus emblica* (amala), *Piper longum* (pipla), *Pyrus pashia* (mayal), *Rubus ellipticus* (ainselu), *Tamarindus indica* (imli), *Zanthoxylum armatum* (timur), *Ziziphus mauritiana* (bayar).

Fiber-yielding plants: *Agave cantula* (ketuke), *Bixa orellana* (sindure), *Bombax ceiba* (simal), *Cannabis sativa* (gaanja), *Daphne papyracea* (lokta), *Dioscorea pentaphylla* (mitho gittho), *Edgeworthia gardneri* (argheli), *Engelhardia spicata* (mauwa), *Ficus lacor* (kabro), *Girardinia diversifolia* (allo), *Pandanus furcatus* (bantari).

Fodder plants: *Acer campbellii* (phirphire), *Artocarpus lakoocha* (badhar), *Brassaiopsis hainla* (chuletro), *Ficus nerifolia* (dudhilo), *Ilex dipyrena* (liso), *Leucaenia leucocephala* (ipil-ipil), *Litsea monopetala* (kutmiro), *Machilus odoratissima* (bhate kaulo), *Quercus semecarpifolia* (khasru), *Saurauia napaulensis* (gogan), *Sorbus cuspidata* (chyuli).

Gums/resins/latex yielding plants: *Acacia catechu* (khayar), *Aegle marmelos* (bel), *Calotropis procera* (aank), *Ficus religiosa* (pipal), *Holarrhena pubescens* (ban khirro), *Pinus roxburghii* (khote salla), *Plumeria rubra* (karbir), *Sapium insigne* (khirro), *Semecarpus anacardium* (bhalayo), *Vachellia farnesiana* (ganchaune khayar).

Hedge: *Duranta repens* (nil kanda), *Euphorbia royleana* (siundi), *Ipomoea carnea* var. *fistulosa* (besharam), *Jatropha curcas* (sajiwan), *Justicia adhatoda* (asuro), *Jasminum humile* (jaee), *Lantana camara* (boksi kanda), *Lonicera japonica* (juhi phul), *Malvaviscus arboreus* (ghanti phul), *Thuja orientalis* (dhupi).

Medicinal plants: *Aconitum heterophyllum* (atis), *Asparagus racemosus* (kurilo), *Azadirachta indica* (neem), *Dactylorhiza hatagirea* (paanchaule), *Nardostachys jatamansi* (jatamasi), *Neopicrorhiza scrophulariiflora* (kutki), *Ophiocordyceps sinensis* (yarsagumba), *Paris polyphylla* (satuwa), *Rauwolfia serpentina* (sarpagandha), *Rheum australe* (padamchal), *Swertia chirayita* (chiraito), *Taxus wallichiana* (lauth salla), *Tinospora sinensis* (gurjo ko lahara), *Terminalia chebula* (harro).

Oil-yielding plants (Essential oils): *Cinnamomum tamala* (tejpaat), *Gaultheria fragrantissima* (dhasingare), *Jatropha curcas* (sajiwan), *Juniperus indica* (dhupi), *Olea cuspidata* (jaitun), *Rhododendron anthopogon* (sunpati), *Ricinus communis* (ander), *Valeriana wallichii* (sugandhawal), *Zanthoxylum armatum* (timmur).

Poisonous plants: *Acacia intsia* (arkhu), *Aconitum spicatum* (bikh), *Arisaema tortuosum* (baanko), *Datura metel* (dhaturo), *Boehmeria virgata* var. *macrostachya* (haathi sisnu), *Mucuna pruriens* (kauso), *Ricinus communis* (ander), *Sapium insigne* (khirro), *Schima wallichii* (chilaune), *Urtica dioica* (sisnu).

Roadside/Avenue trees/Shade trees: *Albizia julibrissin* (rato siris), *Albizia lucidior* (padke siris), *Callistemon lanceolatus* (kalki phul), *Cinnamomum camphora* (kapoor), *Eucalyptus lanceolatus* (masala ko bot), *Grevillea robusta* (kangiyo phul), *Jacaranda mimosifolia* (nilo parijat), *Melia azedarach* (bakenu), *Populus euro-americana* (lahare pipal), *Prunus cerasoides* (painyu).

Sacred/Religious plants: *Achyranthes aspera* (datiun), *Aegle marmelos* (bel), *Cynodon dactylon* (dubo), *Desmostachya pinnata* (kush), *Elaeocarpus sphaericus* (rudraachhya), *Ficus religiosa* (pipal), *Ficus benghalensis* (bar), *Jasminum sambac* (chameli), *Nelumbo nucifera* (kamal), *Ocimum tenuiflorum* (tulasi).

Timber and firewood: *Abies spectabilis* (talispatra), *Adina cordifolia* (karma/haldi), *Alnus nepalensis* (utis), *Dalbergia sissoo* (sisau), *Lyonia ovalifolia* (angeri), *Macaranga pustulata* (malato), *Magnolia champaca* (chaanp), *Pinus roxburghii* (khote salla), *Pinus wallichiana* (gobre salla), *Rhododendron arboreum* (lali gurans), *Schima wallichii* (chilaune), *Shorea robusta* (saal).

Besides the above mentioned major categories, several plant species are used for various domestic and commercial purposes. It is suggested to consult some important literature on Useful plant resources and ethnobotany of useful plants, for example, Fodder Trees and Tree Fodder in Nepal (Panday 1982), Wild Edible Plants of Nepal (Malla *et al* 1982), The Useful Plants of India (Ambasta 1992), Ethnobotany of Nepal (Rajbhandari 2001), Medicinal Plants of Dolpo (Lama *et al* 2001), Plants and People of Nepal (Manandhar 2002), Compendium of Medicinal Plants of Nepal (Baral and Kurmi 2006) and Medicinal Plants of Nepal (DPR 2016).

1.9 Other categories of plants

a. Poisonous plants

Some plants or plant parts are poisonous to human being and livestocks, if they are consumed, whereas some plants are harmful to skin, causing blisters and irritation. For example, root of *Aconitum spicatum* (bish), *Roscoea alpina* (bhuin sarro), latex of *Rhus wallichii* (bhalayo), *Sapium insigne* (khirro), stinging hairs in pods of *Mucuna nigricans* (kauso), stinging hairs in entire plant of *Urtica dioica* (sisnu), and *Girardinia diversifolia* (allo) are harmful plants.

Similarly, some species are poisonous to livestocks, if they are consumed, for example, young twigs of *Lyonia ovalifolia* (angeri), *Rhododendron campanulatum* (chimal), entire plant of *Euphorbia wallichii* (dudhe), *Kalanchoe spathulata* (ajammari), seeds of *Abrus precatorius* (rati gedi) and *Ricinus communis* (ander).

Some plant species are specifically used as fish poison, for example, *Acacia intsia* (arkhu), *Agave cantula* (kettuke), *Albizia lucidior* (siris), *Engelhardia spicata* (mauwa), *Litsea cubeba* (siltimmur), *Milletia fruticosa* (gaujo), *Sapium insigne* (khirro) and *Ulmus wallichiana* (dhamina).

b. Invasive species

An invasive species are species, which are non native and introduced from other countries, which may harm the environment, and cause damage to native species. There are at least 219 alien species of flowering plants that are naturalized in Nepal, of which 21 naturalized species are considered as invasive species (Tiwari *et al* 2005, Shrestha 2016). Examples of such invasive species are: *Ageratina adenophora* (banmara), *Ageratum conyzoides* (gandhe jhaar), *Amaranthus spinosus* (kaande lunde), *Argemone mexicana* (thaakal), *Bidens pilosa* (kuro), *Chromolaena odorata* (aule banmara), *Eichhornia crasssipes* (jal kumbhi), *Ipomoea carnea* subsp. *fistulosa* (besharam), *Lantana camara* (boksi kanda), *Leersia hexandra* (karaunte jhaar), *Mikania micrantha* (lahare banmara), *Myriophyllum aquaticum*, *Parthenium hysterophorus* (Paati jhaar), and *Xanthium strumarium* (bhende kurro).

c. Wetland species

Due to diversity in ecosystems and habitats, several species of wetland or marshy plants occur in Nepal. For example, *Acorus calamus* (bojho), *Alisma plantago-*

aquatica (Jal parni), *Eicchornia crassipes* (jal kumbhi), *Euryale ferox* (makhaana), *Ipomoea carnea* subsp. *fistulosa* (besharam), *Myriophyllum aquaticum*, *Nelumbo nucifera* (kamal), *Nymphaea nouchali* (jal kamal), *Oryza rufipogon* (jungali dhan), *Pistia stratioites* (paani banda), *Trapa quadrispinosa* (singada), and *Typha angustifolia* (pater).

1.10 Flora of Nepal Initiative

Publication of Flora of Nepal, in other words the encyclopaedia of the plants of Nepal, is a long-awaited dream project of Nepal. The Department of Plant Resources (then called Department of Medicinal Plant), under the aegis of Ministry of Forest and Soil Conservation, Government of Nepal, was established in 1960, with the main objective to publish Flora of Nepal. To fulfil this objective, the National Herbarium and Plant Laboratories was founded in 1961, focussing on extensive plant collections from different corners of Nepal, maintenance of herbarium specimens, and publications related to Flora of Nepal. A team of Nepali botanists, especially Samar Bahadur Malla, Tirtha Bahadur Shrestha, and Puspa Ratna Shakya, with the help of a handful of botanists, had initiated to work in the publication of Flora of Nepal in early 1960's. As the outcome of their contributions, the DPR published series of taxonomic publications such as Key to Dicot and Monocot genera of flowering plants of Nepal (1967, 1968, 1981), Flora of Rajnikunj/Gokarna (1967), Flora of Phulchoki and Godavari (1969), Flora of Nagarjun (1973), Flora of Langtang (1976), Catalogue of Vascular Plants of Nepal (1976) and Flora of Kathmandu Valley (1986).

One of the main contributions of the Department of Plant Resources is the publications of series of Fascicles on Flora of Nepal (1995-2010). Some of the important fascicles are: Aceraceae, Amaranthaceae, Apocynaceae, Capparaceae, Caprifoliaceae, Fabaceae, Fagaceae, Geraniaceae, Hypericaceae, Malvaceae, Melastomataceae, Myrsinaceae, Rosaceae, Theaceae, and Verbenaceae. Several occasional publications related to Flora of Nepal have been published by the Department of Plant Resources, since its establishment.

The main basis of writing taxonomic treatments for the 'Flora' is the herbarium specimens, collected since early 19th century till date, which are deposited in the national and international herbaria. The specimens are mainly deposited in the British and Japanese herbaria, and Nepal houses nearly 60 percent species representing the flora of the country. Based on the Nepalese collections in British and Japanese Herbaria, especially in BM, K (UK), and TI, KYO (Japan), with the initiation of Hiroshi Hara (University of Tokyo, Japan), with the cooperation of British taxonomists (AO Chater, WT Stearn, and LHJ Williams) published 'Enumeration of Flowering Plants of Nepal' (Vol. 1-3, 1978-1982). These books are considered as a precursor for the publication of Flora of Nepal.

It has been greatly realized that the publication of Flora of Nepal, with a full description of families, genera, species (up to infraspecific taxa), along with keys for identification, would not be possible without collaboration and contrib-utions from the national and international institutions, and taxonomists. To mate-rialize this initiation, international organizations like Royal Botanic Garden Edinburgh (UK) and University of Tokyo (Japan) have joined hands with national organizations like Nepal Academy of Science and Technology, Ministry of Forest and Soil Conservation (Department of Plant Resources), and Tribhuvan University (Central Department of Botany) to publish the Flora of Nepal.

Following major activities have been initiated and accomplished related to the explorations, documentation (Table 3) and publications on Flora of Nepal (Table 4).

Table 3. Important Explorations and Activities related to Flora of Nepal (in chronological order: 1793-2014)

Year	Explorers/ Activities	Explorations/ achievements	Specimens/Herbaria
1793	William J. Kirkpatrick (UK)	Birgunj – Kathmandu	No plant collection
1802-1803	Francis Buchanan-Hamilton (UK)	Raxaul – Kathmandu	Collection of >1,100 plant species, probably over 800 species new to world (BM)
1817-1819	Edward Gardner (UK)	Kathmandu, Rasuwa, Sindhupalchok area	*ca* 1000 specimens (K)
1820-1821	Nathaniel Wallich (Denmark/UK)	Raxaul – Kathmandu (sent collectors to Gosainkunda area)	1,834 specimens (K)
1848	Joseph Dalton Hooker (UK)	Eastern Nepal (Taplejung)	Olangchung Gola, Yangma, Ghunsa (K)
1884-1886	John Firminger Duthie (UK)	Far-west Nepal (Darchula)	65 species (DD)
1905	Jules Jacot-Guillar-mod (Switzerland)	Eastern Nepal (Taplejung)	66 specimens (G)
1907	Isaac Henry Burkill (UK)	Raxaul to Kathmandu, Trishuli Valley (Nuwakot area)	470 specimens (K/BM)
1927-1931	Capt. Lal Dhwoj and Khadananda Sharma Nepal)	West, Central, East Nepal	N/A (BM)
1929	Basant Lal Gupta (India)	West Nepal: Karnali Valley (Khaptad area)	*ca* 500 Specimens (DD, KATH)
1932-1937	Khadanda Sharma (Nepal)	Central Nepal (Ramechhap, Dolakha)	167 specimens (BM, E)

Year	Explorers/ Activities	Explorations/ achievements	Specimens/Herbaria
1948-1957	Mohan Lal Banerji (India)	East Nepal	583 species reported (CAL)
1949	Oleg Polunin (UK)	Central Nepal (Langtang, Ganesh, Jugal Himal)	*ca* 1900 specimens (BM)
1950	Donald Lowndes (UK)	Central Nepal (Marsyangdi Valley)	656 specimens (BM)
1952	Oleg Polunin, WR Sykes and LHJ Williams (UK)	West, Central Nepal (Karnali, Kali Gandaki Rivers)	17,000 specimens (BM)
1952-53	Sasuke Nakao (Japan)	Kathmandu – Trishuli – Manaslu area	5,000 specimens (KYO)
1952, 1954	Albert Zimmermann (Switzerland)	Central, East Nepal	2,000 specimens (G)
1954	JDA Stainton, WR Sykes and LHJ Williams (UK)	West, Central Nepal	9,314 specimens (BM)
1956-1975	John David Adam Stainton (UK)	West, Central, East Nepal	*ca* 5,000 specimens (BM)
1960	Establishment of Department of Plant Resources	Then, Department of Medicinal Plants, GoN	
1961	Establishment of National Herbarium and Plant Laboratories, Godawari	Department of Plant Resources, MoFSC	One of the main purposes was to publish Flora of Nepal
1962, 1968	Samar Bahadur Malla (Nepal)	West, Central Nepal	928 specimens (KATH)
1963-1972	Hiroshi Hara (Japan)	Central, East Nepal	*ca* 60,000 specimens (TI)
1966-1967	Dan Henry Nicolson (USA)	West, Central, East Nepal	*ca* 1,200 specimens (US)
1968-1984	Jean-Francois Dobremez (France)	West, Central, East Nepal	N/A (GRE, E)
1972	Madhu Sudan Bista (Nepal)	West Nepal	3,000 specimens (KATH)
1972, 1975	Tirtha Bahadur Shrestha (Nepal)	West, Central, East Nepal	4,200 specimens (KATH, BM)
1972-1985	Puspa Ratna Shakya (Nepal)	West, Central, East Nepal	*ca* 17,000 specimens (KATH)
1973-1982	Damodar Prasad Joshi (Nepal)	West, Central Nepal	*ca* 12,000 specimens (KATH)
1973, 1981	Christopher Grey-Wilson (UK)	West, Central, East Nepal	1100^{+} specimens (K)

Year	Explorers/ Activities	Explorations/ achievements	Specimens/Herbaria
1974-1984	Prabha Pradhan (Nepal)	West, Central, East Nepal	4,200 specimens (KATH)
1975-1976	Saman Bahadur Rajbhandary (Nepal)	Central, East Nepal	1,216 specimens (KATH)
1975-2014	Keshab Raj Rajbhandari (Nepal)	West, Central, East Nepal	N/A (KATH, TI)
1976-1980	Vidya Laxmi Gurung (Nepal)	Central, East Nepal	3,190 specimens (KATH)
1976-1987	Narayan P. Manandhar (Nepal)	West, Central, East Nepal	8100 specimens (KATH)
1976-1986	George Miehe (Germany)	Central, East Nepal	2,740 specimens (BM)
1978, 1981	Michael A. Farille (France)	Central Nepal	N/A
1983-1991	Hideaki Ohba (incl. S. Akiyama, H. Ikeda, H. Kanai, M. Suzuki, M. Wakabayashi, etc.); Japan	Central, East Nepal	52,500 specimens (TI)
1985-1988	Nirmal Kumar Bhattarai (Nepal)	West, Central Nepal	6,642 specimens (KATH)
1986	George Miehe	Central Nepal	16,904 specimens (BM)
1988-1997	Mitsuo Suzuki (Japan)	West, Central, East Nepal	*ca* 65,000 specimens (TI)
1989	KEKE: K, RBGE Team UK)	East Nepal	2,000 specimens (K, E)
1991, 2001	David G. Long (UK)	East Nepal	1,246 specimens of flowering plants + 500 specimens of bryophytes
1994-1995	Futoshi Miyamoto (Japan)	Central, East Nepal	N/A (TI, KATH)
1995	Registration of Tribhuvan University Central Herbarium (Central Department of Botany, TU)	Index Herbariorum, New York Botanical Garden, USA	Curator: KK Shrestha
1997-1999	Repatriation of Plant Information for Flora of Nepal	UK Darwin Initiative Project (Natural History Museum, London)	JR Press and KK Shrestha (Coordinators)
1999	International MOU for Flora of Nepal	Royal Botanic Garden Edinburgh, University of Tokyo, and NAST	Signed by Regius Keeper, President, Vice Chancellor

Year	Explorers/ Activities	Explorations/ achievements	Specimens/Herbaria
2001	National MOU for Flora of Nepal	NAST, MoFSC (DPR) and TU (CDB)	Signatories: NAST, MoFSC, TU authorities
2003-2006	Building Capacity for Flora of Nepal	UK Darwin Initiative Project (RBGE, UK)	Collaborators: NAST, CDBTU, DPR
2003-2014	Mark Watson, Colin Pendry (UK)	West, Central, East Nepal	N/A (E, KATH, TI)
2004	On-line version of 'Annotated Checklist of Flowering Plants of Nepal: with updates'	www.efloras.org (Missouri Botanical Garden, USA)	Coordinator: KK Shrestha (CDBTU)
2007	Shuichi Noshiro (Japan)	Central Nepal	N/A
2008-2014	Hiroshi Ikeda (Japan)	West, Central, East Nepal	N/A (TI, E, KATH)

Table 4. Important publications on Flora of Nepal (in chronological order: 1811-2016)

Year	Publication Title	Authors	Publishers/Remarks
1811	An Account of the Kingdom of Nepaul	WJ Kirkpatrick	London
1819	An Account of the Kingdom of Nepal and of the territories Annexed to this domination by the House of Gorkha	FB Hamilton	Manjushri Publishing House, New Delhi, India (Rep. ed. 1971)
1824-1826	Tentamen Florae Nepalensis Illustratae	N. Wallich	London
1825	Prodromus Florae Nepalensis	D. Don	London (Described 766 species of flowering plants), collected by Hamilton and Wallich)
1828-1849	A Numerical List of dried specimens of plants in East India Company's Museum	N. Wallich	London (Catalogued >20,000 specimens)
1830-1832	Plantae Asiaticae Rariores (Vols. 1-3)	N. Wallich	London
1854	Himalayan Journals (Vols. 1-2)	JD Hooker	J. Murray, London
1872-1897	The Flora of British India (Vols. 1-7)	JD Hooker (ed)	London
1886	Account of a botanical tour in Kuamaon	JF Duthie	*Gardeners' Chronicle* **31**: 276-458
1910	Notes from a journey to Nepal	IH Burkill	*Rec. Bot. Surv. India* **4**(4): 59-140

Year	Publication Title	Authors	Publishers/Remarks
1928	Flora of Nepal. Appendix in 'Nepal Vol. 1'	P. Landon	Listed 131 families, 758 genera and 1702 species. [Constable and Co. Ltd. London, pp. 334-358]
1931	List of plants collected in West Nepal	RN Parker	*For. Bull.* **76**: 1-9
1934	Plants of Nepal	T. Hay	*J. Roy. Hort. Soc.* **59**: 459-462 (Based on collections of L. Dhwoj and KN Sharma)
1950	Plant hunting in the Nepal Himalayas	O. Polunin	*Geog. Mag.* **23**: 132-147
1953	The 1952 expedition to Western Nepal	LHJ Williams	*J. Roy. Hort. Soc.* **78**: 323-337
1955	Flowering Plants and Ferns	S. Kitamura	In H. Kihara (Ed.) Fauna and Flora of Nepal Himalaya Vol. 1, 73-290. Kyoto University Press, Kyoto
1956	Resultats des expeditions scientifiques Genevoises au Nepal en 1952 et 1954 (Partie botanique)	A. Zimmer-mann	*Candollea* **15**: 127-147
1963	Chapters on the history of Botany in India	IH Burkill	Government of India Press, Delhi, India
1965	Contributions to the Flora of East Nepal	ML Banerji	*Rec. Bot. Surv. India* **19**(2): 1-90
1966	The Flora of Eastern Himalaya	H. Hara (ed)	University of Tokyo, Tokyo, Japan
1967	Flora of Rajnikunj (Gokarna forest)	PR Pande (ed)	Department of Med. Plants, Thapathali
1969	Flora of Phulchoki and Godawari	PN Suwal (ed)	Department of Med. Plants, Thapathali
1970	Natural History and Economic Botany of Nepal	DD Bhatt	Orient Longman Ltd., New Delhi
1971	The Flora of Eastern Himalaya, Second Report	H. Hara (ed)	*Univ. Mus. Univ. Tokyo Bull.* No. 2, Tokyo, Japan
1972	Forests of Ncpal	JDA Stainton	John Murray, London
1973	Flora of Nagarjun	SB Malla *et al* (eds)	Department of Med. Plants, Thapathali
1974	Supplement to the Flora of Phulchoki and Godawari	SB Malla *et al* (eds)	Department of Med. Plants, Thapathali
1975	The Flora of Eastern Himalaya, Third Report	H. Ohashi (ed)	*Univ. Mus. Univ. Tokyo Bull.* No. 8, Tokyo, Japan
1976	Catalogue of Vascular Plants of Nepal	SB Malla *et al* (eds)	Department of Med. Plants, Thapathali

Year	Publication Title	Authors	Publishers/Remarks
1976	Flora of Langtang and cross section vegetation survey (Central Nepal)	SB Malla *et al* (eds)	Department of Med. Plants, Thapathali
1978	An Enumeration of Flowering Plants of Nepal (Vol. 1)	H. Hara, WT Stearn, LHJ Williams (eds)	British Museum (Natural History) London
1979	An Enumeration of Flowering Plants of Nepal (Vol. 2)	H. Hara, LHJ Williams (eds)	British Museum (Natural History) London
1982	An Enumeration of Flowering Plants of Nepal (Vol. 3)	H. Hara, AO Chater, LHJ Williams (eds)	British Museum (Natural History) London
1984	Flowers of the Himalaya	O. Polunin, JDA Stainton	Oxford University Press, New Delhi
1984	The Orchids of Nepal Himalaya	ML Banerji, P. Pradhan	J. Cramer, Germany
1986	Flora of Kathmandu Valley	SB Malla *et al* (eds)	Department of Medicinal Plants, Thapathali, KTM
1987	An annotated list of vascular plants collected in the valleys south of Mt. Everest	G. Miehe	*Bull. Brit. Mus. Nat. Hist. (Bot.)* **16**: 225-268
1988	Flowers of the Himalaya: a Supplement	A. Stainton	Oxford University Press, New Delhi
1988	The Himalayan Plants (Vol. 1)	H. Ohba, SB Malla (eds)	University Museum, University of Tokyo, Bull. No. 31
1991	The Himalayan Plants (Vol. 2)	H. Ohba, SB Malla (eds)	University Museum, University of Tokyo, Bull. No. 34
1992	The Alpine Flora of the Jaljale Himal, East Nepal	H. Ohba and S. Akiyama	Univ. Mus., Univ. Tokyo, Nature and Culture No. 4
1994	Name List of the Flowering Plants of Nepal	H. Koba, S. Akiyama, Y. Endo, H. Ohba	University of Tokyo, Tokyo, Japan
1998	Biodiversity in Nepal: Status and Conservation	RP Chaudhary	S. Devi, Saharanpur, India and Tecpress Books, Bangkok
1999	A Contribution to the Flora of Ganesh Himal, Central Nepal	H. Ohba and H. Ikeda (eds)	Univ. Mus., Univ. Tokyo, Nature and Culture No. 5
1999	Plant Diversity of Eastern Nepal – Flora of Plains of Eastern Nepal	M. Siwakoti, SK Varma	Bishen Singh Mahendra Pal Singh, Dehradun, India
1999	The Himalayan Plants, Volume 3	H. Ohba (ed)	University Museum, University of Tokyo, Bull. No. 39, Tokyo, Japan

Year	Publication Title	Authors	Publishers/Remarks
2000	Flora of Hinku and Hunku Valleys, East Nepal	H Ohba, H Ikeda (eds)	University Museum, University of Tokyo, Nature and Culture No. 6, Tokyo, Japan
2000	Annotated Checklist of Flowering Plants of Nepal	JR Press, KK Shrestha, DA Sutton	The Natural History Museum, London
2000	Catalogue of Type Specimens from Nepal	KK Shrestha, JR Press	The Natural History Museum, London
2001	Flowering Plants of Nepal (Phanerogams)	MS Bista, MK Adhikari, KR Rajbhandari (eds)	Department of Plant Resources, Thapathali
2002	Himalayan Botany in the Twentieth and Twenty-first Centuries	S. Noshiro, KR Rajbhandari (eds)	The Society of Himalayan Botany, University of Tokyo
2002	Pteridophytes of Nepal	MS Bista, MK Adhikari, KR Rajbhandari (eds)	Department of Plant Resources, Thapathali, Kathmandu
2006	The Himalayan Plants, Volume 4	H. Ohba (ed)	University Museum, University of Tokyo, Bull. No. 42, Tokyo, Japan
2008	Flora of Mustang, Nepal	H. Ohba, Y. Iokawa, LR Sharma (eds)	Kodansha Scientific Ltd., Tokyo
2009	Building Capacity for plant biodiversity, inventory and conservation in Nepal: Activities of the recent Darwin Initiative Project in Nepal	CA Pendry, MF Watson, KK Shrestha and KR Rajbhandari	*Newsletter of Himalayan Botany* (Tokyo) 42: 16-27
2009	Orchids of Nepal	BB Raskoti	Bhakta B Raskoti and Rita Ale, Kathmandu
2009-2011	Endemic Flowering Plants of Nepal (Parts 1-3)	KR Rajbhandari *et al*	Department of Plant Resources
2010-2012	Catalogue of Nepalese Flowering Plants (Part 1-3)	KR Rajbhandari *et al*	National Herbarium and Plant Laboratories, DPR
2011	Flora of Nepal, Vol. 3 (Magnoliaceae – Rosaceae)	MF Watson, S. Akiyama, H. Ikeda, KR Rajbhandari and KK Shrestha	Royal Botanic Garden Edinburgh, UK
2015	Nepal (Companion volume of Flora of Nepal)	G. Miehe, C. Pendry, RP Chaudhary (eds)	Royal Botanic Garden Edinburgh, UK

Year	Publication Title	Authors	Publishers/Remarks
2015	Ferns and Fern-allies of Nepal Vol. 1	CR Fraser-Jenkins, DR Kandel, S. Pariyar	National Herbarium and Plant Laboratories, DPR, Lalitpur, Nepal
2015	Catalogue of Nepalese Flowering Plants. Supplement 1	KR Rajbhandari *et al*	National Herbarium and Plant Laboratories
2015	A Handbook of the Orchids of Nepal	KR Rajbhandari	Department of Plant Resources, Thapathali
2016	History of Botanical Explorations in Nepal: 1802-2015	KR Rajbhandari	Jha PK, Siwakoti M, and Rajbhandary S (eds) Frontiers of Botany, CDB TU, pp. 1-99.
2016	Endemic Plants of Nepal: Update	KR Rajbhandari, SK Rai, GD Bhatt	*Bull. Dept. Pl. Res.* No. **38**: 106-144
2016	Catalogue of herbarium specimens deposited at TUCH	KR Rajbhandari, M Siwakoti, S Rajbhandary, and SK Ghimire	Central Department of Botany, TU

THE WAY FORWARD

As per the mandate of Convention on Biological Diversity (CBD), Nepal has also committed to explore and document all biological resources of the country, and to publish Flora and Fauna of Nepal within 2020. Being the focal point of GEF (Global Environment Facility) and CBD, the Department of Plant Resources, Ministry of Forest and Soil Conservation, in collaboration with national and international organizations, are trying to publish the Flora of Nepal within the proposed deadline. However, at the present pace, it seems to be a herculean task to publish remaining volumes of Nepal within the proposed deadline. It has been anticipated that the Government of Nepal should give high priority for publishing the Flora of Nepal series, and facilitate cooperation and collaboration of national and international organizations to achieve the goal.

To fulfil the gap, the present attempt has been made to prepare the Handbook of Flowering Plants of Nepal, documenting all reported species from Nepal, which will be published in two volumes (Volume 1 and Volume 2). The present volume (Handbook of Flowering Plants of Nepal, Vol. 1) comprises 91 families of gymnosperms and angiosperms (Cycadaceae – Betulaceae), 696 genera and 3,004 taxa (2,857 species, 33 subspecies, 113 varieties and 1 forma), encompassing nearly 40 percent of Nepal Flora; as well as 106 species of exotic or introduced species, and 137 species of doubtful or uncertain (Data Deficient) species (Appendix 1, 3).

The previous version of the Checklist (Press *et al* 2000), *pro parte*, comprises 86 families, 635 genera and 2547 species, whereas the present version of book Vol. 1 (Shrestha *et al* 2018) comprises 91 families, 696 genera and 2857 species, which indicates nearly 12% increase in species number. For example, increase in the species number in the three largest families is as follows: Orchidaceae (323 species to 480 species), Poaceae (376 species to 446 species), and Fabaceae (299 species to 322 species).

REFERENCES

Ambasta, SP 1992 *The Useful Plants of India*, New Delhi: Publications and Information Directorate, CSIR

APG I 1998 An ordinal classification for the families of flowering plants, *Annals of the Missouri Botanical Garden* **85** 531-553

APG IV 2016 An Update of the Angiosperm Phylogeny Group Classification for the Orders and Families of Flowering Plants: APG IV, *Botanical Journal of the Linnean Society* **181** 1–20

Baral SR 1995 Enumeration of the Algae of Nepal, *Biodiversity Profile Project, Publication No. 11*, Kathmandu: Department of National Parks and Wildlife Conservation

Baral SR and Kurmi PP 2006 *A Compendium of Medicinal Plants of Nepal,* Kathmandu: Mrs. Rachana Sharma

Bhuju U, Shakya PR, Basnet TB, and Shrestha S 2007 *Biodiversity Resource Book: Protected Areas, Ramsar sites, and World Heritage Sites*, Kathmandu: ICIMOD, and Ministry of Science and Technology, Government of Nepal

Bista M 2006 Gymnosperms of Nepal, in *Environment and Plants: Glimpses of Research in South Asia* (eds) PK Jha, RP Chaudhary, SB Karmacharya and V Prasad, Kathmandu: Ecological Society

BPP 1995 An Assessment of Representation of the Terrestrial Ecosystems in the Protected Areas System of Nepal, in *Biodiversity Profiles Project,* Kathmandu: HMG/N

Byng JW 2014 The Flowering Plants Handbook: A practical guide to families and genera of the world, Hertford: Plant Gateway Ltd.

Chapman AD 2009 *Numbers of Living Species in Australia and the World* 2nd Edition, Australia: Department of the Environment, Water, Heritage and the Arts, Australia Government

Chaudhary RP 1998 *Biodiversity in Nepal – Status and Conservation*, Bangkok: S. Devi Sharanpur (U.P.) India and Tecpress Books

Chevallier A 1996 *Encyclopdia of Medicinal Plants*, London: A Dorling Kindersley Book

Connor S 2009 The Independent: 23 May 2009, http://www.independent.co.uk/news/science/a-seahorsethe-size-of-a-pea-and-the-other-bizarre-species-new-toscience-1689776.html

Dobremez JF 1970 Biogeographie du centre Nepal, *Bull. Ass. Geographes Fr.* **379–380** 79–90

Dobremez JF 1972 Les grandes divisions phytogeographiques du Nepal et de l'Himalaya, *Bulletin de la Société botanique de France* **119** 111–120

Don D 1825 Prodromus Florae Nepalensis, London

DPR 2016 *Medicinal Plants of Nepal Second Edition*, Kathmandu: Department of Plant Resources, Ministry of Forest and Soil Conservation

Fraser-Jenkins CR, Kandel DR, and Pariyar S 2015 *Ferns and Fern Allies of Nepal. Vol. 1*, Lalitpur: Department of Plant Resources, Ministry of Forest and Soil Conservation, National Herbarium and Plant Laboratories

Frodin DG 2001 *Guide to Standard Floras of the World*, Cambridge: Cambridge University Press

Gaston KG and Spicer JI 1998 *Biodiversity: An Introduction*, Blackwell Science Ltd.

Ghimire SK 2008 Medicinal plants in the Nepal Himalaya: Current issues, sustainable harvesting, Knowledge gaps and Research priorities, in *Medicinal Plants in Nepal: An Anthology of Contemporary Research* (eds) PK Jha, SB Karmacharya, MK Chettri, CB Thapa and BB Shrestha, Kathmandu: Ecological Society (ECOS), 25–42 pages

Govaerts R 2001 How many species of seed plants are there?, *Taxon* **50** 1085–1090

Groombridge B and Jenkins MD 2002 *World Atlas of Biodiversity*. Prepared by the UNEP World Conservation Monitoring Centre, Berkeley: University of California Press

Hara H (eds) 1966, 1971 *Flora of Eastern Himalaya* (Vol. 1 & 2), Tokyo: University Museum, University of Tokyo

Hara H, Chater AO, and Williams LHJ 1982 *An Enumeration of the Flowering Plants of Nepal* Vol. III, London: The British Museum (Natural History)

Hara H, Stearn WT, and Williams LHJ 1978 *An Enumeration of the Flowering Plants of Nepal* Vol. I, London: The British Museum (Natural History)

Hara H and Williams LHJ 1979 *An Enumeration of the Flowering Plants of Nepal* Vol. II, London: The British Museum (Natural History)

Hooker JD 1854 *Himalayan Journals* Vols. 1–2, Lodon: J. Murray

Hooker JD 1872–1897 *The Flora of British India* 7 vols, London: Reeve

Howell E 2015 How long have humans been on Earth?, *Universe Today*, 23 December 2015

IUCN Nepal [International Union for Conservation of Nature Nepal] 2004 *National Register of Medicinal and Aromatic Plants* (Revised and updated) Xiii + 202 pp. IUCN – The World Conservation Union, Nepal

Jeffries MJ 2005 *Biodiversity and Conservation* Second Edition, London and New York: Routledge Taylor & Francis Group

Kier G, Kreft H, Lee TM, Jetz W, Ibisch PL, Nowicki C, Mutke J, and Barthlott W 2009 A global assessment of endemism and species richness across island and mainland regions, *Proceedings of the National Academy of Sciences* **106** 9322–9327

Kihara H (ed) 1955 *Fauna and Flora of Nepal Himalaya* Vol. I, Kyoto: Fauna and Flora Research Society, Kyoto University

Kost G and Adhikari MK 2015 Mycota, in *Nepal: An introduction to the natural history, ecology and human environment of the Himalayas* (eds) G Miehe, CA Pendry and RP Chaudhary, Edinburgh: Royal Botanic Garden Edinburgh, 203–210 pages

Kunin WE and Lawton JH 1996 Does biodiversity matters? Evaluating the case for conserving species, in *Biodiversity: a biology of numbers and difference* (ed) KJ Gaston, Oxford: Blackwell Science 283–308 pages

Kunwar RM, Shrestha K, Dhungana SK, Shrestha PR, and Shrestha KK 2010 Floral Biodiversity of Nepal: an Update, *Journal of the Natural History Museum* **25** 295–311

Lama YC, Ghimire SK, and Aumeeruddy-Thomas Y 2001 *Medicinal Plants of Dolpo: Amchis' Knowledge and Conservation*, Kathamandu: WWF Nepal Program

Lovejoy TE 1997 Biodiversity: Why is it? in *Biodiversity II: Understanding and protecting our biological resources* (eds) ML Reaka-Kudla, DE Wilson and OE Wilson, Washington DC: Joseph Henry Press, 7–14 pages

Malla SB, Rajbhandari AB, Rajbhandari SB, Shrestha TB, Adhikari PM, and Adhikari SR 1976 *Flora of Langtang and Cross Section Vegetation Survey (Central Nepal)*, Bulletin of Department of Medicinal of Plants, Number 6, Kathmandu: Department of Medicinal Plants, HMG

Malla SB, Rajbhandari SB, Shrestha TB, Adhikari PM, and Adhikari SR 1982 *Wild edible plants of Nepal,* Bulletin of Department of Medicinal Plants of Nepal, Number 9, Kathmandu: Department of Medicinal Plants

Malla SB, Rajbhandari SB, Shrestha TB, Adhikari PM, Adhikari SR, and Shakya PR (eds) 1986 *Flora of Kathmandu Valley*, Bulletin of Department of Medicinal of Plants, Number 11, Kathmandu: Department of Medicinal Plants, HMG

Manandhar NP 2002 *Plants and People of Nepal*, Portland: Timber Press

Miehe G, Pendry CA, and Chaudhary R (eds) 2015 Nepal: An Introduction to the natural history, ecology and human environment of the Himalayas, Edinburgh: Royal Botanic Garden Edinburgh

MoFSC [Ministry of Forest and Soil Conservation] 2014a *Nepal fifth National Report on Convention on Biological diversity*, Kathmandu: Ministry of Forest and Soil Conservation, Government of Nepal

MoFSC [Ministry of Forest and Soil Conservation] 2014b *Nepal Biodiversity Strategy and Action Plan: 2014-2020*, Kathmandu: Ministry of Forest and Soil Conservation, Government of Nepal

Mora C, Tittensor DP, Adl S, Simpson AGB, and Worm B 2011 How many species are there on earth and in the ocean?, *PLoS Biology* **9** e1001127

Ohashi H 1975 *Flora of Eastern Himalaya*, Vol. 3, Tokyo: University of Tokyo Press

Ohba H and Ikeda H (eds) 1999. *A Contribution to the Flora of Ganesh Himal, Central Nepal*, Tokyo: The University Museum, The University of Tokyo, Nature and Culture No. 5

Ohba H and Ikeda H (eds) 2000 *The Flora of Hinku and Hunku Valleys, East Nepal*, Tokyo: The University Museum, The University of Tokyo, Nature and Culture No. 6

Ohba H, Iokawa Y, and Sharma LR 2008 *Flora of Mustang, Nepal*, Tokyo: Kodansha Scientific

Olson DM, Dinerstein E, Wikramanayake ED, Burgess ND, Powell GVN, Underwood EC, D'Amico JA, Itoua I, Strand HE, Morrison JC, Loucks CJ, Allnutt TF, Ricketts TH, Kura Y, Lamoreux JF, Wettengel WW, Hedao P, and Kassem KR 2001 Terrestrial ecoregions of the world: New map of life on earth, *Bioscience* **51** 933–938

Panday KK 1982 *Fodder Trees and Tree Fodder in Nepal*, Berne and Birmensdorf: Swiss Development Cooperation, Switzerland, and Swiss Federal Institute of Forestry Research

Polunin O and Stainton A 1984, Oxford: *Flowers of the Himalaya*, Oxford Press

Pradhan N 2016 Bryophytes of Nepal, in *Frontiers of Botany* (eds) PK Jha, M Siwakoti and S Rajbhandary, Kathmandu: Central Department of Botany, Tribhuvan University, 100–123 pages

Pradhan N and Joshi SD 2008 A diversity account of Bryaceae (Bryophyte: Musci) of Nepal, *Journal of the Natural History Museum* **23** 19–26

Prance GT 2001 *Systematics: Relevance to the Twenty-first Century,* John Wiley and Sons, Ltd., U.K.

Prasad V 2013 Biodiversity: Algae, in *Biological Diversity and Conservation* (eds) PK Jha, FP Neupane, ML Shrestha and IP Khanal, Lalitpur: Nepal Academy of Science and Technology, 97–103 pages

Press JR and Shrestha KK 1999 Collections of Flowering Plants by Francis Buchanan–Hamilton from Nepal, 1802–1803, *Bulletin of the Natural History Museum, London (Botany)* **30** 101–130

Press JR, Shrestha KK, and Sutton DA 2000 *Annotated Checklist of the Flowering Plants of Nepal*, London: The Natural History Museum

Rajbhandari KR 2001 *Ethnobotany of Nepal*, Kathmandu: Ethnobotanical Society of Nepal

Rajbhandari KR 2016 History of Botanical Explorations in Nepal: 1802-2015, in *Frontiers of Botany* (eds) PK Jha, M Siwakoti and S Rajbhandary, Kathmandu: Central Department of Botany, Tribhuvan University, 1–99 pages

Rajbhandari KR and Adhikari MK 2009 *Endemic Flowering Plants of Nepal* Part 1 (Acanthaceae – Euphorbiaceae), Kathmandu: Department of Plant Resources, Ministry of Forest and Soil Conservation

Rajbhandari KR and Baral SR (eds) 2010 *Catalogue of Nepalese Flowering Plants* I, Lalitpur: National Herbarium and Plant Laboratories, Department of Plant Resources, Ministry of Forest and Soil Conservation

Rajbhandari KR, Bhatt GD, and Chhetri R (eds) 2015 *Catalogue of Nepalese Flowering Plants* Supplement 1, Lalitpur: National Herbarium and Plant Laboratories, Department of Plant Resources, Ministry of Forest and Soil Conservation

Rajbhandari KR, Bhattarai KR, and Baral SR (eds) 2011 *Catalogue of Nepalese Flowering Plants* Part II, Lalitpur: National Herbarium and Plant Laboratories, Department of Plant Resources, Ministry of Forest and Soil Conservation

Rajbhandari KR, Bhattarai KR, and Baral SR (eds) 2012 *Catalogue of Nepalese Flowering Plants* III, Lalitpur: National Herbarium and Plant Laboratories, Department of Plant Resources, Ministry of Forest and Soil Conservation

Rajbhandari KR and Dhungana SK 2010 *Endemic Flowering Plants of Nepal* Part 2 (Gentianaceae – Primulaceae), Kathmandu: Department of Plant Resources, Ministry of Forest and Soil Conservation

Rajbhandari KR and Dhungana SK 2011 *Endemic Flowering Plants of Nepal* Part 3 (Ranunculaceae – Zingiberaceae), Kathmandu: Department of Plant Resources, Ministry of Forest and Soil Conservation

Rajbhandari KR, Rai SK, and Bhatt GD 2016 Endemic Flowering Plants of Nepal: An Update, *Plant Resources (A Scientific Publication), Bulletin of Department of Plant Resources* **38** 106-144

Rajbhandari KR, Siwakoti M, Rajbhandary S, and Ghimire SK 2016 *Catalogue of Herbarium Specimens deposited in TUCH*, Kathmandu: Central Department of Botany, Tribhuvan University

RBG Kew 2016 *The State of the World's Plants Report – 2016*, Kew: Royal Botanic Gardens

Scotland RW and Wortley AH 2003 How many species of seed plants are there?, *Taxon* **52** 101–104

Sharma LR 1995 *Enumeration of the Lichens of Nepal*, Kathmandu: Biodiversity Profile Project, Department of National Park and Wildlife Conservation/Ministry of Forest and Soil Conservation

Shrestha BB 2016 Invasive Alien Plant Species in Nepal, in *Frontiers of Botany* (eds) PK Jha, M Siwakoti and S Rajbhandary, Kathmandu: Central Department of Botany, Tribhuvan University, 269–284 pages

Shrestha GL and Shrestha B 1999 An Overview of Wild Relatives of Cultivated Plants in Nepal, in *Wild Relatives of Cultivated Plants in Nepal* (eds) R Shrestha and B Shrestha, Kathmandu: The Green Energy Mission/ Nepal, 19–23 pages

Shrestha KK 2016 Global Biodiversity and Taxonomy Initiatives in Nepal, in *Frontiers of Botany* (eds) PK Jha, M Siwakoti and S Rajbhandary, Kathmandu: Central Department of Botany, Tribhuvan University, 177–223 pages

Shrestha KK and Press JR 2000 Catalogue of Type specimens from Nepal: a catalogue of type specimens of flowering plants described from Nepal, London: The Natural History Museum, 123 pages

Shrestha KK, Tiwari NN, and Ghimire SK 2001 MAPDON – Medicinal and Aromatic Plant database of Nepal, 53-74. *In Proceedings of Nepal-Japan Joint Symposium on Conservation and Utilization of Himalayan Medicinal Resources*, Department of Plant Resources and SCDHMR, Japan

Shrestha TB and Joshi RM 1996 *Rare endemic and endangered plants of Nepal*, Kathmandu: WWF Nepal Program

Stainton A 1988 *Flowers of the Himalaya – a Supplement*, London: Oxford Press

Stainton JDA 1972 *Forests of Nepal*, London: John Murray

Stearn WT 1960 *Allium* and *Milula* in the Central and Eastern Himalaya. *Bulletin of British Museum and Natural History* (Botany) **2** 159–191

Thapa N 2002 *Pteridophytes of Nepal*, Bulletin of Department of Plant Resources Number 19, Kathmandu: Department of Plant Resources

Thorne RF 2002 How many species of seed plants are there?, *Taxon* **51** 511–522

Tiwari S, Adhikari B, Siwakoti M, and Subedi K 2005 *An Inventory and Assessment of Invasive Alien Plant Species of Nepal*, Kathmandu: IUCN Nepal

Vetaas OR and Grytnes JA 2002 Distribution of vascular plant species richness and endemic richness along the Himalayan elevation gradient in Nepal, *Global Ecology and Biogeography* **11** 291–301

Wallich N 1824–1826 *Tentamen Florae Nepalensis*, London: Asiatic Lithographical Press

Wallich N 1829–1832 A Numerical List of dried specimens of plants in the East India Company and Museum, London

Watson MF, Akiyama S, Ikeda H, Pendry C, Rajbhandari KR, and Shrestha KK (eds) 2011 *Flora of Nepal. Vol. 3* (Magnoliaceae – Rosaceae), Edinburgh: Royal Botanic Garden Edinburgh, U.K.

WCMC 1992 *Global Biodiversity: Status of Earth's Living Resources*, London: World Conservation Monitoring Centre, Chapman & Hall

Wikramanayake ED, Dinerstein E, Loucks CJ, Olson DM, Morrison J, Lamoreux J, McKnight M, and Hedao P 2002 *Terrestrial Ecoregions of the Indo-Pacific: A Conservation Assessment*, Washington DC: Island Press

Zeigler D 2007 *Understanding Biodiversity*, London: Praeger Publishers

Weblinks

Catalogue of Life: www.catalogueoflife.org

CITES (Convention on International Trade in Endangered Species of Wild Fauna and Flora): https://www.cites.org/eng/app/appendices.php

Encyclopaedia of Life: www.eol.org

IUCN Red List of Threatened Species: http://www.iucnredlist.org/search

Royal Botanic Garden Edinburgh: www.rbge.org.uk

Royal Botanic Gardens Kew: www.kew.org

The Natural History Museum London: www.nhm.ac.uk

The Plant List: www.theplantlist.org

The University Museum, The University of Tokyo: www.um.u-tokyo.ac.jp

TROPICOS: www.tropicos.org

www.bas.ac.uk

www.ecology.com

www.humanorigins.si.edu

■■■■

CHAPTER 2

SPERMATOPHYTA (Seed Plants)

GYMNOSPERMS

Arrangement of orders and families in Gymnosperms follows the classification system of Christenhusz *et al* 2011 [136].

CYCADIDAE

CYCADALES

CYCADACEAE

Cycas L.

Cycas pectinata Buch.-Ham. [1, 4, 5]

Cycas circinalis var. *pectinata* (Griff.) Schuster; *Cycas pectinata* Griff.

थाकल, कलवल (Thaakal, Kalwal); Assam Cycas

Evergreen dwarf trees. E Nepal: 300-700 m. Nepal, E Himalaya, Assam-Burma, E Asia and SE Asia.

Use(s): Food (fronds, fruit), Ornamental; Yields coarse sago.

IUCN Category: **Vulnerable** A2c (ver 3.1).

CITES Appendix: II.

Cycas revoluta Thunb. [5]

कलवल (Kalwal); Sago palm

Evergreen dwarf trees. C, E Nepal: 100-1400 m. Commonly planted in the garden. Native to E Asia.

Use(s): Ornamental, Food (seeds).

GINKGOIDAE

GINKGOALES

GINKGOACEAE

Ginkgo L.

Ginkgo biloba L. [1, 4, 5]

बालकुमारी (Baal kumari); Ginkgo, Maiden hair tree

Deciduous medium-sized trees. C Nepal: *ca.* 1400 m. Commonly planted as road side avenue trees. Native to E Asia.

Use(s): Ornamental, Medicine (leaves, seeds).

GNETIDAE

GNETALES

GNETACEAE

Gnetum L.

Gnetum montanum Markgr. [1, 4, 5]

भोटे लहरा (Bhote laharaa)

Evergreen woody climbers. C, E Nepal: 300-1800 m. Nepal, E Himalaya, Assam-Burma, E Asia and SE Asia.

Use(s): Medicine (seeds), Food (seeds), Fiber (bark).

CITES Appendix: III.

EPHEDRALES

EPHEDRACEAE

Ephedra L.

Ephedra gerardiana Wall. ex Stapf [1, 4, 5, 71]

सोमलता, काग चरो; छेदुम् (Somlataa, Kaag charo; Tshedum: Tibetan); Jointfir

Perennial undershrubs. W, C, E Nepal: 2000-5200 m. Nepal, W Himalaya, E Himalaya, Tibetan Plateau and C Asia.

Use(s): Medicine (whole plant), Firewood, Fodder.

Ephedra intermedia Schrenk ex C.A.Mey. [1, 4]

Ephedra intermedia var. *tibetica* Stapf; *Ephedra tibetica* (Stapf) V.V.Nikitin

सोमलता (Somlataa); Blue joint fir

Evergreen shrubs. W, C Nepal: 2700-4000 m. Nepal, W Himalaya, Tibetan Plateau, SW Asia and N Asia.

Use: Medicine (whole plant).

Ephedra pachyclada Boiss. [1, 4, 5]

Small dense shrubs. W, C Nepal: 2000-4100 m. Nepal, W Himalaya and SW Asia.

Ephedra saxatilis (Stapf) Royle ex Florin [4]

Ephedra gerardinia var. *saxatilis* Stapf; *Ephedra gerardiana* var. *sikkimensis* Stapf (1, 4); *Ephedra saxatilis* var. *sikkimensis* (Stapf) Florin

Small shrubs. W, C, E Nepal: 2300-5200 m. Nepal and E Himalaya.

Use(s): Medicine (stem), Firewood.

PINIDAE

PINALES

PINACEAE

Abies Juss.

Abies densa Griff. [1, 4, 5, 71]

Abies spectabilis var. *densa* (Griff.) Silba

गोब्रे सल्ला (Gobre sallaa); Bhutan fir, Sikkim fir

Semi-evergreen trees. E Nepal: 3000-4000 m. Nepal, E Himalaya, Tibetan Plateau and Assam-Burma.

Use(s): Timber, Firewood.

Abies pindrow (Royle ex D.Don) Royle [1, 4, 5, 71]

Pinus pindrow Royle ex D.Don; *Abies webbiana* var. *pindrow* (Royle ex D.Don) Brandis

ठिन्ग्रे सल्ला (Thingre sallaa); West Himalayan fir

Large evergreen trees. W Nepal: 2100-3600 m. Nepal and W Himalaya.

Use(s): Timber, Firewood, Medicine (leaves), Fiber (pulp).

Abies spectabilis (D.Don) Spach [1, 4, 5, 71]

Pinus spectabilis D.Don; *Abies webbiana* (Wall. ex D.Don) Lindl.; *Pinus webbiana* Wall. ex D.Don

var. **spectabilis**

गोब्रे सल्ला, तालिस पत्र (Gobre sallaa, Taalis patra); East Himalayan fir

Large evergreen trees. W, C, E Nepal: 2400-3800 m. Nepal, W Himalaya, E Himalaya, Tibetan Plateau and SW Asia.

Use(s): Timber, Medicine (leaves), Essential oil (Himalayan silver fir oil), Ritual.

IUCN Category: Near Threatened (ver 3.1).

var. **langtangensis** Silba, *Phytologia* **68**(1): 22. 1990. [4]

Abies spectabilis subsp. *langtangensis* (Silba) Silba

लांगटांगे सल्ला (Lantange salla); Langtang fir

Evergreen trees. C Nepal: *ca.* 3100 m. Endemic to Nepal (?)

Central: *Rasuwa*, Ghora Tabela, South of Langtang Khola, 3048 m, T. Cotter *et al* N94 (Holotype: A).

Use(s): Firewood, Construction materials.

Cedrus Trew.

Cedrus deodara (Roxb. ex D.Don) G.Don [1, 4, 5, 71]

Pinus deodara Roxb. ex D.Don

देवदार (Devdar); Himalayan Cedar

Evergreen trees. W, C Nepal: 1100-3000 m. Nepal, W Himalaya, Tibetan Plateau and SW Asia.

Use(s): Timber, Medicine (heart-wood, bark), Essential oil (Cedar wood oil).

Larix Mill.

Larix griffithii Hook.f. [4]

Larix griffithiana Carriere (1,5,71); *Pinus griffithiana* (Carriere) Voss, *Pinus griffithii* (Hook.f.) Parl.

लेक सल्ला (Lek sallaa); East Himalayan Larch, Sikkim Larch

Medium-sized deciduous trees. C, E Nepal: 2400-3800 m. Nepal, E Himalaya and Tibetan Plateau.

Use(s): Timber, Firewood.

Larix himalaica W.C.Cheng and L.K.Fu [1, 4, 5, 71]

Larix potaninii var. *himalaica* (W.C.Cheng & L.K.Fu) Farjon & Silba (11)

लांगटांग सल्ला (Langtang sallaa); Langtang Larch

Deciduous trees. C Nepal: 2400-4000 m. Nepal and Tibetan Plateau.

Use(s): Timber, Firewood.

Picea Link

Picea smithiana (Wall.) Boiss. [1, 4, 5, 71]

Pinus smithiana Wall.

var. **smithiana**

झुले सल्ला, जुरे सल्ला (Jhule sallaa, Jure sallaa); West Himalayan Spruce

Evergreen trees. W, C Nepal: 2100-3600 m. Nepal, W Himalaya and SW Asia.

Use(s): Timber, Firewood, Medicine (leaves, young cone), Fiber (pulp).

var. **nepalensis** Franco, *Enum. Fl. Pl. Nepal* **1**: 26. 1978. [1, 4]

Picea smithiana subsp. *nepalensis* (Franco) Silba

Evergreen trees. W Nepal: *ca.* 2400 m. Endemic to Nepal (?)

West: *Kalikot*, Dillikot, *ca.* 2400 m. O. Polunin, WR Sykes & LHJ Williams 3920 (Holotype: BM).

Picea spinulosa (Griff.) A.Henry [4]

Abies spinulosa Griff.; *Picea spinulosa* var. *yatungensis* Silba

East Himalayan Spruce, Sikkim Spruce

Large evergreen trees. W, C Nepal: 2900-3600 m. Nepal, E Himalaya and Tibetan Plateau.

Use(s): Timber, Firewood.

Pinus L.

Pinus roxburghii Sarg. [1, 4, 5]

Pinus longifolia Salisb.

खोटे सल्ला, रानी सल्ला (Khote sallaa, Raani sallaa); Chir Pine

Large evergreen trees. W, C, E Nepal: 500-2700 m. Nepal, W Himalaya, E Himalaya and Tibetan Plateau.

Use(s): Timber, Essential oil (turpentine oil), Medicine (resin, seeds), Tannin (bark), Food (seeds).

Pinus wallichiana A.B.Jacks. [1, 4, 5, 71]

Pinus excelsa Wall. ex D.Don; *Pinus nepalensis* Chambray

var. **wallichiana**

गोब्रे सल्ला (Gobre sallaa); Blue pine

Large evergreen trees. W, C, E Nepal: 1800-4300 m. Nepal, W Himalaya, E Himalaya, E Asia and SW Asia.

Use(s): Timber, Firewood, Medicine (resin), Turpentine (resin).

var. **manangensis** H.Ohba & M.Suzuki, *Bull. Univ. Mus. Univ. Tokyo*, **31**: 350. 1988. [4]

Pinus wallichiana subsp. *manangensis* (H.Ohba & M. Suzuki) Silba

Evergreen trees. C Nepal: 3500-3700 m. Endemic to Nepal (?)

Central: *Manang*, Manang-Khangsar, 3550-3650 m, H. Ohba *et al* 8340255 (Holotype: TI).

Tsuga (Endl.) Carriere

Tsuga dumosa (D.Don) Eichler [1, 4, 71]

Pinus dumosa D.Don

ठिंगे सल्ला, ठिन्ग्रे सल्ला (Thinge sallaa, Thingre sallaa); Himalayan Hemlock

Evergreen trees. W, C, E Nepal: 2000-3600 m. Nepal, E Himalaya, Tibetan Plateau, Assam-Burma, E Asia and SE Asia.

Use(s): Timber, Firewood, Medicine (bark), Incense (leaves), Dye (bark).

ARAUCARIALES

ARAUCARIACEAE

Araucaria Juss.

Araucaria bidwillii Hook. [1, 4, 5]

काँडे सल्ला (Kaande salla); Bunya pine, Monkey puzzle tree

Evergreen large trees. C Nepal: *ca.* 1400 m. Commonly planted in the garden. Native to Australasia.

Use(s): Ornamental, Food (seeds).

PODOCARPACEAE

Podocarpus L'Her. ex Pers.

Podocarpus neriifolius D.Don [1, 4, 5]

Margbensonia neriifolia (D.Don) A.V.Bobrov & Melikyan; *Podocarpus discolor* Blume

गुन्सी (Gunsi); Brown Pine

Evergreen trees. C, E Nepal: 800-1500 m. Nepal, Assam-Burma, S Asia, E Asia and Australasia.

Use(s): Ornamental, Medicine (leaves), Timber.

CITES Appendix: III.

CUPRESSALES

CUPRESSACEAE (incl. TAXODIACEAE)

Cryptomeria D.Don

Cryptomeria japonica (Thunb. ex L.f.) D.Don [1, 4, 5]

कत्ले सल्ला (Katle sallaa); Japanese red Cedar

Evergreen trees. C, E Nepal: 1300-2600 m. Widely planted in E Himalaya, Assam-Burma and SE Asia. Native to E Asia.

Use(s): Timber, Construction materials, Essential oil (Japanese Cedar wood oil), Incense (leaves).

Cupressus L.

Cupressus cashmeriana Royle ex Carriere. [8]

Kashmir Cypress

Evergreen trees. C Nepal: *ca.* 1200 m. Probably planted. Native to Eastern Himalaya.

Cupressus sempervirens L. [1, 4]

Mediterranean Cypress

Evergreen trees. C Nepal: *ca.* 1350 m. Planted as ornamental tree. Native to W Asia and E Mediterranean region.

Use(s): Timber, Construction materials, Medicine (wood, fruit), Essential oil (leaves).

Cupressus torulosa D.Don [1, 4, 5, 71]

Cupressus corneyana Carriere; Tree (1, 4); *Cupressus torulosa* var. *corneana* Carriere

राज सल्ला (Raaj sallaa); Himalayan Cypress

Evergreen trees. W, C Nepal: 1300-3300 m. Nepal, W Himalaya and Tibetan Plateau.

Use(s): Timber, Medicine (leaves, fruit), Incense (leaves), Essential oil (leaves), Ornamental.

Juniperus L.

Juniperus communis L. [71]

धुपी (Dhupi); Common Juniper

Evergreen shrubs or small trees. W, C Nepal: 1800-4100 m. Nepal and W Himalaya.

Use(s): Ritual, Essential oil (juniper oil), Medicine (leaves, fruit), Firewood.

Juniperus indica Bertol. [1, 4, 71]

Juniperus wallichiana Hook.f. & Thomson ex Brandis; *Sabina indica* (Bertol.) W.C.Cheng & L.K.Fu; *Sabina wallichiana* (Hook.f. & Thomson ex Brandis) W.C.Cheng & L.K.Fu

धुपी; सुक्पा (Dhupi; Shuk pa: Amchi); Black Juniper

Evergreen small trees or shrubs. W, C, E Nepal: 2100-4500 m. Nepal, W Himalaya, E Himalaya, Tibetan Plateau and E Asia.

Use(s): Ritual (incense: leaves), Essential oil (Juniper oil), Firewood, Medicine (leaves, fruit).

Juniperus recurva Buch.-Ham. ex D.Don [1, 4, 5, 71]

भैरुंग धुपी (Bhairung dhupi); Drooping Juniper

Evergreen small trees or shrubs. W, C, E Nepal: 1900-4600 m. Nepal, W Himalaya, E Himalaya, Tibetan Plateau, E Asia and SW Asia.

Use(s): Ritual (incense: leaves), Essential oil (Juniper oil), Firewood, Medicine (stem).

Juniperus sibirica Burgsd. [4]

Juniperus communis var. *saxatilis* Pall. (1, 4, 5)

धुपी (Dhupi); Himalayan pencil Cedar

Evergreen shrubs. W, C Nepal: 2700-3200(-4100) m. Nepal, W Himalaya, W Asia, C Asia, E Asia, Europe and N America.

Use(s): Ritual, Firewood, Medicine (leaves, fruit).

Juniperus squamata Buch.-Ham. ex D.Don [1, 4, 5, 71]

Juniperus recurva var. *squamata* (D.Don) Parl.

धुपी (Dhupi); Himalayan Juniper

Evergreen prostrate shrubs. W, C, E Nepal: 3000-4500 m. Nepal, W Himalaya, E Himalaya, Assam-Burma and E Asia.

Use(s): Firewood, Medicine (whole plant), Ritual (incense: leaves).

TAXACEAE

Taxus L.

Taxus contorta Griff. [5, 19, 80]

Taxus fuana N.Li & R.R.Mill (4); *Taxus wallichiana* subsp. *contorta* (Griff.) Silba

काँडे लोटी, लौठ सल्ला (Kaande loti, Lauth sallaa); West Himalayan Yew

Evergreen trees. W, C Nepal: (1700-)2100-2700(-3500) m. Nepal, W Himalaya and Tibetan Plateau.

Use(s): Timber, Firewood, Medicine (bark, leaves).

IUCN Category: **Endangered** A2acd (ver 3.1).

CITES Appendix: II.

Taxus mairei (Lemee & H.Lev.) S.Y.Hu ex T.S.Liu [8, 19, 80]

Tsuga mairei Lemee & H.Lev.; *Taxus chinensis* var. *mairei* (Lemee & H.Lev.) W.C.Cheng & L.K.Fu, *Taxus wallichiana* var. *mairei* (Lemee & H.Lev.) L.K.Fu & N.Li

लौठ सल्ला, बर्मे सल्ला (Lauth sallaa, Burme sallaa); Chinese Yew

Evergreen trees. C Nepal: 1500-2200 m. Nepal, E Himalaya, Assam-Burma and E Asia.

Use(s): Medicine (bark, leaves), Timber, Firewood.

IUCN Category: **Vulnerable** A2d (ver 3.1).

CITES Appendix: II.

Taxus wallichiana Zucc. [4, 5, 15, 19]

Taxus baccata subsp. *wallichiana* (Zucc.) Pilg. (1, 71); *Taxus yunnanensis* W.C.Cheng & L.K.Fu

लौठ सल्ला, बर्मे सल्ला, ढेंग्रे सल्ला (Lauth sallaa, Barme sallaa, Dhengre sallaa); East Himalayan Yew

Medium-sized evergreen trees. C, E Nepal: 1900-3300(-3500) m. Nepal, E Himalaya, Tibetan Plateau, Assam-Burma and E Asia.

Use(s): Timber, Medicine (bark, leaves), Food (fruit), Fodder.

IUCN Category: **Endangered** A2acd (ver 3.1).

CITES Appendix: II.

Exotic/Cultivated Taxa

ARAUCARIACEAE

Araucaria columnaris (G.Forst.) Hook. (Coral reef Araucaria). C Nepal: *ca.* 1400 m. Native to Australasia. [5]

Araucaria cunninghamii Aiton ex D.Don (Hoop Pine). C Nepal: *ca.* 1400 m. Native to Australasia. [5]

Araucaria heterophylla (Salisb.) Franco (Norfolk island Pine). C Nepal: *ca.* 1400 m. Native to Australasia. [5]

CUPRESSACEAE

Juniperus horizontalis Moench (Creeping Juniper). C Nepal: *ca.* 1400 m. Native to America.

Metasequoia glyptostroboides Hu & W.C.Cheng (Dawn red wood). C Nepal: *ca.* 1400 m. Native to E Asia [5]

Platycladus orientalis (L.) Franco (*Thuja orientalis* L.); (**मयुर पंखी,** Mayur pankhi; Oriental Arborvitae). C Nepal: *ca.* 1400 m. Native to N Asia and C Asia.

Thuja occidentalis L. (**धुपी मयुर पंखी,** Dhupi, Mayur pankhi; Eastern arborvitae). C Nepal: *ca.* 1400 m. Native to N America.

PINACEAE

Pinus patula Schltdt. & Cham. (Mexican weeping pine). Native to N America. [5]

ANGIOSPERMS

Arrangement of orders and families in Angiosperms follows the Angiosperm Phylogeny Group classification system, APG IV [137].

NYMPHAEALES

NYMPHAEACEAE

Euryale Salisb.

Euryale ferox Salisb. [6, 9]

मखाना (Makhanaa); Prickly water lily

Perennial aquatic herbs. E Nepal: 100-500 m (Naturalized). Nepal, S Asia and E Asia.

Use(s): Food (seeds), Medicine (seeds).

Nymphaea L.

Nymphaea lotus L.

var. **pubescens** (Willd.) Hook.f. & Thomson [8]

Nymphaea pubescens Willd.

कमल (Kamal); Egyptian lotus

Perennial aquatic herbs. W Nepal: 100-200 m. Nepal, Tropical Asia and Australasia.

Use(s): Ornamental, Ritual, Medicine (seeds, flowers).

Nymphaea nouchali Burm.f. [6]

Nymphaea stellata Willd.

var. **nouchali**

निल कमल (Nil kamal); Blue water lily

Perennial aquatic herbs. W Nepal: 100-300 m. Nepal, W Himalaya, E Himalaya, Tibetan Plateau, Assam-Burma, S Asia, SE Asia and Australasia.

Use(s): Ornamental, Ritual, Medicine (rhizome, leaves, flowers, seeds), Food (all parts).

var. **versicolor** (Sims) R.Ansari & Jeeja

Nymphaea versicolor Sims; *Nymphaea stellata* var. *versicolor* (Sims) Hook.f. & Thomson (2, 4)

निल कमल (Nil kamal)

Perennial aquatic herbs. E Nepal: *ca.* 200 m. Nepal, S Asia, E Asia and Africa.

Use(s): Ornamental, Ritual, Medicine (root, flowers, seeds).

Nymphaea odorata Aiton [39]

Fragrant water lily

Perennial aquatic herbs. E Nepal: *ca.* 1600 m (Naturalized). Native to N America and S America.

Nymphaea rubra Roxb. ex Andrews [6]

Red water lily

Perennial aquatic herbs. E Nepal: *ca.* 300 m (Naturalized). Native to S Asia.

Use: Ornamental.

Nymphaea tetragona Georgi [8, 36]

Nymphaea pygmaea Ait.

Pygmy water lily

Perennial aquatic herbs. W Nepal: 100-300 m. Nepal, S Asia, E Asia, Europe and N America.

Use(s): Ornamental, Food (leaf buds, seeds).

Exotic/Cultivated Taxa

Nymphaea mexicana Zucc. (**जल कमल,** Jal kamal; Yellow water lily). Native to S USA and Mexico. [14]

AUSTROBAILEYALES

SCHISANDRACEAE

Kadsura Juss.

Kadsura heteroclita (Roxb.) Craib. [8, 23]

Uvaria heteroclita Roxb.

Evergreen climbing shrubs. C, E Nepal: 1500-2000 m. Nepal, E Himalaya, Assam-Burma, S Asia and SE Asia.

Use(s): Medicine (essential oil), Food (fruit).

Schisandra Michx.

Schisandra grandiflora (Wall.) Hook.f. & Thomson [2, 4, 6, 10]

Kadsura grandiflora Wall.; *Sphaerostema grandiflorum* (Wall.) Blume

सिंगटो, सिगुन्टा लहरा (Singato, Sigunta laharaa)

Deciduous woody climbers. W, C, E Nepal: (1500-)2100-3400 m. Nepal, W Himalaya, E Himalaya and Tibetan Plateau.

Use(s): Food (fruit), Medicine (root).

Schisandra neglecta A.C.Sm. [2, 4, 6, 10]

Schisandra elongata Hook.f. & Thomson

Deciduous woody climbers. C, E Nepal: 1700-3100 m. Nepal, E Himalaya, Assam-Burma and E Asia.

Use(s): Food (fruit), Medicine (fruit), Fodder.

Schisandra propinqua (Wall.) Baill. [2, 4, 6, 10]

Kadsura propinqua Wall.; *Sphaerostema propinquum* (Wall.) Blume

पहेंलो सिंगटो (Pahelo singato)

Deciduous woody climbers. W, C, E Nepal: 1600-2200 m. Nepal, W Himalaya, E Himalaya, Tibetan Plateau, Assam-Burma, E Asia and SE Asia.

Use(s): Food (fruit), Medicine (root).

MESANGIOSPERMS

MAGNOLIIDS

PIPERALES

SAURURACEAE

Houttuynia Thunb.

Houttuynia cordata Thunb. [3, 4, 7]

गने (Gane); Fish mint

Perennial herbs. W, C, E Nepal: 1300-2500 m. Nepal, E Himalaya, Tibetan Plateau, Assam-Burma, E Asia and SE Asia.

Use(s): Medicine (root, stem, leaves), Food.

PIPERACEAE

Peperomia Ruiz & Pav.

Peperomia heyneana Miq. [3, 4, 7]

Perennial herbs. C, E Nepal: 900-2600 m. Nepal, E Himalaya, Tibetan Plateau, Assam-Burma and S Asia.

Use: Medicine (whole plant).

Peperomia pellucida (L.) Kunth [3, 4, 7]

Piper pellucidum L.; *Peperomia exigua* (Blume) Miq.

लतपते (Latpate)

Annual herbs. W, C, E Nepal: 100-2000 m. Nepal, E Himalaya, Assam-Burma, S Asia and SE Asia.

Use: Food (fruit).

Peperomia tetraphylla (G.Forst.) Hook. & Arn. [3, 4, 7]

Piper tetraphyllum G.Forst.; *Piper reflexum* L.f.

Perennial herbs. W, C, E Nepal: 1000-2500 m. Nepal, Tibetan Plateau, E Himalaya, S Asia, SE Asia, Africa and America.

Use: Medicine (whole plant).

Piper L.

Piper boehmeriifolium (Miq.) Wall. ex C.DC. [3, 4]

Chavica boehmeriifolia Miq.

ठुलो पिपला, लामो पिपला (Thulo piplaa, Lamo piplaa)

Subshrubs. C, E Nepal: 300-1500 m. Nepal, E Himalaya, Assam-Burma, E Asia and SE Asia.

Use(s): Medicine (fruit), Food (condiments).

Piper longum L. [3, 4, 7]

पिपला, गज पिपला (Piplaa, Gaj piplaa)

Climbing shrubs. W, C, E Nepal: 100-800 m. Nepal, E Himalaya, S Asia, E Asia and SE Asia.

Use(s): Food (fruit), Medicine (root, fruit).

Piper mullesua Buch.-Ham. ex D.Don [3, 4, 7]

Piper guigual Buch.-Ham. ex D.Don

पहाडी पान (Pahaadi paan)

Shrubs or climbers. W, C, E Nepal: 400-2500 m. Nepal, E Himalaya, Tibetan Plateau and S Asia.

Use(s): Medicine (root, fruit), Food (fruit).

Piper peepuloides Roxb. [3, 4, 7]

बन पिपला (Ban piplaa)

Climbing shrubs. E Nepal: 400-1500 m. Nepal, E Himalaya, Assam-Burma and S Asia.

Use(s): Medicine (fruit), Food (fruit).

Piper rhytidocarpum Hook.f. [7]

Climbing shrubs. E Nepal: *ca.* 1600 m. Nepal, E Himalaya, Tibetan Plateau and S Asia.

Piper suipigua Buch.-Ham. ex D.Don [3, 4, 7]

Piper nepalense Miq. (3, 4)

Climbing shrubs. W, C, E Nepal: 200-2000 m. Nepal, E Himalaya and E Asia.

Use: Medicine (fruit).

Piper wallichii (Miq.) Hand.-Mazz. [3, 4, 7]

Chavica wallichii Miq.; *Piper aurantiacum* Wall. ex DC.

टांग (Taang: Chepang)

Woody climbers. C, E Nepal: 100-2800 m. Nepal, E Himalaya, Assam-Burma and E Asia.

Use: Medicine (root, fruit).

Exotic/Cultivated Taxa

Piper betle L. (**पान,** Paan; Betel). Native to S Asia and SE Asia. *Use*: Medicine (root, leaves).

Piper nigrum L. (**मरिच,** Marich; Black pepper). Native to S Asia (S India). *Use(s)*: Food/Medicine (fruit). [15]

Doubtful/Uncertain Taxa

Piper chaba Blume (Jungali paan; wild betel leaves). Native to S Asia and SE Asia.

Piper saxatile Wall. (Wallich, Cat. no. 6664: Kew (K-W), K001124444). [107]

ARISTOLOCHIACEAE

Aristolochia L.

Aristolochia griffithii Hook.f. & Thomson ex Duch. [3, 4, 7]

Isotrema griffithii (Hook.f. & Thomson ex Duch.) C.E.C.Fisch.

इश्वरी, शिवलिंगी (Ishwari, Sibalingi)

Climbing shrubs. C, E Nepal: 2000-2900 m. Nepal, E Himalaya, Tibetan Plateau and Assam-Burma.

Use: Medicine (root).

Aristolochia nakaoi F.Maek. [3, 4]

Isotrema nakaoi (F.Maek.) A.R.Sm.

Climbers. C Nepal: *ca.* 2200 m. Nepal and E Himalaya.

Aristolochia platanifolia (Klotzsch) Duch. [3, 4]

Siphisia platanifolia Klotzsch

Stout climbers. E Nepal. Nepal and E Himalaya.

Aristolochia saccata Wall. [3, 4]

Siphisia saccata (Wall.) Klotzsch

Evergreen climbing shrubs. W, C Nepal: *ca.* 2100 m. Nepal, E Himalaya, Tibetan Plateau and E Asia.

Aristolochia tagala Cham. [3, 4, 7]

Aristolochia roxburghiana Klotzsch

Indian birthwort

Evergreen climbers. E Nepal: 1000-2400 m. Nepal, E Himalaya, Assam-Burma, S Asia, E Asia and SE Asia.

Use: Medicine (root).

Asarum L.

Asarum himalaicum Hook.f. & Thomson ex Klotzsch [3, 4, 7]

Perennial herbs. E Nepal: 1300-3100 m. Nepal, E Himalaya, Tibetan Plateau and E Asia.

Exotic/Cultivated Taxa

Aristolochia littoralis D.Parodi (*Aristolochia elegans* Mast); (**हैचा-पाली स्वाँ,** Haincha-paali swaan: Newari); Native to S America (Brazil). [14]

Doubtful/Uncertain Taxa

Aristolochia indica L. (Ishamul). *Use*: Medicine (root, leaves, seeds). [106]

MAGNOLIALES

MYRISTICACEAE

Horsfieldia Willd.

Horsfieldia kingii (Hook.f.) Warb. [3, 4, 10]

Myristica kingii Hook.f.

रुन्चे पात (Runche paat)

Evergreen trees. E Nepal: 300-600(-1300) m. Nepal, E Himalaya, Assam-Burma, E Asia and SE Asia.

Use: Medicine (stem, bark, leaves, seeds).

Knema Lour.

Knema tenuinervia W.J. de Wilde [3, 4, 10]

रातो भलायो (Raato bhalaayo)

Evergreen large trees. E Nepal: 100-1000(-1500) m. Nepal, E Himalaya, Assam-Burma and SE Asia.

Exotic/Cultivated Taxa

Myristica fragrans Houtt. (**जाइफल,** Jaiphal; Nutmeg). Native to SE Asia (Indonesia). Seeds imported for commercial purpose. *Use(s)*: Medicine/ Condiments (seeds).

MAGNOLIACEAE

Magnolia L.

Magnolia campbellii Hook.f. & Thomson [2, 4, 6, 10]

Yulania campbellii (Hook.f. & Thomson) D.L.Fu

पुर्बेली चाँप, घोगे चाँप (Purbeli chaanp, Ghoge chaanp)

Deciduous trees. C, E Nepal: 2200-2800 m. Nepal, E Himalaya, Tibetan Plateau, Assam-Burma and E Asia.

Use(s): Timber, Furniture.

Magnolia champaca (L.) Baill. ex Pierre [10]

Michelia champaca L. (2, 4, 6); *Michelia aurantiaca* Wall.

सुन चाँप, औले चाँप (Sun chaanp, Aule chaanp); Champak

Evergreen trees. C, E Nepal: 900-1900 m. Nepal, E Himalaya, Assam-Burma, S Asia, E Asia and SE Asia.

Use(s): Timber, Medicine (bark, flowers, seeds), Fodder, Ritual, Dye (flowers).

Magnolia doltsopa (Buch.-Ham. ex DC.) Figlar [10]

Michelia doltsopa Buch.-Ham. ex DC. (2, 4, 6); *Magnolia excelsa* Wall.; *Michelia excelsa* (Wall.) Blume

रानी चाँप (Raani chaanp); Sweet Michelia

Deciduous trees. C, E Nepal: 1900-2600 m. Nepal, E Himalaya, Tibetan Plateau, Assam-Burma and E Asia.

Use(s): Timber, Firewood, Fodder, Tannin (bark).

Magnolia globosa Hook.f. & Thomson [2, 4, 6, 10]

Oyama globosa (Hook.f. & Thomson) N.H.Xia & C.Y.Wu

खुकि चाँप (Khuki chaanp)

Deciduous trees. E Nepal: 2800-3400 m. Nepal, E Himalaya, Tibetan Plateau, Assam-Burma and E Asia.

Use: Timber.

Magnolia grandiflora L. [4, 14, 71]

रुख कमल (Rukh kamal); Bull bay

Evergreen trees. C Nepal: *ca.* 1400 m. Planted in the old gardens. Native to N America.

Use: Ornamental.

Magnolia hodgsonii (Hook.f. & Thomson) H.Keng [10]

Talauma hodgsonii Hook.f. & Thomson (2, 4, 6)

भालु काठ (Bhaalu kaath)

Evergreen trees. C, E Nepal: 900-1800 m. Nepal, E Himalaya, Tibetan Plateau, Assam-Burma, E Asia and SE Asia.

Use(s): Firewood, Handle of khukuri (Nepalese knife).

Magnolia insignis Wall. [10]

Manglietia insignis (Wall.) Blume (2, 4, 6)

बन कमल (Ban kamal); Red lotus tree

Evergreen trees. C Nepal: 1500-2700 m. Nepal, W Himalaya, Assam-Burma, E Asia and SE Asia.

Use: Firewood.

Magnolia kisopa (Buch.-Ham. ex DC.) Figlar [10]

Michelia kisopa Buch.-Ham. ex DC. (2, 4, 6); *Sampacca kisopa* (Buch.-Ham. ex DC.) Kuntze

बन चाँप (Ban chaanp)

Deciduous trees. W, C, E Nepal: 1000-3000 m. Nepal, W Himalaya, E Himalaya and Tibetan Plateau.

Use(s): Timber, Food (bark), Fodder.

Magnolia lanuginosa (Wall.) Figlar & Noot. [10]

Michelia lanuginosa Wall.; *Michelia velutina* DC. (2, 4, 6); *Sampacca lanuginosa* (Wall.) Kuntze

गोगल चाँप, सुन चम्पा (Gogal chaanp, Sun champa)

Deciduous trees. C, E Nepal: 300-2100 m. Nepal, E Himalaya, Tibetan Plateau, Assam-Burma and E Asia.

Use(s): Timber, Firewood.

Exotic/Cultivated Taxa

Magnolia coco (Lour.) DC. (*Lirianthe coco* (Lour.) N.H.Xia & C.Y.Wu); (Dwarf Magnolia). Native to E Asia [4, 98]

Magnolia liliiflora Desr. (Tulip Magnolia). Native to E Asia (SW China). [2, 4, 98]

Magnolia × soulangeana Soul.-Bad. (रक्त कमल, Rakta kamal; Saucer Magnolia) [14]

Magnolia figo (Lour.) DC. (*Michelia figo* (Lour.) Spreng.); (Banana Magnolia). [4, 98]

Doubtful/Uncertain Taxa

Magnolia pterocarpa Roxb. [2, 4]

ANNONACEAE

Alphonsea Roxb.

Alphonsea ventricosa (Roxb.) Hook.f. & Thomson [6, 131]

Uvaria ventricosa Roxb.

ठेकी फल (Theki phal)

Large trees. E Nepal: 100-200 m. Nepal, Assam-Burma and S Asia.

Use(s): Timber, Food (fruit pulp), Medicine (root).

Annona L.

Annona reticulata L. [2, 4, 6, 10]

रामफल (Ramphal); Custard apple, Bullock's heart

Small semi-evergreen trees. C Nepal: 200-1100 m. Cultivated or naturalized in disturbed habitats. Asia, Africa, N America, S America and Australasia.

Use(s): Medicine (root, bark, leaves, seeds), Food (fruit), Insecticide (seed oil).

Annona squamosa L. [2, 4, 6, 10]

सरिफा, सीताफल (Saripha, Sita phal); Sugar apple; Custard apple

Small evergreen trees or shrubs. W, C, E Nepal: 100-1000 m. Cultivated, sometimes naturalized in disturbed habitats. Asia, Africa, N America, S America and Australasia.

Use(s): Medicine (root, leaves, seeds), Food (fruit), Insecticide (seeds), Fish poison (leaves).

Artabotrys R.Br.

Artabotrys hexapetalus (L.f.) Bhandari [6, 10]

Annona hexapetala L.f.; *Annona uncinata* Lam.; *Artabotrys uncinatus* (Lam.) Merr.

कनक चम्पा, कटहर चम्पा (Kanak champaa, Katahar champaa); Tail grape

Scandent or climbing shrubs. W, C Nepal: 100-1300 m (Cultivated and naturalized). Nepal, Assam-Burma, S Asia and SE Asia.

Use(s): Medicine (oil), Ornamental, Ritual.

Desmos Lour.

Desmos chinensis Lour. [2, 4, 6, 10]

माले लहरा (Maale laharaa)

Woody climbing shrubs. E Nepal: 300-600 m. Nepal, E Himalaya, Assam-Burma, S Asia, E Asia and SE Asia.

Use: Medicine (root).

Miliusa Lesch. ex A.DC.

Miliusa globosa (A.DC.) Panigrahi & S.C.Mishra [6, 10]

Guatteria globosa A.DC.; *Miliusa roxburghiana* (Wall.) Hook.f. & Thomson (2, 4)

Small trees or large shrubs. E Nepal: 100-300 m. Nepal, E Himalaya, Assam-Burma and S Asia.

Use(s): Household utensils (wood), Medicine (leaves).

Miliusa macrocarpa Hook.f. & Thomson [6, 10]

काली काठ (Kaali kaath)

Small trees. E Nepal:1200-1700 m. Nepal, E Himalaya and Assam-Burma.

Miliusa tomentosa (Roxb.) J.Sinclair [2, 4, 6, 10]

Uvaria tomentosa Roxb.; *Saccopetalum tomentosum* (Roxb.) Hook.f. & Thomson

करी (Kari)

Large deciduous trees. W Nepal: 200-400 m. Nepal and S Asia.

Use(s): Timber, Construction materials, Food (fruit), Fodder, Fiber (bark), Gum.

Miliusa velutina (Dunal) Hook.f. & Thomson [2, 4, 6, 10]

Uvaria velutina Dunal

काली (Kaali); Velvety Miliusa

Small deciduous trees. C, E Nepal: 100-500 m. Nepal, Assam-Burma, S Asia and SE Asia.

Use(s): Timber, Construction materials, Fodder, Food (fruit), Medicine (bark), Fiber (stem).

Polyalthia Blume

Polyalthia fragrans (Dalzell) Benth. & Hook.f. [10]

Guatteria fragrans Dalzell

Evergreen trees. C, E Nepal: 100-500 m. Planted as avenue tree in Tarai. Nepal, S Asia and E Asia.

Use: Timber.

Polyalthia longifolia (Sonn.) Thwaites [6, 10]

Uvaria longifolia Sonn.

नक्कली अशोक (Nakkali Ashok); Mast tree

Evergreen trees. C Nepal: 100-300 m. Widely planted as avenue tree in Tarai. Nepal, S Asia and E Asia.

Use(s): Ornamental (road trees), Timber, Food (fruit), Medicine (bark).

Uvaria L.

Uvaria hamiltonii Hook.f. & Thomson [2, 4, 6, 10]

रबु लहरा (Rabu laharaa)

Large scandent shrubs. E Nepal: 300-600 m. Nepal, Assam-Burma, S Asia and SE Asia.

Use: Medicine (bark).

Doubtful/Uncertain Taxa

Melodorum bicolor (Roxb.) Hook.f. & Thomson [2, 4]

LAURALES

LAURACEAE

Actinodaphne Nees

Actinodaphne angustifolia (Blume) Nees [3, 4, 7, 10]

Litsea angustifolia Blume; *Tetranthera angustifolia* (Blume) Nees

झाँक्री काठ, लामपाते (Jhaankri kaath, Laampate)

Evergreen trees. C, E Nepal: 400-700 m. Nepal, E Himalaya and Assam-Burma.

Use(s): Fodder, Medicine (leaves, seeds), Timber.

Actinodaphne longipes Kosterm. [3, 4, 7, 10]

Actinodaphne reticulata var. *glabra* Meisn.

Evergreen trees. C, E Nepal: 1600-2600 m. Nepal, E Himalaya and Assam-Burma.

Actinodaphne obovata (Nees) Blume [3, 4, 10]

Tetradenia obovata Nees; *Tetranthera obovata* Buch.-Ham. ex Wall.

Evergreen trees. E Nepal: 300-1600 m. Nepal, E Himalaya, Tibetan Plateau, Assam-Burma and E Asia.

Use: Medicine (bark).

Actinodaphne sikkimensis Meisn. [3, 4, 7, 10]

खपटे (Khapate)

Evergreen trees. C, E Nepal: 1600-2700 m. Nepal, E Himalaya and Assam-Burma.

Beilschmiedia Nees

Beilschmiedia gammieana King ex Hook.f. [3, 4, 8, 10]

Large evergreen trees. C, E Nepal: 1000-2000 m. Nepal and E Himalaya.

Beilschmiedia roxburghiana Nees [3, 4, 7, 10]

Laurus bilocularis Roxb.

Large evergreen trees. E Nepal: 200-400 m. Nepal, E Himalaya, Tibetan Plateau, Assam-Burma, S Asia and SE Asia.

Use: Construction materials.

Cinnamomum Schaeffer

Cinnamomum bejolghota (Buch.-Ham.) Sweet [3, 4, 7, 10]

Laurus bejolghota Buch.-Ham.; *Cinnamomum obtusifolium* Roxb. ex Nees; *Laurus bazania* Buch.-Ham.

सिन्कौली, बडा सिंगोली, मरिचे मालागिरी (Sinkaulee, Badaa singolee, Mariche maalagiri)

Evergreen shrubs or small trees. E Nepal: 1200-1900 m. Nepal, E Himalaya, Assam-Burma, S Asia, E Asia and SE Asia.

Use(s): Medicine (bark), D-camphore a substitute for Sassafras.

Cinnamomum camphora (L.) J.Presl [3, 4, 7]

Laurus camphora L.

कपूर (Kapoor); Camphor tree

Evergreen trees. C Nepal: *ca.* 1400 m. Widely planted as avenue tree. Native to Asia.

Use(s): Medicine (root, stem, leaves), Volatile oil.

Cinnamomum glanduliferum (Wall.) Meisn. [3, 4, 10]

Laurus glandulifera Wall.; *Camphora gladulifera* (Wall.) Nees

सिन्कौली, मालागिरी, सुगन्ध कोकिला (Sinkaulee, Maalagiri, Sugandha kokilaa); Nepal camphor tree

Evergreen shrubs or small trees. W, C, E Nepal: 700-2600 m. Nepal, E Himalaya, Tibetan Plateau, Assam-Burma, E Asia and SE Asia.

Use(s): Medicine (wood, leaves, fruit), Essential oil (Sugandha kokila oil).

Note: *Cinnamomum glaucescens* (Buch.-Ham. ex Nees) Hand.-Mazz. (3, 4) and *Cinnamomum parthenoxylon* (Jack) Meisn. (3,4) are misapplied names for **Cinnamomum glanduliferum** (Wall.) Meisn, so both names are excluded (10).

Cinnamomum impressinervium Meisn. [3, 4, 8, 10]

Evergreen shrubs or small trees. C, E Nepal: 1700-2100 m. Nepal, E Himalaya and Assam-Burma.

Use: Medicine (leaves).

Cinnamomum tamala (Buch.-Ham.) Nees & Eberm. [3, 4, 7, 10]

Laurus tamala Buch.-Ham.; *Cinnamomum albiflorum* Nees

तेज पात (Tejpaat); Nepali Cinnamon

Evergreen shrubs or small trees. W, C, E Nepal: 400-2300 m. Nepal, W Himalaya, E Himalaya, Assam-Burma and E Asia.

Use(s): Food and food additives (bark, leaves), Essential oils (Cinnamon oil), Medicine (bark, leaves).

Cinnamomum tenuipile Kosterm. [7, 10, 23]

Alseodaphne mollis W.W.Sm.

सुगन्ध कोकिला (Sugandha kokilaa)

Deciduous trees. W, E Nepal: 400-1600 m. Nepal, E Himalaya and E Asia.

Cryptocarya R.Br.

Cryptocarya amygdalina Nees [3, 4, 7, 10]

Cryptocarya floribunda Nees; *Laurus amygdalina* Buch.-Ham. ex Wall., nom. nud.; *Laurus floribunda* Wall., nom. nud.

Large evergreen trees. C, E Nepal: 100-1600 m. Nepal, E Himalaya, Tibetan Plateau and Assam-Burma.

Use: Timber.

Dodecadenia Nees

Dodecadenia grandiflora Nees [3, 4, 7, 10]

Dodecadenia grandiflora var. *griffithii* (Hook.f.) D.G.Long; *Dodecadenia griffithii* Hook.f. (3, 4); *Laurus macrophylla* D.Don; *Tetranthera grandiflora* Wall., nom. nud.

नेपाली दालचीनी, काउले (Nepali daalchini, Kaaule)

Evergreen trees. W, C, E Nepal: 600-3400 m. Nepal, E Himalaya, Tibetan Plateau, Assam-Burma and E Asia.

Use: Fodder.

Lindera Thunb.

Lindera assamica (Meisn.) Kurz [3, 4, 7, 10]

Aperula assamica Meisn.

Evergreen trees. E Nepal: 2000-2800 m. Nepal, E Himalaya and Assam-Burma.

Use(s): Timber, Fodder.

Lindera bootanica Meisn. [4, 7, 10]

Daphnidium venosum Meisn.; *Lindera venosa* (Meisn.) Hook.f. (3, 4)

Deciduous trees. E Nepal: 1800-2000 m. Nepal and E Himalaya.

Lindera heterophylla Meisn. [3, 4, 7, 10]

Deciduous shrubs or small trees. C, E Nepal: 1900-3200 m. Nepal and E Himalaya.

Lindera nacusua (D.Don) Merr. [3, 4, 7, 10]

Laurus nacusua D.Don; *Laurus nacushia* Buch.-Ham., nom. nud.; *Laurus umbellata* Buch.-Ham. ex D.Don

पहेंलो खपटे (Pahenlo khapate)

Evergreen shrubs or small trees. W, C, E Nepal: 1200-2500 m. Nepal, W Himalaya, E Himalaya, Tibetan Plateau, Assam-Burma and E Asia.

Lindera neesiana (Wall. ex Nees) Kurz [3, 4, 7, 10]

Benzoin neesianum Wall. ex Nees; *Aperula neesiana* (Wall. ex Nees) Blume

सिल टिम्मुर, ठेकी फूल (Sil timur, Theki phul)

Deciduous shrubs or small trees. C, E Nepal: 800-2900 m. Nepal, E Himalaya, Assam-Burma and E Asia.

Use(s): Medicine (root, bark, leaves, seeds), Food, Essential oil (Lindera oil).

Lindera pulcherrima (Nees) Hook.f. [3, 4, 7, 10]

Daphnidium pulcherrimum Nees; *Lindera pulcherrima* var. *attenuata* C.K.Allen (3, 4)

फुसुरे, खराने (Phusure, Kharaane)

Evergreen shrubs or small trees. W, C, E Nepal: 700-3600 m. Nepal, W Himalaya, E Himalaya, Tibetan Plateau, Assam-Burma and E Asia.

Use(s): Timber, Fodder, Firewood.

Litsea Lam.

Litsea chartacea (Wall. ex Nees) Hook.f. [3, 4, 7, 10]

Tetranthera chartacea Wall. ex Nees

Evergreen trees. C Nepal: 1500-1800 m. Nepal, E Himalaya and E Asia.

Use(s): Fodder, Firewood.

Litsea cubeba (Lour.) Pers. [3, 4, 7, 10]

Laurus cubeba Lour.; *Persea cubeba* (Lour.) Spreng.; *Tetranthera cubeba* (Lour.) Kostel.

सिल टिम्मुर (Sil timur); Mountain pepper

Evergreen trees. C, E Nepal: 1000-2600 m. Nepal, E Himalaya, Assam-Burma and E Asia.

Use(s): Medicine (root, bark, stem, leaves, seeds), Essential oil (Litsea oil), Fish poison.

Litsea doshia (D.Don) Kosterm., *J. Sci. Res. [Jakarta]* **1**: 90. 1952. [3, 4, 7, 10, 12, 29]

Tetranthera doshia D.Don; *Litsea oblonga* (Wall. ex Nees) Hook.f.; *Tetranthera oblonga* Wall. ex Nees

पहेंली, काठे काउलो (Pahenli, Kaathe kaaulo)

Medium-sized evergreen trees. W, C, E Nepal: 300-2900 m. **Endemic to Nepal**.

Central: *Kathmandu*, Narainhetty, F. Buchanan-Hamilton s.n. (Holotype: BM; Isotype: LINN).

Use: Fodder.

Litsea elongata (Nees) Hook.f. [3, 4, 7, 10]

Daphnidium elongatum Nees; *Tetranthera elongata* Wall., nom. nud.

Evergreen trees. C, E Nepal: 1500-3700 m. Nepal, W Himalaya, E Himalaya, Tibetan Plateau, Assam-Burma and E Asia.

Use: Fodder.

Litsea glutinosa (Lour.) C.B.Rob. [3, 4, 10]

Sebifera glutinosa Lour.; *Tetranthera laurifolia* Jacq.; *Tetranthera polycephala* Wall. ex Meisn.

मेदा (Medaa); Bolly beech

Deciduous trees. C, E Nepal: 100-1000 m. Nepal, W Himalaya, E Himalaya, Assam-Burma, S Asia, E Asia and SE Asia.

Use(s): Medicine (root, bark, fruit), Food (fruit), Construction materials.

Litsea hookeri (Meisn.) D.G.Long [10]

Cylicodaphne hookeri Meisn.; *Litsea khasyana* Meisn. (3, 4); *Tetranthera khasyana* Meisn.

Evergreen trees. C, E Nepal: 200-800 m. Nepal, E Himalaya and Assam-Burma.

Litsea kingii Hook.f. [4, 8, 10, 24]

Deciduous trees. E Nepal: 1900-2900 m. Nepal, E Himalaya, Tibetan Plateau, Assam-Burma and E Asia.

Use: Fodder.

Litsea monopetala (Roxb.) Pers. [3, 4, 7, 10]

Tetranthera monopetala Roxb.; *Litsea polyantha* Juss.

कुटमिरो, कदमेरो (Kutmiro, Kadmero)

Evergreen trees. C, E Nepal: 80-2100 m. Nepal, W Himalaya, E Himalaya, Assam-Burma, S Asia, E Asia and SE Asia.

Use(s): Fodder, Medicine (root, bark), Household construction, Agricultural implements.

Litsea panamanja (Buch.-Ham. ex Nees) Hook.f. [7, 10]

Tetranthera panamanja Buch.-Ham. ex Nees

Evergreen trees. E Nepal: 700-800 m. Nepal, E Himalaya, Assam-Burma, E Asia and SE Asia.

Litsea salicifolia (J.Roxb. ex Nees) Hook.f. [3, 4, 7, 10]

Tetranthera salicifolia J.Roxb. ex Nees; *Tetranthera glauca* Wall. ex Nees

पहेंली (Pahenli)

Evergreen trees. C, E Nepal: 100-800 m. Nepal, E Himalaya, Assam-Burma, S Asia, E Asia and SE Asia.

Use: Fodder.

Litsea sericea (Wall. ex Nees) Hook.f. [3, 4, 7, 10]

Tetranthera sericea Wall. ex Nees

पहेले (Pahele)

Deciduous shrubs or small trees. C, E Nepal: 1300-3900 m. Nepal, E Himalaya, Tibetan Plateau, S Asia and E Asia.

Use: Fodder.

Machilus Nees

Machilus clarkeana King ex Hook.f. [7, 10]

Machilus gammieana King ex Hook.f.; *Persea clarkeana* (King ex Hook.f.) Kosterm. (3, 4); *Persea gammieana* (Hook.f.) Kosterm. (3, 4)

सेतो काउलो (Seto kaaulo)

Evergreen trees. W, C, E Nepal: 1600-2900 m. Nepal, E Himalaya and Assam-Burma.

Use: Fodder.

Machilus duthiei King ex Hook.f. [3, 4, 7, 10]

Persea duthiei (King ex Hook.f.) Kosterm. (3, 4)

काठे काउलो, माहिलो काउलो (Kaathe kaaulo, Maahilo kaaulo)

Evergreen large trees. W, C, E Nepal: 1500-3200 m. Nepal, W Himalaya, E Himalaya, Tibetan Plateau and Assam-Burma.

Use(s): Medicine (twig), Fodder, Firewood.

Machilus gamblei King ex Hook.f. [10]

Machilus bombycina King ex Hook.f.; *Persea bombycina* (King ex Hook.f.) Kosterm. (3, 4); *Persea gamblei* (King ex Hook.f.) Kosterm. (3, 4)

चिप्ले काउलो, काठे काउलो (Chiple kaaulo, Kaathe kaaulo)

Evergreen trees. W, C, E Nepal: 700-2400 m. Nepal, E Himalaya, Tibetan Plateau, Assam-Burma, E Asia and SE Asia.

Use(s): Construction materials, Fodder, Medicine (bark, leaves).

Machilus glaucescens (Nees) Wight [7, 10]

Ocotea glaucescens Nees; *Laurus villosa* Roxb.; *Persea glaucescens* (Nees) D.G.Long; *Persea villosa* (Roxb.) Kosterm. (3, 4); *Phoebe glaucescens* (Nees) Nees

Evergreen trees. C, E Nepal: 300-2400 m. Nepal, E Himalaya and Assam-Burma.

Machilus odoratissima Nees [7, 10]

Laurus odoratissima Wall., nom. nud.; *Persea odoratissima* (Nees) Kosterm. (3, 4)

भाटे काउलो, सेतो काउलो (Bhate kaulo, Seto kaaulo); Fragrant bay tree

Evergreen trees. W, C, E Nepal: 300-2200 m. Nepal, W Himalaya, E Himalaya, Assam-Burma and SE Asia.

Use(s): Timber, Fodder.

Machilus pubescens Blume, *Mus. Bot.* **1**: 330. 1851. [10, 29]

Persea blumei Kosterm. (3, 4); *Cinnamomum tomentosum* D.Don; *Persea tomentosa* (D.Don) Spreng. (4, 27)

Evergreen trees. C Nepal: 100-2500 m. **Endemic to Nepal**.

Central: Nepalia, N. Wallich s.n. (Type: K-W); Suembu, Kathmandu, Buchanan-Hamilton s.n. (Holotype: BM), [type of *Persea tomentosa* (D.Don) Spreng.]

Machilus sericea Blume [7, 10]

Laurus sericea Wall., nom. nud.; *Persea wallichii* D.G.Long (4); *Phoebe sericea* Nees, nom. illegit.

Evergreen trees. W, C Nepal: 200-2500 m. Nepal and W Himalaya.

Neocinnamomum H.Liu

Neocinnamomum caudatum (Nees) Merr. [3, 4, 7, 10]

Cinnamomum caudatum Nees; *Laurus caudata* Wall. ex Nees

Evergreen trees. C, E Nepal: 900-2600 m. Nepal, E Himalaya, Assam-Burma, E Asia and SE Asia.

Neolitsea (Benth.) Merr.

Neolitsea cuipala (D.Don) Kosterm. [3, 4, 7, 10]

Tetranthera cuipala D.Don; *Litsea cuipala* (D.Don) Nees; *Tetradenia lanuginosa* Nees

कल्चे (Kalche: Gurung)

Medium-sized trees. C, E Nepal: 900-2100 m. Nepal, W Himalaya, E Himalaya and Assam-Burma.

Use: Medicine (seed oil).

Neolitsea foliosa (Nees) Gamble [7, 10]

Tetradenia foliosa Nees; *Litsea foliosa* (Nees) Nees; *Tetranthera foliosa* Wall., nom.nud.

Evergreen trees. C, E Nepal: 1500-3500 m. Nepal, E Himalaya and Assam-Burma.

Neolitsea pallens (D.Don) Momiy. & H.Hara [3, 4, 7, 10]

Tetranthera pallens D.Don; *Litsea consimilis* (Nees) Nees; *Neolitsea umbrosa* (Nees) Gamble; *Tetradenia consimilis* Nees

मकै काठ, खपटे (Makai kaath, Khapate)

Evergreen small trees. W, C, E Nepal: 1700-3500 m. Nepal, W Himalaya and Tibetan Plateau.

Use(s): Construction materials, Fodder, Medicine (fruit).

Phoebe Nees

Phoebe attenuata (Nees) Nees [3, 4, 8, 10]

Ocotea attenuata Nees; *Laurus attenuata* Wall., nom. nud.

Evergreen trees. E Nepal: 1000-1400 m. Nepal, E Himalaya and Assam-Burma.

Use: Timber.

Phoebe cathia (D.Don) Kosterm. [3, 4, 8, 10, 12]

Cinnamomum cathia D.Don; *Laurus cathia* Buch.-Ham., nom.nud.

Evergreen trees. C Nepal: 1300-1900 m. Nepal, E Himalaya, Assam-Burma and S Asia.

Phoebe hainesiana Brandis [10]

Evergreen trees. C, E Nepal: 1300-1600 m. Nepal and E Himalaya.

Use: Construction materials.

Phoebe lanceolata (Nees) Nees [3, 4, 7, 10]

Ocotea lanceolata Nees; *Laurus lanceolaria* Roxb.; *Laurus salicifolia* Buch.-Ham. ex Nees

झाँक्री काठ (Jhaankri kaath)

Evergreen trees. W, C, E Nepal: 100-2900 m. Nepal, E Himalaya, Assam-Burma, S Asia, E Asia and SE Asia.

Use(s): Construction materials, Fodder, Medicine (fruit).

Phoebe pallida (Nees) Nees [3, 4, 7, 10]

Ocotea pallida Nees; *Laurus pubescens* Wall.; *Persea pallida* (Nees) Oliv.

Evergreen trees. C Nepal: 1800-2000 m. Nepal, W Himalaya and Assam-Burma.

Exotic/Cultivated Taxa

Cinnamomum verum J.Presl. (Syn. *Cinnamomum zeylanicum* Blume); (**दाल चीनी**, Daalchini; True cinnamon). Native to Sri Lanka. *Use(s)*: Medicine/ Condiments (bark).

Persea americana Mill. (Avocado). Native to Mexico. Cultivated for edible fruit. *Use(s)*: Food (fruit), Medicine (bark, leaves, fruit).

Doubtful/Uncertain Taxa

Actinodaphne reticulata Meisn. [3, 4]

Cinnadenia paniculata (Hook.f.) Kosterm. [3, 4]

Litsea confertiflora (Meisn.) Kosterm. [3, 4]

Litsea glabrata (Wall. ex Nees) Hook.f. [3, 4]

Litsea lancifolia (Roxb. ex Nees) Benth. & Hook.f. ex Fern.-Vill. [3, 4]

CHLORANTHALES

CHLORANTHACEAE

Chloranthus Sw.

Chloranthus erectus (Buch.-Ham.) Verdc. [4]

Cryphaea erecta Buch.-Ham.; *Chloranthus elatior* Link (3); *Chloranthus officinalis* Blume

Undershrubs. E Nepal: 100-2000 m. Nepal, E Himalaya, E Asia and SE Asia.

Use(s): Medicine (whole plant), Food/Beverage (twigs).

MONOCOTS

ACORALES

ACORACEAE

Acorus L.

Acorus calamus L. [1, 4, 5]

Acorus angustatus Raf.; *Acorus angustifolius* Schott; *Acorus asiaticus* Nakai

बोझो (Bojho); Sweet flag

Perennial marshy herbs. W, C, E Nepal: 100-2300 m. Nepal, Asia, Europe and N America.

Use(s): Medicine (rhizome), Essential oil (rhizome), Ornamental.

ALISMATALES

ARACEAE

Alocasia (Schott) G.Don

Alocasia fornicata (Roxb.) Schott [5]

Arum fornicatum Roxb.; *Alocasia fornicata* (Kunth) Schott

Elephant ear

Perennial herbs. E Nepal: 200-300 m. Nepal, S Asia and SE Asia.

Alocasia macrorrhizos (L.) G.Don [8]

Arum macrorrhizon L.; *Philodendron punctatum* Kunth

घ्याम्पे तरुल, जाँते तरुल (Ghyaampe tarul, Jaante tarul); Giant Taro

Perennial herbs. C Nepal: 700-800 m. Nepal, Tibetan Plateau, S Asia, E Asia, SE Asia and Australasia.

Use: Medicine (tuber, leaves).

Alocasia navicularis (K.Koch & C.D.Bouché) K.Koch & C.D.Bouché [1, 4]

Colocasia navicularis K.Koch & C.D.Bouché

पिंडालु, माने (Pindaalu, Maane); Taro

Perennial herbs. E Nepal: 400-500 m. Nepal, Assam-Burma, S Asia, E Asia and SE Asia.

Use: Use to obtain starches and cellulose products.

Amorphophallus Blume ex Decne.

Amorphophallus bulbifer (Roxb.) Blume [1, 4, 5]

Arum bulbiferum Roxb.

ओल (Ol); White yam

Perennial herbs. W, C, E Nepal: 300-900 m. Nepal, E Himalaya, Assam-Burma and S Asia.

Use: Medicine (corm).

Amorphophallus napalensis (Wall.) Bogner & Mayo [5]

Thomsonia napalensis Wall.

Perennial herbs. W, C Nepal: 700-1500 m. Nepal and E Himalaya.

Amorphophallus paeoniifolius (Dennst.) Nicolson [67]

Dracontium paeoniifolium Dennst.; *Amorphophallus campanulatus* Blume ex Decne.

Elephant foot yam

Perennial herbs. W, C Nepal: 100-300 m. Nepal, Assam-Burma, S Asia, E Asia, SE Asia and Australasia.

Use(s): Medicine (corm), Food (corm).

Ariopsis Nimmo

Ariopsis peltata Nimmo [5]

Perennial herbs. W, C Nepal: 600-2000 m. Nepal, E Himalaya and Assam-Burma.

Use: Food (leaf as vegetable).

Arisaema Mart. (**बाँको, सर्पको मकै,** Baanko, Sarpako makai); Cobra lily

Arisaema concinnum Schott [1, 4, 5]

Perennial herbs. C, E Nepal: 800-2500 m. Nepal, W Himalaya, E Himalaya, Tibetan Plateau, Assam-Burma and SE Asia.

Use: Medicine (tuber).

Arisaema costatum (Wall.) Mart. [1, 4, 5]

Arum costatum Wall.

काल (Kaal)

Perennial herbs. W, C, E Nepal: 1200-2800 m. Nepal and Tibetan Plateau.

Use(s): Medicine (rhizome), Food (rhizome).

Arisaema echinatum (Wall.) Schott [1, 4, 5]

Arum echinatum Wall.

Perennial herbs. C, E Nepal: 1500-3400 m. Nepal, E Himalaya, Tibetan Plateau and E Asia.

Arisaema erubescens (Wall.) Schott [1, 4, 5]

Arum erubescens Wall.; *Arisaema consanguineum* Schott (1, 4, 5)

रकाया बाँको (Rakaayaa baanko)

Perennial herbs. W, C, E Nepal: 1600-3300 m. Nepal, E Himalaya, Tibetan Plateau, Assam-Burma, E Asia and SE Asia.

Use(s): Food (tender shoot), Medicine (rhizome).

Arisaema exappendiculatum H.Hara [1, 4, 5]

Dragon lily

Perennial tuberous herbs. C, E Nepal: 2300-2900 m. Nepal and Tibetan Plateau.

Arisaema flavum (Forssk.) Schott [1, 4]

Arum flavum Forssk.; *Arisaema flavum* var. *abbreviatum* Schott ex J.Murata (1, 4)

तिन्चु; धावा (Tinchu; Dhawa: Amchi)

Perennial tuberous herbs. W, C Nepal: 2000-4300 m. Nepal, W Himalaya, Tibetan Plateau, S Asia, SW Asia and Africa.

Use(s): Medicine (root/tuber, flowers), Food (tender shoot).

Arisaema griffithii Schott [1, 4]

var. **griffithii**

ढोकायो साग (Dhokaayo saag)

Perennial herbs. C, E Nepal: 2400-3600 m. Nepal, E Himalaya, Tibetan Plateau and Assam-Burma.

Use(s): Medicine (leaves), Food (tender shoot), Ornamental.

var. **verrucosum** (Schott) H.Hara [1, 4, 5]

Arisaema verrucosum Schott; *Arisaema pradhanii* C.E.C.Fisch.

Perennial herbs. E Nepal: *ca.* 300 m. Nepal, E Himalaya and Assam-Burma.

Arisaema intermedium Blume [1, 4, 5]

var. **intermedium**

Perennial herbs. W, C, E Nepal: 2000-3500 m. Nepal, W Himalaya and Tibetan Plateau and E Asia.

Use: Medicine (spadix, leaves).

var. **biflagellatum** (H.Hara) Pradhan [1, 4]

Arisaema biflagellatum H.Hara

Perennial herbs. W Nepal: *ca.* 3500 m. Nepal and E Himalaya.

Arisaema jacquemontii Blume [1, 4, 5]

Arisaema cornutum Schott; *Arisaema exile* Schott

बाँको; धावा (Baanko; Dhawa: Amchi)

Perennial tuberous herbs. W, C, E Nepal: 2400-4300 m. Nepal, W Himalaya, E Himalaya, Tibetan Plateau, S Asia and SW Asia.

Use(s): Medicine (root/tuber, leaves, fruit), Food.

Arisaema nepenthoides (Wall.) Mart. [1, 4, 5]

Arum nepenthoides Wall.; *Arisaema ochraceum* Schott

तुवा (Tuwaa)

Perennial herbs. C, E Nepal: 1900-3600 m. Nepal, E Himalaya, Tibetan Plateau, Assam-Burma and E Asia.

Arisaema propinquum Schott [1, 4, 5]

Arisaema sikkimense Stapf ex Chatterjee; *Arisaema wallichianum* Hook.f.

Perennial herbs. W, C, E Nepal: 2400-3800 m. Nepal, W Himalaya, E Himalaya, Tibetan Plateau, S Asia and SW Asia.

Use: Medicine (tuber).

Arisaema speciosum (Wall.) Mart.

Arum speciosum Wall.

var. **speciosum** [1, 4, 5]

बिष झार्ने (Bish jhaarne)

Perennial herbs. C, E Nepal: 2200-3500 m. Nepal, E Himalaya and Tibetan Plateau.

Use: Medicine (tuber).

var. **mirabile** (Schott) Engl. [1, 4, 5]

Arisaema mirabile Schott

बिष झार्ने (Bish jhaarne)

Perennial herbs. C, E Nepal: 2200-3500 m. Nepal and E Himalaya.

Arisaema tortuosum (Wall.) Schott [1, 4]

Arum tortuosum Wall.

var. **tortuosum**

बीर बाँको (Bir baanko); Whipcord lily

Perennial herbs. W, C, E Nepal: 900-2900 m. Nepal, W Himalaya, E Himalaya, Tibetan Plateau, Assam-Burma, S Asia, E Asia and SW Asia.

Use(s): Food (corm), Medicine (corm), Insecticide.

var. **curvatum** (Roxb.) Engl. [1, 4, 5]

Arum curvatum Roxb.

Perennial herbs. C Nepal: 1500-2500 m. Nepal, W Himalaya, E Himalaya and Assam-Burma.

Arisaema utile Hook.f. ex Schott [1, 4, 5]

ढोकाया, तिन पाते (Dhokaya, Tinpate); Sikkim cobra lily

Perennial herbs. W, C, E Nepal: 2500-3800 m. Nepal, E Himalaya, Tibetan Plateau, S Asia, E Asia and SW Asia.

Use: Food (leaves, spathe).

Arisaema vexillatum H.Hara & H.Ohashi [1, 4, 5]

Perennial herbs. C, E Nepal: 3300-4000 m. Nepal and Tibetan Plateau.

Colocasia Schott

Colocasia affinis Schott [1, 4, 8]

Colocasia bicolor C.L.Long & L.M.Cao

Perennial marshy herbs. C Nepal: *ca.* 1200 m. Nepal, E Himalaya, S Asia and E Asia.

Colocasia esculenta (L.) Schott [1, 4, 5]

Arum esculentum L.; *Arum chinense* L.; *Arum colocasia* L.; *Caladium esculentum* (L.) Vent.

गाभा, कर्कलो (Gaavaa, Karkalo); Coco yam, Wild taro

Perennial marshy herbs. W, C, E Nepal: 300-1200 m. Nepal, native to Asia.

Use(s): Medicine (corm, leaves), Food (corm, leaves).

Colocasia fallax Schott [1, 4, 5]

Colocasia yunnanensis C.L.Long & X.Z.Cai

Dwarf elephant ear

Perennial marshy herbs. W, C, E Nepal: 300-2000 m. Nepal, E Himalaya, S Asia and E Asia.

Use: Food (corm, leaves)

Gonatanthus Klotzsch

Gonatanthus pumilus (D.Don) Engl. & K.Krause [1, 4]

Caladium pumilum D.Don; *Remusatia pumila* (D.Don) H.Li & A.Hay (5)

जलुको, माने (Jaluko, Maane); Pygmy elephant ear

Perennial epiphytic or terrestrial herbs. W, C, E Nepal: 1000-2800 m. Nepal, E Himalaya, Tibetan Plateau, S Asia, E Asia and SE Asia.

Use: Food (tender leaves).

Lasia Lour.

Lasia spinosa (L.) Thwaites [1, 4, 5]

Dracontium spinosum L.; *Pothos heterophyllus* Roxb.; *Pothos spinosus* (L.) Buch.-Ham. ex Wall.

काँडे कर्कलो (Kaande karkalo)

Perennial herbs. E Nepal: 300-700 m. Nepal, E Himalaya, Tibetan Plateau, Assam-Burma, S Asia E Asia and SE Asia.

Use(s): Medicine (root), Food (leaves, fruit).

Lemna L.

Lemna aequinoctialis Welw.

Lemna perpusilla Torr. (5)

Perennial or annual aquatic herbs. W, C, E Nepal: 600-2300 m. Nepal, native to temperate and tropical regions.

Lemna minor L. (1, 4)

हाँसको चिउरा (Haansko chiuraa); Common duckweed

Perennial aquatic herbs. W Nepal: *ca.* 3000 m. Nepal, S Asia, SW Asia, Europe, Africa, N America and Australasia.

Use: Forage (for water birds and fishes).

Pistia L.

Pistia stratiotes L. [1, 4, 5]

Pistia obcordata Schleid.

जल कुम्भी, पानी बन्दा (Jal kumbhi, Paani bandaa); Water lettuce

Perennial aquatic herbs. W, C, E Nepal: 200-600 m. Invasive species in Nepal. Native to S America.

Use(s): Medicine (root, leaves), Food (leaves), Green manure.

Pothos L.

Pothos chinensis (Raf.) Merr.

Tapanava chinensis Raf.; *Pothos cathcartii* Schott (1, 4, 5)

Perennial climbers. W, C, E Nepal: 200-1500 m. Nepal, E Himalaya, Tibetan Plateau, Assam-Burma, S Asia, E Asia and SE Asia.

Use: Medicine (leaves).

Remusatia Schott

Remusatia hookeriana Schott [1, 4, 5]

Gonatanthus ornatus Schott

Perennial epiphytic herbs. C, E Nepal: 1500-2400 m. Nepal, E Himalaya, Tibetan Plateau, E Asia and SE Asia.

Remusatia vivipara (Roxb.) Schott [1, 4, 5]

Arum viviparum Roxb.; *Caladium viviparum* (Roxb.) Nees; *Colocasia vivipara* (Roxb.) Thwaites

कालो पिंडालु, जलुका (Kaalo pidaalu, Jalukaa)

Perennial epiphytic herbs. W, C, E Nepal: 1500-2400 m. Nepal, E Himalaya, Tibetan Plateau, S Asia, E Asia, SE Asia, SW Asia, Africa and Australasia.

Use(s): Medicine (root), Food (young leaves, tubers).

Rhaphidophora Hassk.

Rhaphidophora decursiva (Roxb.) Schott [1, 4, 5]

Pothos decursivus Roxb.; *Rhaphidophora grandis* Schott. (5); *Scindapsus decursivus* (Roxb.) Schott

Creeping Philodendron

Epiphytic or terrestrial climbers. C, E Nepal: 1100-2000 m. Nepal, E Himalaya, Tibetan Plateau, Assam-Burma, S Asia, E Asia and SE Asia.

Rhaphidophora glauca (Wall.) Schott [1, 4, 5]

Pothos glaucus Wall.; *Monstera glauca* K.Koch ex Ender; *Rhaphidophora glauca* var. *khasiana* Hook.f.; *Scindapsus glaucus* (Wall.) Schott

हड्डी जोर, कान चिरुवा (Haddi jor, Kaan chiruwaa)

Epiphytic climbers. W, C, E Nepal: 400-1800 m. Nepal, E Himalaya, Tibetan Plateau, Assam-Burma, S Asia and SE Asia.

Use: Medicine (whole plant).

Sauromatum Schott

Sauromatum brevipes (Hook.f.) N.E.Br. [1, 4, 5]

Typhonium brevipes Hook.f.

बन कर्कलो (Ban karkalo)

Perennial epiphytic herbs. W, C, E Nepal: 1400-2800 m. Nepal, E Himalaya, Tibetan Plateau and S Asia.

Use: Food (spathe).

Sauromatum diversifolium (Wall. ex Schott) Cusimano & Hett.

Typhonium diversifolium Wall. ex Schott (1, 4); *Typhonium alpinum* C.Y.Wu ex H.Li

तालु (Taalu)

Perennial herbs. W, C Nepal: 2400-4300 m. Nepal, W Himalaya, E Himalaya, Tibetan Plateau, Assam-Burma, E Asia and SE Asia.

Use(s): Medicine (root, flowers, fruit), Food (young leaves).

Sauromatum venosum (Dryand. ex Aiton) Kunth [1, 4]

Arum venosum Dryand. ex Aiton; *Arisaema venosum* (Aiton) Blume

Perennial herbs. W, C Nepal: 900-2500 m. Nepal, W Himalaya, E Himalaya, Tibetan Plateau, Assam-Burma, E Asia and Africa.

Use(s): Medicine (tuber, stem), Food (stem, leaves).

Scindapsus Schott

Scindapsus officinalis (Roxb.) Schott [1, 4, 5]

Pothos officinalis Roxb.

कान चिर्नु, गज पिपला, ठुलो पिपला (Kaan chirnu, Gaj pipalaa, Thulo pipalaa)

Perennial climbers. W, C, E Nepal: 200-1300 m. Nepal, E Himalaya, Assam-Burma, S Asia and SE Asia.

Use(s): Medicine (stem, leaves, fruit), Food (seeds).

Spirodela Schleid.

Spirodela polyrhiza (L.) Schleid. [5]

Lemna polyrhiza L.

Greater duck weed

Perennial aquatic herbs. W, C, E Nepal: 60-1300 m. Cosmopolitan.

Use: Medicine (whole plant).

Typhonium Schott

Typhonium trilobatum (L.) Schott [1, 4, 5]

Arum trilobatum L.

Bengal Arum

Perennial herbs. C Nepal: 300-2500 m. Nepal, E Himalaya, Assam-Burma, S Asia, E Asia and SE Asia.

Use: Medicine (root).

Wolffia Horkel ex Schleid.

Wolffia globosa (Roxb.) Hartog & Plas [1, 4]

Lemna globosa Roxb.

Asian watermeal

Perennial aquatic herbs. C, E Nepal: *ca.* 1400 m. Nepal, Tibetan Plateau, S Asia, SE Asia, Africa and Australasia.

Use: Food (whole plant).

Note: Smallest flowering plant of the world (0.1-0.2 mm in diameter).

Exotic/Cultivated Taxa

Caladium bicolor (Aiton) Vent. (**टिके कर्कलो,** Tike karkalo). Native to S America.

Epipremnum aureum (Linden & André) G.S.Bunting, (*Pothos aureus* Linden & André); (Money plant). Native to French Polynesia.

Monstera deliciosa Liebm. (**लहरे कर्कलो;** Split leaves Philodendron); Native to tropical America.

Xanthosoma sagittifolium (L.) Schott (**दर्शन पिपल;** Darshan peepal); Native to S America.

Zantedeschia aethiopica (L.) Spreng. (14, 54); (**शंख फूल;** Shankha phul); Calla lily. Native to S Africa.

TOFIELDIACEAE

Tofieldia Huds.

Tofieldia himalaica Baker [1, 4, 8]

Asphodeliris himalaica (Baker) Kuntze

Perennial herbs. C, E Nepal: 2100-3500 m. Nepal and E Himalaya.

ALISMATACEAE (incl. BUTOMACEAE)

Alisma L.

Alisma orientale (Sam.) Juz

Alisma plantago-aquatica var. *orientale* Sam.; *Alisma plantago-aquatica* subsp. *orientale* (Sam.) Sam. (1, 4)

जल पर्नी (Jal parnee); Water plantain

Perennial aquatic herbs. W Nepal: *ca.* 1000 m. Nepal, Tibetan Plateau, Assam-Burma, S Asia, E Asia, N Asia, C Asia, SW Asia, Europe, Africa, N America and Australasia.

Use: Medicine (root, leaves, fruit).

Butomopsis Kunth

Butomopsis latifolia (D.Don) Kunth [5]

Butomus latifolius D.Don; *Butomopsis lanceolata* (Roxb.) Kunth; *Butomus lanceolatus* Roxb.

कर्कले झार (Karkale jhaar); Bladder wort

Annual aquatic herbs. W, C, E Nepal: 200-1400 m. Nepal, Assam-Burma, S Asia, E Asia, SE Asia, Africa and Australasia.

Use: Medicine (whole plant).

Caldesia Parl.

Caldesia parnassifolia (L.) Parl.

Alisma parnassifolium L.; *Alisma reniforme* D.Don; *Caldesis reniformis* (D.Don) Makino (1, 4)

Perennial or annual herbs. C Nepal: 100-700 m. Nepal, S Asia, E Asia, N Asia, SW Asia, Europe, Africa and Australasia.

Sagittaria L.

Sagittaria guayanensis Kunth

subsp. **lappula** (D.Don) Bogin [5]

Sagittaria lappula D.Don; *Lophotocarpus guayanensis* var. *lappula* (D.Don) Buchenau

लाफ (Laaph); Guayanese arrow head

Annual or perennial aquatic herbs. W, C, E Nepal: 100-1300 m. Nepal, S Asia, E Asia, SE Asia, SW Asia and Africa.

Use: Medicine (whole plant).

Sagittaria trifolia L. [1, 4, 5]

Sagittaria hastata D.Don

लाफ (Laaph); Water plantain

Perennial aquatic herbs. W, C Nepal: 200-600 m. Nepal, Tibetan Plateau, S Asia, E Asia, N Asia, C Asia, SW Asia, Europe and Africa.

Use: Medicine (whole plant).

Doubtful/Uncertain Taxa

Sagittaria montevidensis Cham. & Schltdl., W Nepal: *ca.* 200 m. Native of N America and S America. [124]

Sagittaria sagittifolia L., C Nepal: *ca.* 200 m. Nepal, Tibetan Plateau, S Asia, E Asia, SE Asia, Europe and Australasia.

HYDROCHARITACEAE (incl. NAJADACEAE)

Blyxa Noronha ex Thouars.

Blyxa aubertii Rich. [1, 4, 5]

Blyxa ecaudata Hayata

लसुने (Lasune); Bamboo plant

Annual aquatic herbs. W, C, E Nepal: 200-1700 m. Nepal, E Himalaya, Assam-Burma, S Asia, E Asia, SE Asia, Africa and Australasia.

Use: Food.

Blyxa echinosperma (C.B.Clarke) Hook.f. [1, 4]

Hydrotrophus echinospermus C.B.Clarke; *Blyxa aubertii* var. *echinosperma* (C.B.Clarke) C.D.K.Cook & Luond (5)

Annual aquatic herbs. W, C, E Nepal: 700-2000 m. Nepal, Assam-Burma, S Asia, E Asia, SE Asia and Australasia.

Blyxa japonica (Miq.) Maxim. ex Asch. & Gürke [1, 4]

Hydrilla japonica Miq.

Japanese rush

Annual aquatic herbs. C Nepal: *ca.* 800 m. Nepal, Assam-Burma, S Asia, E Asia, SE Asia and Europe.

Hydrilla Rich.

Hydrilla verticillata (L.f.) Royle [1, 4, 5]

Serpicula verticillata L.f.

पानि उन्यु (Paani unyu); Hydrilla

Perennial or annual aquatic herbs. W, C, E Nepal: 100-1600 m. Nepal, Tibetan Plateau, Assam-Burma, S Asia, E Asia, SE Asia, C Asia, SW Asia, Europe, Africa and Australasia.

Use: Medicine (whole plant).

Najas L.

Najas graminea Delile [1, 4]

Caulinia graminea (Delile) Tzvelev

Ricefield waternymph

Annual aquatic herbs. W, C Nepal: 300-2300 m. Nepal, Assam-Burma, S Asia, E Asia, SE Asia, C Asia, SW Asia, Europe, Africa and Australasia.

Use: Medicine (whole plant).

Najas minor All. [1, 4]

Caulinia minor (All.) Coss. & Germ.

Brittle waternymph

Annual aquatic herbs. C Nepal: 700-1000 m. Nepal, S Asia, E Asia, SE Asia, N Asia, C Asia, SW Asia, Europe and Africa.

Use(s): Medicine (whole plant), Food (for water birds).

Nechamandra Planch.

Nechamandra alternifolia (Roxb. ex Wight) Thwaites [1, 4]

Vallisneria alternifolia Roxb. ex Wight; *Nechamandra roxburghii* Planch.

Indian Elodea

Annual or long lived aquatic herbs. W, C Nepal: 600-900 m. Nepal, Assam-Burma, S Asia E Asia and SE Asia.

Ottelia Pers.

Ottelia alismoides (L.) Pers. [1, 4, 5]

Stratiotes alismoides L.; *Ottelia japonica* Miq.

Duck lettuce

Annual or long lived aquatic herbs. W, C, E Nepal: 100-300 m. Nepal, Assam-Burma, S Asia, E Asia, SE Asia, Africa and Australasia.

Use: Medicine (leaves).

Vallisneria L.

Vallisneria natans (Lour.) H.Hara [1, 4]

Physkium natans Lour.; *Vallisneria asiatica* Niki

सलिल कुन्तला, फुल्के कमल (Salil kuntalaa, Phulke kamal); Eel grass

Perennial or annual aquatic herbs. W Nepal: *ca.* 300 m. Nepal, Tibetan Plateau, S Asia, E Asia, N Asia, SE Asia, SW Asia and Australasia.

Use: Medicine (whole plant).

Vallisneria spiralis L. [5]

फुल्के कमल (Phulke kamal); Tape grass, Eel grass

Perennial or annual aquatic herbs. W Nepal: 100-200 m. Nepal, Pantropical and Subtropical.

Use: Medicine (root).

APONOGETONACEAE

Aponogeton L.f.

Aponogeton natans (L.) Engl. & K.Krause [4, 38]

Saururus natans L.

Drifting sword plant

Perennial aquatic herbs. E Nepal: 100-800 m. Nepal, native to S Asia.

Use: Food (root).

Doubtful/Uncertain Taxa

Aponogeton appendiculatus H.Bruggen, W Nepal: *ca.* 200 m. Nepal and S Asia (?)[124]

Aponogeton crispus Thunb., W Nepal: *ca.* 200 m. Nepal and S Asia (?) [124]

Aponogeton undulates Roxb., W Nepal: *ca.* 200 m. Nepal, Assam-Burma and S Asia. [124]

JUNCAGINACEAE

Triglochin L.

Triglochin maritima L. [1, 4, 5]

Hexaglochin maritima (L.) Nieuwl.

ना-रम (Naa-ram: Tibetan); Sea arrow grass

Perennial aquatic herbs. W, C Nepal: 2600-4600 m. Nepal, E Himalaya, Tibetan Plateau, E Asia, N Asia, SW Asia and temperate and cold regions of N Hemisphere.

Use: Medicine (whole plant).

Triglochin palustris L. [1, 4, 5]

Tristemon palustris (L.) Raf.

Marsh arrow grass

Perennial aquatic herbs. W, C, E Nepal: 2300-4700 m. Nepal, E Himalaya, Tibetan Plateau, S Asia, E Asia, N Asia, C Asia and SW Asia.

POTAMOGETONACEAE

Potamogeton L.

Potamogeton crispus L. [1, 4, 5]

Potamogeton crenulatus D.Don

Curled pondweed

Perennial aquatic herbs. W, C, E Nepal: 100-2400 m. Nepal, Asia, Europe, Africa, N America and Australasia.

Use: Forage.

Potamogeton distinctus A.Benn [5]

Potamogeton longipetiolatus E.G.Camus

पानी झार (Paani jhaar); Round-leaved pondweed

Perennial aquatic herbs. W, C Nepal: 900-3000 m. Nepal, E Himalaya, Tibetan Plateau, E Asia, SE Asia, N Asia and Australasia.

Potamogeton lucens L. [1, 4, 5]

Potamogeton sinicus Migo

Shining pondweed

Perennial aquatic herbs. W, C, E Nepal: 200-3000 m. Nepal, Tibetan Plateau, Assam-Burma, S Asia, E Asia, N Asia, C Asia, SW Asia, Europe and Africa.

Use: Forage.

Potamogeton natans L. [1, 4, 5]

Potamogeton linguatus Hagstr.

पानी झार (Paani jhaar); Broad-leaved pondweed

Perennial aquatic herbs. C Nepal: 2600-4400 m. Nepal, Tibetan Plateau, Assam-Burma, E Asia, C Asia, N Asia, SW Asia, Europe, Africa and N America.

Use: Forage.

Potamogeton nodosus Poir. [1, 4, 5]

Potamogeton indicus Roxb.

पानी झार (Paani jhaar); Long-leaved pondweed

Perennial aquatic herbs. W, C, E Nepal: 200-3000 m. Nepal, Asia, Europe, Africa, America and Australasia.

Potamogeton octandrus Poir. [1, 4]

Potamogeton asiaticus A.Benn.

Pondweed

Perennial aquatic herbs. C Nepal: 700-3000 m. Nepal, Assam-Burma, S Asia, E Asia, SE Asia, N Asia, Africa and Australasia.

Potamogeton perfoliatus L. [5]

Spirillus perfoliatus (L.) Nieuwn.

Clasping-leaved pondweed

Perennial aquatic herbs. C Nepal: 2800-3300 m. Nepal, Asia, Europe, Africa, N America and Australasia.

Potamogeton pusillus L. [1, 4, 5]

Spirillus pusillus (L.) Nieuwn.

Lesser pondweed

Perennial aquatic herbs. W, C Nepal: 2100-3000 m. Nepal, Asia, Europe, Africa and N America.

Stuckenia Borner

Stuckenia filiformis (Pers.) Borner

Potamogeton filiformis Pers. [1, 4]

Perennial aquatic herbs. W, C Nepal: 2000-4700 m. Nepal, Asia, Europe, N America and S America.

Stuckenia pectinata (L.) Borner [5]

Potamogeton pectinatus L. (1, 4)

Sago pondweed

Perennial aquatic herbs. C Nepal: 200-3000 m. Nepal, Asia, Europe, Africa, America and Australasia.

Zannichellia L.

Zannichellia palustris L. [1, 4, 5]

Zannichellia pedicellata (Wahlenb. & Rosen) Fr.; *Zannichellia palustris* var. *pedicellata* Wahlenb. & Rosen

Horned pondweed

Perennial aquatic herbs. W, C, E Nepal: 100-1100 m. Nepal, S Asia, SW Asia, Europe and America.

DIOSCOREALES

NARTHECIACEAE

Aletris L.

Aletris glabra Bureau & Franch. [5]

Aletris sikkimensis Hook.f. (1, 4)

Perennial herbs. C, E Nepal: 3200-4600 m. Nepal, E Himalaya, Tibetan Plateau and E Asia.

Aletris gracilis Rendle [1, 4, 5]

Aletris stelliflora Hand.-Mazz.

Perennial herbs. E Nepal: 3000-3600 m. Nepal, E Himalaya, Tibetan Plateau, Assam-Burma and E Asia.

Aletris nana S.C.Chen [5]

Aletris alpestris var. *occidentalis* H.Hara (1, 4)

Perennial herbs. C, E Nepal: 3200-4600 m. Nepal, Tibetan Plateau and E Asia.

Aletris pauciflora (Klotzsch) Hand.-Mazz. [1, 4, 5]

Stachyopogon pauciflorus Klotzsch; *Aletris nepalensis* var. *delavayi* Franch.

पांग (Paang -Tamang)

Perennial herbs. W, C, E Nepal: 2500-4900 m. Nepal, W Himalaya, E Himalaya, Tibetan Plateau, Assam-Burma, S Asia and E Asia.

Use: Medicine (tuber).

BURMANNIACEAE

Burmannia L.

Burmannia coelestis D.Don [1, 4]

Burmannia chinensis Gand.

निलमणि (Neelmani); Indian bluethread

Annual saprophytic swampy herbs. C Nepal: 300-1300 m. Nepal, Assam-Burma, S Asia, E Asia, SE Asia and Australasia.

Burmannia disticha L. [1, 4]

Annual saprophytic swampy herbs. C, E Nepal: 400-3000 m. Nepal, Assam-Burma, S Asia, E Asia, SE Asia and Australasia.

Burmannia nepalensis (Miers) Hook.f. [1, 4]

Gonianthes nepalensis Miers

Annual saprophytic herbs. C, E Nepal: 400-1600 m. Nepal, S Asia, E Asia and SE Asia.

DIOSCOREACEAE

Dioscorea L.

Dioscorea alata L. [1, 4, 5]

Dioscorea purpurea Roxb.

घर तरुल, कुकुर तरुल (Ghar tarul, Kukur tarul); White yam

Perennial climbers. C, E Nepal: 100-1800 m. Nepal, E Asia, cultivated Pantropically.

Use(s): Medicine (tuber), Forage.

Dioscorea belophylla (Prain) Voigt ex Haines [1, 4, 5]

Dioscorea nummularia var. *belophylla* Prain

Perennial climbers. W, C, E Nepal: 200-1600 m. Nepal, W Himalaya, E Himalaya, Assam-Burma and S Asia.

Use(s): Food (tuber), Forage.

Dioscorea bulbifera L. [1, 4, 5]

Helmia bulbifera (L.) Kunth

बन तरुल, भ्याकुर, गिट्ठा, कुकुर तरुल (Ban tarul, Bhyaakur, Gitthaa, Kukur tarul); Potato yam

Perennial climbers. W, C, E Nepal: 100-2600 m. Nepal, E Himalaya, Tibetan Plateau, Assam-Burma, S Asia, E Asia, SE Asia, Africa and Australasia.

Use(s): Medicine (tuber), Food (tuber), Forage.

Dioscorea deltoidea Wall. ex Griseb. [1, 4, 5]

काँडे भ्याकुर (Kaande bhyaakur); Cushcush yam

Perennial climbers. W, C, E Nepal: 200-3900 m. Nepal, E Himalaya, Tibetan Plateau, Assam-Burma, S Asia, E Asia and SE Asia.

Use(s): Medicine (tuber, axillary bulb), Food, Forage.

CITES Appendix: II.

Dioscorea esculenta (Lour.) Burkill [1, 4, 8]

Oncus esculentus Lour.; *Dioscorea spinosa* Roxb. ex Wall.

सुथानी तरुल (Suthaani tarul); Chinese yam

Perennial climbers. C, E Nepal: 2100-2600 m. Nepal, native to S Asia and SE Asia, cultivated in tropical Asia.

Use(s): Medicine (tuber), Fodder.

Dioscorea glabra Roxb. [1, 4, 5]

Dioscorea glabra var. *longifolia* Prain & Burkill

पातली तरुल (Paatali tarul)

Perennial climbers. W, C, E Nepal: 100-2400 m. Nepal, E Himalaya, Assam-Burma, S Asia, E Asia and SE Asia.

Use: Fodder.

Dioscorea hamiltonii Hook.f. [1, 4, 5]

Dioscorea persimilis Prain & Burkill

बन तरुल (Ban tarul)

Perennial climbers. W, C, E Nepal: 200-2400 m. Nepal, E Himalaya, Assam-Burma and SE Asia.

Use(s): Food, Forage.

Dioscorea hispida Dennst. [1, 4, 5]

Dioscorea daemona Roxb.

गीट्ठो (Gittho); Asiatic bitter yam

Perennial climbers. W, C, E Nepal: 100-2000 m. Nepal, E Himalaya, Tibetan Plateau, S Asia, E Asia and SE Asia.

Use(s): Medicine (tuber), Forage.

Dioscorea kamoonensis Kunth [1, 4, 5]

Dioscorea bonatiana Prain & Burkill

रानी भ्याकुर (Raani bhyaakur); Kamoon yam

Perennial climbers. W, C, E Nepal: 100-2200 m. Nepal, E Himalaya, Tibetan Plateau, E Asia and SE Asia.

Use: Forage.

Dioscorea melanophyma Prain & Burkill [1, 4, 5]

Dioscorea tenii R.Knuth

Perennial climbers. W, C Nepal: 1300-2500 m. Nepal, Tibetan Plateau and E Asia.

Use: Forage.

Dioscorea pentaphylla L. [1, 4, 5]

Dioscorea quartiniana A.Rich.

मिठे तरुल (Mithe tarul)

Perennial climbers. W, C, E Nepal: 100-2700 m. Nepal, Tibetan Plateau, Assam-Burma, S Asia, E Asia, SE Asia, Africa and Australasia.

Use(s): Medicine (whole plant), Food, Forage.

Dioscorea prazeri Prain & Burkill [1, 4, 5]

Dioscorea sikkimensis Prain & Burkill

कुकुर तरुल, भ्याकुर (Kukur tarul, Bhyaakur); Deltoid yam

Perennial climbers. W, C, E Nepal: 100-1600 m. Nepal, E Himalaya, Assam-Burma and SE Asia.

Use(s): Medicine (tuber), Food, Forage.

Dioscorea pubera Blume [1, 4, 5]

Dioscorea anguina Roxb.

पान्गला तरुल (Paangla tarul)

Perennial climbers. W, C, E Nepal: 100-1000 m. Nepal, E Himalaya, Assam-Burma, S Asia and SE Asia.

Exotic/Cultivated Taxa

Dioscorea polystachya Turcz. (*Dioscorea batatas* Decne.); (**तरुल,** Tarul; Cinnamon vine). Native to E Asia.

PANDANALES

PANDANACEAE

Pandanus Parkinson

Pandanus furcatus Roxb. [5]

केउरा, तारिका (Keuraa, Taarikaa)

Evergreen shrubs or small trees. C, E Nepal: 1100-1600 m. Nepal, E Himalaya, SE Asia, Africa and Australasia.

Use(s): Construction materials, Medicine (young twig).

Pandanus nepalensis H.St. John [1, 4]

केउरा, बनतरी (Keuraa, Bantari); Himalayan screw pine

Evergreen shrubs or small trees. C, E Nepal: 700-1600 m. Nepal, Himalaya, SE Asia, Africa and Australasia.

Use(s): Medicine (leaves, flowers), Food (fruit), Utensils and handicrafts, Construction materials.

Pandanus tectorius Parkinson

Pandanus fascicularis Lam. (5); *Pandanus sinensis* (Warb.) Martelli

बन घिउकुमारी, केतकी (Ban ghiukumari, Ketaki); Screw pine

Shrubs or small trees. C, E Nepal: *ca.* 1000 m. Native to S Asia, E Asia and SE Asia.

Use: Medicine (root, flowers).

LILIALES

MELANTHIACEAE (incl. LILIACEAE)

Paris L.

Paris mairei H.Lev. [5]

Daiswa pubescens (Hand.-Mazz.) Takht.; *Paris violacea* H.Lev.

Perennial herbs. E Nepal: 1800-3500 m. Nepal and E Asia.

Paris marmorata Stearn

Paris polyphylla subsp. *marmorata* (Stearn) H.Hara (1, 4)

Perennial herbs. C, E Nepal: 2400-3100 m. Nepal, E Himalaya, Tibetan Plateau and E Asia.

Paris polyphylla Sm. [1, 4]

Daiswa polyphylla (Sm.) Raf.; *Paris polyphylla* var. *wallichii* H.Hara (1,4,5)

सतुवा (Satuwaa); Love apple

Perennial herbs. W, C, E Nepal: 1800-3500 m. Nepal, E Himalaya, Tibetan Plateau, Assam-Burma, S Asia, E Asia and SE Asia.

Use: Medicine (rhizome).

Paris thibetica Franch. [8]

Paris polyphylla var. *appendiculata* H.Hara (1, 4)

सतुवा (Satuwaa)

Perennial herbs. E Nepal: 2500-3500 m. Nepal, E Himalaya, Tibetan Plateau, Assam-Burma and E Asia.

Use: Medicine (rhizome).

Trillium L.

Trillium govanianum Wall. ex D.Don

Trillidium govanianum (Wall. ex Royle) Kunth (1, 4, 5)

Wood lily

Perennial herbs. W, C, E Nepal: 2700-4000 m. Nepal, E Himalaya, Tibetan Plateau and S Asia.

Use: Medicine (root).

Ypsilandra Franch.

Ypsilandra yunnanensis W.W.Sm. & Jeffrey

Helonias yunnanensis (W.W.Sm. & Jeffrey) N.Tanaka (4); *Ypsilandra yunnanensis* var. *himalaica* H.Hara

Perennial herbs. C, E Nepal: 2700-4500 m. Nepal, E Himalaya, Tibetan Plateau, Assam-Burma and E Asia.

COLCHICACEAE

Disporum Salisb.

Disporum calcaratum D.Don [1, 4]

Disporum hamiltonianum D.Don

Perennial herbs. C, E Nepal: 1100-2900 m. Nepal, E Himalaya, Assam-Burma, E Asia and SE Asia.

Disporum cantoniense (Lour.) Merr.

Fritillaria cantoniensis Lour.; *Disporum cantoniense* var. *parviflorum* (Wall.) H.Hara (1, 4, 5)

महजरी, सानो कुकुरडाइनो (Mahajari, Sano kukurdaino); Fairy bells

Perennial herbs. W, C, E Nepal: 1100-2900 m. Nepal, E Himalaya, Tibetan Plateau, Assam-Burma, S Asia, E Asia and SE Asia.

Use(s): Medicine (root), Food.

Gloriosa L.

Gloriosa superba L. [1, 4, 5, 14]

केवरी, हरितालि फूल (Kewari, Haritali phul); Glory lily

Perennial scandent climbers. W, C, E Nepal: 400-2200 m. Nepal, Assam-Burma, S Asia, E Asia, SE Asia and Africa.

Use(s): Ornamental, Medicine (tuberous root, leaves).

Iphigenia Kunth

Iphigenia indica (L.) A.Gray ex Kunth [1, 4, 8]

Melanthium indicum L.; *Lloydia melanantha* H.Lev.

Indian grass lily

Perennial herbs. W, C Nepal: 1300-2000 m. Nepal, E Himalaya, Assam-Burma, S Asia, E Asia, SE Asia and Australasia.

Use: Medicine (corm).

SMILACACEAE

Heterosmilax Kunth

Heterosmilax japonica Kunth [8]

Heterosmilax indica A.DC.

Woody climbers. E Nepal: 100-1800 m. Nepal, E Himalaya, Assam-Burma, E Asia and SE Asia.

Smilax L. (**कुकुर डाइनो,** Kukurdaaino)

Smilax aspera L. [1, 4, 5]

Smilax maculata Roxb.

चोपचिनी (Chopchini)

Perennial climbers. W, C, E Nepal: 400-4300 m. Nepal, E Himalaya, Tibetan Plateau, Assam-Burma, S Asia, E Asia, SW Asia, Europe and Africa.

Use(s): Medicine (root, fruit), Food (tender shoot, leaves), Construction materials.

Smilax elegans Wall. ex Kunth

Smilax parvifolia Wall. ex Hook.f.

subsp. **elegans** [4, 5]

Perennial climbers. W, C, E Nepal: 1600-3100 m. Nepal, E Himalaya, Tibetan Plateau, Assam-Burma, S Asia and SW Asia.

Use(s): Medicine (root, stem), Food (young shoot), Fodder, Construction materials.

subsp. **glaucophylla** (Klotzsch) Noltie [4, 5]

Smilax glaucophylla Klotzsch

Perennial climbers. W, C, E Nepal: 1800-3100 m. Nepal, W Himalaya, E Himalaya, Tibetan Plateau and Assam-Burma.

subsp. **osmastonii** (F.T.Wang & Tang) Noltie [4]

Smilax osmastonii F.T.Wang & Tang (1, 4)

Perennial climbers. C Nepal: *ca.* 2700 m. Nepal, E Himalaya and Assam-Burma.

Smilax ferox Wall. ex Kunth [1, 4, 5]

चटलांग (Chatlaang)

Perennial climbers. W, C, E Nepal: 1100-2700 m. Nepal, E Himalaya, Assam-Burma, S Asia, E Asia and SE Asia.

Use(s): Food (young shoot, leaves), Construction materials.

Smilax lanceifolia Roxb.

Smilax micropoda A.DC.

चोप चिनी, सानो कुकुरडाइनो (Chopchini, Saano kukurdaaino)

Perennial climbers. C, E Nepal: 1500-2100 m. Nepal, E Himalaya, Assam-Burma, E Asia and SE Asia.

Use(s): Medicine (root), Food (young shoot, leaves), Construction materials.

Smilax longebracteolata Hook.f.

Smilax elegans subsp. *suberecta* Noltie (4)

Perennial climbers. C Nepal: *ca.* 2400 m. Nepal, E Himalaya, Assam-Burma, Tibetan Plateau and E Asia.

Smilax menispermoidea A.DC. [1, 4, 5]

Smilax rubriflora Rehder

हाडे कुकुरडाइनो (Haade kukurdaaino)

Perennial climbers. W, C, E Nepal: 1100-3400 m. Nepal, E Himalaya, Tibetan Plateau, Assam-Burma, E Asia and SE Asia.

Use(s): Food (tender shoot, leaves), Construction materials.

Smilax minutiflora A.DC. [1, 4, 5]

Perennial climbers. W, C, E Nepal: 2300-3500 m. Nepal, E Himalaya, Assam-Burma and E Asia.

Use: Construction materials.

Smilax munita S.C.Chen [5]

Smilax myrtillus var. *rigida* Noltie (4); *Smilax rigida* Wall. ex Kunth

Perennial climbers. C, E Nepal: 2100-2900 m. Nepal, E Himalaya, Tibetan Plateau, Assam-Burma and E Asia.

Use: Food (tender shoot).

Smilax myrtillus A.DC. [5]

Smilax rigida subsp. *myrtillus* (A.DC.) T.Koyama; *Smilax rigida* var. *myrtillus* (A.DC.) T.Koyama

Perennial climbers. C Nepal: 1800-2800 m. Nepal, E Himalaya, Tibetan Plateau, Assam-Burma, S Asia and E Asia.

Use: Construction materials.

Smilax ocreata A.DC. [1, 4]

Smilax perulata H.Lev. & Vaniot

Perennial climbers. C, E Nepal: 1000-2700 m. Nepal, E Himalaya, Tibetan Plateau, Assam-Burma, S Asia, E Asia and SE Asia.

Use: Construction materials.

Smilax orthoptera A.DC. [1, 4, 5]

Perennial climbers. W, C, E Nepal: 100-1500 m. Nepal, E Himalaya and Assam-Burma.

Use: Construction materials.

Smilax ovalifolia Roxb. ex D.Don [1, 4, 5]

Smilax macrophylla Roxb.

नादर (Naadar)

Perennial climbers. C, E Nepal: 200-2200 m. Nepal, Assam-Burma, S Asia, E Asia and SE Asia.

Use(s): Medicine (root, berries), Food (tender shoot, leaves), Fodder.

Smilax perfoliata Lour. [1, 4, 5]

Smilax prolifera Roxb. (4)

Perennial climbers. W, C, E Nepal: 400-3000 m. Nepal, Assam-Burma, S Asia, E Asia and SE Asia.

Use(s): Medicine (root), Food (tender leaves, shoot).

Smilax wallichii Kunth [4, 5]

Perennial climbers. C Nepal: 1300-2600 m. Nepal, E Himalaya and S Asia.

Use: Construction materials.

Smilax wightii A.DC. [1, 4]

Perennial climbers. C, E Nepal: 300-1000 m. Nepal, E Himalaya and S Asia.

Use: Construction materials.

Smilax zeylanica L. [1, 4]

Smilax elliptica Desv. ex Ham.

चोपचिनी (Chopchini)

Perennial climbers. C, E Nepal: 100-1500 m. Nepal, E Himalaya, Assam-Burma, S Asia and SE Asia.

Use: Medicine (root, tender shoot).

LILIACEAE

Cardiocrinum Lindl.

Cardiocrinum giganteum (Wall.) Makino [1, 4, 5]

Lilium giganteum Wall.

घिउपाते (Ghiupaate), भोगटि (Bhogati-Gurung); Giant lily

Perennial herbs. W, C, E Nepal: 1200-3700 m. Nepal, E Himalaya, Tibetan Plateau, Assam-Burma and E Asia.

Use: Medicine (root, leaves).

Clintonia Raf.

Clintonia udensis Trautv. & C.A.Mey.

Clintonia alpina Kunth ex Baker; *Clintonia udensis* var. *alpina* (Kunth ex Baker) H.Hara (1, 4, 5)

रेछेमा (Rechhema-Sherpa); Bead lily

Perennial herbs. W, C, E Nepal: 3000-4000 m. Nepal, E Himalaya, Tibetan Plateau and Assam-Burma.

Use: Food.

Fritillaria L.

Fritillaria cirrhosa D.Don [1, 4, 5]

Lilium bonatii H.Lev.

काकोली (Kaakoli); Snakehead fritillary

Perennial herbs. W, C, E Nepal: 3000-4700 m. Nepal, E Himalaya, Tibetan Plateau, Assam-Burma, S Asia and E Asia.

Use(s): Medicine (bulb), Food (root, fresh gynoecium).

Gagea Salisb.

Gagea lutea (L.) Ker Gawl. [5]

Ornithogalum luteum L.; *Gagea elegans* Wall. ex G.Don (1, 4)

Perennial herbs. W, C Nepal: 2800-4300 m. Nepal, W Himalaya, E Asia, N Asia and SW Asia.

Gagea noltiei Peruzzi, J.-M.Tison, A.Peterson & J.Peterson

Lloydia delicatula Noltie (4)

Perennial herbs. E Nepal: 3600-4600 m. Nepal and E Himalaya.

Lilium L. (Lily)

Lilium bakerianum Collett & Hemsl. [1, 4]

Perennial herbs. E Nepal: 1500-1800 m. Nepal, Assam-Burma and E Asia.

Lilium nanum Klotzsch

var. **nanum**

Lilium nanum Klotzsch forma *nanum* (1, 4, 5); *Nomocharis nana* (Klotzsch) E.H.Wilson

खिरौला (Khiraulaa)

Perennial herbs. W, C, E Nepal: 3700-4700 m. Nepal, E Himalaya, Tibetan Plateau, Assam-Burma and E Asia.

var. **flavidum** (Rendle) Sealy

Fritillaria flavida Rendle; *Lilium nanum* forma *flavidum* (Rendle) H.Hara (1, 4, 5)

Perennial herbs. W, C, E Nepal: 3700-4800 m. Nepal, W Himalaya and Tibetan Plateau.

Lilium nepalense D.Don [1, 4, 5]

खिरौला, बन लसुन (Khiraulaa, Ban lasun); Nepalese lily

Perennial herbs. W, C, E Nepal: 2000-3800 m. Nepal, E Himalaya, Tibetan Plateau, Assam-Burma and E Asia.

Use(s): Medicine (bulb), Food (bulb).

Lilium sherriffiae Stearn [1, 4]

Perennial herbs. E Nepal: 2700-4000 m. Nepal and E Himalaya.

Lilium wallichianum Schult. & Schult.f.

var. **wallichianum** [1, 4, 5]

बन लसुन (Ban lasun)

Perennial herbs. W, C, E Nepal: 1100-2000 m. Nepal, E Himalaya, Assam-Burma and S Asia.

Use(s): Medicine/Food (bulb).

var. **neilgherrense** (Wight) H.Hara [1, 4]

Lilium neilgherrense Wight

तिते पिंडालु (Tite pindaalu)

Perennial herbs. W, C Nepal: 1100-2400 m. Nepal and S Asia.

Use(s): Medicine/Food (bulb).

Lloydia Salisb. ex Rchb.

Lloydia flavonutans H.Hara [1, 4]

Perennial herbs. C, E Nepal: 3700-5200 m. Nepal, E Himalaya, Tibetan Plateau and Assam-Burma.

Lloydia longiscapa Hook. [1, 4, 5]

Perennial herbs. W, C, E Nepal: 2500-5000 m. Nepal, W Himalaya, E Himalaya and Tibetan Plateau.

Lloydia serotina (L.) Rchb. [1, 4]

Bulbocodium serotinum L.; *Lloydia himalensis* Royle

var. **serotina**

Perennial herbs. W, C, E Nepal: 3500-4900 m. Nepal, E Himalaya and Tibetan Plateau.

var. **parva** (C.Marquand & Airy Shaw) H.Hara [1, 4]

Lloydia serotina forma *parva* C.Marquand & Airy Shaw

Perennial herbs. C, E Nepal: 4100-4700 m. Nepal, E Himalaya and Tibetan Plateau.

Lloydia tibetica Baker ex Oliv. [1, 4]

Lloydia montana (Dammer) P.C.Kuo

Perennial herbs. C Nepal: 2300-4100 m. Nepal, E Himalaya, Tibetan Plateau and E Asia.

Lloydia yunnanensis Franch. [4, 5]

Lloydia filiformis Franch.

Perennial herbs. C, E Nepal: 2300-4300 m. Nepal, E Himalaya, Assam-Burma and E Asia.

Nomocharis Franch.

Nomocharis oxypetala (D.Don) E.H.Wilson

Fritillaria oxypetala D.Don; *Lilium oxypetalum* (D.Don) Baker (1, 4, 5)

Yellow Himalayan lily

Perennial herbs. W Nepal: 3100-3900(-5100) m. Nepal, W Himalaya and Assam-Burma.

Use: Medicine (bulb).

Notholirion Wall.

Notholirion bulbuliferum (Lingelsh.) Stearn [1, 4, 5]

Paradisea bulbulifera Lingelsh.; *Lilium hyacinthinum* E.H.Wilson

Perennial herbs. C, E Nepal: 1800-4500 m. Nepal, E Himalaya, Tibetan Plateau and E Asia.

Notholirion macrophyllum (D.Don) Boiss. [1, 4, 5]

Fritillaria macrophylla D.Don; *Lilium macrophyllum* (D.Don) Voss

Perennial herbs. W, C, E Nepal: 2500-4400 m. Nepal, E Himalaya, Tibetan Plateau and E Asia.

Use: Medicine (whole plant).

Notholirion thomsonianum (Royle) Stapf [1, 4]

Fritillaria thomsoniana Royle

Rosy Himalayan lily

Perennial herbs. C, E Nepal: 800-1800 m. Nepal, W Himalaya and SW Asia.

Streptopus Michx.

Streptopus parasimplex H.Hara & H.Ohashi [1, 4, 5]

Perennial herbs. C, E Nepal: 3100-4200 m. Nepal and E Himalaya.

Streptopus simplex D.Don [1, 4, 5]

Perennial herbs. W, C, E Nepal: 1800-4000 m. Nepal, E Himalaya, Tibetan Plateau, Assam-Burma and E Asia.

Tricyrtis Wall.

Tricyrtis pilosa Wall.

Compsoa maculata D.Don; *Tricyrtis maculata* (D.Don) J.F.Macbr. (1, 4, 5)

Spotted toadlily

Perennial herbs. C, E Nepal: 1300-3900 m. Nepal, E Himalaya, Tibetan Plateau, S Asia and E Asia.

Exotic/Cultivated Taxa

Clivia miniata (Lindl.) Bosse (Kafir lily). Native to S Africa. [54]

Lilium tigrinum Ker Gawl., Native to E Asia. [14]

Tulipa gesneriana L. (Garden Tulip). Native to SW Asia.

Doubtful/Uncertain Taxa

Gagea nakaiana Kitag. [104]

ASPARAGALES

ORCHIDACEAE

The whole family Orchidaceae is listed in the CITES Appendix II. However, *Paphiopedilum* is listed in the Appendix I.

Acampe Lindl.

Acampe papillosa (Lindl.) Lindl. [1, 4, 5, 13]

Saccolabium papillosum Lindl.

रसना (Rasanaa)

Epiphytic herbs. C, E Nepal: 200-1500 m. Nepal, E Himalaya, Assam-Burma and SE Asia.

Use(s): Medicine (root), Food, Ornamental.

Acampe rigida (Buch.-Ham. ex Sm.) P.F.Hunt [1, 4, 5, 13]

Aerides rigida Buch.-Ham. ex Sm.; *Acampe multiflora* (Lindl.) Lindl. (1, 4); *Vanda multiflora* Lindl.

Epiphytic herbs. C, E Nepal: 200-1500 m. Nepal, E Himalaya, Assam-Burma, S Asia, E Asia, SE Asia and Africa.

Use: Ornamental.

Acanthephippium Blume

Acanthephippium striatum Lindl. [13, 58]

Acanthephippium simplex Aver.; *Acanthephippium sinense* Rolfe

Terrestrial herbs. E Nepal: 400-1500 m. Nepal, E Himalaya, E Asia and SE Asia.

Use: Ornamental.

Acrochaene Lindl.

Acrochaene punctata Lindl. [8, 13, 89]

Bulbophyllum kingii Hook.f.; *Monomeria punctata* (Lindl.) Schltr.

Spotted Acrochaene

Epiphytic herbs. E Nepal: *ca.* 1700 m. Nepal, E Himalaya, Assam-Burma and SE Asia.

Aerides Lour.

Aerides multiflora Roxb. [1, 4, 13, 14]

Aerides affinis Wall. ex Lindl.; *Epidendrum geniculatum* Hook.f.

प्याजी सुनाखरी (Pyaaji sunaakhari); Cat's-tail orchid

Epiphytic herbs. W, C, E Nepal: 200-1200 m. Nepal, W Himalaya, E Himalaya, Assam-Burma and SE Asia.

Use(s): Medicine (leaves, bulb, root), Ornamental.

Aerides odorata Lour. [1, 4, 13]

Aerides cornuta Roxb.; *Epidendrum odoratum* (Lour.) Poir.

Epiphytic herbs. W, C, E Nepal: 200-1500 m. Nepal, E Himalaya, Assam-Burma, E Asia and SE Asia.

Use(s): Medicine (leaves), Ornamental.

Agrostophyllum Blume

Agrostophyllum callosum Rchb.f. [1, 4, 5, 13]

Epiphytic herbs. C, E Nepal: 1900-2500 m. Nepal, E Himalaya, Tibetan Plateau, Assam-Burma, E Asia and SE Asia.

Use: Medicine (leaves).

Agrostophyllum planicaule Rchb.f. [5, 13]

Agrostophyllum khasianum Griff.; *Eria planicaulis* Wall. ex Lindl.

Epiphytic herbs. E Nepal: 500-800 m. Nepal, E Himalaya, Assam-Burma, E Asia and SE Asia.

Androcorys Blume

Androcorys pugioniformis (Lindl. ex Hook.f.) K.Y.Lang.

Herminium pugioniforme Lindl. ex Hook.f. (1, 4); *Monorchis pugioniformis* (Lindl. ex Hook.f.) O.Schwarz

Terrestrial herbs. W, C, E Nepal: 2700-5200 m. Nepal, W Himalaya, E Himalaya and Tibetan Plateau.

Androcorys pusillus (Ohwi & Fukuy.) Masam.

Herminium pusillum Ohwi & Fukuy. (41)

Terrestrial herbs. C Nepal: *ca.* 3700 m. Nepal and E Asia.

Anoectochilus Blume

Anoectochilus brevilabris Lindl. [13, 40]

Anoectochilus griffithii Hook.f.; *Anoectochilus sikkimensis* King & Pantl.

Terrestrial herbs. C Nepal: 900-1500 m. Nepal, E Himalaya and SE Asia.

Anoectochilus roxburghii (Wall.) Lindl. [1, 4]

Chrysobaphus roxburghii Wall.

Terrestrial herbs. C Nepal: 300-1500 m. Nepal, W Himalaya, E Himalaya, Tibetan Plateau, S Asia, E Asia and SE Asia.

Use: Medicine (whole plant).

Anoectochilus setaceus Blume [13, 40]

Terrestrial herbs. C Nepal: 300-1500 m. Nepal, E Himalaya, Assam-Burma, S Asia and SE Asia.

Anthogonium Lindl.

Anthogonium gracile Lindl. [1, 4, 5, 13]

Anthogonium corydaloides Schltr.; *Anthogonium griffithii* Rchb.f.

Terrestrial herbs. C, E Nepal: 1500-2400 m. Nepal, E Himalaya, Tibetan Plateau, Assam-Burma, S Asia, E Asia and SE Asia.

Aphyllorchis Blume

Aphyllorchis alpina King & Pantl. [13, 40]

Terrestrial herbs. C, E Nepal: 2100-2600 m. Nepal, E Himalaya and Tibetan Plateau.

Apostasia Blume

Apostasia wallichii R.Br. [1, 4, 13]

Apostasia alba Rolfe; *Apostasia gracilis* Rolfe

Yellow-grass orchid

Terrestrial herbs. C Nepal: 600-1000 m. Nepal, E Himalaya, Tibetan Plateau, Assam-Burma, S Asia, E Asia, SE Asia and Australasia.

Arundina Blume

Arundina graminifolia (D.Don) Hochr. [1, 4, 5, 13]

Bletia graminifolia D.Don; *Arundina bambusifolia* Lindl.; *Cymbidium bambusifolium* Roxb.

गाभा (Gaabhaa)

Terrestrial herbs. C, E Nepal: 200-2300 m. Nepal, E Himalaya, Tibetan Plateau, Assam-Burma, S Asia, E Asia and SE Asia.

Use(s): Medicine (root), Fodder, Ornamental.

Ascocentrum Schltr. ex J.J.Sm.

Ascocentrum ampullaceum (Roxb.) Schltr. [1, 4, 5]

Aerides ampullacea Roxb.; *Saccolabium ampullaceum* (Roxb.) Lindl.

Epiphytic herbs. C, E Nepal: 100-900 m. Nepal, W Himalaya, E Himalaya, Assam-Burma, S Asia, E Asia and SE Asia.

Use: Ornamental.

Bhutanthera Renz

Bhutanthera albomarginata (King & Pantl.) Renz [5, 13, 40]

Habenaria albomarginata King & Pantl. (1, 4); *Peristylus albomarginatus* (King & Pantl.) K.Y.Lang

Terrestrial herbs. C, E Nepal: 3400-4300 m. Nepal and E Himalaya.

Bhutanthera alpina (Hand.-Mazz.) Renz [13, 40, 89]

Habenaria alpina Hand.-Mazz.; *Peristylus alpinus* (Hand.-Mazz.) K.Y.Lang

Terrestrial herbs. C Nepal: 4200-4300 m. Nepal, E Himalaya and Tibetan Plateau.

Bhutanthera fimbriata Raskoti, *Phytotaxa* **62**: 57. 2012. [8, 13, 108]

Terrestrial herbs. C Nepal: *ca.* 3800 m. **Endemic to Nepal**.

Central: *Rasuwa*, Langtang National Park, *ca.* 3800 m, BB Raskoti 234 (Holotype: KATH).

Brachycorythis Lindl.

Brachycorythis obcordata (Lindl.) Summerh. [1, 4, 5, 13]

Platanthera obcordata Lindl.

गामडोल (Gaamdol)

Terrestrial herbs. W, C, E Nepal: 900-2600 m. Nepal, W Himalaya and E Himalaya.

Use(s): Medicine (tuber), Food.

Bulbophyllum Thouars

Bulbophyllum affine Wall. ex Lindl. [1, 4, 5, 13]

Sarcopodium affine (Wall. ex Lindl.) Lindl. & Paxton

Epiphytic herbs. C, E Nepal: 600-1800 m. Nepal, E Himalaya, E Asia and SE Asia.

Use: Ornamental.

Bulbophyllum ambrosia (Hance) Schltr.

Eria ambrosia Hance

subsp. **nepalense** J.J.Wood [4, 13]

Epiphytic herbs. C Nepal: 1400-1500 m. Nepal and E Asia (Yunnan).

Bulbophyllum bisetum Lindl. [8, 13]

Epiphytic or lithophytic herbs. C, E Nepal: 1500-2000 m. Nepal, E Himalaya and SE Asia.

Bulbophyllum careyanum (Hook.) Spreng. [1, 4, 5, 13]

Anisopetalum careyanum Hook.; *Pleurothallis purpurea* D.Don

Epiphytic herbs. C, E Nepal: 600-2100 m. Nepal, E Himalaya, Assam-Burma and SE Asia.

Use(s): Medicine (leaves, pseudobulb), Ornamental.

Bulbophyllum cariniflorum Rchb.f. [13]

Bulbophyllum densiflorum Rolfe

Epiphytic herbs. C Nepal: 1300-1800 m. Nepal, E Himalaya, Tibetan Plateau and SE Asia.

Use: Ornamental.

Bulbophyllum caudatum Lindl. [1, 4, 13]

Cirrhopetalum caudatum (Lindl.) King & Pantl.

Epiphytic herbs. C, E Nepal: 500-2000 m. Nepal, E Himalaya and Tibetan Plateau.

Bulbophyllum cauliflorum Hook.f. [13]

Phyllorkis cauliflora (Hook.f.) Kuntze

Epiphytic or lithophytic herbs. E Nepal: 800-1800 m. Nepal, E Himalaya and Tibetan Plateau.

Bulbophyllum cylindraceum Wall. ex Lindl. [1, 4, 5, 13]

Phyllorkis cylindracea (Wall. ex Lindl.) Kuntze

Epiphytic herbs. C, E Nepal: 1600-2400 m. Nepal, E Himalaya and E Asia.

Use: Ornamental.

Bulbophyllum elatum (Hook.f.) J.J.Sm. [1, 4, 5, 13]

Cirrhopetalum elatum Hook.f.

Epiphytic herbs. C Nepal: 900-2600 m. Nepal, E Himalaya, Tibetan Plateau and E Asia.

Use: Ornamental.

Bulbophyllum emarginatum (Finet) J.J.Sm. [13, 40]

Cirrhopetalum emarginatum Finet

Epiphytic herbs. C Nepal: 2000-2600 m. Nepal, E Himalaya, Tibetan Plateau, Assam-Burma, E Asia and SE Asia.

Bulbophyllum eublepharum Rchb.f. [13, 40]

Bulbophyllum yuanyangense Z.H.Tsi.

Epiphytic herbs. E Nepal: 2000-2500 m. Nepal, E Himalaya, Tibetan Plateau and E Asia.

Bulbophyllum gamblei (Hook.f.) Hook.f. [13, 40]

Bulbophyllum leptanthum var. *gamblei* Hook.f.

Epiphytic or lithophytic herbs. E Nepal: 200-2000 m. Nepal and E Himalaya.

Bulbophyllum griffithii (Lindl.) Rchb.f. [8, 13, 59]

Sarcopodium griffithii Lindl.

Epiphytic herbs. C, E Nepal: 1000-2600 m. Nepal, E Himalaya, E Asia and SE Asia.

Bulbophyllum guttulatum (Hook.f.) N.P.Balakr. [1, 4, 5, 13]

Cirrhopetalum guttulatum Hook.f.

Epiphytic herbs. C, E Nepal: 1000-2300 m. Nepal, E Himalaya, Tibetan Plateau and SE Asia.

Use: Ornamental.

Bulbophyllum helenae (Kuntze) J.J.Sm. [1, 4, 13]

Phyllorkis helenae Kuntze

Epiphytic herbs. E Nepal: 900-1200 m. Nepal, E Himalaya, Assam-Burma, E Asia and SE Asia.

Use: Ornamental.

Bulbophyllum hirtum (Sm.) Lindl. ex Wall. [1, 4, 5, 13]

Stelis hirta Sm.; *Phyllorkis hirta* (Sm.) Kuntze

Epiphytic herbs. C Nepal: 1000-2600 m. Nepal, W Himalaya, E Himalaya, Assam-Burma, E Asia and SE Asia.

Use: Ornamental.

Bulbophyllum leopardinum (Wall.) Lindl. ex Wall. [1, 4, 5, 13]

Dendrobium leopardinum Wall.

Epiphytic or lithophytic herbs. C, E Nepal: 1500-3200 m. Nepal, W Himalaya, E Himalaya, Tibetan Plateau, E Asia and SE Asia.

Use(s): Medicine (whole plant), Ornamental.

Bulbophyllum moniliforme C.S.P.Parish & Rchb.f. [13, 40]

Epiphytic herbs. C Nepal: 1400-1500 m. Nepal, Assam-Burma and SE Asia.

Bulbophyllum muscicola Rchb.f. [5, 13, 40]

Bulbophyllum hookeri (Duthie) J.J.Sm. (1, 4)

Epiphytic herbs. C Nepal: 2100-2500 m. Nepal, W Himalaya, E Himalaya and Assam-Burma.

Use: Ornamental.

Bulbophyllum nepalense Raskoti & Ale, *Edinburgh J. Bot.* **70**(2): 381. 2013. [8, 13]

Epiphytic herbs. C Nepal: *ca.* 2300 m. **Endemic to Nepal**.

Central: *Kathmandu*, Shivapuri Nagarjun National Park, *ca.* 2300 m, BB Raskoti 325 (Holotype: KATH).

Bulbophyllum odoratissimum (Sm.) Lindl. [1, 4, 5, 13]

Stelis odoratissima Sm.

Epiphytic herbs. C Nepal: 1000-2500 m. Nepal, E Himalaya, Tibetan Plateau, Assam-Burma, E Asia and SE Asia.

Use(s): Medicine (whole plant), Ornamental.

Bulbophyllum polyrrhizum Lindl. [1, 4, 5, 13]

Phyllorchis polyrrhiza (Lindl.) Kuntze

Epiphytic herbs. C, E Nepal: 900-2100 m. Nepal, W Himalaya, E Himalaya, Assam-Burma, E Asia and SE Asia.

Bulbophyllum reptans (Lindl.) Lindl. [1, 4, 5, 13]

Tribrachia reptans Lindl.; *Bulbophyllum reptans* var. *subracemosa* Hook.f.

Epiphytic herbs. W, C, E Nepal: 1500-2600 m. Nepal, W Himalaya, E Himalaya, Tibetan Plateau, Assam-Burma, E Asia and SE Asia.

Bulbophyllum retusiusculum Rchb.f. [1, 4, 5, 13]

Cirrhopetalum retusiusculum (Rchb.f.) Hemsl.

Epiphytic herbs. C, E Nepal: 2100-3000 m. Nepal, E Himalaya, Tibetan Plateau, Assam-Burma, E Asia and SE Asia.

Bulbophyllum rigidum King & Pantl. [5, 13, 40]

Epiphytic herbs. C Nepal: 1000-2000 m. Nepal and E Himalaya.

Use: Ornamental.

Bulbophyllum rolfei (Kuntze) Seidenf. [5, 13, 40]

Phyllorkis rolfei Kuntze; *Bulbophyllum parvulum* (Hook.f.) J.J. Sm.; *Cirrhopetalum parvulum* Hook.f.

Epiphytic herbs. C, E Nepal: 2000-2600 m. Nepal, E Himalaya, Assam-Burma and E Asia.

Bulbophyllum roxburghii (Lindl.) Rchb.f. [8, 13, 40]

Cirrhopetalum roxburghii Lindl.

Epiphytic herbs. E Nepal: 300-600 m. Nepal and E Himalaya.

Bulbophyllum sarcophyllum (King & Pantl.) J.J.Sm. [13, 40]

Cirrhopetalum sarcophyllum King & Pantl.

Epiphytic herbs. E Nepal: 1000-1200 m. Nepal and E Himalaya.

Bulbophyllum scabratum Rchb.f. [1, 4, 13]

Bulbophyllum confertum Hook.f.; *Cirrhopetalum caespitosum* Wall. ex Lindl.

Epiphytic herbs. E Nepal: 2000-2200 m. Nepal, E Himalaya and S Asia.

Bulbophyllum secundum Hook.f. [1, 4, 5, 13]

Phyllorkis secunda (Hook.f.) Kuntze

Epiphytic herbs. C Nepal: 2000-2200 m. Nepal, E Himalaya, Assam-Burma, E Asia and SE Asia.

Bulbophyllum striatum (Griff.) Rchb.f. [1, 4, 13]

Dendrobium striatum Griff.

Epiphytic herbs. C Nepal: 2000-2300 m. Nepal, E Himalaya, E Asia and SE Asia.

Bulbophyllum triste Rchb.f. [1, 4, 5, 13]

Phyllorkis tristis (Rchb.f.) Kuntze

Epiphytic herbs. C, E Nepal: 1200-2300 m. Nepal, W Himalaya, E Himalaya, Assam-Burma, S Asia, E Asia and SE Asia.

Use: Ornamental.

Bulbophyllum umbellatum Lindl.

var. **umbellatum** [1, 4, 5, 13]

Epiphytic herbs. C, E Nepal: 300-3100 m. Nepal, W Himalaya, E Himalaya, Tibetan Plateau, Assam-Burma, E Asia and SE Asia.

Use(s): Medicine (whole plant), Ornamental.

var. **fuscescens** (Hook.f.) P.K.Sarkar [40]

Cirrhopetalum maculosum var. *frutescens* Hook.f.

Epiphytic herbs. Nepal: 1000-2000 m. Nepal and E Himalaya.

Bulbophyllum viridiflorum (Hook.f.) Schltr. [1, 4, 5, 13]

Cirrhopetalum viridiflorum Hook.f.

Epiphytic herbs. C, E Nepal: 1000-2600 m. Nepal, E Himalaya, S Asia and E Asia.

Bulbophyllum wallichii (Lindl.) Rchb.f. [1, 4, 5, 13]

Cirrhopetalum wallichii Lindl.; *Phyllorkis wallichii* (Lindl.) Kuntze

Epiphytic herbs. W, C, E Nepal: 1000-2500 m. Nepal, W Himalaya, E Himalaya, Assam-Burma, E Asia and SE Asia.

Bulbophyllum xylophyllum C.S.P.Parish & Rchb.f. [13]

Woody-leaved Bulbophyllum

Epiphytic herbs. E Nepal: 400-900 m. Nepal, E Himalaya, Assam-Burma, E Asia and SE Asia.

Bulbophyllum yoksunense J.J.Sm. [1, 4, 40]

Epiphytic herbs. C, E Nepal: 2000-2600 m. Nepal, E Himalaya and SE Asia.

Bulbophyllum yunnanense Rolfe [13]

Bulbophyllum otoglossum Tuyama

Epiphytic or lithophytic herbs. E Nepal: 1400-2800 m. Nepal, E Himalaya and E Asia.

Calanthe R.Br.

Calanthe alpina Hook.f. ex Lindl. [1, 4, 5, 13]

Alismorkis alpina (Hook.f. ex Lindl.) Kuntze; *Calanthe schlechteri* Hara

Terrestrial herbs. W, C, E Nepal: 1900-3500 m. Nepal, W Himalaya, E Himalaya, Tibetan Plateau and E Asia.

Calanthe biloba Lindl. [1, 4, 13]

Alismorkis biloba (Lindl.) Kuntze.

Terrestrial herbs. E Nepal: 1000-2800 m. Nepal, E Himalaya, Assam-Burma, E Asia and SE Asia.

Calanthe brevicornu Lindl. [1, 4, 5, 13]

Alismorkis brevicornu (Lindl.) Kuntze; *Calanthe lamellosa* Rolfe; *Calanthe scaposa* Z.H.Tsi & K.Y.Lang

Terrestrial herbs. C, E Nepal: 1500-3100 m. Nepal, W Himalaya, E Himalaya, Tibetan Plateau, Assam-Burma and E Asia.

Use: Ornamental.

Calanthe chloroleuca Lindl. [5, 13]

Terrestrial herbs. E Nepal: 1500-2500 m. Nepal and E Himalaya.

Calanthe davidii Franch. [13]

Calanthe pachystalix Rchb.f. ex Hook.f. (1, 4)

Terrestrial or lithophytic herbs. C Nepal: 500-3300 m. Nepal, W Himalaya, Tibetan Plateau, E Asia and SE Asia.

Calanthe densiflora Lindl. [5, 13, 61]

Alismorkis densiflora (Lindl.) Kuntze; *Calanthe kazuoi* K.Yamam.

Terrestrial herbs. C Nepal: 100-3000 m. Nepal, E Himalaya, Tibetan Plateau, E Asia and SE Asia.

Calanthe griffithii Lindl. [5, 13, 40]

Calanthe anjanae Lucksom

Terrestrial herbs. E Nepal: 2000-2300 m. Nepal, E Himalaya, Tibetan Plateau and Assam-Burma.

Calanthe mannii Hook.f. [1, 4, 5, 13]

Alismorkis mannii (Hook.f.) Kuntze

बाँदरको धोति (Baandarko dhoti)

Terrestrial herbs. W, C, E Nepal: 1500-2500 m. Nepal, W Himalaya, E Himalaya, Tibetan Plateau, Assam-Burma, E Asia and SE Asia.

Calanthe plantaginea Lindl. [1, 4, 5, 13]

Terrestrial herbs. W, C Nepal: 1500-2500 m. Nepal, W Himalaya, E Himalaya, Tibetan Plateau and E Asia.

Use(s): Medicine (rhizome), Ornamental.

Calanthe puberula Lindl. [1, 4, 5, 13]

Alismorkis puberula (Lindl.) Kuntze

Terrestrial herbs. C, E Nepal: 1800-2500 m. Nepal, W Himalaya, E Himalaya, Tibetan Plateau, E Asia and SE Asia.

Use(s): Medicine (rhizome), Ornamental.

Calanthe sylvatica (Thouars) Lindl.

Centrosis sylvatica Thouars; *Calanthe masuca* (D.Don) Lindl. (1, 4)

पंखा फूल (Pankhaa phool)

Terrestrial herbs. W, C, E Nepal: 1500-2800 m. Nepal, E Himalaya, Tibetan Plateau, Assam-Burma, S Asia, E Asia, SE Asia and Africa.

Use(s): Medicine (stem, flowers), Ornamental.

Calanthe tricarinata Lindl. [1, 4, 5, 13]

Alismorkis tricarinata (Lindl.) Kuntze

Terrestrial herbs. W, C, E Nepal: 1500-3200 m. Nepal, W Himalaya, E Himalaya, Tibetan Plateau and E Asia.

Use(s): Medicine (leaves, pseudobulb), Ornamental.

Calanthe trulliformis King & Pantl. [1, 4, 5, 13]

Terrestrial herbs. C, E Nepal: 2100-2600 m. Nepal, W Himalaya, E Himalaya and Assam-Burma.

Callostylis Blume

Callostylis rigida Blume [13]

Callostylis rigida subsp. *discolor* (Lindl.) Brieger; *Eria discolor* Lindl. (40)

Epiphytic herbs. C, E Nepal: 100-1600 m. Nepal, Assam-Burma, S Asia, E Asia and SE Asia.

Cephalanthera Rich.

Cephalanthera longifolia (L.) Fritsch [1, 4, 5, 13]

Serapias helleborine var. *longifolia* L.; *Cephalanthera ensifolia* var. *acuminata* (Lindl.) T.Tang & F.T.Wang

Terrestrial herbs. W, C Nepal: 1200-3200 m. Nepal, W Himalaya, E Himalaya, Tibetan Plateau, Assam-Burma, E Asia, N Asia, SW Asia, Europe and Africa.

Use: Medicine (rhizome).

Ceratostylis Blume

Ceratostylis himalaica Hook.f. [1, 4, 13]

Ritaia himalaica (Hook.f.) King & Pantl.

Epiphytic herbs. E Nepal: 600-1900 m. Nepal, E Himalaya, Tibetan Plateau, Assam-Burma, S Asia, E Asia and SE Asia.

Cheirostylis Blume

Cheirostylis griffithii Lindl. [1, 4, 13]

Cheirostylis macrantha Schltr.

Saprophytic/epiphytic or terrestrial herbs. C Nepal: 1500-2200 m. Nepal, W Himalaya, E Himalaya, Assam-Burma, S Asia, E Asia and SE Asia.

Chiloschista Lindl.

Chiloschista lunifera (Rchb.f.) J.J.Sm. [1, 4, 14]

Thrixspermum luniferum Rchb.f.; *Sarcochilus luniferus* (Rchb.f.) Benth. ex Hook.f.

Epiphytic herbs. C, E Nepal: 100-1800 m. Nepal, E Himalaya and Assam-Burma.

Use: Ornamental.

Chiloschista parishii Seidenf. [13, 40]

Epiphytic herbs. E Nepal: 1000-1800 m. Nepal, E Himalaya, Assam-Burma, E Asia and SE Asia.

Chiloschista usneoides (D.Don) Lindl. [1, 4, 13]

Epidendrum usneoides D.Don

Epiphytic herbs. C Nepal: 900-2500 m. Nepal, W Himalaya and E Himalaya.

Chiloschista viridiflava Seidenf. [13]

Epiphytic herbs. W, C, E Nepal: 1600-1700 m. Nepal and SE Asia.

Chrysoglossum Blume

Chrysoglossum ornatum Blume [5, 13, 40]

Chrysoglossum erraticum Hook.f.

Terrestrial herbs. C Nepal: 700-1700 m. Nepal, E Himalaya, S Asia, E Asia and SE Asia.

Use: Ornamental.

Cleisostoma Blume

Cleisostoma aspersum (Rchb.f.) Garay [5, 13, 40]

Sarcanthus asperses Rchb.f.; *Stereochilus bicuspidatus* (Hook.f.) King & Pantl. (1, 4)

Epiphytic herbs. W, C, E Nepal: 600-1900 m. Nepal, W Himalaya, E Himalaya, Assam-Burma, E Asia and SE Asia.

Cleisostoma filiforme (Lindl.) Garay [1, 4, 5, 13]

Sarcanthus filiformis Lindl.

Epiphytic herbs. C, E Nepal: 100-1400 m. Nepal, E Himalaya, Assam-Burma, E Asia and SE Asia.

Use: Ornamental.

Cleisostoma racemiferum (Lindl.) Garay [1, 4, 5, 13]

Saccolabium racemiferum Lindl.

Epiphytic herbs. C, E Nepal: 1100-2000 m. Nepal, E Himalaya, Assam-Burma, E Asia and SE Asia.

Cleisostoma simondii (Gagnep.) Seidenf. [5, 13, 40]

Vanda simondii Gagnep.

Epiphytic herbs. C, E Nepal: 400-600 m. Nepal, E Himalaya, E Asia and SE Asia.

Coelogyne Lindl.

Coelogyne barbata Lindl. ex Griff. [13, 40]

Epiphytic herbs. W, C, E Nepal: 1000-1800 m. Nepal, E Himalaya, Tibetan Plateau and E Asia.

Coelogyne corymbosa Lindl. [1, 4, 5, 13, 14]

Pleione corymbosa (Lindl.) Kuntze

चांदी गाभा (Chaandi gaavaa)

Epiphytic herbs. C, E Nepal: 1500-2900 m. Nepal, E Himalaya, Tibetan Plateau, Assam-Burma, S Asia and E Asia.

Use(s): Medicine (pseudobulb), Ornamental.

Coelogyne cristata Lindl. [1, 4, 5, 13, 14]

Cymbidium speciosissimum D.Don; *Pleione speciosissima* (D.Don) Kuntze

चांदी गाभा, भालु केरा (Chaandi gaavaa, Bhaalu kera)

Epiphytic herbs. W, C, E Nepal: 1000-2500 m. Nepal, W Himalaya, E Himalaya, Tibetan Plateau and S Asia.

Use(s): Medicine (pseudobulb), Ornamental.

Coelogyne fimbriata Lindl. [5, 13, 40]

Coelogyne fuliginosa Lodd. ex Hook. (1, 4); *Pleione fimbriata* (Lindl.) Kuntze

Epiphytic herbs. C, E Nepal: 500-2600 m. Nepal, E Himalaya, Tibetan Plateau, Assam-Burma, E Asia and SE Asia.

Use: Ornamental.

Coelogyne flaccida Lindl. [1, 4, 13]

Coelogyne lactea Rchb.f.; *Pleione flaccida* (Lindl.) Kuntze

थुर गाभा (Thur gaavaa)

Epiphytic herbs. C Nepal: 900-1100 m. Nepal, E Himalaya, Assam-Burma, E Asia and SE Asia.

Use(s): Medicine (pseudobulb), Ornamental.

Coelogyne fuscescens Lindl. [1, 4, 5, 13]

Coelogyne fuscescens var. *viridiflorum* Pradhan

Epiphytic or lithophytic herbs. C, E Nepal: 1200-2000 m. Nepal, E Himalaya, Assam-Burma, E Asia and SE Asia.

Use(s): Medicine (pseudobulb), Ornamental.

Coelogyne longipes Lindl. [1, 4, 5, 13]

Pleione longipes (Lindl.) Kuntze

Epiphytic herbs. C, E Nepal: 1500-2300 m. Nepal, E Himalaya, Tibetan Plateau, Assam-Burma, E Asia and SE Asia.

Use: Ornamental.

Coelogyne nitida (Wall. ex D.Don) Lindl. [5, 13, 40]

Cymbidium nitidum Wall. ex D.Don; *Coelogyne ochracea* Lindl.; *Pleione nitida* (Lindl.) Kuntze

चांदी गाभा (Chaandi gaavaa)

Epiphytic herbs. W, C, E Nepal: 1300-2400 m. Nepal, E Himalaya, Assam-Burma, E Asia and SE Asia.

Use(s): Medicine (pseudobulb), Ornamental.

Coelogyne ovalis Lindl. [1, 4, 13]

Coelogyne decora Wall. ex Voigt; *Coelogyne pilosissima* Planch.

Epiphytic herbs. W, C Nepal: 900-1700 m. Nepal, W Himalaya, E Himalaya, Tibetan Plateau, Assam-Burma, E Asia and SE Asia.

Use(s): Medicine (pseudobulb), Ornamental.

Coelogyne prolifera Lindl. [1, 4, 5, 13]

Coelogyne flavida Hook.f. ex Lindl. (1, 4); *Pleione flavida* (Hook.f. ex Lindl.) Kuntze; *Pleione prolifera* (Lindl.) Kuntze

थुरगाउजो (Thurgaujo)

Epiphytic herbs. C, E Nepal: 1000-2300 m. Nepal, E Himalaya, Assam-Burma, E Asia and SE Asia.

Use: Medicine (pseudobulb).

Coelogyne punctulata Lindl. [1, 4, 13]

Coelogyne brevifolia Lindl.; *Coelogyne nitida* (Roxb.) Hook.f.

Epiphytic or lithophytic herbs. E Nepal: 200-2900 m. Nepal, E Himalaya, Tibetan Plateau, Assam-Burma and E Asia.

Use: Ornamental.

Coelogyne raizadae S.K.Jain & S.Das [13, 40]

Epiphytic herbs. E Nepal: 1500-2300 m. Nepal, E Himalaya, Tibetan Plateau, E Asia and SE Asia.

Coelogyne schultesii S.K.Jain & S.Das [4, 13]

Coelogyne longipes var. *verruculata* S.C.Chen

Epiphytic herbs. C, E Nepal: 1000-1900 m. Nepal, E Himalaya, Assam-Burma, S Asia, E Asia and SE Asia.

Coelogyne stricta (D.Don) Schltr. [1, 4, 5, 13]

Cymbidium strictum D.Don; *Coelogyne elata* Lindl.

Epiphytic or lithophytic herbs. C, E Nepal: 1200-2000 m. Nepal, W Himalaya, E Himalaya, Tibetan Plateau, Assam-Burma, E Asia and SE Asia.

Use(s): Medicine (pseudobulb), Ornamental.

Conchidium Griff.

Conchidium extinctorium (Lindl.) Y.P.Ng & P.J.Cribb [13, 40]

Dendrobium extinctorium Lindl.

Epiphytic herbs. C Nepal: 500-600 m. Nepal, Assam-Burma and S Asia.

Conchidium muscicola (Lindl.) Rauschert [13, 40]

Dendrobium muscicola Lindl.; *Dendrobium parviflorum* D.Don; *Eria muscicola* (Lindl.) Lindl. (1, 4)

Epiphytic or lithophytic herbs. C, E Nepal: 300-3000 m. Nepal, W Himalaya, E Himalaya, Assam-Burma, S Asia, E Asia and SE Asia.

Use: Medicine (whole plant).

Corallorhiza Gagnebin.

Corallorhiza trifida Châtel. [1, 4, 13]

Ophrys corallorhiza L.

Coralroot orchid

Saprophytic herbs. W, C Nepal: 3800-3900 m. Nepal, W Himalaya, E Asia, N Asia, Europe and N America.

Cremastra Lindl.

Cremastra appendiculata (D.Don) Makino [1, 4, 13]

Cymbidium appendiculatum D.Don

Terrestrial herbs. C Nepal: 1800-2300 m. Nepal, E Himalaya, Tibetan Plateau and E Asia.

Crepidium Tausch

Crepidium acuminatum (D.Don) Szlach. [13]

Malaxis acuminata D.Don (1, 4, 5); *Microstylis biloba* Lindl.; *Microstylis wallichii* Lindl.

Terrestrial herbs. W, C, E Nepal: 400-3100 m. Nepal, W Himalaya, E Himalaya, Tibetan Plateau, Assam-Burma, E Asia, SE Asia and Australasia.

Use: Medicine (root, pseudobulb).

Crepidium biauritum (Lindl.) Szlach. [13, 40]

Microstylis biaurita Lindl.; *Malaxis biaurita* (Lindl.) Kuntze (92)

Terrestrial herbs. C Nepal: 900-1400 m. Nepal, W Himalaya, E Himalaya, Assam-Burma, E Asia and SE Asia.

Crepidium calophyllum (Rchb.f.) Szlach. [13, 40]

Microstylis calophylla Rchb.f.; *Malaxis calophylla* (Rchb.f.) Kuntze

Terrestrial or lithophytic herbs. E Nepal: 500-600 m. Nepal, E Himalaya, Assam-Burma, E Asia and SE Asia.

Crepidium josephianum (Rchb.f.) Marg. [13, 40]

Microstylis josephiana Rchb.f.

Terrestrial herbs. E Nepal: 500-800 m. Nepal and S Asia.

Crepidium khasianum (Hook.f.) Szlach. [13, 40]

Microstylis khasiana Hook.f.; *Malaxis khasiana* (Hook.f.) Kuntze (1, 4)

Terrestrial or lithophytic herbs. E Nepal: 1000-3000 m. Nepal, Assam-Burma, E Asia and SE Asia.

Crepidium mackinnonii (Duthie) Szlach. [13, 40]

Microstylis mackinnonii Duthie

Terrestrial herbs. W, C Nepal: 800-2100 m. Nepal, W Himalaya, S Asia, E Asia and SE Asia.

Cryptochilus Wall.

Cryptochilus luteus Lindl. [1, 4, 13]

Cryptochilus farreri Schltr.

Epiphytic herbs. W, C, E Nepal: 1200-2400 m. Nepal, W Himalaya, E Himalaya, Assam-Burma, E Asia and SE Asia.

Use: Ornamental.

Cryptochilus sanguineus Wall. [1, 4, 13]

Epiphytic herbs. C, E Nepal: 1600-2400 m. Nepal, E Himalaya, Tibetan Plateau, Assam-Burma and E Asia.

Use: Ornamental.

Cymbidium Sw.

Cymbidium aloifolium (L.) Sw. [1, 4, 5, 13]

Epidendrum aloifolium L.; *Cymbidium pendulum* (Roxb.) Sw. (1, 4)

हरजोर (Harjor)

Epiphytic or lithophytic herbs. W, C, E Nepal: 100-1600 m. Nepal, W Himalaya, E Himalaya, Assam-Burma, S Asia, E Asia and SE Asia.

Use(s): Medicine (root, pseudobulb), Forage, Ornamental.

Cymbidium bicolor Lindl. [13, 40]

Epiphytic herbs. W, C, E Nepal: 100-1600 m. Nepal, Assam-Burma, S Asia and SE Asia.

Cymbidium cyperifolium Wall. ex Lindl.

var. **szechuanicum** (Y.S.Wu & S.C.Chen) S.C.Chen & Z.J.Liu [13, 40]

Cymbidium szechuanicum Y.S.Wu & S.C.Chen

Epiphytic herbs. C, E Nepal: 1000-2900 m. Nepal, E Himalaya and E Asia.

Cymbidium dayanum Rchb.f. [13, 40]

Epiphytic herbs. C Nepal: *ca.* 1400 m. Nepal, E Himalaya, Assam-Burma, E Asia and SE Asia.

Cymbidium devonianum Paxton [1, 4, 5, 13]

Cymbidium simonsianum King & Pantl.

थिर गाभा (Thir gaavaa)

Epiphytic or lithophytic herbs. E Nepal: 1500-1800 m. Nepal, E Himalaya, Assam-Burma, S Asia, E Asia and SE Asia.

Use(s): Medicine (whole plant), Ornamental.

Cymbidium eburneum Lindl. [1, 4, 13]

Cyperorchis eburnea (Lindl.) Schltr.

Epiphytic herbs. W, C, E Nepal: 300-1700 m. Nepal, E Himalaya, Assam-Burma, S Asia and E Asia.

Use: Ornamental.

Cymbidium elegans Lindl. [13, 40]

Cymbidium longifolium D.Don (1, 4); *Cyperorchis elegans* (Lindl.) Blume

Epiphytic herbs. C, E Nepal: 1500-2800 m. Nepal, E Himalaya, Tibetan Plateau, Assam-Burma and E Asia.

Use(s): Medicine (leaves, pseudobulb, root), Ornamental.

Cymbidium erythraeum Lindl. [1, 4, 5, 13]

Epiphytic herbs. C, E Nepal: 1500-2400 m. Nepal, E Himalaya, Tibetan Plateau, Assam-Burma, S Asia, E Asia and SE Asia.

Use: Ornamental.

Cymbidium faberi Rolfe [13, 40]

Cymbidium cerinum Schltr.

Epiphytic herbs. E Nepal: 700-3000 m. Nepal, E Himalaya and E Asia.

Cymbidium × gammieanum King & Pantl. [5, 13]

Cyperorchis gammieana (King & Pantl.) Schltr.

Epiphytic herbs. C, E Nepal: 500-200 m. Nepal and E Himalaya.

Cymbidium hookerianum Rchb.f. [1, 4, 5, 13]

Cymbidium giganteum var. *hookerianum* (Rchb.f.) Desbois

Epiphytic herbs. E Nepal: 1600-2600 m. Nepal, E Himalaya, Tibetan Plateau, E Asia and SE Asia.

Use: Ornamental.

Cymbidium iridioides D.Don [1, 4, 5, 13]

Cymbidium giganteum Wall. ex Lindl.; *Cyperorchis gigantea* Schltr.

Epiphytic or lithophytic herbs. C, E Nepal: 1500-2800 m. Nepal, W Himalaya, E Himalaya, Tibetan Plateau, Assam-Burma, S Asia, E Asia and SE Asia.

Use(s): Medicine (root, leaves, pseudobulb), Ornamental.

Cymbidium lancifolium Hook. [1, 4, 5, 13]

Cymbidium javanicum Blume

Lithophytic or terrestrial/epiphytic herbs. C Nepal: 1300-2400 m. Nepal, E Himalaya, Tibetan Plateau, Assam-Burma, E Asia and SE Asia.

Use: Ornamental.

Cymbidium macrorhizon Lindl. [13, 40]

Pachyrhizanthe macrorhizon (Lindl.) Nakai

Saprophytic herbs. W, C Nepal: 200-2500 m. Nepal, W Himalaya, E Himalaya, Assam-Burma, E Asia and SE Asia.

Cymbidium mannii Rchb.f. [13, 40]

Cymbidium bicolor subsp. *obtusum* DuPuy & P.J.Cribb (4); *Cymbidium flaccidum* Schltr.

Epiphytic herbs. W, C, E Nepal: 900-1500 m. Nepal, E Himalaya, Assam-Burma, S Asia, E Asia and SE Asia.

Cymbidium micranthum Z.J.Liu & S.C.Chen [13, 40]

Epiphytic or terrestrial herbs. C Nepal: *ca.* 1500 m. Nepal and E Asia.

Cypripedium L.

Cypripedium cordigerum D.Don [1, 4, 5]

जिब्रे (Jibre)

Terrestrial herbs. W, C Nepal: 2800-3900 m. Nepal, W Himalaya, E Himalaya and Tibetan Plateau.

Use(s): Medicine (root), Ornamental, Food (tender leaves).

IUCN Category: **Vulnerable** B2ab (ver 3.1).

Cypripedium elegans Rchb.f. [1, 4, 13]

Terrestrial herbs. C, E Nepal: 3300-4200 m. Nepal, W Himalaya, E Himalaya, Tibetan Plateau and E Asia.

Use(s): Medicine (root), Ornamental.

IUCN Category: **Endangered** B2ab (ver 3.1).

Cypripedium himalaicum Rolfe [1, 4, 5, 13]

Cypripedium macranthos var. *himalaicum* (Rolfe ex Hemsl.) Kraenzl.

खुजुक्पा (Khujukpaa: Amchi)

Terrestrial herbs. W, C, E Nepal: 3000-4800 m. Nepal, E Himalaya and Tibetan Plateau.

Use(s): Medicine (whole plant), Food, Ornamental.

IUCN Category: **Endangered** B2ab (ver 3.1).

Dactylorhiza Neck. ex Nevski

Dactylorhiza hatagirea (D.Don) Soó [1, 4, 5]

Orchis hatagirea D.Don

पांचऔंले, हाथेजरा; वांगपो-लागपा (Paanchaunle, Hathejara; Wangpo-lagkpa: Amchi); Himalayan marsh orchid, Salep

Terrestrial herbs. W, C, E Nepal: 2800-4600 m. Nepal, W Himalaya, E Himalaya, Tibetan Plateau, E Asia and N Asia.

Use(s): Medicine (root/tuber), Food (leaves), Ornamental.

Dactylorhiza viridis (L.) R.M.Bateman, Pridgeon & M.W.Chase [13]

Satyrium viride L.; *Coeloglossum viride* (L.) Hartm. (1, 4); *Platanthera viridis* (L.) Lindl.

Terrestrial herbs. W, C Nepal: 3100-3900 m. Nepal, W Himalaya, E Himalaya, Tibetan Plateau, E Asia, N Asia, C Asia, SW Asia, Europe and N America.

Dendrobium Sw.

Dendrobium amoenum Wall. ex Lindl. [1, 4, 5, 13]

Callista amoena (Wall. ex Lindl.) Kuntze

Epiphytic herbs. C, E Nepal: 1100-2900 m. Nepal, W Himalaya, E Himalaya, Assam-Burma and S Asia.

Use(s): Medicine (pseudobulb), Ornamental.

Dendrobium anceps Sw. [1, 4, 5, 13]

Epiphytic herbs. E Nepal: 500-1400 m. Nepal, W Himalaya, E Himalaya, Assam-Burma and SE Asia.

Use: Ornamental.

Dendrobium aphyllum (Roxb.) C.E.C.Fisch. [1, 4, 5, 13]

Limodorum aphyllum Roxb.

Epiphytic herbs. C, E Nepal: 100-2200 m. Nepal, E Himalaya, Assam-Burma, S Asia, SE Asia and Australasia.

Use: Ornamental.

Dendrobium bicameratum Lindl. [1, 4, 5, 13]

Callista bicamerata (Lindl.) Kuntze; *Dendrobium breviflorum* Lindl.

Epiphytic herbs. W, C Nepal: 1400-2400 m. Nepal, W Himalaya, E Himalaya, Assam-Burma and E Asia.

Use: Medicine (whole plant).

Dendrobium chrysanthum Wall. ex Lindl. [1, 4, 5, 13]

Callista chrysantha (Wall. ex Lindl.) Kuntze

Epiphytic herbs. C, E Nepal: 1300-2000 m. Nepal, W Himalaya, E Himalaya, Tibetan Plateau, Assam-Burma, E Asia and SE Asia.

Use(s): Medicine (stem), Ornamental.

Dendrobium chryseum Rolfe [5, 13, 40]

Callista aurantiaca Kuntze

Epiphytic herbs. C Nepal: 1200-2200 m. Nepal, W Himalaya, E Himalaya, Assam-Burma, E Asia and SE Asia.

Use: Ornamental.

Dendrobium crepidatum Lindl. & Paxton [1, 4, 13]

Callista crepidata (Lindl. & Paxton) Kuntze

Epiphytic herbs. C Nepal: 600-1400 m. Nepal, W Himalaya, E Himalaya, Assam-Burma, E Asia and SE Asia.

Use(s): Medicine (pseudobulb), Ornamental.

Dendrobium cumulatum Lindl. [13, 40]

Callista cumulata (Lindl.) Kuntze

Epiphytic herbs. W, C, E Nepal: 100-700 m. Nepal, E Himalaya, Assam-Burma and SE Asia.

Use: Ornamental.

Dendrobium darjeelingense Pradhan [13, 40]

Epiphytic herbs. C Nepal: *ca.* 2000 m. Nepal and E Himalaya.

Dendrobium denneanum Kerr [1, 4, 13]

Callista clavata Kuntze; *Dendrobium clavatum* Wall. ex Lindl.

Epiphytic herbs. C Nepal: 600-2500 m. Nepal, E Himalaya, Assam-Burma, S Asia, E Asia and SE Asia.

Dendrobium densiflorum Wall. [1, 4, 5, 13, 14]

Callista densiflora (Lindl.) Kuntze

सुन गाभा (Sun gaabhaa)

Epiphytic herbs. C, E Nepal: 900-2900 m. Nepal, E Himalaya, Tibetan Plateau, Assam-Burma, E Asia and SE Asia.

Use(s): Medicine (pseudobulb), Ornamental.

Dendrobium denudans D.Don [1, 4, 5, 13]

Callista denudans (D.Don) Kuntze

Epiphytic herbs. W, C, E Nepal: 1000-2200 m. Nepal, W Himalaya and E Himalaya.

Dendrobium eriiflorum Griff. [1, 4, 5, 13]

Callista eriiflora (Griff.) Kuntze

Epiphytic or lithophytic herbs. W, C, E Nepal: 1400-2200 m. Nepal, E Himalaya, Assam-Burma, S Asia and SE Asia.

Use(s): Medicine (pseudobulb), Ornamental.

Dendrobium farmeri Paxton [13, 40]

Callista farmeri (Paxton) Kuntze

Epiphytic herbs. E Nepal: 100-700 m. Nepal, E Himalaya, Assam-Burma and SE Asia.

Use: Ornamental.

Dendrobium fimbriatum Hook. [1, 4, 5, 13]

Callista fimbriata (Hook.) Kuntze

Epiphytic herbs. C, E Nepal: 200-2100 m. Nepal, W Himalaya, E Himalaya, Assam-Burma, E Asia and SE Asia.

Use: Medicine (whole plant).

Dendrobium formosum Roxb. ex Lindl. [1, 4, 5, 13]

Callista formosa (Roxb. ex Lindl.) Kuntze

Epiphytic herbs. C, E Nepal: 400-1500 m. Nepal, E Himalaya, Assam-Burma, S Asia and SE Asia.

Use: Ornamental.

Dendrobium gibsonii Paxton [1, 4, 13]

Callista gibsonii (Lindl.) Kuntze

Epiphytic herbs. E Nepal: 900-2000 m. Nepal, E Himalaya, Assam-Burma, E Asia and SE Asia.

Use: Ornamental.

Dendrobium heterocarpum Wall. ex Lindl. [1, 4, 5, 13, 14]

Callista heterocarpa (Wall. ex Lindl.) Kuntze; *Dendrobium aureum* Lindl.

Epiphytic herbs. W, C, E Nepal: 1000-2200 m. Nepal, W Himalaya, E Himalaya, Assam-Burma, S Asia, E Asia and SE Asia.

Use(s): Medicine (pseudobulb), Ornamental.

Dendrobium hookerianum Lindl. [13, 40]

Callista hookeriana (Lindl.) Kuntze

Epiphytic herbs. C, E Nepal: 300-2000 m. Nepal, E Himalaya, Tibetan Plateau, Assam-Burma and E Asia.

Dendrobium longicornu Lindl. [1, 4, 5, 13]

Callista longicornis (Lindl.) Kuntze; *Dendrobium longicornu* var. *hirsuta* (Griff.) Hook.f.

बावर (Baawar)

Epiphytic herbs. C, E Nepal: 1200-2900 m. Nepal, W Himalaya, E Himalaya, Tibetan Plateau, Assam-Burma, E Asia and SE Asia.

Use(s): Medicine (whole plant), Ornamental.

Dendrobium moniliforme (L.) Sw. [13, 40]

Epidendrum moniliforme L.; *Dendrobium candidum* Wall. ex Lindl. (1,4,40)

Epiphytic or lithophytic herbs. C, E Nepal: 600-2500 m. Nepal, W Himalaya, E Himalaya, Assam-Burma, E Asia and SE Asia.

Dendrobium monticola P.F.Hunt & Summerh. [1, 4, 5, 13]

Callista alpestris Kuntze; *Dendrobium alpestre* Royle

जिवन्ति (Jiwanti)

Epiphytic or lithophytic herbs. C Nepal: 1700-3000 m. Nepal, W Himalaya, E Himalaya, Tibetan Plateau, E Asia and SE Asia.

Use(s): Medicine (whole plant), Ornamental.

Dendrobium moschatum (Buch.-Ham.) Sw. [1, 4, 5, 13]

Epidendrum moschatum Buch.-Ham.; *Cymbidium moschatum* (Buch.-Ham.) Willd.

Epiphytic herbs. C, E Nepal: 200-1200 m. Nepal, W Himalaya, E Himalaya, Assam-Burma, E Asia and SE Asia.

Use(s): Medicine (pseudobulb), Ornamental.

Dendrobium nobile Lindl. [1, 4, 5, 13]

Callista nobilis (Lindl.) Kuntze; *Dendrobium lindleyanum* Griff.

Epiphytic herbs. C, E Nepal: 400-3400 m. Nepal, E Himalaya, Tibetan Plateau, Assam-Burma, E Asia and SE Asia.

Use(s): Medicine (stem), Ornamental.

Dendrobium ochreatum Lindl. [13, 40]

Callista ochreata (Lindl.) Kuntze

Epiphytic herbs. C, E Nepal: 1200-1600 m. Nepal, Assam-Burma and SE Asia.

Use: Ornamental.

Dendrobium peguanum Lindl. [1, 4, 13]

Callista pygmaea Kuntze

Epiphytic herbs. E Nepal: 300-1200 m. Nepal, E Himalaya and Assam-Burma.

Dendrobium polyanthum Wall. ex Lindl. [13]

Callista primulina (Lindl.) Kuntze; *Dendrobium primulinum* Lindl. (1, 4, 5)

Epiphytic herbs. W, C Nepal: 1200-1400 m. Nepal, W Himalaya, Assam-Burma, E Asia and SE Asia.

Dendrobium porphyrochilum Lindl. [1, 4, 5, 13]

Callista porphyrochila (Lindl.) Kuntze; *Dendrobium caespitosum* King & Pantl.

Epiphytic or lithophytic herbs. C, E Nepal: 1800-2600 m. Nepal, E Himalaya, Assam-Burma, S Asia, E Asia and SE Asia.

Dendrobium pulchellum Roxb. ex Lindl. [1, 4, 13]

Dendrobium brevifolium Lindl.; *Dendrobium dalhousieanum* Wall. ex Paxton

Epiphytic herbs. C, E Nepal: 1200-1800 m. Nepal, E Himalaya, S Asia and SE Asia.

Use: Ornamental.

Dendrobium transparens Wall. ex Lindl. [1, 4, 8, 13]

Callista transparens (Wall. ex Lindl.) Kuntze

Epiphytic herbs. W, C, E Nepal: 600-2000 m. Nepal, W Himalaya, E Himalaya, Assam-Burma, E Asia and SE Asia.

Use(s): Medicine (pseudobulb), Ornamental.

Dendrolirium Blume

Dendrolirium lasiopetalum (Willd.) S.C.Chen & J.J.Wood [13]

Aerides lasiopetala Willd.; *Eria lasiopetala* (Willd.) Ormerod (5, 40); *Eria pubescens* (Hook.) Lindl. ex Loudon (1, 4)

Epiphytic herbs. W, C, E Nepal: 100-1600 m. Nepal, W Himalaya, E Himalaya, Assam-Burma, E Asia and SE Asia.

Didymoplexis Griff.

Didymoplexis pallens Griff. [13, 40, 84]

Gastrodia pallens (Griff.) F.Muell.

Saprophytic herbs. E Nepal: 100-200 m. Nepal, E Himalaya, E Asia, SE Asia, SW Asia and Australasia.

Dienia Lindl.

Dienia cylindrostachya Lindl. [13, 40]

Malaxis cylindrostachya (Lindl.) Kuntze (1,4,5); *Microstylis cylindrostachya* (Lindl.) Rchb.f.

टुकुना (Tukunaa)

Terrestrial herbs. W, C, E Nepal: 2000-4000 m. Nepal, W Himalaya, E Himalaya and Tibetan Plateau.

Use(s): Medicine (pseudobulb), Food (root).

Dienia ophrydis (J.Koenig) Seidenf. [13, 40]

Epidendrum ophrydis J.Koenig; *Malaxis latifolia* Sm. (1, 4, 5); *Malaxis ophrydis* (J.Koenig) Ormerod

Terrestrial or epiphytic herbs. C, E Nepal: 1400-2200 m. Nepal, W Himalaya, E Himalaya, Assam-Burma, S Asia, E Asia, SE Asia and Australasia.

Diphylax Hook.f.

Diphylax urceolata Hook.f. [1, 4, 5]

Habenaria urceolata C.B.Clarke

Terrestrial or epiphytic/lithophytic herbs. C, E Nepal: 3000-4200 m. Nepal, W Himalaya, E Himalaya, Assam-Burma and E Asia.

Diplomeris D.Don

Diplomeris hirsuta (Lindl.) Lindl. [1, 4, 5, 13]

Diplochilos hirsutus Lindl.

Terrestrial or lithophytic herbs. C, E Nepal: 200-1900 m. Nepal, W Himalaya, E Himalaya and E Asia.

Diplomeris pulchella D.Don [1, 4, 13]

Habenaria uniflora (Roxb.) Griff.; *Orchis uniflora* Roxb.

Terrestrial or lithophytic herbs. C, E Nepal: 600-2600 m. Nepal, E Himalaya, Tibetan Plateau, Assam-Burma, E Asia and SE Asia.

Epigeneium Gagnep.

Epigeneium amplum (Lindl.) Summerh. [1, 4, 5, 13]

Dendrobium amplum Lindl.; *Bulbophyllum aplum* (Lindl.) Rchb.f.

Epiphytic or lithophytic herbs. C, E Nepal: 1300-2000 m. Nepal, E Himalaya, Tibetan Plateau, E Asia and SE Asia.

Use: Ornamental.

Epigeneium fuscescens (Griff.) Summerh. [1, 4, 5, 13]

Dendrobium fuscescens Griff.; *Bulbophyllum fuscescens* (Griff.) Rchb.f.

Epiphytic herbs. E Nepal: 600-2100 m. Nepal, E Himalaya, Tibetan Plateau and E Asia.

Use: Ornamental.

Epigeneium rotundatum (Lindl.) Summerh. [1, 4, 13]

Sarcopodium rotundatum Lindl.; *Dendrobium rotundatum* (Lindl.) Hook.f.

Epiphytic herbs. E Nepal: 1500-2000 m. Nepal, E Himalaya, Tibetan Plateau, Assam-Burma and E Asia.

Use: Ornamental.

Epipactis Zinn

Epipactis helleborine (L.) Crantz [1, 4, 5, 13]

Serapias helleborine L.; *Epipactis latifolia* (L.) All.

Terrestrial herbs. W, C, E Nepal: 200-3700 m. Nepal, W Himalaya, E Himalaya, Tibetan Plateau, E Asia, N Asia, C Asia, SW Asia, Europe and Africa.

Use: Medicine (tuber).

Epipactis royleana Lindl. [1, 4, 5, 13]

Helleborine royleana (Lindl.) Soó; *Limodorum royleanum* (Lindl.) Kuntze

छासाक्रुंगगाइ (Chhasaakrungai: Chepang)

Terrestrial herbs. W, C Nepal: 1200-3700 m. Nepal, W Himalaya, E Himalaya, C Asia and SW Asia.

Use(s): Medicine (root/tuber), Food (root).

Epipactis veratrifolia Boiss. & Hohen. [1, 4, 5, 13]

Epipactis amoena Buch.-Ham. ex Wall.; *Epipactis wallichii* Schltr.

Terrestrial herbs. W, C, E Nepal: 600-1900 m. Nepal, W Himalaya, E Himalaya, Tibetan Plateau, E Asia, SW Asia and Africa.

Use: Ornamental.

Epipogium Borkh.

Epipogium aphyllum Sw. [1, 4, 8, 13]

Epipactis epipogium (L.) All.; *Satyrium epipogium* L.

Saprophytic herbs. W, C Nepal: 2500-3600 m. Nepal, W Himalaya, E Himalaya, Tibetan Plateau, E Asia and N Asia.

Epipogium japonicum Makino [41]

Galera japonica (Makino) Makino

Terrestrial herbs. C Nepal: *ca.* 2300 m. Nepal and E Asia.

Epipogium roseum (D.Don) Lindl. [1, 4, 13]

Limodorum roseum D.Don; *Epipogium tuberosum* Duthie

Saprophytic herbs. C Nepal: 500-2000 m. Nepal, W Himalaya, E Himalaya, Tibetan Plateau, S Asia, E Asia, SE Asia, Africa and Australasia.

Eria Lindl.

Eria annapurnensis L.R.Shakya & M.R.Shrestha, *Edinburgh J. Bot.* **64**(1): 1. 2007. [5, 13, 27]

Epiphytic herbs. C Nepal: *ca.* 2000 m. **Endemic to Nepal**.

Central: *Kaski*, Chhomrong, *ca.* 2000 m, LR Shakya, MR Shrestha & S. Shakya, 74 (Holotype: KATH; Isotype: TUCH).

Eria apertiflora Summerh. [5, 13, 90]

Eria rivesii Gagnep.

Epiphytic herbs. W, C, E Nepal: 1400-1800 m. Nepal, E Himalaya, Assam-Burma and SE Asia.

Use: Ornamental.

Eria baniae Bajrach., L.R.Shakya & Chettri, *J. Orchid Soc. India* **16**(1-2): 91. 2002. [13, 27]

Epiphytic herbs. C Nepal: *ca.* 1600 m. **Endemic to Nepal**.

Central: *Lalitpur*, E. North facing slope Khani Gaon, Godawari, *ca.* 1600 m, D. Bajracharya, LR Shakya & M. Chettri, CBS150 (Holotype: TUCH).

Eria carinata Gibson [5, 13, 40]

Epiphytic herbs. C, E Nepal: 1300-2000 m. Nepal, E Himalaya and SE Asia.

Use: Ornamental.

Eria clausa King & Pantl. [8, 13, 40, 91]

Dendrobium subsessile Gagnep.

Epiphytic or lithophytic herbs. C Nepal: 500-2000 m. Nepal, E Himalaya, Tibetan Plateau, Assam-Burma, E Asia and SE Asia.

Eria coronaria (Lindl.) Rchb.f. [1, 4, 5, 13]

Coelogyne coronaria Lindl.; *Trichosma suavis* Lindl. (1, 4)

Epiphytic or lithophytic herbs. C, E Nepal: 1500-2300 m. Nepal, E Himalaya, Tibetan Plateau, E Asia and SE Asia.

Use: Ornamental.

Eria nepalensis Bajrach. & K.K.Shrestha, *J. Jpn. Bot.* **78**: 158. 2003. [5, 13, 27]

Epiphytic herbs. C Nepal: *ca.* 200 m. **Endemic to Nepal**.

Central: *Chitwan*, Sauraha, *ca.* 200 m, DM Bajracharya CN 260 (Holotype: KATH; Isotype: TUCH).

Eria pokharensis Bajrach., Subedi & K.K.Shrestha, *J. Orchid Soc. India* **17**(1-2): 1. 2003. [13, 27]

Epiphytic herbs. C Nepal: 900-1600 m. **Endemic to Nepal**.

Central: *Kaski*, Lumle, Pokhara, 900-1000 m, DM Bajracharya & A. Subedi CN220 (Holotype: KATH).

Esmeralda Rchb.f

Esmeralda cathcarti (Lindl.) Rchb.f. [13]

Vanda cathcartii Lindl.; *Arachnanthe cathcartii* (Lindl.) Benth. & Hook.f.; *Arachnis cathcartii* (Lindl.) J.J.Sm. (1, 4)

Epiphytic herbs. E Nepal: 600-2000 m. Nepal and E Himalaya.

Use: Ornamental.

Esmeralda clarkei Rchb.f. [5, 13, 40]

Arachnanthe clarkei (Rchb.f) Rolfe; *Arachnis clarkei* (Rchb.f.) J.J.Sm.; *Vanda clarkei* (Rchb.f.) N.E.Br.

Epiphytic herbs. E Nepal: 500-2100 m. Nepal, E Himalaya, Assam-Burma, E Asia and SE Asia.

Use: Ornamental.

Eulophia R.Br. ex Lindl.

Eulophia bicallosa (D.Don) P.F.Hunt & Summerh. [1, 4, 13]

Bletia bicallosa D.Don; *Eulophia bicarinata* (Lindl.) Hook.f.

हात्तिपाईले (Hattipaile)

Terrestrial herbs. C Nepal: 200-300 m. Nepal, E Himalaya, Assam-Burma, E Asia, SE Asia and Australasia.

Eulophia dabia (D.Don) Hochr. [1, 4, 5, 13]

Bletia dabia D.Don; *Eulophia campestris* Wall. ex Lindl.; *Limodorum dabium* (D.Don) Buch.-Ham. ex D.Don

हात्तिपाईले (Haattipaile)

Terrestrial herbs. W, C Nepal: 300-2500 m. Nepal, W Himalaya, E Himalaya, S Asia, E Asia, C Asia and SW Asia.

Use: Medicine (root).

Eulophia densiflora Lindl. [13, 40]

Cyrtopera densiflora (Lindl.) Rchb.f.

Terrestrial herbs. W, C Nepal: *ca.* 1000 m. Nepal and E Himalaya.

Eulophia explanata Lindl. [1, 4, 5, 13]

Lissochilus explanatus (Lindl.) Schltr.

Terrestrial herbs. C Nepal: 200-2800 m. Nepal, W Himalaya and S Asia.

Eulophia flava (Lindl.) Hook.f. [1, 4, 13]

Cyrtopera flava Lindl.; *Lissochilus flavus* (Lindl.) Schltr.

Terrestrial herbs. C Nepal: 100-400 m. Nepal, W Himalaya, Assam-Burma, E Asia and SE Asia.

Eulophia graminea Lindl. [1, 4, 13]

Eulophia inconspicua Griff.; *Eulophia sinensis* Miq.

Terrestrial herbs. C Nepal: 900-1100 m. Nepal, W Himalaya, E Himalaya, Assam-Burma, S Asia, E Asia and SE Asia.

Use: Ornamental.

Eulophia herbacea Lindl. [1, 4, 5, 13]

Eulophia albiflora Edgew. ex Lindl.; *Eulophia brachypetala* Lindl.

Terrestrial herbs. W, C Nepal: 100-1400 m. Nepal, W Himalaya, S Asia, E Asia and SE Asia.

Use: Medicine (tuber).

Eulophia mackinnonii Duthie [5, 13, 40]

Terrestrial herbs. C Nepal: 800-1500 m. Nepal, W Himalaya and E Himalaya.

Eulophia obtusa (Lindl.) Hook.f. [13, 40]

Cyrtopera obtusa Lindl.; *Lissochilus obtusus* (Lindl.) Schltr.

Terrestrial herbs. C Nepal: 800-1500 m. Nepal and W Himalaya.

Eulophia spectabilis (Dennst.) Suresh [13, 40]

Wolfia spectabilis Dennst.; *Eulophia nuda* Lindl. (1, 4)

Terrestrial herbs. C, E Nepal: 400-1800 m. Nepal, E Himalaya, Assam-Burma, S Asia, E Asia, SE Asia and Australasia.

Use(s): Medicine (tuber), Ornamental.

Flickingeria A.D.Hawkes

Flickingeria fugax (Rchb.f.) Seidenf. [5, 13, 40]

Dendrobium fugax Rchb.f.

Epiphytic herbs. C, E Nepal: 400-1600 m. Nepal, W Himalaya, E Himalaya, Assam-Burma and SE Asia.

Use(s): Medicine (whole plant), Ornamental.

Flickingeria macraei (Lindl.) Seidenf.

Dendrobium macraei Lindl. (13); *Ephemerantha macraei* (Lindl.) P.F.Hunt & Sumerrh. (1, 4)

जिवन्ती (Jiwanti)

Epiphytic herbs. E Nepal: 500-2400 m. Nepal, E Himalaya and S Asia.

Use: Medicine (whole plant)

Galearis Raf.

Galearis roborowskyi (Maxim.) S.C.Chen, P.J.Cribb & S.W.Gale [8, 13]

Orchis roborovskyi Maxim.; *Galearis stracheyi* (Hook.f.) P.F.Hunt (1, 4); *Orchis stracheyi* Hook.f.

Epiphytic herbs. W, C, E Nepal: 3300-4800 m. Nepal, W Himalaya, E Himalaya and E Asia.

Use: Medicine (tuber).

Galearis spathulata (Lindl.) P.F.Hunt [1, 4, 13]

Gymnadenia spathulata Lindl.; *Aorchis spathulata* (Lindl.) Verm. (5)

Epiphytic herbs. W, C, E Nepal: 3000-5000 m. Nepal, W Himalaya, E Himalaya, Tibetan Plateau and E Asia.

Galeola Lour.

Galeola lindleyana (Hook.f. & J.W.Thomson) Rchb.f. [1, 4, 5, 13]

Cyrtosia lindleyana Hook.f. & J.W.Thomson

Saprophytic herbs. W, C, E Nepal: 1300-2800 m. Nepal, W Himalaya, E Himalaya, Tibetan Plateau, E Asia and SE Asia.

Gastrochilus D.Don

Gastrochilus acutifolius (Lindl.) Kuntze [5, 13, 40]

Saccolabium acutifolium Lindl.

Epiphytic herbs. C, E Nepal: 1200-2100 m. Nepal, W Himalaya and E Himalaya.

Use: Ornamental.

Gastrochilus affinis (King & Pantl.) Schltr. [1, 4, 13]

Saccolabium affine King & Pantl.

Epiphytic herbs. C, E Nepal: 2400-2900 m. Nepal, E Himalaya and E Asia.

Gastrochilus calceolaris (Buch.-Ham. ex Sm.) D.Don [1, 4, 13]

Aerides calceolaris Buch.-Ham. ex Sm.; *Epidendrum calceolare* (Buch.-Ham. ex Sm.) D.Don; *Gastrochilus calceolaris* var. *biflora* L.R.Shakya & M.R.Shrestha (5)

Epiphytic herbs. W, C, E Nepal: 900-2300 m. Nepal, W Himalaya, E Himalaya, Tibetan Plateau, Assam-Burma, E Asia and SE Asia.

Use: Ornamental.

IUCN Category: **Critically Endangered** A4acd (ver 3.1).

Gastrochilus dasypogon (Sm.) Kuntze [1, 4, 13]

Aerides dasypogon Sm.; *Epidendrum umbellatum* Buch.-Ham. ex Lindl.

Epiphytic herbs. C, E Nepal: 100-1400 m. Nepal and E Himalaya.

Use: Ornamental.

Gastrochilus distichus (Lindl.) Kuntze [1, 4, 5, 13]

Saccolabium distichum Lindl.

Epiphytic herbs. C, E Nepal: 1700-2800 m. Nepal, W Himalaya, E Himalaya, Tibetan Plateau and E Asia.

Gastrochilus inconspicuus (Hook.f.) Kuntze [4, 13]

Saccolabium inconspicuum Hook.f.; *Luisia micrantha* Hook.f.

Epiphytic herbs. C Nepal: 500-1500 m. Nepal, W Himalaya and Assam-Burma.

Gastrochilus nepalensis Raskoti, *Phytotaxa* **233**(2): 180. 2015. [41]

Epiphytic herbs. C Nepal: *ca.* 2350 m. **Endemic to Nepal**.

Central: *Kaski*, above Deurali on the way to Kande, 2350 m, BB Raskoti 00745 (Holotype: KATH).

Gastrochilus obliquus (Lindl.) Kuntze [13, 40]

Saccolabium obliquum Lindl.; *Gastrochilus bigibbus* (Rchb.f. ex Hook.f.) Kuntze (5)

Epiphytic herbs. C Nepal: 200-800 m. Nepal, E Himalaya, Assam-Burma, E Asia and SE Asia.

Use: Ornamental.

Gastrochilus pseudodistichus (King & Pantl.) Schltr. [5, 13, 40]

Saccolabium pseudodistichum King & Pantl.

Epiphytic herbs. C Nepal: 1700-2500 m. Nepal, E Himalaya, E Asia and SE Asia.

Gastrodia R.Br.

Gastrodia dyeriana King & Pantl. [13]

Saprophytic herbs. W, C, E Nepal: 1500-2200 m. Nepal, E Himalaya and Tibetan Plateau.

Gastrodia elata Blume [13, 40]

Gastrodia viridis Makino

Saprophytic herbs. W, C, E Nepal: 400-3200 m. Nepal, E Himalaya, E Asia and N Asia.

IUCN Category: **Vulnerable** A2c (ver 3.1).

Gastrodia falconeri D.L.Jones & M.A.Clem. [5, 13, 40]

Gastrodia orobanchoides (Falc.) Hook.f.

Saprophytic herbs. C Nepal: 2300-3000 m. Nepal and W Himalaya.

Geodorum Jacks.

Geodorum densiflorum (Lam.) Schltr. [5, 13, 40]

Limodorum densiflorum Lam.

Terrestrial herbs. C Nepal: 100-1400 m. Nepal, W Himalaya, Assam-Burma, S Asia, E Asia, SE Asia and Australasia.

Geodorum pallidum D.Don [1, 4, 13]

Cymbidium nutans (Roxb.) Sw.; *Malaxis nutans* (Roxb.) Willd.

Terrestrial herbs. C Nepal: 800-1000 m. Nepal and E Himalaya.

Goodyera R.Br.

Goodyera biflora (Lindl.) Hook.f. [1, 4, 13]

Georchis biflora Lindl.

Terrestrial herbs. W, C Nepal: 1900-2200 m. Nepal, W Himalaya, Tibetan Plateau, E Asia and SE Asia.

Goodyera foliosa (Lindl.) Benth. ex C.B.Clarke [1, 4, 5, 13]

Georchis foliosa Lindl.; *Goodyera commelinoides* Fukuy.

Terrestrial herbs. C, E Nepal: 2000-2500 m. Nepal, W Himalaya, E Himalaya, Tibetan Plateau, Assam-Burma, E Asia and SE Asia.

Goodyera fusca (Lindl.) Hook.f. [1, 4, 5, 13]

Hetaeria fusca Lindl.; *Epipactis fusca* (Lindl.) A.A.Eaton

Terrestrial herbs. W, C, E Nepal: 1500-4700 m. Nepal, W Himalaya, E Himalaya, Tibetan Plateau, Assam-Burma, E Asia and SE Asia.

Goodyera hemsleyana King & Pantl. [1, 4, 13]

Epipactis hemsleyana (King & Pantl.) A.A.Eaton

Terrestrial herbs. C, E Nepal: 2000-2200 m. Nepal, E Himalaya and E Asia.

Goodyera hispida Lindl. [13, 40]

Epipactis hispida (Lindl.) A.A.Eaton

Terrestrial herbs. C, E Nepal: 600-2000 m. Nepal, E Himalaya, E Asia and SE Asia.

Goodyera nankoensis Fukuy. [41]

Terrestrial herbs. E Nepal: *ca.* 3200 m. Nepal, Tibetan Plateau and E Asia.

Goodyera procera (Ker Gawl.) Hook. [1, 4, 13]

Neottia procera Ker Gawl.; *Epipactis procera* (Ker Gawl.) A.A.Eaton

Terrestrial herbs. W, C Nepal: 700-1500 m. Nepal, W Himalaya, E Himalaya, Tibetan Plateau, Assam-Burma, S Asia, E Asia and SE Asia.

Goodyera recurva Lindl. [8, 13, 62]

Epipactis recurva (Lindl.) A.A.Eaton; *Goodyera prainii* Hook.f.; *Goodyera recurva* var. *prainii* (Hook.f.) Pradhan

Epiphytic herbs. C Nepal: 2300-2500 m. Nepal, E Himalaya and E Asia.

Goodyera repens (L.) R.Br. [1, 4, 5, 13]

Satyrium repens L.

Terrestrial herbs. W, C, E Nepal: 1000-4200 m. Nepal, W Himalaya, E Himalaya, Tibetan Plateau, Assam-Burma, E Asia, SE Asia, N Asia, Europe and N America.

Use: Medicine (tuber).

Goodyera schlechtendaliana Rchb.f. [5, 13]

Epipactis schlechtendaliana (Rchb.f.) A.A.Eaton

Terrestrial herbs. W, C, E Nepal: 1500-2800 m. Nepal, E Himalaya, Tibetan Plateau, Assam-Burma, S Asia and SE Asia.

Use: Medicine (whole plant).

Goodyera viridiflora (Blume) Lindl. ex D.Dietr. [13]

Neottia viridiflora Blume; *Goodyera cordata* (Lindl.) Benth. & Hook.f. (1, 4)

Terrestrial herbs. C, E Nepal: 1500-1600 m. Nepal, W Himalaya, E Himalaya, E Asia, SE Asia and Australasia.

Goodyera vittata (Lindl.) Benth. ex Hook.f. [1, 4, 8, 13]

Georchis vittata Lindl.; *Orchiodes vittata* (Lindl.) Kuntze

Terrestrial herbs. C, E Nepal: 2000-3100 m. Nepal, W Himalaya, E Himalaya and Tibetan Plateau.

Gymnadenia R.Br.

Gymnadenia conopsea (L.) R.Br. [1, 4, 13]

Orchis conopsea L.

Terrestrial herbs. C Nepal: 2000-4700 m. Nepal, Tibetan Plateau, E Asia, N Asia and Europe.

Gymnadenia orchidis Lindl. [1, 4, 5, 13]

Gymnadenia cylindrostachya Lindl.; *Habenaria orchidis* (Lindl.) Hook.f.

हात्तिजरा (Hattijara)

Terrestrial herbs. W, C, E Nepal: 2100-4700 m. Nepal, W Himalaya, E Himalaya, Tibetan Plateau and E Asia.

Use: Medicine (root/pseudobulb).

Habenaria Willd.

Habenaria aitchisonii Rchb.f. [1, 4, 5, 13]

Habenaria diceras Schltr. (1, 4)

Terrestrial herbs. W, C, E Nepal: 2400-3900 m. Nepal, W Himalaya, E Himalaya, Tibetan Plateau, E Asia and SW Asia.

Habenaria arietina Hook.f. [1, 4, 5, 13]

Habenaria intermedia var. *arietana* (Hook.f.) Finet; *Habenaria pectinata* var. *arietina* (Hook.f.) Kraenzl.

Terrestrial herbs. W, C, E Nepal: 2400-3900 m. Nepal, W Himalaya, E Himalaya, Tibetan Plateau and SW Asia.

Use: Medicine (root, leaves).

Habenaria commelinifolia (Roxb.) Wall. ex Lindl. [1, 4, 5, 13]

Orchis commelinifolia Roxb.

Terrestrial herbs. W, C Nepal: 200-3200 m. Nepal, Assam-Burma, E Asia and SE Asia.

Use: Medicine (tuber).

Habenaria davidii Franch. [41]

Habenaria pectinata var. *davidii* (Franch.) Finet

Terrestrial herbs. E Nepal: 2100-2200 m. Nepal, Tibetan Plateau and E Asia.

Habenaria dentata (Sw.) Schltr. [1, 4, 5, 13]

Orchis dentata Sw.; *Platanthera dentata* (Sw.) Lindl.

Terrestrial herbs. C, E Nepal: 500-2300 m. Nepal, E Himalaya, Tibetan Plateau, Assam-Burma, E Asia and SE Asia.

Habenaria digitata Lindl. [1, 4, 5, 13]

Habenaria graveolens A.V.Duthie

Terrestrial herbs. C, E Nepal: 200-1500 m. Nepal, W Himalaya, Assam-Burma and SE Asia.

Habenaria diphylla Dalzell [5, 13]

Habenaria humistrata Rolfe ex Downie; *Liparis diphyllos* Nimmo

Terrestrial herbs. W, C, E Nepal: 200-1700 m. Nepal, W Himalaya, E Asia and SE Asia.

Habenaria furcifera Lindl. [1, 4, 5, 13]

Habenaria ovalifolia Wight

संकलो (Sankalo)

Terrestrial herbs. W, C Nepal: 100-800 m. Nepal, W Himalaya, E Himalaya, Assam-Burma, E Asia and SE Asia.

Use: Medicine (tuber).

Habenaria intermedia D.Don [1, 4, 5, 13]

Kryptostoma intermedium (D.Don) Olszewski & Szlach.; *Ochyrorchis intermedia* (D.Don) Szlach.

थुनमा (Thunma)

Terrestrial herbs. W, C Nepal: 1800-3000 m. Nepal, W Himalaya and Tibetan Plateau.

Use(s): Medicine (tuber), Food (tuber, tender leaves).

Habenaria longifolia (Lindl.) Buch.-Ham. ex Lindl. [1, 4, 13]

Gymnadenia longifolia Lindl.; *Plantaginorchis longifolia* (Buch.-Ham. ex Lindl.) Szlach.

Terrestrial herbs. C Nepal: 100-2000 m. Nepal and W Himalaya.

Habenaria malintana (Blanco) Merr. [13, 40]

Thelymitra malintana Blanco

Terrestrial herbs. E Nepal: 500-1300 m. Nepal, E Himalaya, Assam-Burma, E Asia and SE Asia.

Habenaria malleifera Hook.f. [5, 13]

Habenaria furfuracea Hook.f.

Terrestrial herbs. E Nepal: 1500-1900 m. Nepal, E Himalaya and SE Asia.

Habenaria marginata Colebr. [1, 4, 5, 13]

Platanthera marginata (Colebr.) Lindl.

Terrestrial herbs. W, C Nepal: 100-1500 m. Nepal, W Himalaya, E Himalaya, Assam-Burma, E Asia and SE Asia.

Use: Medicine (tuber).

Habenaria nepalensis Kolan., *Phytotaxa* **261**(1): 92. 2016. [45]

Terrestrial herbs. C Nepal: *ca.* 300 m. **Endemic to Nepal**.

Central: Ramnagar/Chikana, *ca.* 300 m, Wesche 3139 (Holotype: BM; Isotype: BM).

Habenaria pantlingiana Kraenzl. (13)

Habenaria stenopetala var. *polytricha* Hook.f. [1, 4]

Terrestrial herbs. E Nepal: 2000-2100 m. Nepal, E Himalaya, E Asia and SE Asia.

Habenaria pectinata D.Don [1, 4, 5, 13]

Habenaria ensifolia Lindl.

Terrestrial herbs. W, C, E Nepal: 500-3500 m. Nepal, W Himalaya, E Himalaya and E Asia.

Use(s): Medicine (tuber), Ornamental.

Habenaria plantaginea (Lindl.) Lindl. [1, 4, 5, 13]

Gymnadenia plantaginea Lindl.

Terrestrial herbs. W, C, E Nepal: 200-2500 m. Nepal, W Himalaya, Assam-Burma and S Asia.

Habenaria pubescens Lindl. [1, 4, 13]

Terrestrial herbs. C Nepal: 600-1200 m. Nepal, W Himalaya and Assam-Burma.

Habenaria reniformis (D.Don) Hook.f. [1, 4, 13]

Listera reniformis D.Don; *Neottia reniformis* (D.Don) Spreng.

Terrestrial herbs. C, E Nepal: 400-1200 m. Nepal, E Himalaya, E Asia and SE Asia.

Habenaria stenopetala Lindl. [1, 4, 5, 13]

Habenaria amanoana Ohwi

Terrestrial herbs. W, C, E Nepal: 400-2500 m. Nepal, W Himalaya, E Himalaya, Tibetan Plateau, Assam-Burma, S Asia, E Asia and SE Asia.

Habenaria wolongensis K.Y.Lang [41]

Terrestrial herbs. E Nepal: *ca.* 2800 m. Nepal and E Asia.

Habenella Small

Habenella wallichii Kolan., *Polish Botanical Journal* **60**(1): 27. 2015. [43]

Terrestrial herbs. C Nepal. **Endemic to Nepal**.

Central: *Makawanpur*, Hetauda, BB Sharma s.n. (Holotype: K).

Hemipilia Lindl.

Hemipilia cordifolia Lindl. [1, 4, 5, 13]

Hemipilia bulleyi Rolfe

Terrestrial herbs. W, C Nepal: 1800-2400 m. Nepal, W Himalaya, E Himalaya, Tibetan Plateau, Assam-Burma and E Asia.

Herminium L.

Herminium alaschanicum Maxim. [8, 13, 81]

Peristylus alaschanica (Maxim.) N.Pearce & P.J.Cribb

Terrestrial herbs. C Nepal: 1800-4500 m. Nepal, Tibetan Plateau, E Asia and N Asia.

Herminium angustilabre King & Pantl. [41]

Monorchis angustilabris (King & Pantl.) O.Schwarz

Terrestrial herbs. E Nepal: *ca.* 3400 m. Nepal, E Himalaya and E Asia.

Herminium chloranthum Tang & F.T.Wang [41, 81]

Terrestrial herbs. C Nepal: *ca.* 3800 m. Nepal, Tibetan Plateau and E Asia.

Herminium hongdeyuanii Raskoti, *Phytotaxa* **98**(1): 23. 2013. [8, 13, 81]

Terrestrial herbs. C Nepal: *ca.* 2200 m. **Endemic to Nepal**.

Central: *Makawanpur*, Chandragiri, *ca.* 2200 m, BB Raskoti 571 (Holotype: KATH).

Herminium jaffreyanum King & Pantl. [13, 81]

Monorchis jaffreyana (King & Pantl.) O.Schwarz

Terrestrial herbs. E Nepal: 2300-3000 m. Nepal and E Himalaya.

Herminium josephi Rchb.f. [1, 4, 5, 81]

Herminium forrestii Schltr.; *Herminium grandiflorum* Lindl. ex Hook.f.; *Monorchis josephi* (Rchb.f.) O.Schwarz

Terrestrial herbs. W, C, E Nepal: 2700-4400 m. Nepal, E Himalaya, Tibetan Plateau and E Asia.

Herminium kamengense A.N.Rao [41, 81]

Terrestrial herbs. E Nepal: *ca.* 2200 m. Nepal, E Himalaya and Tibetan Plateau.

Herminium lanceum (Thunb. ex Sw.) Vuijk [1, 4, 5, 13, 81]

Ophrys lancea Thunb. ex Sw.; *Herminium minutiflorum* Schltr.; *Platanthera angustifolia* (Lindl.) O.Schwarz

Terrestrial herbs. W, C, E Nepal: 1100-3500 m. Nepal, W Himalaya, E Himalaya, Assam-Burma, Assam-Burma, E Asia and SE Asia.

Use: Medicine (whole plant).

Herminium longilobatum S.N.Hedge & A.N.Rao [41, 81]

Terrestrial herbs. E Nepal: *ca.* 2500 m. Nepal and E Himalaya.

Herminium mackinonii Duthie [1, 4, 5, 13, 81]

Monorchis mackinonii (Duthie) O.Schwarz

Terrestrial or epiphytic herbs. W, C, E Nepal: 2100-2700 m. Nepal, W Himalaya and E Himalaya.

Herminium macrophyllum (D.Don) Dandy [1, 4, 5, 13, 81]

Neottia macrophylla D.Don; *Spiranthes macrophylla* (D.Don) Spreng.

Terrestrial herbs. W, C, E Nepal: 3700-4600 m. Nepal, W Himalaya, E Himalaya and Tibetan Plateau.

Herminium monophyllum (D.Don) P.F.Hunt & Summerh. [1, 4, 5, 13, 81]

Neottia monophylla D.Don; *Spiranthes monophylla* (D.Don) Spreng.

Terrestrial herbs. W, C Nepal: 1000-4500 m. Nepal and E Himalaya.

Herminium monorchis (L.) R.Br. [1, 4, 5, 13, 81]

Ophrys monorchis L.; *Satyrium monorchis* (L.) Pers.

Musk orchid

Terrestrial herbs. W, C, E Nepal: 3400-4600 m. Nepal, W Himalaya, Tibetan Plateau, E Asia, N Asia, C Asia, SW Asia and Europe.

Use: Medicine (root).

Herminium orbiculare Hook.f. [13, 62, 81]

Monorchis orbicularis (Hook.f.) O.Schwarz

Terrestrial herbs. E Nepal: *ca.* 3000 m. Nepal, E Himalaya and Tibetan Plateau.

Herminium quinquelobum King & Pantl. [1, 4, 5, 13, 81]

Monorchis quinqueloba (King & Pantl.) O.Schwarz

Terrestrial or epiphytic herbs. C, E Nepal: 100-2900 m. Nepal, E Himalaya and E Asia.

Herminium souliei (Finet) Rolfe [13, 81]

Herminium angustifolium var. *souliei* Finet; *Monorchis soulei* (Rolfe) O. Schwarz

Terrestrial herbs. W, C, E Nepal: 1400-4200 m. Nepal, Tibetan Plateau and E Asia.

Herpysma Lindl.

Herpysma longicaulis Lindl. [1, 4, 8, 13]

Herpysma bracteata (Blume) J.J.Sm.

Terrestrial herbs. C Nepal: 900-1700 m. Nepal, E Himalaya, Assam-Burma, E Asia and SE Asia.

Liparis Rich.

Liparis bootanensis Griff. [5, 13]

Liparis lancifolia Hook.f.

Terrestrial herbs. C, E Nepal: 1400-2500 m. Nepal, E Himalaya, Tibetan Plateau, Assam-Burma, E Asia and SE Asia.

Liparis cespitosa (Lam.) Lindl. [5, 13]

Epidendrum cespitosum Lam.; *Malaxis angustifolia* Blume

Terrestrial or epiphytic herbs. C, E Nepal: 500-1400 m. Nepal, W Himalaya, E Himalaya, Tibetan Plateau, Assam-Burma, S Asia, E Asia, SE Asia, Africa and Australasia.

Liparis campylostalix Rchb.f. [41]

Liparis pauciflora Rolfe

Terrestrial herbs. W, C, E Nepal: 1100-3800 m. Nepal, Tibetan Plateau, E Asia and N Asia.

Liparis cathcartii Hook.f. [1, 4, 5, 13]

Leptorkis cathcartii (Hook.f.) Kuntze

Terrestrial herbs. W, C, E Nepal: 2100-3100 m. Nepal, E Himalaya and E Asia.

Liparis cordifolia Hook.f. [1, 4, 13]

Leptorkis cordifolia (Hook.f.) Kuntze

Terrestrial herbs. C, E Nepal: 1400-2000 m. Nepal, E Himalaya, Tibetan Plateau, S Asia, E Asia and SE Asia.

Liparis deflexa Hook.f. [5, 13]

Leptorkis deflexa (Hook.f.) Kuntze; *Liparis prazeri* King & Pantl.

Terrestrial herbs. C, E Nepal: 300-1400 m. Nepal, W Himalaya and E Himalaya.

Liparis elliptica Wight [13]

Liparis hookeri Ridl.

Terrestrial or epiphytic herbs. E Nepal: 200-1600 m. Nepal, E Himalaya, Tibetan Plateau, Assam-Burma, S Asia, E Asia, SE Asia and Australasia.

Liparis ferruginea Lindl. [8, 13, 63]

Leptorkis ferruginea (Lindl.) Kuntze

Terrestrial herbs. C Nepal: *ca.* 400 m. Nepal, E Asia and SE Asia.

Liparis glossula Rchb.f. [1, 4, 5, 13]

Leptorkis glossula (Rchb.f.) Kuntze

Terrestrial herbs. W, C, E Nepal: 1800-4300 m. Nepal, W Himalaya, E Himalaya, Tibetan Plateau and E Asia.

Liparis langtangensis Raskoti & Ale, *Novon* **23**: 83. 2014. [8, 13]

Terrestrial herbs. C Nepal: 3000-3900 m. **Endemic to Nepal.**

Central: *Rasuwa*, Kyangin Kharka, Langtang National Park, 3700-3900 m, BB Raskoti 300 (Holotype: KATH; Isotype: TUCH).

Liparis nervosa (Thunb.) Lindl. [1, 4, 5, 13]

Ophrys nervosa Thunb.; *Malaxis nervosa* (Thunb.) Sw.

var. **nervosa**

Terrestrial herbs. W, C, E Nepal: 400-2800 m. Nepal, W Himalaya, E Himalaya, Tibetan Plateau, S Asia, E Asia, SE Asia; wide spread in Old and New world tropics.

Use: Medicine (tuber).

var. **khasiana** (Hook.f.) P.K.Sarkar [40]

Liparis bituberculata var. *khasiana* Hook.f.; *Liparis breviscapa* A.P.Das & Lama; *Liparis khasiana* (Hook.f.) Tang & F.T.Wang

Terrestrial herbs. C Nepal: *ca.* 1800 m. Nepal and E Himalaya.

Liparis odorata (Willd.) Lindl. [13]

Malaxis odorata Willd.; *Malaxis lancifolia* Sm.

Terrestrial or lithophytic herbs. C Nepal: 900-2000 m. Nepal, W Himalaya, E Himalaya, Tibetan Plateau, Assam-Burma, E Asia, SE Asia and Australasia.

Liparis olivacea Lindl., *Gen. Sp. Orchid. Pl.* **1**: 27. 1830. [1, 4, 13, 27]

Leptorkis olivacea (Lindl.) Kuntze

Terrestrial herbs. C Nepal: *ca.* 2300 m. **Endemic to Nepal.**

Central: Nepalia, N. Wallich, 1942 (Syntype: K-W).

Liparis perpusilla Hook.f. [1, 4, 5, 13]

Leptorkis perpusilla (Hook.f.) Kuntze; *Liparis togashii* Tuyama

Terrestrial or epiphytic herbs. C, E Nepal: 2300-2900 m. Nepal, E Himalaya and E Asia.

Liparis petiolata (D.Don) P.F.Hunt & Summerh. [1, 4, 5, 13]

Acianthus petiolatus D.Don; *Liparis nepalensis* Lindl.; *Liparis pulchella* Hook.f.; *Malaxis cordifolia* Sm.

Terrestrial herbs. C Nepal: 1300-2800 m. Nepal, E Himalaya, Tibetan Plateau, E Asia and SE Asia.

Liparis plantaginea Lindl. [13]

Leptorkis plantaginea (Lindl.) Kuntze

Terrestrial herbs. C Nepal: 800-2000 m. Nepal, E Himalaya and SE Asia.

Liparis platyrachis Hook.f. [1, 4, 13]

Leptorkis platyrachis (Hook.f.) Kuntze

Terrestrial herbs. E Nepal: 1400-1500 m. Nepal, W Himalaya, E Himalaya and E Asia.

Liparis pygmaea King & Pantl. [1, 4, 13]

Terrestrial or lithophytic herbs. W, C Nepal: 3200-3500 m. Nepal, E Himalaya and Tibetan Plateau.

Liparis resupinata Ridl. [1, 4, 5, 13]

Leptorkis resupinata (Ridl.) Kuntze

Terrestrial or epiphytic herbs. C, E Nepal: 1500-2600 m. Nepal, W Himalaya, E Himalaya, Tibetan Plateau and E Asia.

Liparis rostrata Rchb.f. [1, 4, 13]

Liparis diodon Rchb.f.; *Leptorkis diodon* (Rchb.f.) Kuntze

Terrestrial herbs. W Nepal: 2000-3000 m. Nepal, W Himalaya, E Himalaya, Tibetan Plateau and E Asia.

Use: Medicine (tuber).

Liparis stricklandiana Rchb.f. [13]

Leptorkis stricklandiana (Rchb.f.) Kuntze

Epiphytic herbs. C, E Nepal: 600-2400 m. Nepal, E Himalaya, Tibetan Plateau, E Asia and SE Asia.

Liparis viridiflora (Blume) Lindl. [1, 4, 5, 13]

Malaxis viridiflora Blume; *Leptorkis viridiflora* (Blume) Kuntze; *Liparis pendula* Lindl.

Epiphytic or lithophytic herbs. C, E Nepal: 600-2500 m. Nepal, W Himalaya, E Himalaya, Tibetan Plateau, Assam-Burma, S Asia, E Asia, SE Asia and Australasia.

Luisia Gaudich.

Luisia brachystachys (Lindl.) Blume [13]

Mesoclastes brachystachys Lindl.; *Luisia siamensis* Rolfe ex Downie

Epiphytic herbs. W, C Nepal: 1300-1900 m. Nepal, W Himalaya, E Himalaya, Assam-Burma, S Asia, E Asia and SE Asia.

Luisia trichorrhiza (Hook.) Blume [1, 4, 13]

Vanda trichorhiza Hook.

अर्जोना (Arjona)

Epiphytic herbs. C, E Nepal: 700-1400 m. Nepal, W Himalaya, E Himalaya, Assam-Burma and SE Asia.

Use(s): Medicine (root), Ornamental.

Luisia tristis (G.Forst.) Hook.f. [13]

Epidendrum triste G.Forst.; *Luisia zeylanica* Lindl. (1, 4, 5)

कौवाको केरा, बोरि झार (Kauwako Kera, Bori jhaar)

Epiphytic herbs. W, C, E Nepal: 300-3100 m. Nepal, W Himalaya, E Himalaya and S Asia.

Use(s): Medicine (leaves), Ornamental.

Malaxis Sol. ex Sw.

Malaxis dolpensis M.R.Shrestha, L.R.Shakya & Ghimire, *Pleione* **1**: 6. 2007. [5, 13, 27, 86]

Terrestrial herbs. W Nepal: *ca.* 4200 m. **Endemic to Nepal**.

Central: *Dolpa*, Khorakchhang, *ca.* 4200 m, MR Shreshta, SK Ghimire & L. Shahi 873 (Holotype: KATH).

Malaxis monophyllos (L.) Sw. [5, 13]

Ophrys monophyllos L.; *Microstylis monophyllos* (L.) Lindl.

var. **monophyllos**

Terrestrial or lithophytic herbs. C, E Nepal: 2600-4500 m. Nepal, E Himalaya, Tibetan Plateau, E Asia, Europe and N America.

var. **obtusa** Tsukaya & H.Okada, *J. Jap. Bot.* **75**(5): 314. 2000. [13]

Microstylis monophyllos var. *obtusa* (Tsukaya & H.Okada) Marg.

Terrestrial herbs. E Nepal: 3300-3400 m. **Endemic to Nepal**.

East: *Solukhumbu*, Thasing Dingma to Saure Kharka through Sanu Khola, 3350 m, F. Miyamoto, M. Amano, H. Ikeda, CM Joshi, K. Arai & T. Komatsu, 9584106 (Holotype: TI; Isotype: KATH).

Malaxis muscifera (Lindl.) Kuntze [1, 4, 5, 13]

Dienia muscifera Lindl.; *Microstylis muscifera* (Lindl.) Ridl.

रसाभक (Rsaabhak)

Terrestrial herbs. W, C, E Nepal: 2000-4300 m. Nepal, W Himalaya, E Himalaya and Assam-Burma.

Use: Medicine (pseudobulb).

IUCN Category: **Vulnerable** A2cd (ver 3.1).

Malaxis tamurensis Tuyama, *Fl. E. Himalaya* 444, f.29a,b. 1966. [1, 4, 13, 27]

Terrestrial herbs. E Nepal: *ca.* 1200 m. **Endemic to Nepal**.

East: *Dhankuta*, Dhankuta, *ca.* 1200 m, TI6305500 (Holotype: TI).

Micropera Lindl.

Micropera obtusa (Lindl.) T.Tang & F.T.Wang [13]

Camarotis obtusa Lindl.

Epiphytic herbs. C, E Nepal: 100-500 m. Nepal, E Himalaya, Assam-Burma and SE Asia.

Monomeria Lindl.

Monomeria barbata Lindl. [1, 4, 5, 13]

Epicranthes barbata (Lindl.) Rchb.f.

Epiphytic herbs. C Nepal: 1800-3200 m. Nepal, E Himalaya, Tibetan Plateau, Assam-Burma, E Asia and SE Asia.

Mycaranthes Blume

Mycaranthes floribunda (D.Don) S.C.Chen & J.J.Wood. [13]

Dendrobium floribundum D.Don (1, 4); *Eria paniculata* Lindl. (1, 4); *Pinalia paniculata* (Lindl.) Kuntze

Epiphytic herbs. C, E Nepal: 600-1800 m. Nepal, E Himalaya, Assam-Burma, E Asia and SE Asia.

Myrmechis Blume

Myrmechis pumila (Hook.f.) Tang & F.T.Wang [1, 4, 8, 13]

Odontochilus pumilus Hook.f.; *Zeuxine pumila* (Hook.f.) King & Pantl.

Terrestrial herbs. C Nepal: 2500-3200 m. Nepal, E Himalaya, Assam-Burma, E Asia and SE Asia.

Neogyna Rchb.f.

Neogyna gardneriana (Lindl.) Rchb.f. [1, 4, 13]

Coelogyne gardneriana Lindl.; *Pleione gardneriana* (Lindl.) Kuntze

Saprophytic or epiphytic herbs. C Nepal: 1500-3000 m. Nepal, E Himalaya, Tibetan Plateau, Assam-Burma, S Asia, E Asia and SE Asia.

Use: Ornamental.

Neottia Guett.

Neottia acuminata Schltr.

Aphyllorchis parviflora King & Pantl. (1, 4, 13); *Neottia oblonga* Tang & F.T.Wang

Saprophytic herbs. W, C Nepal: 3000-4000 m. Nepal, W Himalaya, E Himalaya, Tibetan Plateau, E Asia and N Asia.

Neottia chandrae Raskoti, J.J.Wood & Ale, *Nord. J. Bot.* **30**:187.2012. [8, 13]

Terrestrial herbs. C Nepal: *ca.* 3800 m. **Endemic to Nepal**.

Central: *Rasuwa*, Lauribinayak, Langtang National Park, *ca.* 3800 m, BB Raskoti 280 (Holotype: KATH; Isotype: TUCH).

Neottia listeroides Lindl. [1, 4, 5, 13]

Listera lindleyana (Decne.) King & Pantl.; *Neottia lindleyana* Decne.

Terrestrial or saprophytic herbs. W, C Nepal: 2700-3400 m. Nepal, W Himalaya, E Himalaya, Tibetan Plateau and E Asia.

Neottia mucronata (Panigrahi & J.J.Wood) Szlach. [13]

Listera mucronata Panigrahi & J.J.Wood (5)

Saprophytic herbs. C, E Nepal: 2400-2500 m. Nepal, E Himalaya and E Asia.

Neottia nepalensis (N.P.Balakr.) Szlach., *Fragm. Florist. Geobot.* **3**: 118. 1995. [13]

Listera nepalensis N.P.Balakr. (1, 4, 5, 27)

Terrestrial saprophytic herbs. W, C Nepal: 2600-4000 m. **Endemic to Nepal**.

Central: *Baglung*, *ca.* 3800 m, JDA Stainton, WR Sykes & LHJ Williams, 3431 (Holotype: CAL; Isotype: BM).

Neottia pinetorum (Lindl.) Szlach. [13]

Listera pinetorum Lindl. (1, 4, 5)

Saprophytic herbs. C, E Nepal: 2200-3700 m. Nepal, W Himalaya, E Himalaya, Tibetan Plateau and E Asia.

Neottia tenuis (Lindl.) Szlach. [13]

Listera tenuis Lindl. (1, 4)

Saprophytic herbs. E Nepal: 3000-4000 m. Nepal, W Himalaya and E Himalaya.

Neottianthe (Rchb.) Schltr.

Neottianthe cucullata (L.) Schltr. [1, 4, 5, 13]

Orchis cucullata L.; *Gymnadenia cucullata* (L.) Rich.

var. **cucullata**

Terrestrial herbs. W, C, E Nepal: 2300-5000 m. Nepal, E Himalaya, Tibetan Plateau, E Asia, N Asia and Europe.

Use(s): Medicine (root), Ornamental.

var. **calcicola** (W.W.Sm.) Soó [5, 13]

Gymnadenia calcicola W.W.Sm.; *Neottianthe calcicola* (W.W.Sm.) Schltr. (1, 4)

Terrestrial herbs. W, C, E Nepal: 3700-5000 m. Nepal, W Himalaya, E Himalaya, Tibetan Plateau and E Asia.

Neottianthe secundiflora (Hook.f.) Schltr. [1, 4, 5, 13]

Habenaria secundiflorus Hook.f.; *Gymnadenia secundiflora* (Hook.f.) Kraenzl.

Terrestrial herbs. W, C, E Nepal: 2200-3200 m. Nepal, W Himalaya, E Himalaya, Tibetan Plateau, Assam-Burma and E Asia.

Nervilia Comm. ex Gaudich.

Nervilia aragoana Gaudich. [1, 4, 13]

Nervilia scottii (Rchb.f.) Schltr. (1, 4)

स्थल कमल (Sthal kamal)

Terrestrial herbs. C, E Nepal: 500-1300 m. Nepal, W Himalaya, E Himalaya, Tibetan Plateau, Assam-Burma, S Asia, E Asia, SE Asia and Australasia.

Use: Medicine (whole plant).

Nervilia crociformis (Zoll. & Moritzi) Seidenf. [5, 13]

Bolborchis crociformis Zoll. & Moritzi; *Nervilia crispata* (Blume) Schltr. ex K.Schum. & Lauterb. (1, 4)

Terrestrial herbs. E Nepal: 400-2100 m. Nepal, W Himalaya, E Himalaya, E Asia, SE Asia, Africa and Australasia.

Nervilia gammieana (Hook.f.) Pfitzer [1, 4, 13]

Pogonia gammieana Hook.f.

Terrestrial herbs. C, E Nepal: 400-800 m. Nepal, W Himalaya and E Himalaya.

Nervilia mackinnonii (Duthie) Schltr. [13, 64]

Pogonia mackinnonii Duthie

Terrestrial herbs. C Nepal: 900-1700 m. Nepal, W Himalaya, E Himalaya, Assam-Burma and E Asia.

Nervilia macroglossa (Hook.f.) Schltr. [13, 40]

Pogonia macroglossa Hook.f.

Terrestrial herbs. E Nepal: 600-1800 m. Nepal, E Himalaya, Assam-Burma and SE Asia.

Nervilia plicata (Andrews) Schltr. [13, 64]

Arethusa plicata Andrews; *Epipactis plicata* Roxb.

Terrestrial herbs. W, C, E Nepal: 300-1000 m. Nepal, W Himalaya, E Himalaya, Assam-Burma, S Asia, E Asia, SE Asia and Australasia.

Oberonia Lindl.

Oberonia acaulis Griff. [1, 4, 5, 13]

Oberonia myriantha Lindl.; *Oberonia sikkimensis* Lindl.

Epiphytic herbs. C, E Nepal: 600-2000 m. Nepal, W Himalaya, E Himalaya, Tibetan Plateau, Assam-Burma, E Asia and SE Asia.

Oberonia brachystachys Lindl. [5, 13]

Malaxis brachystachya (Lindl.) Rchb.f.; *Oberonia demissa* Lindl.

Epiphytic herbs. E Nepal: 200-800 m. Nepal, E Himalaya, Assam-Burma and SE Asia.

Oberonia caulescens Lindl. [1, 4, 5, 13]

Malaxis caulescens (Lindl.) Rchb.f.; *Oberonia auriculata* King & Pantl.

Epiphytic herbs. C, E Nepal: 600-2500 m. Nepal, W Himalaya, E Himalaya, Tibetan Plateau, E Asia and SE Asia.

Use: Medicine (tuber).

Oberonia cavaleriei Finet [13]

Epiphytic herbs. C Nepal: 1000-1500 m. Nepal, S Asia, E Asia and SE Asia.

Oberonia emarginata King & Pantl. [13]

Epiphytic herbs. E Nepal: 1600-2700 m. Nepal, E Himalaya and SE Asia.

Oberonia ensiformis (Sm.) Lindl. [1, 4, 5, 13]

Malaxis ensiformis Sm.; *Oberonia iridifolia* Lindl. (1, 4); *Oberonia trilobata* Griff.

Epiphytic herbs. C, E Nepal: 800-1700 m. Nepal, W Himalaya, E Himalaya, S Asia, E Asia and SE Asia.

Oberonia falcata King & Pantl. [1, 4, 5, 13]

Oberonia caudata King & Pantl.

Epiphytic herbs. C, E Nepal: 1300-2500 m. Nepal, E Himalaya, E Asia and SE Asia.

Oberonia falconeri Hook.f. [1, 4, 5, 13]

Oberonia siamensis Schltr.

Epiphytic herbs. W, C, E Nepal: 100-1300 m. Nepal, W Himalaya, E Asia and SE Asia.

Oberonia jenkinsiana Griff. ex Lindl. [5, 13]

Malaxis jenkinsiana (Griff. ex Lindl.) Rchb.f.

Epiphytic herbs. C Nepal: 400-1400 m. Nepal, E Himalaya, Assam-Burma, E Asia and SE Asia.

Oberonia lobulata King & Pantl. [13]

Epiphytic herbs. W, C, E Nepal: *ca.* 300 m. Nepal and E Himalaya.

Oberonia mucronata (D.Don) Ormerod & Seidenf. [5, 13]

Stelis mucronata D.Don; *Cymbidium iridifolium* Roxb.; *Oberonia denticulata* Wight

Epiphytic herbs. E Nepal: 200-2000 m. Nepal, E Himalaya, Assam-Burma, S Asia, E Asia and SE Asia.

Oberonia myosurus (G.Forst.) Lindl. [4, 40]

Epidendrum myosurus G.Forst.

Epiphytic herbs. W, C Nepal: 1000-1500 m. Nepal and SE Asia.

Oberonia nepalensis L.R.Shakya & R.P.Chaudhary, *Harvard Pap. Bot.* **4**(1): 357. 1999. [4, 8, 13, 27]

Epiphytic herbs. C Nepal: 900-1100 m. **Endemic to Nepal**.

Central: *Kaski*, Pokhara (around Annapurna Conservation Zone), *ca.* 1100 m, LR Shakya, RL Singh & RP Chaudhary 27 (Holotype: KATH; Isotype: TUCH).

Oberonia obcordata Lindl. [1, 4, 5, 13]

Malaxis obcordata (Lindl.) Rchb.f.; *Oberonia treutleri* Hook.f.

Epiphytic herbs. C, E Nepal: 1100-2000 m. Nepal, E Himalaya, Tibetan Plateau, S Asia, E Asia and SE Asia.

Oberonia pachyphylla King & Pantl. [5, 13]

Oberonia apiculata Kerr

Epiphytic herbs. E Nepal: 200-800 m. Nepal, S Asia and SE Asia.

Oberonia pachyrachis Rchb.f. ex Hook.f. [1, 4, 5, 13]

Iridorkis pachyrachis (Rchb.f. ex Hook.f.) Kuntze

Epiphytic herbs. W, C, E Nepal: 800-3500 m. Nepal, W Himalaya, E Himalaya, Tibetan Plateau, Assam-Burma, E Asia and SE Asia.

Oberonia prainiana King & Pantl. [5, 13]

Oberonia brunnescens Ridl.

Epiphytic herbs. C, E Nepal: 500-1500 m. Nepal, W Himalaya, E Himalaya, E Asia and SE Asia.

Oberonia pyrulifera Lindl. [5, 13]

Malaxis pyrulifera (Lindl.) Rchb.f.

Epiphytic herbs. C, E Nepal: 300-2400 m. Nepal, W Himalaya, E Himalaya, E Asia and SE Asia.

Oberonia recurva Lindl. [8, 13]

Oberonia parvula King & Pantl.; *Oberonia setifera* Lindl.

Epiphytic herbs. C Nepal: 200-1500 m. Nepal, E Himalaya and E Asia.

Oberonia rufilabris Lindl. [1, 4, 13]

Malaxis rufilabris (Lindl.) Rchb.f.; *Oberonia pantlingiana* L.R.Shakya & R.P.Chaudhary

Epiphytic herbs. C Nepal: 300-1500 m. Nepal, E Himalaya, Assam-Burma, S Asia, E Asia and SE Asia.

Odontochilus Blume

Odontochilus crispus (Lindl.) Hook.f. [1, 4, 8, 13]

Anoectochilus crispus Lindl.

Terrestrial herbs. C, E Nepal: 400-2400 m. Nepal, E Himalaya, Tibetan Plateau and E Asia.

Odontochilus lanceolatus (Lindl.) Blume [1, 4, 13]

Anoectochilus lanceolatus Lindl.

Terrestrial herbs. C, E Nepal: 800-2200 m. Nepal, E Himalaya, Assam-Burma, S Asia, E Asia and SE Asia.

Odontochilus nandae Raskoti & H.Kurzweil, *Phytotaxa* **233**(3): 293. 2015. [44]

Terrestrial herbs. C Nepal: 2100-2400 m. **Endemic to Nepal**.

Central: *Kaski*, Panchase, *ca.* 2400 m, BB Raskoti 119 (Holotype: KATH).

Oreorchis Lindl.

Oreorchis foliosa (Lindl.) Lindl. [4, 13]

Corallorhiza foliosa Lindl.

var. **foliosa**

Terrestrial herbs. W, C, E Nepal: 2500-4500 m. Nepal, W Himalaya, E Himalaya and E Asia.

var. **indica** (Lindl.) N.Pearce & P.J.Cribb [4, 13]

Corallorhiza indica Lindl.; *Oreorchis indica* (Lindl.) Hook.f.

Terrestrial or saprophytic herbs. C, E Nepal: 2700-3800 m. Nepal, W Himalaya, E Himalaya, Tibetan Plateau, Assam-Burma and E Asia.

Oreorchis micrantha Lindl. [1, 4, 5, 13]

Oreorchis rolfei Duthie

Terrestrial or saprophytic herbs. C, E Nepal: 2400-3000 m. Nepal, W Himalaya, E Himalaya, Tibetan Plateau, Assam-Burma and E Asia.

Oreorchis nepalensis N.Pearce & P.J.Cribb [4, 13]

Terrestrial or saprophytic herbs. C Nepal: 2700-3100 m. Nepal and Tibetan Plateau.

Oreorchis porphyranthes Tuyama, *J. Jap. Bot.* **50**: 69. 1975. [1, 4, 13, 27]

Terrestrial or saprophytic herbs. C Nepal: 3000-3700 m. **Endemic to Nepal**.

Central: *Myagdi*, S. Gurjakhani, 10800 ft. (3290 m), JDA Stainton, WR Sykes & LHJ Williams 3048 (Holotype: BM).

Ornithochilus (Wall. ex Lindl.) Benth. & Hook.f.

Ornithochilus difformis (Wall. ex Lindl.) Schltr. [1, 4]

Aerides difformis Wall. ex Lindl.; *Sarcochilus difformis* (Wall. ex Lindl.) Tang & F.T.Wang

Epiphytic herbs. C, E Nepal: 1400-2400 m. Nepal, W Himalaya, E Himalaya, Assam-Burma, E Asia and SE Asia.

Otochilus Lindl.

Otochilus albus Lindl. [1, 4, 5, 13]

Coelogyne alba (Lindl.) Rchb.f.

Epiphytic herbs. C, E Nepal: 1500-2400 m. Nepal, E Himalaya, Tibetan Plateau, Assam-Burma and SE Asia.

Use(s): Medicine (whole plant), Ornamental.

Otochilus fuscus Lindl. [1, 4, 5, 13]

Coelogyne fusca (Lindl.) Rchb.f.; *Otochilus lancifolius* Griff.

Epiphytic herbs. C, E Nepal: 750-2500 m. Nepal, E Himalaya, Assam-Burma, E Asia and SE Asia.

Use: Ornamental.

Otochilus lancilabius Seidenf. [4, 5, 13]

Otochilus albus var. *lancilabius* (Seidenf.) Pradhan

Epiphytic herbs. W, C, E Nepal: 1400-2500 m. Nepal, W Himalaya, E Himalaya, Tibetan Plateau, Assam-Burma and SE Asia.

Use(s): Medicine (pseudobulb), Ornamental.

Otochilus porrectus Lindl. [1, 4, 5, 13]

Coelogyne porrecta (Lindl.) Rchb.f.; *Otochilus forrestii* W.W.Sm.; *Tetrapeltis fragrans* Wall. ex Lindl.

जिवन्ति (Jiwanti)

Epiphytic herbs. W, C, E Nepal: 1000-2400 m. Nepal, E Himalaya, Assam-Burma, E Asia and SE Asia.

Use(s): Medicine (whole plant), Ornamental.

Pachystoma Blume

Pachystoma pubescens Blume [13]

Pachychilus pubescens (Blume) Blume

Terrestrial herbs. C Nepal: 500-1800 m. Nepal, W Himalaya, E Himalaya, Assam-Burma, S Asia, E Asia, SE Asia and Australasia.

Panisea (Lindl.) Lindl.

Panisea demissa (D.Don) Pfitzer [1, 4, 5, 13]

Dendrobium demissum D.Don; *Coelogyne parviflora* Lindl.; *Panisea parviflora* (Lindl.) Lindl.

हात्ती टाउके (Haatti tauke)

Epiphytic herbs. C, E Nepal: 1500-2400 m. Nepal, E Himalaya, Assam-Burma, S Asia, E Asia and SE Asia.

Panisea panchaseensis Subedi [13, 77]

Epiphytic herbs. C Nepal: 2100-2500 m. Nepal and E Himalaya.

Panisea tricallosa Rolfe. [13]

Panisea pantlingii (Pfitzer) Schltr.; *Panisea unifolia* S.C.Chen

Epiphytic herbs. W, C, E Nepal: 100-2100 m. Nepal, E Himalaya, E Asia and SE Asia.

Panisea uniflora (Lindl.) Lindl. [1, 4, 5, 13]

Coelogyne uniflora Lindl.; *Pleione uniflora* (Lindl.) Kuntze

Epiphytic herbs. C, E Nepal: 400-2300 m. Nepal, E Himalaya, S Asia, E Asia and SE Asia.

Use: Ornamental.

Paphiopedilum Pfitzer

Paphiopedilum fairrieanum (Lindl.) Stein. [13]

Cypripedium fairrieanum Lindl.; *Cypripdium assamicum* Linden ex K.Koch & Fintelm.

Terrestrial herbs. W Nepal: 1500-1800 m. Nepal and E Himalaya.

IUCN Category: **Critically Endangered** A2acd+3cd+4acd (ver 3.1).

CITES Appendix: I.

Paphiopedilum venustum (Wall. ex Sims) Pfitzer [13]

Cypripedium venustum Wall. ex Sims

Terrestrial herbs. E Nepal: 1100-1600 m. Nepal, E Himalaya, Tibetan Plateau and S Asia.

IUCN Category: **Endangered** A2acd (ver 3.1).

CITES Appendix: I.

Papilionanthe Schltr.

Papilionanthe teres (Roxb.) Schltr. [13]

Dendrobium teres Roxb.; *Vanda teres* (Roxb.) Lindl. (1, 4)

हरजोर, थुर्जो (Harjor, Thurjo)

Epiphytic herbs. C, E Nepal: 200-2100 m. Nepal, E Himalaya, Assam-Burma, S Asia, E Asia and SE Asia.

Use(s): Medicine (whole plant), Ornamental.

Papilionanthe uniflora (Lindl.) Garay [5, 13]

Mesoclastes uniflora Lindl.; *Aerides longicornu* Hook.f.; *Aerides uniflora* (Lindl.) Summerh. (1, 4)

Epiphytic herbs. C, E Nepal: 400-2100 m. Nepal and E Himalaya.

Use: Ornamental.

Papilionanthe vandarum (Rchb.f.) Garay [13]

Aerides vandarum Rchb.f. (1, 4)

Epiphytic herbs. E Nepal: 1600-1700 m. Nepal, E Himalaya, Assam-Burma and E Asia.

Pecteilis Raf.

Pecteilis susannae (L.) Raf. [1, 4, 5, 13]

Orchis susannae L.; *Habenaria susannae* (L.) R.Br. ex Spreng.; *Platanthera susannae* (L.) Lindl.

Terrestrial herbs. W, C, E Nepal: 100-1800 m. Nepal, Assam-Burma, E Asia and SE Asia.

Use(s): Medicine (tuber), Ornamental.

Pecteilis triflora (D.Don) Tang & F.T.Wang [1, 4, 5, 13]

Habenaria triflora D.Don; *Habenaria uniflora* Buch.-Ham. ex D.Don; *Platanthera uniflora* (Buch.-Ham. ex D.Don) Lindl.

Terrestrial herbs. W, C, E Nepal: 400-1200 m. Nepal and W Himalaya.

Pelatantheria Ridl.

Pelatantheria insectifera (Rchb.f.) Ridl. [13]

Sarcanthus insectifer Rchb.f.

Epiphytic herbs. W, C, E Nepal: 200-1000 m. Nepal, W Himalaya, Assam-Burma and SE Asia.

Use: Ornamental.

Peristylus Blume

Peristylus affinis (D.Don) Seidenf. [13]

Habenaria affinis D.Don; *Gymnadenia affinis* (D.Don) Rchb.f.

Epiphytic herbs. W, C, E Nepal: 200-1000 m. Nepal, E Himalaya, Assam-Burma, E Asia and SE Asia.

Use: Ornamental.

Peristylus aristatus Lindl. [1, 4, 5, 13]

Habenaria aristata (Lindl.) Hook.f.

Terrestrial herbs. C, E Nepal: 1700-2200 m. Nepal, E Himalaya and S Asia.

Peristylus calcaratus (Rolfe) S.Y.Hu [41]

Glossula calcarata Rolfe

Terrestrial herbs. E Nepal: *ca.* 2000 m. Nepal, E Asia and SE Asia.

Peristylus constrictus (Lindl.) Lindl. [1, 4, 5, 13]

Herminium constrictum Lindl.; *Habenaria constricta* (Lindl.) Hook.f.

Terrestrial herbs. W, C, E Nepal: 300-2300 m. Nepal, E Himalaya, Assam-Burma, E Asia and SE Asia.

Use(s): Medicine (tuber), Food (tuber).

Peristylus densus (Lindl.) Santapau & Kapadia [5, 13]

Coeloglossum densum Lindl.; *Habenaria stenostachya* (Lindl. ex Benth.) Benth.

Terrestrial herbs. C, E Nepal: 2100-2700 m. Nepal, Assam-Burma, S Asia, E Asia and SE Asia.

Peristylus duthiei (Hook.f.) Deva & H.B.Naithani [13]

Herminium duthiei Hook.f. (1, 4, 5); *Monorchis duthiei* (Hook.f.) O.Schwarz

Terrestrial herbs. W, C, E Nepal: 2800-4600 m. Nepal and W Himalaya.

Peristylus elisabethae (Duthie) R.K.Gupta [1, 4, 5, 13]

Habenaria elisabethae Duthie; *Herminium elisabethae* (Duthie) Tang & F.T.Wang

Terrestrial or lithophytic herbs. W, C, E Nepal: 900-3200 m. Nepal, W Himalaya, E Himalaya and Tibetan Plateau.

Peristylus fallax Lindl. [1, 4, 5, 13]

Habenaria fallax (Lindl.) King & Pantl.; *Herminium fallax* (Lindl.) Hook.f.; *Platanthera fallax* (Lindl.) Schltr.

Terrestrial herbs. W, C, E Nepal: 2100-3700 m. Nepal, W Himalaya, E Himalaya, Tibetan Plateau and E Asia.

Peristylus goodyeroides (D.Don) Lindl. [1, 4, 5, 13]

Habenaria goodyeroides D.Don

Terrestrial herbs. W, C, E Nepal: 500-1500 m. Nepal, W Himalaya, E Himalaya, Assam-Burma, E Asia and SE Asia.

Use: Medicine (whole plant).

Peristylus hamiltonianus (Lindl.) Lindl. [1, 4, 8, 13]

Herminium hamiltonianum Lindl.; *Habenaria hamiltoniana* (Lindl.) Hook.f.

Terrestrial herbs. C, E Nepal: 1500-2200 m. Nepal, E Himalaya, S Asia and SE Asia.

Peristylus intrudens (Ames) Ormerod [8, 13, 93]

Habenaria intrudens Ames

Terrestrial herbs. C Nepal: 800-900 m. Nepal and E Asia.

Peristylus lacertifer (Lindl.) J.J.Sm. [13, 40]

Coeloglossum lacertiferum Lindl.; *Habenaria lacertifera* (Lindl.) Benth.

Terrestrial herbs. C Nepal: *ca.* 300 m. Nepal, E Himalaya, Assam-Burma, E Asia and SE Asia.

Peristylus lawii Wight [1, 4, 5, 13]

Habenaria lawii (Wight) Hook.f.

Terrestrial herbs. W, C Nepal: 400-1000 m. Nepal, W Himalaya and Assam-Burma.

Peristylus mannii (Rchb.f.) Mukerjee [8, 13, 60]

Coeloglossum mannii Rchb.f.; *Herminium mannii* (Rchb.f.) Tang & F.T.Wang (81); *Platanthera mannii* (Rchb.f.) Schltr.

Terrestrial herbs. C, E Nepal: 1700-3000 m. Nepal, E Himalaya, Assam-Burma, S Asia, E Asia and SE Asia.

Peristylus nematocaulon (Hook.f.) Banerji & P.Pradhan [5, 13]

Habenaria nematocaulon Hook.f.; *Platanthera juncea* (King & Pantl.) Kraenzl. (1, 4); *Platanthera nematocaulon* (Hook.f.) Kraenzl. (52)

Terrestrial or lithophytic herbs. C, E Nepal: 2700-3600 m. Nepal, E Himalaya and Tibetan Plateau.

Peristylus parishii Rchb.f. [5, 13]

Habenaria parishii (Rchb.f.) Hook.f.

Terrestrial or lithophytic herbs. C, E Nepal: 600-1900 m. Nepal, E Himalaya, Assam-Burma, E Asia and SE Asia.

Peristylus prainii (Hook.f.) Kraenzl. [5, 13]

Habenaria prainii Hook.f.

Terrestrial herbs. W, C Nepal: 700-2100 m. Nepal, E Himalaya, Assam-Burma and SE Asia.

Peristylus richardianus Wight [1, 4, 13]

Habenaria bicornuta Hook.f.

Terrestrial herbs. E Nepal: 2000-2500 m. Nepal and S Asia.

Peristylus superanthus J.J.Wood [4, 13]

Terrestrial herbs. E Nepal: 3000-3600 m. Nepal, E Himalaya and E Asia.

Peristylus tentaculatus (Lindl.) J.J.Sm. [13]

Glossula tentaculata Lindl.; *Habenaria tentaculata* (Lindl.) Rchb.f.

Terrestrial herbs. E Nepal: 100-2000 m. Nepal, E Asia and SE Asia.

Peristylus tipuliferus (C.S.P.Parish & Rchb.f.) Mukerjee [5]

Habenaria tipulifera C.S.P.Parish & Rchb.f.; *Habenaria brandisii* Hook.f.

Terrestrial herbs. E Nepal: 1900-2000 m. Nepal and E Himalaya.

Phaius Lour.

Phaius flavus (Blume) Lindl. [1, 4, 13]

Limodorum flavum Blume; *Phaius maculatus* Lindl.

Terrestrial herbs. C Nepal: 200-1500 m. Nepal, E Himalaya, Tibetan Plateau, S Asia, E Asia and SE Asia.

Use: Ornamental.

Phaius tancarvilleae (L'Hér.) Blume [1, 4, 5, 13]

Limodorum tancarvilleae L'Hér.

Terrestrial herbs. W, C, E Nepal: 200-2000 m. Nepal, W Himalaya, E Himalaya, Tibetan Plateau, Assam-Burma, S Asia, E Asia, SE Asia and Australasia.

Use(s): Medicine (tuber), Ornamental.

Phalaenopsis Blume

Phalaenopsis deliciosa Rchb.f. [13]

Doritis deliciosa (Rchb.f.) T.Yukawa & K.Kita; *Kingidium deliciosum* (Rchb.f.) H.R.Sweet

Epiphytic herbs. E Nepal: 200-500 m. Nepal, E Himalaya, Assam-Burma, S Asia, E Asia and SE Asia.

Use: Ornamental.

Phalaenopsis mannii Rchb.f. [1, 4, 5, 13]

Polychilos mannii (Rchb.f.) Shim

Epiphytic herbs. C, E Nepal: 100-600 m. Nepal, E Himalaya, Assam-Burma, E Asia and SE Asia.

Use: Ornamental.

Phalaenopsis parishii Rchb.f. [13]

Kingidium decumbens (Griff.) P.F.Hunt; *Phalaenopsis decumbens* (Griff.) Holttum (1, 4)

Terrestrial or epiphytic herbs. C, E Nepal: 200-1900 m. Nepal, E Himalaya, Assam-Burma and SE Asia.

Phalaenopsis taenialis (Lindl.) Christenson & Pradhan [5, 13]

Aerides taenialis Lindl.; *Kingidium taenialis* (Lindl.) P.F.Hunt

Epiphytic herbs. W, C, E Nepal: 1100-2300 m. Nepal, W Himalaya, E Himalaya, Tibetan Plateau, Assam-Burma, E Asia and SE Asia.

Use: Ornamental.

Pholidota Lindl.

Pholidota articulata Lindl. [1, 4, 5, 13]

Coelogyne articulata (Lindl.) Rchb.f.; *Pholidota griffithii* Hook.f. (1, 4)

स्याबेलाम्दा, तिमपुनो (Syabe lamdaa, Timpuno: Gurung)

Epiphytic herbs. W, C, E Nepal: 500-2500 m. Nepal, W Himalaya, E Himalaya, Tibetan Plateau, Assam-Burma, E Asia and SE Asia.

Use(s): Medicine (pseudobulb), Ornamental.

Pholidota imbricata Hook. [1, 4, 5, 13]

Coelogyne imbricata (Hook.) Rchb.f.; *Cymbidium imbricatum* (Hook.) Roxb.

पत्थर केरा, भालु केरा (Patthar keraa, Bhalu keraa)

Epiphytic or lithophytic herbs. C, E Nepal: 300-2900 m. Nepal, W Himalaya, E Himalaya, Tibetan Plateau, Assam-Burma, S Asia, E Asia and SE Asia.

Use(s): Medicine (pseudobulb), Ornamental.

Pholidota pallida Lindl. [13]

Coelogyne pallida (Lindl.) Rchb.f.

Epiphytic or lithophytic herbs. W, C, E Nepal: 800-2800 m. Nepal, E Himalaya, Assam-Burma, E Asia and SE Asia.

Use: Medicine (root, pseudobulb).

Pholidota protracta Hook.f. [1, 4, 5, 13]

Epiphytic herbs. C, E Nepal: 500-2200 m. Nepal, E Himalaya, Tibetan Plateau, Assam-Burma, E Asia and SE Asia.

Use: Ornamental.

Pholidota recurva Lindl. [1, 4, 5, 13]

Coelogyne recurva (Lindl.) Rchb.f.

Epiphytic herbs. C Nepal: 700-1800 m. Nepal, E Himalaya, Assam-Burma, E Asia and SE Asia.

Use: Ornamental.

Phreatia Lindl.

Phreatia elegans Lindl. [42]

Eria elegans (Lindl.) Rchb.f.; *Thelasis elegans* (Lindl.) Blume

Epiphytic herbs. C Nepal: *ca.* 1600 m. Nepal, E Himalaya, Tibetan Plateau, S Asia and SE Asia.

Pinalia Lindl.

Pinalia acervata (Lindl.) Kuntze [13]

Eria acervata Lindl. (5)

Epiphytic herbs. E Nepal: 300-3000 m. Nepal, E Himalaya, Tibetan Plateau, Assam-Burma, E Asia and SE Asia.

Pinalia amica (Rchb.f.) Kuntze [13]

Eria amica Rchb.f.; *Eria confusa* Hook.f. (1, 4)

Epiphytic herbs. C Nepal: 600-2100 m. Nepal, W Himalaya, E Himalaya, Assam-Burma, E Asia and SE Asia.

Pinalia bipunctata (Lindl.) Kuntze [13]

Eria bipunctata Lindl. (5, 90)

Epiphytic herbs. C Nepal: 1100-1700 m. Nepal, W Himalaya, E Himalaya, E Asia and SE Asia.

Pinalia bractescens (Lindl.) Kuntze

Eria bractescens Lindl. (1, 4, 5, 13); *Eria griffithii* Rchb.f.

Epiphytic herbs. C, E Nepal: 300-1800 m. Nepal, Assam-Burma and SE Asia.

Use: Ornamental.

Pinalia concolor (C.S.P.Parish & Rchb.f.) Kuntze [13]

Eria concolor C.S.P.Parish & Rchb.f. (97)

Epiphytic herbs. E Nepal: *ca.* 500 m. Nepal, Assam-Burma and SE Asia.

Pinalia excavata (Lindl.) Kuntze [13]

Eria excavata Lindl. (1, 4, 5)

Epiphytic or lithophytic herbs. C Nepal: 500-2400 m. Nepal, E Himalaya and Tibetan Plateau.

Pinalia graminifolia (Lindl.) Kuntze [13]

Eria graminifolia Lindl. (1, 4, 5)

Epiphytic herbs. C, E Nepal: 1500-2500 m. Nepal, E Himalaya, Tibetan Plateau, Assam-Burma and E Asia.

Pinalia leucantha Kuntze

Eria alba Lindl. (1, 4, 13)

Epiphytic herbs. W, C, E Nepal: 1500-2500 m. Nepal, W Himalaya and E Himalaya.

Pinalia obesa (Lindl.) Kuntze

Eria obesa Lindl. (13, 40, 97)

Epiphytic herbs. E Nepal: *ca.* 500 m. Nepal, Assam-Burma and SE Asia.

Pinalia spicata (D.Don) S.C.Chen & J.J.Wood [13]

Octomeria spicata D.Don; *Eria spicata* (D.Don) Hand.-Mazz. (1, 4, 5)

Epiphytic or lithophytic herbs. C Nepal: 300-1800 m. Nepal, W Himalaya, E Himalaya, Tibetan Plateau, Assam-Burma, E Asia and SE Asia.

Use(s): Medicine (whole plant), Ornamental.

Pinalia stricta (Lindl.) Kuntze

Eria stricta Lindl. (1, 4); *Mycaranthes stricta* (Lindl.) Lindl. (13)

Epiphytic herbs. W, C, E Nepal: 300-1800 m. Nepal, E Himalaya, Tibetan Plateau, Assam-Burma, E Asia and SE Asia.

Platanthera Rich.

Platanthera arcuata Lindl. [1, 4, 5, 13]

Habenaria arcuata (Lindl.) Hook.f.

Terrestrial herbs. W, C Nepal: 2500-3100 m. Nepal, W Himalaya, E Himalaya and Tibetan Plateau.

Use: Medicine (root).

Platanthera bakeriana (King & Pantl.) Kraenzl. [1, 4, 5, 13]

Habenaria bakeriana King & Pantl.

Terrestrial or epiphytic or lithophytic herbs. W, C, E Nepal: 1500-3400 m. Nepal, E Himalaya, Tibetan Plateau and E Asia.

Platanthera biermanniana (King & Pantl.) Kraenzl. [1, 4, 5, 13]

Habenaria biermanniana King & Pantl.

Terrestrial herbs. C, E Nepal: 2100-3200 m. Nepal and E Himalaya.

Platanthera clavigera Lindl. [1, 4, 5, 13]

Habenaria clavigera (Lindl.) Dandy; *Habenaria densa* Wall. ex Lindl.; *Platantheroides clavigera* (Lindl.) Szlach.

कंगा (Kangaa)

Terrestrial herbs. W, C, E Nepal: 1300-4600 m. Nepal, W Himalaya, E Himalaya and Tibetan Plateau.

Use(s): Medicine (tuber), Food (tuber).

Platanthera cumminsiana (King & Pantl.) Renz. [13, 59]

Habenaria cumminsiana King & Pantl.; *Habenella cumminsiana* (King & Pantl.) Szlach. & Kras-Lap.

Terrestrial herbs. E Nepal: *ca.* 2400 m. Nepal and E Himalaya.

Platanthera dulongensis X.H.Jin & Efimov [52]

Terrestrial herbs. E Nepal: 2100-2800 m. Nepal, E Himalaya and E Asia.

Platanthera edgeworthii (Hook.f. ex Collett) R.K.Gupta [5, 13]

Habenaria edgeworthii Hook.f. ex Collett; *Habenella edgeworthii* (Hook.f. ex Collett) Szlach. & Kras-Lap.

कंगाछा (Kangachha)

Terrestrial herbs. W, C Nepal: 1200-3600 m. Nepal, W Himalaya and E Himalaya.

Use: Medicine (root, leaves).

Platanthera exelliana Soó [1, 4, 5, 13]

Habenaria oligantha Hook.f.; *Platanthera elachyantha* Tang & F.T.Wang

Terrestrial herbs. C, E Nepal: 2700-4200 m. Nepal, E Himalaya, Tibetan Plateau and E Asia.

Platanthera japonica (Thunb.) Lindl. [13, 52, 62]

Orchis japonica Thunb.; *Habenaria japonica* (Thunb.) A.Gray

Terrestrial herbs. C Nepal: 600-2600 m. Nepal and E Asia.

Platanthera latilabris Lindl. [1, 4, 5, 13]

Habenaria latilabris (Lindl.) Hook.f.; *Habenella latilabris* (Lindl.) Szlach. & Kras-Lap.; *Platanthera acuminata* Lindl.

Terrestrial herbs. W, C, E Nepal: 900-4000 m. Nepal, W Himalaya, E Himalaya, Tibetan Plateau and E Asia.

Platanthera leptocaulon (Hook.f.) Soó [1, 4, 5, 13, 52]

Habenaria lepotocaulon Hook.f.

Terrestrial herbs. C, E Nepal: 2800-4800 m. Nepal, W Himalaya, E Himalaya, Tibetan Plateau and E Asia.

Platanthera pachycaulon (Hook.f.) Soó [52]

Habenaria pachycaulon Hook.f.

Terrestrial herbs. W, C, E Nepal: 4000-4100 m. Nepal, W Himalaya, E Himalaya, Tibetan Plateau and E Asia.

Platanthera roseotincta (W.W.Sm.) Tang & F.T.Wang [41]

Habenaria roseotincta W.W.Sm.

Terrestrial herbs. E Nepal: *ca.* 3400 m. Nepal, Tibetan Plateau and Assam-Burma.

Platanthera sikkimensis (Hook.f.) Kraenzl. [1, 4, 5, 13, 52]

Habenaria sikkimensis Hook.f.

Terrestrial or epiphytic herbs. C, E Nepal: 1500-3600 m. Nepal, E Himalaya and Tibetan Plateau.

Use: Medicine (pseudobulb).

Platanthera stenantha (Hook.f.) Soó [1, 4, 5, 13, 52]

Habenaria stenantha Hook.f.; *Hemihabenaria stenantha* (Hook.f.) Finet

Terrestrial or epiphytic herbs. W, C, E Nepal: 1200-3100 m. Nepal, W Himalaya, E Himalaya, Tibetan Plateau, Assam-Burma and E Asia.

Platanthera yadongensis Jin & Jin [41]

Terrestrial herbs. E Nepal: *ca.* 2800 m. Nepal and Tibetan Plateau.

Pleione D.Don

Pleione coronaria P.J.Cribb & C.Z.Tang, *Curtis's Bot. Mag.* **184**: 123. 1983. [4, 5, 13, 27]

Epiphytic herbs. C Nepal: 2500-3500 m. **Endemic to Nepal**.

Central: *Dhading*, Ganesh Himal, 2850 m, JDA Stainton 3654 (Holotype: BM).

Pleione hookeriana (Lindl.) Rollisson [1, 4, 5, 13]

Coelogyne hookeriana Lindl.; *Pleione laotica* Kerr

Epiphytic herbs. W, C, E Nepal: 2200-4200 m. Nepal, W Himalaya, E Himalaya, Tibetan Plateau, Assam-Burma, E Asia and SE Asia.

Use: Ornamental.

Pleione humilis (Sm.) D.Don [1, 4, 5, 13]

Epidendrum humile Sm.; *Coelogyne humilis* (Sm.) Lindl.

हात्ति टाउके, बत्ति टाउके (Haatti taauke, Batti taauke)

Epiphytic herbs. W, C, E Nepal: 1800-3000 m. Nepal, W Himalaya, E Himalaya, Tibetan Plateau, Assam-Burma, E Asia and SE Asia.

Use(s): Medicine (pseudobulb), Ornamental.

Pleione maculata (Lindl.) Lindl. & Paxton [1, 4, 13]

Coelogyne maculata Lindl.; *Pleione diphylla* Lindl. & Paxton

लसुन पाते (Lasun paate)

Epiphytic or lithophytic herbs. C Nepal: 1400-2700 m. Nepal, E Himalaya, Assam-Burma, E Asia and SE Asia.

Use(s): Medicine (root), Ornamental.

Pleione praecox (Sm.) D.Don [1, 4, 5, 13]

Epidendrum praecox Sm.; *Coelogyne praecox* (Sm.) Lindl.

लसुन पाते, शक्ति गुम्बा (Lasun paate, Shakti gumbaa)

Epiphytic or lithophytic herbs. C, E Nepal: 1500-2500 m. Nepal, W Himalaya, E Himalaya, Tibetan Plateau, Assam-Burma, S Asia, E Asia and SE Asia.

Use(s): Medicine (pseudobulb), Ornamental.

Podochilus Blume

Podochilus cultratus Lindl. [1, 4, 5, 13]

Epiphytic herbs. C, E Nepal: 400-800 m. Nepal, E Himalaya, Assam-Burma and SE Asia.

Ponerorchis Rchb.f.

Ponerorchis chusua (D.Don) Soó [1, 4, 13]

Orchis chusua D.Don; *Chusua pauciflora* (Lindl.) P.F.Hunt (5); *Habenaria chusua* (D.Don) Benth.

Terrestrial herbs. W, C, E Nepal: 2400-4900 m. Nepal, W Himalaya, E Himalaya, Tibetan Plateau, Assam-Burma, E Asia and N Asia.

Use(s): Medicine (tuber), Ornamental.

Porpax Lindl.

Porpax elwesii (Rchb.f.) Rolfe [5, 13]

Eria elwesii Rchb.f.

Epiphytic or lithophytic herbs. C, E Nepal: 800-1800 m. Nepal, E Himalaya, E Asia and SE Asia.

Porpax meirax (Parish & Rchb.f.) King & Pantl. [1, 4, 13]

Cryptochilus meirax Parish & Rchb.f.

Epiphytic herbs. E Nepal: 900-1800 m. Nepal, E Himalaya, Assam-Burma and SE Asia.

Pteroceras Hasselt ex Hassk.

Pteroceras teres (Blume) Holttum [5, 13]

Dendrocolla teres Blume

Epiphytic herbs. C, E Nepal: 200-800 m. Nepal, W Himalaya, Assam-Burma, S Asia and SE Asia.

Use: Ornamental.

Rhomboda Lindl.

Rhomboda abbreviata (Lindl.) Ormerod. [13]

Hetaeria abbreviata Lindl.; *Anoectochilus abbreviatus* (Lindl.) Seidenf.; *Odontochilus abbreviatus* (Lindl.) Tang & F.T.Wang (1, 4)

Terrestrial or epiphytic herbs. C Nepal: 500-1200 m. Nepal, E Himalaya, Assam-Burma, E Asia and SE Asia.

Rhynchostylis Blume

Rhynchostylis retusa (L.) Blume [1, 4, 5, 13]

Epidendrum retusum L.; *Aerides retusa* (L.) Sw.; *Epidendrum hippium* Buch.-Ham. ex D.Don

घोडे गाभा (Ghode gaavaa)

Epiphytic herbs. W, C, E Nepal: 300-1800 m. Nepal, W Himalaya, E Himalaya, Assam-Burma, S Asia, E Asia and SE Asia.

Use(s): Medicine (leaves), Ornamental.

Risleya King & Pantl.

Risleya atropurpurea King & Pantl. [8, 13, 94]

Terrestrial herbs. E Nepal: *ca.* 3400 m. Nepal, E Himalaya, Tibetan Plateau and Assam-Burma.

Robiquetia Gaudich.

Robiquetia succisa (Lindl.) Seidenf. & Garay [13, 95]

Sarcanthus succisus Lindl.

Epiphytic herbs. E Nepal: 400-1200 m. Nepal, E Himalaya, Assam-Burma, E Asia and SE Asia.

Satyrium Sw.

Satyrium nepalense D.Don [1, 4, 13]

var. **nepalense**

गामडोल, थाम्नी (Gaamdol, Thaamni)

Terrestrial herbs. W, C, E Nepal: 600-4600 m. Nepal, W Himalaya, E Himalaya, Tibetan Plateau, Assam-Burma, S Asia and E Asia.

Use(s): Medicine (tuber), Food (leaves), Ornamental.

var. **ciliatum** (Lindl.) Hook.f. [5, 13]

Satyrium ciliatum Lindl. (1, 4)

थाम्नी (Thaamni)

Terrestrial herbs. W, C, E Nepal: 1800-4300 m. Nepal, W Himalaya, E Himalaya, Tibetan Plateau and E Asia.

Schoenorchis Reinw. ex Blume

Schoenorchis gemmata (Lindl.) J.J.Sm. [1, 4, 5, 13]

Saccolabium gemmatum Lindl.; *Cleisostoma gemmatum* (Lindl.) King & Pantl.

Epiphytic herbs. E Nepal: 400-2500 m. Nepal, E Himalaya, Tibetan Plateau, Assam-Burma, E Asia and SE Asia.

Smitinandia Holttum

Smitinandia micrantha (Lindl.) Holttum [1, 4, 5, 13]

Saccolabium micranthum Lindl.; *Cleiosostoma micranthum* (Lindl.) King & Pantl.

थिर्जो (Thirjo)

Epiphytic herbs. C, E Nepal: 400-1400 m. Nepal, W Himalaya, E Himalaya, Assam-Burma, E Asia and SE Asia.

Use(s): Medicine (whole plant), Ornamental.

Spathoglottis Blume

Spathoglottis ixioides (D.Don) Lindl. [1, 4, 5, 13]

Cymbidium ixioides D.Don

Terrestrial or lithophytic herbs. C, E Nepal: 1500-4600 m. Nepal, E Himalaya and Tibetan Plateau.

Use: Ornamental.

Spiranthes Rich.

Spiranthes sinensis (Pers.) Ames [5, 13]

Neottia sinensis Pers.; *Spiranthes sinensis* var. *amoena* (M.Bieb.) H.Hara (1,4)

बेरुवे सुनाखरी (Beruwe sunaakhari)

Terrestrial herbs. W, C, E Nepal: 100-3700 m. Nepal, W Himalaya, E Himalaya, Tibetan Plateau, Assam-Burma, E Asia, SE Asia, N Asia, SW Asia and Australasia.

Use: Medicine (tuber).

Spiranthes spiralis (L.) Chevall. [13, 96]

Ophrys spiralis L.

Terrestrial herbs. C Nepal: 800-900 m. Nepal, SW Asia and Europe.

Sunipia Lindl.

Sunipia bicolor Lindl. [1, 4, 13]

Bulbophyllum bicolor (Lindl.) Hook.f.; *Ione bicolor* (Lindl.) Lindl. (5)

Epiphytic herbs. W, C, E Nepal: 900-2500 m. Nepal, W Himalaya, E Himalaya, Tibetan Plateau, Assam-Burma, S Asia, E Asia and SE Asia.

Sunipia cirrhata (Lindl.) P.F.Hunt [13]

Ione cirrhata Lindl. (5); *Bulbophyllum cirrhatum* (Lindl.) Hook.f.; *Sunipia paleacea* (Lindl.) P.F.Hunt (1, 4)

Epiphytic herbs. C Nepal: 800-2300 m. Nepal, E Himalaya, Assam-Burma and E Asia.

Sunipia nepalensis Raskoti & Ale, *Phytotaxa* **21**: 55. 2011. [8, 13]

Epiphytic herbs. C Nepal: *ca.* 2400 m. **Endemic to Nepal**.

Central: *Makawanpur*, *ca.* 2400 m, BB Raskoti 270 (Holotype: KATH).

Sunipia scariosa Lindl. [5, 13]

Ione scariosa (Lindl.) King & Pantl.; *Sunipia racemosa* (Sm.) Tang & F.T.Wang (1, 4)

Epiphytic herbs. C, E Nepal: 1200-1800 m. Nepal, E Himalaya, Assam-Burma, E Asia and SE Asia.

Taeniophyllum Blume

Taeniophyllum scaberulum Hook.f. [13, 85]

Epiphytic herbs. C Nepal: 1500-2000 m. Nepal and S Asia (S India).

Tainia Blume

Tainia minor Hook.f. [5, 13]

Terrestrial herbs. C, E Nepal: 1900-2300 m. Nepal, E Himalaya, Tibetan Plateau, Assam-Burma and E Asia.

Thelasis Blume

Thelasis longifolia Hook.f. [5, 13]

Epiphytic herbs. C, E Nepal: 800-1300 m. Nepal, W Himalaya and E Himalaya.

Thelasis pygmaea (Griff.) Lindl. [1, 4, 13]

Euproboscis pygmaea Griff.

Epiphytic herbs. C Nepal: 800-1500 m. Nepal, E Himalaya, Assam-Burma, E Asia and SE Asia.

Thrixspermum Lour.

Thrixspermum pygmaeum (King & Pantl.) Holttum [13]

Sarcochilus pygmaeum King & Pantl.

Epiphytic herbs. C Nepal: 1300-2000 m. Nepal, E Himalaya, Tibetan Plateau and SE Asia.

Thunia Rchb.f.

Thunia alba (Lindl.) Rchb.f. [1, 4, 13]

Phaius albus Lindl.

var. **alba**

चांदे फूल, गोलाईनो (Chaande phul, Golaino)

Epiphytic/lithophytic or terrestrial herbs. W, C, E Nepal: 500-1800 m. Nepal, W Himalaya, E Himalaya, Tibetan Plateau, Assam-Burma, E Asia and SE Asia.

Use(s): Medicine (whole plant), Forage, Ornamental.

var. **bracteata** (Roxb.) N.Pearce & P.J.Cribb [13]

Limodorum bracteatum Roxb.; *Thunia bracteata* (Roxb.) Schltr.

Epiphytic or lithophytic herbs. C Nepal: 600-1600 m. Nepal, W Himalaya and E Himalaya.

Use(s): Forage, Ornamental.

Tipularia Nutt.

Tipularia josephi Rchb.f. ex Lindl. [1, 4, 5, 13]

Terrestrial herbs. C, E Nepal: 2200-3100 m. Nepal, E Himalaya, Tibetan Plateau and Assam-Burma.

Trichotosia Blume

Trichotosia dasyphylla (C.S.P.Parish & Rchb.f.) Kraenzl. [13]

Eria dasyphylla C.S.P.Parish & Rchb.f. (1, 4); *Pinalia dasyphylla* (C.S.P.Parish & Rchb.f.) Kuntze

Epiphytic herbs. C Nepal: 800-1500 m. Nepal, E Himalaya, Assam-Burma, E Asia and SE Asia.

Use: Ornamental.

Trichotosia pulvinata (Lindl.) Kraenzl. [13]

Eria pulvinata Lindl.; *Trichotsia cristata* Kraenzl.

Epiphytic herbs. W, C, E Nepal: 800-2000 m. Nepal, E Himalaya, Assam-Burma, E Asia and SE Asia.

Uncifera Lindl.

Uncifera acuminata Lindl. [1, 4, 13]

Saccolabium acuminatum (Lindl.) Hook.f.

Epiphytic herbs. E Nepal: 1200-1600 m. Nepal, E Himalaya and E Asia.

Uncifera lancifolia (King & Pantl.) Schltr. [13]

Saccolabium lancifolium King & Pantl.

Epiphytic herbs. C Nepal: 2000-2200 m. Nepal, E Himalaya and SE Asia.

Uncifera obtusifolia Lindl. [1, 4, 13]

Gastrochilus obtusifolius (Lindl.) Kuntze; *Saccolabium obtusifolium* (Lindl.) Hook.f.

Epiphytic herbs. C, E Nepal: 900-1100 m. Nepal and E Himalaya.

Use: Ornamental.

Vanda Jones ex R.Br.

Vanda alpina (Lindl.) Lindl. [1, 4, 13]

Luisia alpina Lindl.; *Trudelia alpina* (Lindl.) Garay (5)

Epiphytic herbs. W, C, E Nepal: 1100-1800 m. Nepal, W Himalaya, E Himalaya, S Asia, E Asia and SE Asia.

Vanda cristata Wall. ex Lindl. [1, 4, 13]

Aerides cristata (Wall. ex Lindl.) Wall. ex Hook.f.; *Luisia striata* (Rchb.f.) Kraenzl.; *Trudelia cristata* (Lindl.) Senghas ex Roeth (5)

भ्यागुते फूल, हरचुर (Bhyaagute phool, Harchur)

Epiphytic herbs. W, C, E Nepal: 1000-2300 m. Nepal, W Himalaya, E Himalaya, Tibetan Plateau, S Asia, E Asia and SE Asia.

Use(s): Medicine (root, leaves), Ornamental.

Vanda pumila Hook.f. [13]

Trudelia pumila (Hook.f.) Senghas

Epiphytic herbs. C Nepal: 500-1500 m. Nepal, E Himalaya, Assam-Burma, E Asia and SE Asia.

Vanda tessellata (Roxb.) Hook. ex G.Don [13]

Epidendrum tesselatum Roxb.; *Cymbidium tessellatum* (Roxb.) Sw.; *Vanda roxburghii* R.Br. (1, 4)

गाइतेबारे, हरजोर (Gaitebaare, Harjor)

Epiphytic herbs. W, C, E Nepal: 100-600 m. Nepal, Assam-Burma and S Asia.

Use(s): Medicine (root, leaves), Ornamental.

Vanda testacea (Lindl.) Rchb.f. [5, 13]

Aerides testaceum Lindl.; *Vanda parviflora* Lindl. (1, 4)

हरजोर (Harjor)

Epiphytic herbs. C, E Nepal: 200-2200 m. Nepal, W Himalaya, E Himalaya, Assam-Burma, S Asia and SE Asia.

Use(s): Medicine (leaves), Ornamental.

Vandopsis Pfitzer

Vandopsis undulata (Lindl.) J.J.Sm. [1, 4, 5, 13]

Vanda undulata Lindl.; *Stauropsis undulatis* (Lindl.) Benth. ex Hook.f.

Epiphytic herbs. W, C, E Nepal: 200-2400 m. Nepal, W Himalaya, E Himalaya, Tibetan Plateau and E Asia.

Use: Ornamental.

Zeuxine Lindl.

Zeuxine affinis (Lindl.) Benth. ex Hook.f. [8, 13, 82]

Monochilus affinis Lindl.

Terrestrial herbs. C Nepal: 700-2200 m. Nepal, W Himalaya, E Himalaya, Assam-Burma, E Asia and SE Asia.

Zeuxine flava (Wall. ex Lindl.) Trimen [1, 4, 5, 13]

Monochilus flavus Wall. ex Lindl.; *Zeuxine aurantiaca* Schltr.

Terrestrial herbs. C, E Nepal: 1200-2300 m. Nepal, W Himalaya, E Himalaya, Assam-Burma, E Asia and SE Asia.

Zeuxine goodyeroides Lindl. [1, 4, 13]

Monochilus goodyeroides Lindl.

Terrestrial herbs. C, E Nepal: 1800-3300 m. Nepal, E Himalaya, Assam-Burma, E Asia and SE Asia.

Zeuxine grandis Seidenf. [8, 13, 83]

Terrestrial herbs. C Nepal: *ca.* 700 m. Nepal, E Asia and SE Asia.

Zeuxine longilabris Trimen [1, 4, 13]

Terrestrial herbs. C Nepal: 700-1100 m. Nepal, Assam-Burma and S Asia.

Zeuxine membranacea Lindl. [46]

Zeuxine evrardii Gagnep.; *Zeuxine godefroyi* Rchb.f.

Terrestrial herbs. C Nepal: *ca.* 700 m. Nepal, W Himalaya, E Himalaya, Assam-Burma, E Asia and SE Asia.

Use: Forage.

Zeuxine nervosa (Wall. ex Lindl.) Benth. ex Trimen [1, 4, 13]

Monochilus nervosus Wall. ex Lindl.; *Haplochilus nervosus* (Wall. ex Lindl.) D.Dietr.

Terrestrial herbs. C Nepal: 700-1200 m. Nepal, E Himalaya, S Asia, E Asia and SE Asia.

Zeuxine reflexa King & Pantl. [5, 13]

Terrestrial herbs. E Nepal: 1300-4300 m. Nepal, E Himalaya, E Asia and SE Asia.

Zeuxine strateumatica (L.) Schltr. [1, 4, 5, 13]

Orchis strateumatica L.; *Neottia strateumatica* (L.) R.Br.

Centepede grass orchid

Terrestrial herbs. W, C, E Nepal: 200-1200 m. Nepal, W Himalaya, E Himalaya, Assam-Burma, S Asia, E Asia, SE Asia, SW Asia, S America and Australasia.

Use: Medicine (root, tuber).

Exotic/Cultivated Taxa

Paphiopedilum insigne (Wall. ex Lindl.) Pfitzer [14]

Doubtful/Uncertain Taxa

Stelis biflora Sm. [1, 4]

HYPOXIDACEAE

Curculigo Gaertn

Curculigo capitulata (Lour.) Kuntze [1, 4]

Leucojum capitulatum Lour.; *Molineria capitulata* (Lour.) Herb. (5)

Palm grass

Perennial stemless herbs. W, C, E Nepal: 100-2100 m. Nepal, Assam-Burma, S Asia, E Asia, SE Asia and Australasia.

Use(s): Ornamental, Fiber.

Curculigo crassifolia (Baker) Hook.f. [1, 4]

Molineria crassifolia Baker (5)

धोति सरो (Dhoti saro)

Perennial herbs. E Nepal: 2000-2800 m. Nepal, E Himalaya, Assam-Burma and E Asia.

Use: Medicine (root).

Curculigo gracilis (Kurz) Wall. ex Hook.f. [1, 4]

Molineria gracilis Kurz (8)

Weevil wort

Perennial herbs. C Nepal: 1300-2000 m. Nepal, E Himalaya, Assam-Burma, E Asia and SE Asia.

Curculigo orchioides Gaertn. [1, 4, 5]

Curculigo orchioides var. *minor* Benth.

कालो मुसली (Kaalo musali); Black gold

Perennial tuberous herbs. W, C, E Nepal: 100-1700 m. Nepal, Assam-Burma, S Asia, E Asia, SE Asia, SW Asia and Australasia.

Use(s): Medicine (root/tuber), Food, Forage, Ritual.

Hypoxis L.

Hypoxis aurea Lour. [1, 4, 5]

Curculigo graminifolia Nimmo, *Hypoxis minor* D.Don

बन सिरु, कर्सुल (Ban siru, Karsul); Golden stargrass

Perennial herbs. W, C, E Nepal: 1500-2900 m. Nepal, W Himalaya, E Himalaya, Assam-Burma, S Asia, E Asia, SE Asia and SW Asia.

Use: Medicine (root stock).

IRIDACEAE

Belamcanda Adans.

Belamcanda chinensis (L.) Redoute [1, 4, 5, 14]

Ixia chinensis L.

ट्यांङ्ग पत्रे, पद्म पुष्कर (Tyaang patre, Padam puskar); Blackberry lily

Perennial rhizomatous herbs. W, C, E Nepal: 700-2300 m. Nepal, Himalaya, S Asia, E Asia and SE Asia.

Use(s): Medicine (rhizome), Ornamental.

Iris L.

Iris clarkei Baker ex Hook.f. [1, 4, 5]

Iris himalaica Dykes

बोझो झार (Bojho jhaar); Sino Siberian Iris

Perennial rhizomatous herbs. C, E Nepal: 3000-4200 m. Nepal, E Himalaya, Tibetan Plateau, Assam-Burma and E Asia.

Use: Medicine (rhizome).

Iris collettii Hook.f. [8]

Perennial herbs. E Nepal: 1700-3700 m. Nepal, E Himalaya, Tibetan Plateau, Assam-Burma, E Asia and SE Asia.

Iris decora Wall. [1, 4, 5]

Iris nepalensis D.Don

पद्म पुष्कर (Padam puskar)

Perennial herbs. W, C, E Nepal: 1800-4000 m. Nepal, E Himalaya, Tibetan Plateau and E Asia.

Use: Medicine (root).

Iris goniocarpa Baker [1, 4, 8]

Iris gracilis Maxim.

पिपेरी; मनिंग-ड्रेमा (Piperi; Maning-drema: Amchi)

Perennial rhizomatous herbs. W, C Nepal: 3000-4400 m. Nepal, E Himalaya, Tibetan Plateau, Assam-Burma and E Asia.

Use: Medicine (whole plant).

Iris kemaonensis Wall. ex Royle [1, 4, 5]

Iris kamaonensis Wall.

पिपेरी, पद्म पुष्कर (Piperi, Padam puskar)

Perennial herbs. W, C, E Nepal: 2000-4500 m. Nepal, E Himalaya, Tibetan Plateau and E Asia.

Use: Medicine (root, leaves).

Iris staintonii H.Hara, *J. Jap. Bot.* **49**: 203, f.5. 1974. [1, 4, 5, 27]

Perennial rhizomatous dwarf herbs. C Nepal: 3400-4300 m. **Endemic to Nepal**.

Central: *Mustang*, 3450 m, JDA Stainton 3960 (Holotype: BM; Isotype: KATH).

Exotic/Cultivated Taxa

Crocus sativus L. (**कुमकुम, केशर,** Kumkum, Keshar; Saffron). Native to India, Mediterranean region. *Use(s)*: Medicine (pistil), Volatile oil, Ornamental.

Gladiolus hybridus Hort. ex E.Rodigas (**तरवारे फूल,** Tarbaare phool; Gladiolus). Native to Europe and Africa. [14]

Iris × germanica L. [*Iris × germanica* var. *nepalensis* (Wall. ex Lindl.) Herb.]; (**भासमूल,** Vasmool; German Iris). Native to Europe. *Use*: Medicine (rhizome, leaves). [1, 4]

Iris tectorum Maxim. (Roof Iris). Native to E Asia. [14]

Tritonia crocata (L.) Ker Gawl., Native to S Africa. [14]

ASPHODELACEAE

Aloe L.

Aloe vera (L.) Burm.f. [1, 4]

Aloe perfoliata var. *vera* L.; *Aloe barbadensis* Mill. (14)

घिउ कुमारी (Ghiu kumari); Barbados Aloe

Perennial succulent herbs. Cultivated in Nepal. Native to Africa, naturalized in Asia and N America.

Use(s): Medicine (leaves), Food (leaves), Ornamental.

Asphodelus L.

Asphodelus tenuifolius Cav. [5]

Asphodelus fistulosus var. *tenuifolius* (Cav.) Baker

Annual or perennial herbs. W Nepal: *ca.* 100 m. Nepal, S Asia, SW Asia and Africa.

Use: Medicine (seeds).

Dianella Lam.

Dianella ensifolia (L.) DC. [1, 4, 9]

Dracaena ensifolia L.

Flax lily

Perennial rhizomatous herbs. W, C, E Nepal: 100-1700 m. Nepal, E Himalaya, Assam-Burma, S Asia, E Asia, SE Asia, Africa and Australasia.

Use(s): Medicine (root, leaves), Cosmetics (root).

Exotic/Cultivated Taxa

Hemerocallis fulva (L.) L. (**दिनशोभा, ध्वाङ्फूल,** Dinshovaa, dhwang phul; Orange day lily). Native to Asia. *Use*: Medicine (flowers). [1, 4, 14]

Kniphofia uvaria (L.) Oken (**घोगे फूल,** Ghoge phul; Torch lily). Native to Africa. [14, 54]

AMARYLLIDACEAE

Allium L.

Allium carolinianum DC. [1, 4, 5]

Allium blandum Wall.; *Allium obtusifolium* Klotzsch; *Allium thomsonii* Baker

जंगली लसुन, लुंग-गोक (Jangali lasun; Lung-gok: Amchi)

Perennial bulbous herbs. W, C Nepal: 2300-5100 m. Nepal, E Himalaya, Tibetan Plateau, S Asia, N Asia, C Asia and SW Asia.

Use(s): Medicine, Food (bulb, leaves).

Allium cepa L. [5]

प्याज (Pyaj); Onion

Perennial bulbous herbs. Widely cultivated in Nepal. Native to SW Asia.

Use(s): Medicine (bulb), Food (bulb, leaves).

Allium fasciculatum Rendle [1, 4, 5]

Allium gageanum W.W.Sm.

लसुने साग, जिम्बु (Lasune saag, Jimbu)

Perennial herbs. W, C, E Nepal: 2700-4500 m. Nepal, E Himalaya, Tibetan Plateau, E Asia and N Asia.

Use: Food (whole plant).

Allium hypsistum Stearn, *Bull. Brit. Mus. (Nat. Hist.), Bot.* **2**: 188. 1960. [1, 4, 26]

जिम्बू (Jimbu)

Perennial herbs. W, C Nepal: 2000-5700 m. **Endemic to Nepal.**

Central: *Dolpa,* SW of Saldanggaon, *ca.* 5500 m, O. Polunin, WR Sykes & LHJ Williams 8 (Holotype: BM).

Use(s): Medicine (whole plant), Food (leaves).

Allium prattii C.H.Wright [1, 4, 5]

Allium cannifolium H.Léveillé; *Allium ellipticum* Wall.

Perennial herbs. W, C, E Nepal: 2400-4500 m. Nepal, E Himalaya, Tibetan Plateau and E Asia.

Use(s): Medicine, Food (leaves).

Allium przewalskianum Regel [1, 4, 5]

Allium jacquemontii Regel; *Allium stoliczkii* Regel

जिम्बू (Jimbu)

Perennial herbs. W, C Nepal: 2500-4200 m. Nepal, Tibetan Plateau, S Asia, E Asia, N Asia and SW Asia.

Use: Medicine (whole plant).

Allium sativum L. [5]

लसुन (Lasun); Garlic

Perennial herbs. Widely cultivated in Nepal. Native to C Asia.

Use(s): Medicine (bulb), Food (whole plant).

Allium sikkimense Baker [1, 4, 5]

Allium kansuense Regel; *Allium tibeticum* Rendle

चिउ गोग (Chiu gog –Tibetan)

Perennial herbs. C Nepal: 3000-4800 m. Nepal, E Himalaya, Tibetan Plateau, S Asia and E Asia.

Use: Medicine (whole plant).

Allium tuberosum Rottler ex Spreng. [1, 4, 5]

Allium chinense Maxim.; *Allium clarkei* Hook.f.

दुंडूको साग (Dunduko saag)

Perennial herbs. W, C Nepal: 2300-2600 m. Nepal, Himalaya, Tibetan Plateau and Assam-Burma.

Use(s): Medicine (root, leaves), Food (whole plant).

Allium wallichii Kunth [1, 4, 5]

Allium coeruleum Wall.

बन लसुन, जिम्बू झार (Ban lasun, Jimbu jhaar)

Perennial herbs. W, C, E Nepal: 2100-4800 m. Nepal, E Himalaya, Tibetan Plateau, Assam-Burma, S Asia and E Asia.

Use(s): Medicine (roots, leaves), Food (whole plant).

Crinum L.

Crinum amoenum Ker Gawl. ex Roxb. [1, 4, 5]

हाडे फूल, हाडे लसुन (Haade phool, Haade lasun)

Perennial herbs. C, E Nepal: 200-1800 m. Nepal, E Himalaya and Assam-Burma.

Use(s): Medicine (rhizome), Ornamental.

Crinum viviparum (Lam.) R.Ansari & V.J.Nair

Amaryllis vivipara Lam.; *Crinum defixum* forma *stephenhassardii* Traub (1, 4)

Perennial herbs. C Nepal: *ca.* 1100 m. Nepal, E Himalaya, Assam-Burma, S Asia and SE Asia.

Use: Ornamental.

Milula Prain

Milula spicata Prain [1, 4]

Allium spicatum (Prain) N.Frieson

Perennial herbs. C Nepal: 3000-4500 m. Nepal, Himalaya and Tibetan Plateau.

Pancratium L.

Pancratium verecundum Aiton [1, 4]

Perennial herbs. E Nepal: 400-500 m. Nepal, W Himalaya and E Himalaya.

Use: Ornamental.

Zephyranthes Herb.

Zephyranthes carinata Herb. [1, 4, 5, 54]

Zephyranthes grandiflora Lindl.

Pink rain-lily

Perennial herbs. C, E Nepal: 700-1900 m (Naturalized). Native to N America.

Use(s): Ornamental, Medicine (bulb).

Zephyranthes citrina Baker [47]

Perennial herbs. E Nepal: *ca.* 100 m. Nepal, Tibetan Plateau, S Asia, E Asia, SE Asia, Africa, N America and S America.

Use: Medicine (whole plant).

Exotic/Cultivated Taxa

Agapanthus africanus (L.) Hoffmanns. (**निरकमल,** Nirkamal; Blue African lily). Native to Africa. [14, 54]

Allium ascalonicum L. (**छ्यापे,** Chhyappe; Shallot). Native to SE Asia and C Asia. *Use*: Medicine (bulb).

Allium fistulosum L. (**छ्यापी,** Chhyapi; Welsh onion). Native to Tibetan Plateau and N Asia.

Crinum latifolium L. (**हाडे फूल,** Haade phool; Trumpet lily). Native to SE Asia. *Use*: Medicine (bulb, leaves).

Hippeastrum vittatum (L'Her.) (**द्वांग फूल,** Dhwaang phool; Amaryllis). Native to S America. [14, 54]

Hymenocallis rotata (Ker Gawl.) (**चम्पा फूल,** Champa phul; Spider lily). Endemic to S America. [14]

Lycoris radiata (L'Hér.) (Red spider lily). Native to Tibetan Plateau and E Asia. [14]

Narcissus tazetta L. (**गुन केशरी,** Gun keshari; Daffodil). Native to Europe. [5]

Zephyranthes candida (Lindl.) (**चुलोचा स्वाँ,** Chulocha swan-Newari; White rain lily). Native to S America. *Use*: Medicine (leaves). [5, 14]

Doubtful/Uncertain Taxa

Allium hookeri Thwaites (**छ्यापी,** Chhyapi)

Allium humile Kunth (**जिम्बू,** Jimbu)

Allium oreoprasum Schrenk (**जिम्बू,** Jimbu)

Allium schoenoprasum L. (**छ्यापी,** Chhyapi; Wild chives)

ASPARAGACEAE

Agave L.

Agave americana L. [4, 5]

केतुकी (Ketuki); Century plant

Perennial arborescent herbs. W, C, E Nepal: 200-1400 m. Nepal, E Asia, native to N America.

Use(s): Medicine (root, sap, leaves), Fiber.

Agave angustifolia Haw. [1, 4]

Agave spectabilis Tod.

केतुकी (Ketuki); Bastard Aloe

Perennial arborescent herbs. Cultivated in Nepal: 800-2000 m. Native to America.

Use(s): Medicine (root, leaves), Fiber.

Agave cantula Roxb. [1, 4]

Agave cantala Roxb. (8)

केतुकी, हाथी बार (Ketuki, Haathi bar); Century plant

Perennial arborescent herbs. W, C, E Nepal: 1000-2000 m. Planted and naturalized in tropical Asia.

Use(s): Medicine (root, leaves), Fiber, Fence, Fish poison (leaves).

Agave sisalana Perrine ex Engelm. [1, 4]

Perennial arborescent herbs. E Nepal: *ca.* 1500 m. Native to N America.

Use: Fiber.

Asparagus L.

Asparagus adscendens Roxb. [5]

मुसली काँडा (Musli kaanda)

Perennial straggling herbs. W Nepal: 100-1900 m. Nepal and W Himalaya.

Use: Medicine (root).

Asparagus curillus Buch.-Ham. ex Roxb. [1, 4, 5]

Asparagus nepalensis Baker

कुरिलो (Kurilo)

Sub-scandent shrubs. W, C Nepal: 1000-2800 m. Nepal and E Himalaya.

Use: Medicine (root).

Asparagus filicinus Buch.-Ham. ex D.Don [1, 4, 5]

Asparagus filicinus var. *brevipes* Baker (5)

बन कुरिलो, सतावरी; न्ये-शिंग (Ban kurilo, Sataawari; Nye-shing: Amchi)

Perennial straggling herbs. W, C, E Nepal: 2000-3400 m. Nepal, E Himalaya, Tibetan Plateau, Assam-Burma, S Asia, E Asia, SE Asia and SW Asia.

Use(s): Medicine (root/tuber), Food (tender shoot).

Asparagus lycopodineus (Baker) F.T.Wang & Tang [5]

Asparagus filicinus var. *lycopodineus* Baker

Perennial straggling herbs. W, C Nepal: 1500-2500 m. Nepal, E Himalaya, Tibetan Plateau, Assam-Burma, S Asia and E Asia.

Asparagus officinalis L. [5]

कुरिलो (Kurilo); Asparagus

Perennial herbs. C Nepal: *ca.* 1400 m (Cultivated, nearly naturalized). Native to Europe, SW Asia and N Africa.

Use(s): Medicine (root, shoot), Food (young twig).

Asparagus penicillatus H.Hara, *J. Jap. Bot.* **49**: 134. 1974. [1, 4, 5, 26]

Perennial straggling herbs or subshrubs. W, C Nepal: 1900-2500 m. **Endemic to Nepal**.

West: *Dolpa*, Dunai, Bheri valley, 8000 ft. (2440 m), O. Polunin, WR Sykes & LHJ Williams 2484 (Holotype: BM).

Use(s): Medicine (root), Food, Fodder, Ornamental.

Asparagus racemosus Willd. [1, 4, 5]

Asparagus racemosus var. *subacerosus* Baker (1, 4, 5)

कुरिलो, सतावरी (Kurilo, Sataawari)

Perennial straggling herbs. W, C, E Nepal: 100-2200 m. Nepal, E Himalaya, Tibetan Plateau, Assam-Burma, S Asia, SE Asia, SW Asia, Africa and Australasia.

Use(s): Medicine (tuberous root), Food (tender shoot), Fodder, Ornamental.

Asparagus tibeticus F.T.Wang & S.C.Chen [5]

Perennial straggling herbs. C Nepal: 2900-3800 m. Nepal, Himalaya and Tibetan Plateau.

Use: Medicine (root).

Chlorophytum Ker Gawl.

Chlorophytum arundinaceum Baker [1, 4, 5]

दति साग (Dati Saag-Danuwar/Tharu); White musali

Perennial herbs. W, C, E Nepal: 100-2000 m. Nepal, E Himalaya and Assam-Burma.

Use: Medicine (root/tuber).

Chlorophytum nepalense (Lindl.) Baker [1, 4, 5]

Phalangium nepalense Lindl.; *Chlorophytum flaccidum* W.W.Sm.; *Chlorophytum khasianum* Hook.f. (1, 4, 5)

बन प्याज, दंति साग (Ban pyaaj, Danti Saag-Danuwar/Tharu)

Perennial herbs. W, C, E Nepal: 1200-3200 m. Nepal, E Himalaya, Tibetan Plateau, Assam-Burma and E Asia.

Use(s): Medicine (root/tuber), Food (tender leaves).

Chlorophytum tuberosum (Roxb.) Baker [1, 4, 5]

Anthericum tuberosum Roxb.

White musali

Perennial herbs. W, C, E Nepal: 200-300 m. Nepal, Assam-Burma, S Asia and Africa.

Dipcadi Medik.

Dipcadi serotinum (L.) Medik.

Hyacinthus serotinus L.; *Dipcadi hysudricum* (Edgew.) Baker (1, 4)

Perennial herbs. W Nepal: 1300-3000 m. Nepal and W Himalaya.

Drimia Jacq. ex Willd.

Drimia indica (Roxb.) Jessop [5]

Scilla indica Roxb.; *Urginea indica* (Roxb.) Kunth (1, 4)

बन प्याज (Ban pyaaj); Indian squill

Perennial herbs. W Nepal: 1300-3000 m. Nepal, E Himalaya, S Asia and Africa.

Use: Medicine (bulb, tuber).

Maianthemum F.H.Wigg.

Maianthemum fuscum (Wall.) LaFrankie [4, 5]

Smilacina fusca Wall.; *Smilacina fusca* var. *pilosa* H.Hara

सेतो जरा (Seto jaraa)

Perennial herbs. C, E Nepal: 2200-3000 m. Nepal, E Himalaya, Tibetan Plateau, Assam-Burma and E Asia.

Use: Medicine (root).

Maianthemum oleraceum (Baker) LaFrankie

Tovaria oleracea Baker; *Smilacina oleracea* (Baker) Hook.f. & Thomson (1, 4); *Smilacina oleracea* forma *acuminata* (F.T.Wang & T.Tang) H.Hara (1, 4); *Smilacina oleracea* var. *acuminata* F.T.Wang & T.Tang

Perennial herbs. E Nepal: 2500-4100 m. Nepal, E Himalaya, Tibetan Plateau, Assam-Burma and E Asia.

Maianthemum purpureum (Wall.) LaFrankie [4, 5]

Smilacina purpurea Wall.; *Tovaria purpurea* (Wall.) Baker

खिरिंग्लो, शिकारी साग (Khiringlo, Sikari saag)

Perennial herbs. W, C, E Nepal: 2600-4300 m. Nepal, E Himalaya, Tibetan Plateau and E Asia.

Use(s): Medicine (root), Food (tender leaves, stem).

Ophiopogon Ker Gawl.

Ophiopogon clarkei Hook.f. [1, 4, 5]

Mondo dracaenoides var. *clarkei* (Hook.f.) Farw.

Perennial stoloniferous herbs. W, C, E Nepal: 1500-3500 m. Nepal, E Himalaya, Tibetan Plateau and E Asia.

Ophiopogon intermedius D.Don [1, 4, 5]

Ophiopogon parviflorus (Hook.f.) H.Hara (1, 4)

Perennial herbs. W, C, E Nepal: 700-3000 m. Nepal, E Himalaya, Tibetan Plateau, Assam-Burma, S Asia, E Asia and SE Asia.

Use: Medicine (tuber).

Peliosanthes Andrews

Peliosanthes griffithii Baker [1, 4]

Peliosanthes bakeri Hook.f.

Perennial herbs. E Nepal: 200-700 m. Nepal, E Himalaya, Assam-Burma and S Asia.

Peliosanthes macrostegia Hance

Peliosanthes devayi Fanch.; *Peliosanthes macrophylla* Wall. ex Baker (1,4,5)

Perennial herbs. E Nepal: 300-700 m. Nepal, E Himalaya and Tibetan Plateau.

Polygonatum Mill.

Polygonatum brevistylum Baker [1, 4, 5]

Perennial herbs. E Nepal: 2400-2900 m. Nepal, E Himalaya, Tibetan Plateau, S Asia and E Asia.

Polygonatum cathcartii Baker [1, 4, 5]

Perennial herbs. E Nepal: 2800-3200 m. Nepal, E Himalaya, Tibetan Plateau and E Asia.

Polygonatum cirrhifolium (Wall.) Royle [1, 4, 5]

Convallaria cirrhifolia Wall.; *Polygonatum bulbosum* H.Lev.

खिरौंला, रम्ने (Khiraunlaa; Ramney: Amchi); Solomon's seal

Perennial herbs. W, C, E Nepal: 1200-4600 m. Nepal, E Himalaya, Tibetan Plateau, S Asia and E Asia.

Use(s): Medicine (root/tuber), Food (tender leaves, shoot).

Polygonatum curvistylum Hua [1, 4, 5]

Perennial herbs. W, C Nepal: 2600-2900 m. Nepal and E Asia.

Polygonatum griffithii Baker [5]

Perennial herbs. C Nepal: 1700-2800 m. Nepal and Tibetan Plateau.

Polygonatum hookeri Baker [1, 4, 5]

Polygonatum pumilum Hua

पंगी रान्ये (Pangi raanye-Tibetan); Dwarf Solomon-seal

Perennial herbs. W, C, E Nepal: 2900-5000 m. Nepal, E Himalaya, Tibetan Plateau and E Asia.

Use(s): Medicine (root), Food.

Polygonatum oppositifolium (Wall.) Royle [1, 4, 5]

Convallaria oppositifolia Wall.

Perennial herbs. C, E Nepal: 1600-2300 m. Nepal, E Himalaya and Tibetan Plateau.

Use: Food (young shoots).

Polygonatum punctatum Royle ex Kunth [1, 4, 5]

Convallaria punctata Wall.; *Polygonatum anomalum* Hua

Perennial herbs. C, E Nepal: 2000-2600 m. Nepal, E Himalaya, Tibetan Plateau, Assam-Burma, S Asia, E Asia and SE Asia.

Polygonatum sibiricum Rodoute [5]

Polygonatum chinense Kunth

Perennial herbs. C, E Nepal: 2900-3900 m. Nepal, Tibetan Plateau, E Asia and N Asia.

Polygonatum singalilense H.Hara (1, 4, 5)

Perennial herbs. W, C, E Nepal: 2800-4400 m. Nepal and E Himalaya.

Polygonatum verticillatum (L.) All. [1, 4, 5]

Convallaria verticillata L.; *Polygonatum kansuense* Maxim. ex Batalin (5); *Polygonatum minutiflorum* H.Lev.

खिरौंला, केरुवा (Khiraunlaa, Keruwaa); Whorled solomon's seal

Perennial herbs. W, C, E Nepal: 2400-4700 m. Nepal, E Himalaya, Tibetan Plateau, S Asia, E Asia, N Asia, SW Asia and Europe.

Use(s): Medicine (tuber), Food (tender shoot, leaves).

Rohdea Roth

Rohdea nepalensis (Raf.) N.Tanaka

Tilcusta nepalensis Raf.; *Campylandra aurantiaca* Baker (1, 4); *Tupistra aurantiaca* (Wall. ex Baker) Hook.f. (5)

तमुर्के (Tamurke-Gurung)

Perennial rhizomatous herbs. C, E Nepal: 2300-2700 m. Nepal, E Himalaya, Tibetan Plateau, Assam-Burma, S Asia and E Asia.

Use(s): Medicine (rhizome), Food (fruit).

Theropogon Maxim.

Theropogon pallidus (Wall. ex Kunth) Maxim. [1,4, 5]

Ophiopogon pallidus Wall. ex Kunth

Perennial herbs. W, C, E Nepal: 1800-2700 m. Nepal, E Himalaya, Tibetan Plateau, S Asia and E Asia.

Tupistra Ker Gawl.

Tupistra nutans Wall. ex Lindl. [1, 4]

Rohdea nutans (Wall. ex Lindl.) Schult. & Schult.f.

Perennial herbs. E Nepal: *ca.* 200 m. Nepal, E Himalaya and Assam-Burma.

Exotic/Cultivated Taxa

Agave lurida Aiton (*Agave vera-cruz* Mill.). Native to N America. *Use*: Fiber.

Cordyline fruticosa (L.) A.Chev. (*Convallaria fruticosa* L.). Native to Australasia.

Hosta plantaginea (Lam.) Asch. (*Hemerocallis plantaginea* Lam.); (Plantain lily). Native to E Asia. [14]

Ornithogalum thyrsoides Jacq. (**छ्यापी फूल,** Chhyapi phul; Wonder flower). Native to Africa. [14]

Ornithogalum umbellatum L. (Grass lily). Native to Europe, Africa and N America. [14]

Polianthes tuberosa L. (**सुगन्ध राज, रजनीगन्धा,** Sugandharaj, Rajani ganda; Tube rose). Native to N America.

Sansevieria roxburghiana Schult. & Schult.f. (**दर्शन पिपल, मरुई लहरो,** Darshan peepal, Maraui laharo; Indian bowstring Hemp). Native to Assam-Burma and S Asia.

Yucca flaccida Haw. (*Yucca smalliana* Fernald); (**केत्तुके,** Kettuke; Golden sword). Native to N America.

Yucca gloriosa L. (**केत्तुके,** Kettuke). Native to S America. *Use(s)*: Construction materials, Fiber, Fish poison. [14]

ARECALES

ARECACEAE (incl. PALMAE)

Areca L.

Areca catechu L. [5]

सुपारी (Supari); Betel nut

Trees. E Nepal: *ca.* 100 m. Native to SE Asia; naturalized in Tibetan plateau, S Asia and E Asia.

Use(s): Medicine (nut), Masticatory, Utensils and handicrafts, Ritual.

Calamus L.

Calamus acanthospathus Griff. [1, 4, 5]

Calamus montanus T.Anderson

बेत, गौरी बेत (Bet, Gauri bet); Yunnan rattan

Climbing shrubs. E Nepal: 400-2400 m. Nepal, E Himalaya, Tibetan Plateau, Assam-Burma, S Asia, E Asia and SE Asia.

Use(s): Medicine (root), Utensils and handicrafts, Construction materials.

Calamus latifolius Roxb. [1, 4, 5]

Calamus humilis Roxb.

फेक्रि बेत (Phekri bet)

Climbing shrubs. E Nepal: 100-600 m. Nepal, E Himalaya and S Asia.

Use(s): Utensils and handicrafts, Construction materials.

Calamus leptospadix Griff. [1, 4, 5]

Palmijuncus leptospadix (Griff.) Kuntze

डाङ्गरे बेत (Dangre bet); Indian dwarf rattan

Scandent climbing shrubs. W, C, E Nepal: 100-300 m. Nepal, E Himalaya and S Asia.

Use(s): Utensils and handicrafts, Construction materials.

Calamus nambariensis Becc. [8]

Calamus khasianus Becc.

Climbing shrubs. C Nepal: 700-2000 m. Nepal, E Himalaya, Assam-Burma, S Asia, E Asia and SE Asia.

Calamus tenuis Roxb. [5]

Calamus heliotropium Buch.-Ham. ex Kunth

पानी बेत, बेतलौरी (Paani bet, Bet lauri)

Climbing shrubs. W, C, E Nepal: 100-900 m. Nepal, E Himalaya, Assam-Burma, S Asia and SE Asia.

Use(s): Medicine (root, stem), Fodder, Utensils and handicrafts, Construction materials.

Caryota L.

Caryota urens L. [1, 4, 5]

बन सुपारी, जगर (Ban supaari, Jagar); Fishtail palm

Trees. W, C, E Nepal: 1200-1400 m. Nepal, Assam-Burma, S Asia and SE Asia.

Use: Fumitory and masticators materials.

Cocos L.

Cocos nucifera L. [5]

Cocos indica Royle; *Cocos nana* Griff.

नरिवल (Nariwal); Coconut

Evergreen trees. E Nepal: *ca.* 100 m. Native to tropical eastern regions, widely cultivated.

Use(s): Medicine (root, kernel/seeds), Food, Utensils and handicrafts, Ornamental, Ritual.

Phoenix L.

Phoenix acaulis Roxb. [1, 4, 5]

थाकल, खजुर (Thaakal, Khajur); Wild date-palm

Shrubs. W, C, E Nepal: 300-1500 m. Nepal, E Himalaya, Assam-Burma and S Asia.

Use(s): Medicine (root, fruit), Food (fruit), Fodder, Utensils and handicrafts, Construction materials.

Phoenix loureiroi Kunth

Phoenix humilis Royle ex Becc. (1, 4); *Phoenix humilis* var. *loureiroi* (Kunth) Becc.

खजुर, थाकल (Khajur, Thaakal); Wild date-palm

Shrubs. W, C, E Nepal: 100-1500 m. Nepal, E Himalaya, Assam-Burma, S Asia, E Asia, SE Asia and SW Asia.

Use(s): Medicine (leaves, fruit, seeds), Fodder, Utensils and handicrafts, Construction materials, Food (Fruit).

Phoenix sylvestris (L.) Roxb. [1, 4, 5]

Elate sylvestris L.

खजुर, ताडी (Khajur, Taadi); Date palm

Evergreen trees. W, C Nepal: *ca.* 1500 m. Nepal, E Himalaya, Assam-Burma and S Asia.

Use(s): Medicine (root, stem juice, fruit, kernels), Food (fruit), Fodder, Utensils and handicrafts, Ornamental.

Pinanga Blume

Pinanga gracilis Blume [1, 4]

Areca gracilis Roxb.; *Nenga gracilis* (Blume) Beccari

Himalayan Pinanga

Dwarf trees. E Nepal: 300-500 m. Nepal, E Himalaya, Tibetan Plateau, Assam-Burma and S Asia.

Plectocomia Mart. ex Schult. & Schult.f.

Plectocomia himalayana Griff. [5]

Plectocomia montana Griff. ex T.Anderson

हिमाली बेत (Himali bet)

Scandent shrubs. E Nepal: 100-200 m. Nepal, E Himalaya, S Asia, E Asia and SE Asia.

Use: Construction materials (stem).

Sabal Adans.

Sabal palmetto (Walter) Lodd. ex Schult. & Schult.f. [5]

Corypha palmetta Walter; *Chamaerops palmetto* (Walter) Michx.

Sabal or Cabbage palm

Evergreen trees. C Nepal: *ca.* 1500 m. Native to N America.

Use(s): Construction materials, Fiber (petiole), Food (leaf bud, fruit).

Trachycarpus H.Wendl.

Trachycarpus martianus (Wall. ex Mart.) H.Wendl. [1, 4, 5]

Chamaerops martiana Wall. ex Mart.; *Trachycarpus khasiana* (Griff.) H.Wendl.

Khasia hills fan palm

Trees. C Nepal: 1500-1600 m. Nepal and Assam-Burma.

Wallichia Roxb.

Wallichia oblongifolia Griff.

Wallichia densiflora Mart. (1, 4, 5)

Foxtail palm

Perennial shrubs. W, C, E Nepal: 200-1400 m. Nepal, E Himalaya, Assam-Burma and S Asia.

Exotic/Cultivated Taxa

Borassus flabellifer L. (**ताडी,** Taadi; Sugar palm, Fan palm). Native to S Asia and SE Asia. *Use*: Medicine (bark, sap fruit).

Roystonea regia (Kunth) O.F.Cook (**कुपीताल,** Kupital; Bottle palm). Native to S America and N America.

Trachycarpus fortunei (Hook.) H.Wendl. (*Chamaerops fortunei* Hook.); Windmill palm. Native to E Asia. [5]

Doubtful/Uncertain Taxa

Trachycarpus takil Becc. (Kumaon palm) [1, 4]

COMMELINALES

COMMELINACEAE

Amischotolype Hassk.

Amischotolype hookeri (Hassk.) H.Hara [1, 4]

Forrestia hookeri Hassk.

Perennial herbs. E Nepal: 200-1200 m. Nepal, E Himalaya, Tibetan Plateau, Assam-Burma, S Asia, E Asia and SE Asia.

Commelina L.

Commelina benghalensis L. [1, 4, 5]

Commelina cavaleriei H.Lév.

काने झार (Kaane jhaar)

Perennial or annual herbs. W, C, E Nepal: 100-1800 m. Nepal, Tibetan Plateau, S Asia, E Asia, SE Asia, SW Asia and Africa.

Use(s): Medicine (whole plant), Food (young leaves).

Commelina caroliniana Walter [5]

Commelina hasskarlii C.B.Clarke (1, 4)

Annual herbs. W, C, E Nepal: 100-1700 m. Nepal, E Himalaya, Assam-Burma and S Asia.

Commelina diffusa Burm.f. [1, 4, 5]

Commelina longicaulis Jacq.

सिम काने (Sim kaane)

Perennial or annual herbs. W, C, E Nepal: 100-3700 m. Nepal, Tibetan Plateau and E Asia.

Commelina maculata Edgew. [1, 4, 5]

Commelina obliqua var. *viscida* C.B.Clarke; *Commelina paludosa* var. *viscida* (C.B.Clarke) R.S.Rao & Kammathy

Spotted day-flower

Perennial herbs. W, C Nepal: 700-2500 m. Nepal, E Himalaya, Tibetan Plateau, Assam-Burma, S Asia, E Asia and SE Asia.

Commelina paludosa Blume [1, 4, 5]

Commelina obliqua Buch.-Ham. ex D.Don

बुकी, काने झार (Buki, Kaane jhaar); May flower

Perennial herbs. W, C, E Nepal: 100-3500 m. Nepal, E Himalaya, Tibetan Plateau, Assam-Burma, S Asia, E Asia and SE Asia.

Use(s): Medicine (root), Food (root).

Commelina suffruticosa Blume [1, 4, 5]

कटस प्रिय (Katas priya)

Perennial herbs. W, C, E Nepal: 100-400 m. Nepal, S Asia, E Asia and SE Asia.

Use: Medicine (root).

Commelina undulata R.Br. [5]

Commelina obliqua var. *mathewii* C.B.Clarke

Perennial herbs. W, C, E Nepal: 1900-2400 m. Nepal, S Asia, E Asia, SE Asia and Australasia.

Cyanotis D.Don

Cyanotis axillaris (L.) D.Don ex Sweet [5]

Commelina axillaris L.; *Amischophacelus axillaris* (L.) R.S.Rao & Kammathy (1, 4)

तेने झार (Tene jhaar)

Annual marshy herbs. W, C, E Nepal: 100-2100 m. Nepal, Assam-Burma, S Asia, E Asia, SE Asia and Australasia.

Use: Medicine (whole plant).

Cyanotis cristata (L.) D.Don [1, 4, 5]

Commelina cristata L.

सानो काने (Saano kaane)

Annual or perennial herbs. W, C, E Nepal: 100-2400 m. Nepal, E Himalaya, Tibetan Plateau, Assam-Burma, S Asia, E Asia, SE Asia and Africa.

Use(s): Forage, Ritual.

Cyanotis fasciculata (B.Heyne ex Roth) Schult. & Schult.f. [8]

Tradescantia fasciculata B.Heyne ex Roth

Annual herbs. C Nepal: 1200-1900 m. Nepal and S Asia.

Cyanotis vaga (Lour.) Schult. & Schult.f. [1, 4, 5]

Tradescantia vaga Lour.; *Cyanotis barbata* D.Don

Wandering dew grass

Perennial herbs. W, C, E Nepal: 800-2700 m. Nepal, E Himalaya, Tibetan Plateau, Assam-Burma, S Asia, E Asia, SE Asia and Africa.

Floscopa Lour.

Floscopa scandens Lour. [1, 4, 5]

Aneilema hispidum D.Don; *Floscopa scandens* var. *vaginivillosa* R.H.Miao

सिमकाने घाँस (Simkaane ghans)

Perennial or annual herbs. W, C, E Nepal: 100-1800 m. Nepal, E Himalaya, Tibetan Plateau, Assam-Burma, S Asia, E Asia, SE Asia and Australasia.

Use: Medicine (whole plant).

Murdannia Royle

Murdannia blumei (Hassk.) Brenan [1, 4]

Dichoespermum blumei Hassk.

Annual or perennial marshy herbs. C Nepal: 1200-1400 m. Nepal, Himalaya, Assam-Burma, S Asia and SE Asia.

Murdannia divergens (C.B.Clarke) G.Brückn. [1, 4, 5]

Aneilema herbaceum var. *divergens* C.B.Clarke

Perennial herbs. W, C Nepal: 1200-2400 m. Nepal, E Himalaya, Assam-Burma, S Asia and E Asia.

Murdannia edulis (Stokes) Faden [5]

Commelina edulis Stokes; *Commelina scapiflora* Roxb.; *Murdannia scapiflora* (Roxb.) Royle (1, 4)

स्याल मुसली, निंगाले गाभा (Syaal musli, Nigaale gaavaa)

Perennial herbs. W, C, E Nepal: 100-2000 m. Nepal, Assam-Burma, S Asia, E Asia and SE Asia.

Use(s): Medicine (root), Food (root).

Murdannia japonica (Thunb.) Faden [1, 4, 5]

Commelina japonica Thunb.

निंगाले गाभा (Nigaale gaavaa)

Perennial herbs. W, C, E Nepal: 400-2100 m. Nepal, E Himalaya, Assam-Burma, S Asia, E Asia and SE Asia.

Use(s): Medicine (root), Food (root).

Murdannia keisak (Hassk.) Hand.-Mazz. [1, 4, 8]

Aneilema keisak Hassk.

Asian spiderwort

Annual aquatic herbs. C Nepal: 1100-1900 m. Nepal, Tibetan Plateau, E Asia, SE Asia, N America and S America.

Murdannia nudiflora (L.) Brenan [1, 4, 5]

Commelina nudiflora L.; *Aneilema nudiflorum* (L.) R.Br.

मसिनो काने, काने झार (Masino kaane, Kaane jhaar); Dove weed

Annual herbs. W, C, E Nepal: 100-2400 m. Nepal, E Himalaya, Assam-Burma, S Asia, E Asia, SE Asia, SW Asia, Africa and Australasia.

Use(s): Medicine (root), Food (tender leaves).

Murdannia spirata (L.) G.Brückn. [1, 4, 5]

Commelina spirata L.; *Aneilema nanum* (Roxb.) Kunth; *Commelina nana* Roxb.

Perennial herbs. C, E Nepal: 100-1900 m. Nepal, E Himalaya, Assam-Burma, S Asia, E Asia, SE Asia and Australasia.

Murdannia vaginata (L.) G.Brückn. [1, 4]

Commelina vaginata L.; *Aneilema vaginatum* (L.) R.Br.

Annual or perennial aquatic herbs. E Nepal: *ca.* 200 m. Nepal, S Asia, E Asia and SE Asia.

Pollia Thunb.

Pollia hasskarlii R.S.Rao [1, 4]

Perennial herbs. E Nepal: 800-900 m. Nepal, E Himalaya, Tibetan Plateau, Assam-Burma, S Asia, E Asia and SE Asia.

Pollia subumbellata (C.B.Clarke) C.B.Clarke [8]

Aclisia subumbellata C.B.Clarke

Perennial herbs. E Nepal: 800-1400 m. Nepal, E Himalaya and E Asia.

Rhopalephora Hassk.

Rhopalephora scaberrima (Blume) Faden [5]

Commelina scaberrima Blume; *Dictyospermum scaberrimum* (Blume) J.K.Morton (1, 4)

Perennial herbs. C, E Nepal: 800-2100 m. Nepal, E Himalaya, Tibetan Plateau, Assam-Burma, S Asia, E Asia and SE Asia.

Streptolirion Edgew.

Streptolirion volubile Edgew. [1, 4, 8]

Streptolirion cordifolium Kuntze

Perennial climbing herbs. W, C, E Nepal: 1200-2400 m. Nepal, E Himalaya, Tibetan Plateau, Assam-Burma, S Asia, E Asia and SE Asia.

Exotic/Cultivated Taxa

Tradescantia zebrina Heynh. ex Bosse. (Wandering Jew). Native to C America. [5]

PONTEDERIACEAE

Eichhornia Kunth

Eichhornia crassipes (Mart.) Solms [1, 4, 5, 14]

Pontederia crassipes Mart.

जल कुम्भी (Jal kumbhee); Water hyacinth

Perennial aquatic herbs. C, E Nepal: 200-1500 m. Invasive species in Nepal. Native to S America.

Use(s): Medicine (flowers), Food (tender shoot), Fodder, Ornamental.

Note: Noxious weed in ponds and lakes of Nepal.

Monochoria C.Presl.

Monochoria hastata (L.) Solms [1, 4, 5]

Pontederia hastata L.; *Monochoria hastifolia* C.Presl

निलो जलुके (Nilo jaluke)

Perennial herbs. W, C, E Nepal: 100-300 m. Nepal, E Himalaya, Assam-Burma, S Asia, E Asia and SE Asia.

Use(s): Medicine (whole plant), Food (tender shoots and leaves).

Monochoria vaginalis (Burm.f.) C.Presl [1, 4, 5]

Pontederia vaginalis Burm.f.

निलो जलुके, कर्कले झार (Nilo jaluke, Karkale jhaar)

Annual or perennial aquatic herbs. W, C, E Nepal: 100-1800 m. Nepal, Asia, Africa and Australasia.

Use(s): Medicine (root, leaves), Food (tender leaves).

ZINGIBERALES

MUSACEAE

Ensete Horan.

Ensete glaucum (Roxb.) Cheesman [1, 4]

Musa glauca Roxb.; *Musa nepalensis* Wall.

Perennial arborescent herbs. C, E Nepal: 200-1100 m. Nepal, Assam-Burma, S Asia, E Asia and SE Asia.

Use(s): Ornamental, Rosaries (seeds).

Musa L.

Musa balbisiana Colla [1, 4]

Musa seminifera Lour.

बन केरा (Ban keraa)

Perennial arborescent herbs. E Nepal: 200-600 m. Nepal, E Himalaya, Assam-Burma, S Asia and SE Asia.

Use(s): Medicine (fruit), Food (buds, flowers, fruit), Fodder.

Musa paradisiaca L. [5]

Musa sapientum L.

केरा (Keraa); Banana, Plantain

Perennial arborescent herbs. C, E Nepal: 300-1500 m. Widely cultivated for fruit. Native to SE Asia.

Use(s): Food (fruit), Ritual, Medicine (root, stem, leaves, fruit).

CANNACEAE

Canna L.

Canna indica L. [1, 4, 5]

Canna chinensis Willd. (1); *Canna edulis* Ker Gawl. (1, 4); *Canna nepalensis* Bouche; *Canna speciosa* Roscoe ex Sims (1, 4)

गने सर्बदा, भुँई चापो (Gane sarbadaa, Bhuinchapo)

Perennial herbs. C, E Nepal: 1000-1400 m. Nepal, E Asia, native to Africa and S America.

Use: Medicine (root, leaves, flowers, seeds).

Exotic/Cultivated taxa

Canna hybrida Hort. (सर्बदा, Sarbada; Canna lily). Native to S America. [14, 54]

MARANTACEAE

Stachyphrynium K.Schum.

Stachyphrynium placentarium (Lour.) Clausager & Borchs.

Phyllodes placentaria Lour.; *Phrynium placentarium* (Lour.) Merr. (1, 4, 5)

Lamb's ears

Perennial herbs or shrubs. C, E Nepal: 500-1500 m. Nepal, E Himalaya, Assam-Burma, E Asia and SE Asia.

Exotic/Cultivated Taxa

Maranta bicolor Ker Gawl. (Prayer plant). Native to Africa, N America and S America. [14]

COSTACEAE

Cheilocostus C.D.Specht

Cheilocostus lacerus (Gagnep.) C.D.Specht

Costus lacerus Gagnep. (1, 4, 8)

Perennial herbs. C, E Nepal: 1400-1500 m. Nepal and E Himalaya.

Hellenia Retz.

Hellenia speciosa (J.Köenig) Govaerts

Banksea speciosa J.Köenig; *Cheilocostus speciosus* (J.Köenig) C.D.Specht; *Costus speciosus* (J.Köenig) Sm. (1, 4, 5)

बेत लौरी (Bet lauri)

Perennial herbs. W, C, E Nepal: 400-2700 m. Nepal, E Himalaya, S Asia, E Asia and SE Asia.

Use(s): Medicine (root), Food (root/rhizome).

ZINGIBERACEAE

Alpinia Roxb.

Alpinia malaccensis (Burm.f.) Roscoe [5]

Maranta malaccensis Burm.f.

Perennial herbs. C Nepal: 300-400 m. Nepal, E Himalaya, Tibetan Plateau, Assam-Burma, S Asia, E Asia and SE Asia.

Use(s): Medicine (rhizome), Essential oils (rhizomes, leaves), Ornamental.

Alpinia nigra (Gaertn.) B.L.Burtt

Zingiber nigrum Gaertn.; *Alpinia allughas* (Retz.) Roscoe (1, 4)

कुरंजन, कुलंजन (Kuranjan, Kulanjan)

Perennial herbs. C Nepal: 200-600 m. Nepal, E Himalaya, S Asia, E Asia and SE Asia.

Use: Medicine (rhizome).

Alpinia roxburghii (Roxb.) Sweet

Alpinia bracteata Roxb. (1, 4)

सेतो बोझो (Seto bojho)

Perennial herbs. E Nepal: *ca.* 300 m. Nepal, E Himalaya, Assam-Burma and S Asia.

Amomum Roxb.

Amomum aromaticum Roxb. [1, 4]

Cardamomum aromaticum (Roxb.) Kuntze

अलैंची (Alainchi); Cardamom

Perennial herbs. C Nepal: 600-2000 m. Nepal, E Himalaya and S Asia.

Use(s): Medicine (seeds), Food, Ornamental.

Amomum dealbatum Roxb. [1, 4]

Cardamomum dealbatum (Roxb.) Kuntze

Perennial herbs. E Nepal: *ca.* 500 m. Nepal, E Himalaya, Assam-Burma, S Asia and SE Asia.

Use: Food (seeds).

Amomum subulatum Roxb. [1, 4, 5]

Cardamomum subulatum (Roxb.) Kuntze

अलैंची, ठुलो अलैंची (Alainchi, Thulo alainchi); Greater Cardamom

Perennial herbs. C, E Nepal: 1000-2000 m. Nepal, E Himalaya, Tibetan Plateau, Assam-Burma, S Asia and E Asia.

Use(s): Medicine (seeds), Food (seeds), Ornamental.

Boesenbergia Kuntze

Boesenbergia longiflora (Wall.) Kuntze

Gastrochilus longiflora Wall.; *Curcumorpha longiflora* (Wall.) A.S.Rao & D.M.Verma (4)

Perennial herbs. E Nepal: 1100-1900 m. Nepal, Assam-Burma, S Asia, E Asia and SE Asia.

Cautleya Royle ex Hook.f.

Cautleya gracilis (Sm.) Dandy [1, 4, 5]

Roscoea gracilis Sm.; *Cautleya lutea* (Royle) Hook.f.; *Roscoea lutea* Royle

var. **gracilis**

Perennial herbs. W, C, E Nepal: 1200-3100 m. Nepal, W Himalaya, E Himalaya, Tibetan Plateau, Assam-Burma, S Asia, E Asia and SE Asia.

var. **robusta** (K.Schum.) Sanjappa

Cautleya lutea var. *robusta* K.Schum.; *Cautleya cathcartii* Baker (1, 4)

Perennial herbs. C, E Nepal: 1700-4000 m. Nepal, E Himalaya and E Asia.

Cautleya spicata (Sm.) Baker [1, 4, 5]

Roscoea spicata Sm.

गग्लेटो, पानि सरो (Gagleto, Paani saro)

Perennial herbs. W, C, E Nepal: 1000-2800 m. Nepal, E Himalaya, Tibetan Plateau and E Asia.

Use(s): Medicine (rhizome, flowers), Food.

Curcuma Roxb.

Curcuma angustifolia Roxb. [1, 4, 5]

बर्खे सरो, कालो बेसार (Barkhe sarro, Kaalo besaar)

Perennial rhizomatous herbs. W, C, E Nepal: 100-2300 m. Nepal and E Himalaya.

Use(s): Medicine (rhizome), Food, Ritual.

Curcuma aromatica Salisb. [1, 4, 5]

बन हलेदो (Ban haledo)

Perennial rhizomatous herbs. W, C, E Nepal: 400-2000 m. Nepal, E Himalaya, Tibetan Plateau, Assam-Burma, S Asia and E Asia.

Use(s): Medicine (rhizome), Food (rhizome).

Curcuma longa L.

Curcuma domestica Valeton

बेसार, हलेदो (Besaar, Haledo); Turmeric

Perennial rhizomatous herbs. W, C, E Nepal: 100-2000 m. Nepal, Tibetan Plateau, S Asia, E Asia and SE Asia.

Use(s): Medicine (rhizome), Food.

Globba L.

Globba clarkei Baker [1, 4, 5]

देवी सरो (Devi saro)

Perennial herbs. C, E Nepal: 600-2300 m. Nepal, E Himalaya, S Asia and SE Asia.

Use: Medicine (root).

Globba macroclada Gagnep. [8]

Perennial herbs. C, E Nepal: 800-1100 m. Nepal, W Himalaya, E Himalaya, Assam-Burma, S Asia and SE Asia.

Globba marantina L. [4]

Globba bulbifera Roxb. (5)

Perennial herbs. W, C, E Nepal: *ca.* 2400 m. Nepal and S Asia.

Use: Food (bulbils).

Globba racemosa Sm.

var. **racemosa** [1, 4, 5]

पानि सरो, देवी सरो (Paani saro, Devi saro)

Perennial herbs. C, E Nepal: 100-2500 m. Nepal, E Himalaya, Tibetan Plateau, Assam-Burma, S Asia and SE Asia.

Use: Medicine (root).

var. **hookeri** (C.B.Clarke ex Baker) S.Kumar

Globba hookeri C.B.Clarke ex Baker (1, 4)

Perennial herbs. C Nepal: 1000-2000 m. Nepal, Assam-Burma and SE Asia.

Hedychium J.Köenig

Hedychium coccineum Buch.-Ham. ex Sm. [1, 4, 5]

Hedychium angustifolium Roxb.; *Hedychium aurantiacum* Roscoe (1, 4, 5)

Perennial herbs. C, E Nepal: 100-2000 m. Nepal, E Himalaya, Tibetan Plateau, Assam-Burma, S Asia, E Asia and SE Asia.

Hedychium coronarium J.Köenig [1, 4]

Hedychium coronarium var. *baimao* Z.Y. Zhu

दुध केवरा (Dudh kewaraa); Common Ginger lily

Perennial herbs. C Nepal: 1000-1900 m. Nepal, E Himalaya, Assam-Burma, S Asia, E Asia, SE Asia and Australasia.

Use(s): Medicine (rhizome), Perfumery (flowers), Ornamental.

Hedychium densiflorum Wall. [1, 4, 5]

Gandasulium densiflorum (Wall.) Kuntze

धुंगलहाप्ती (Dhungal haapti-Tamang)

Perennial herbs. W, C, E Nepal: 400-2800 m. Nepal, E Himalaya, Tibetan Plateau and S Asia.

Use(s): Utensils and handicrafts, Ornamental.

Hedychium elatum R.Br. [1, 4]

Perennial herbs. C Nepal: *ca.* 1700 m. Nepal, E Himalaya and Assam-Burma.

Hedychium ellipticum Buch.-Ham. ex Sm. [1, 4, 5]

Hedychium fastigiatum Wall.

रातो सरो (Raato saro)

Perennial herbs. W, C, E Nepal: 300-3500 m. Nepal and E Himalaya.

Use: Medicine (rhizome).

Hedychium flavescens Carey ex Roscoe [1, 4]

Hedychium emeiense Z.Y.Zhu; *Hedychium panzhuum* Z.Y.Zhu

Perennial herbs. E Nepal: 1200-2000 m. Nepal, S Asia and E Asia.

Hedychium gardnerianum Sheppard ex Ker Gawl. [1, 4, 5, 14]

सुन केवरा, पंखा फूल (Sun kewaraa, Pankhaa phool)

Perennial herbs. C, E Nepal: 1400-2100 m. Nepal and E Himalaya.

Use(s): Medicine (rhizome), Ornamental.

Hedychium glaucum Roscoe [1, 4, 5]

Perennial herbs. E Nepal: 1000-2800 m. Nepal and E Himalaya.

Hedychium spicatum Sm. [1, 4, 5]

var. **spicatum**

Perennial rhizomatous herbs. W, C Nepal: 1900-2100 m. Nepal, Tibetan Plateau, Assam-Burma, E Asia and SE Asia.

Use(s): Medicine (rhizome), Cosmetics (essential oil).

var. **acuminatum** (Roscoe) Wall. [1, 4]

Hedychium acuminatum Roscoe

सेतो सरो, पंखा फूल (Seto saro, Pankhaa phul)

Perennial rhizomatous herbs. W, C, E Nepal: 1300-2800 m. Nepal, E Himalaya, Tibetan Plateau, Assam-Burma, E Asia and SE Asia.

Use: Medicine (rhizome).

var. **trilobum** (Wall. ex Roscoe) Wall. [1, 4]

Hedychium trilobum Wall. ex Roscoe

Perennial rhizomatous herbs. C Nepal. Endemic to Nepal (?)

Hedychium thyrsiforme Sm. [1, 4]

Gandasulium thyrsiforme (Sm.) Kuntze; *Hedychium tocucho* Buch.-Ham. ex Wall.

Perennial herbs. C, E Nepal: 1200-1600 m. Nepal and E Himalaya.

Hedychium villosum Wall. [1, 4]

Perennial herbs. W, C, E Nepal: 100-3400 m. Nepal, Assam-Burma, S Asia, E Asia and SE Asia.

Hemiorchis Kurz

Hemiorchis pantlingii King [1, 4]

Perennial herbs. E Nepal: 200-1100 m. Nepal, E Himalaya, Assam-Burma and S Asia.

Kaempferia L.

Kaempferia rotunda L. [1, 4, 5]

भुइँ चम्पा (Bhuin champaa)

Perennial herbs. C, E Nepal: 500-2100 m. Nepal, Assam-Burma, S Asia, E Asia and SE Asia.

Use: Medicine (whole plant).

Roscoea Sm.

Roscoea alpina Royle [1, 4, 5]

Roscoea intermedia Gagnep.

नक्कली पाँच औंले, भुइँ सरो (Nakkali paanchaunle, Bhuin saro)

Perennial herbs. W, C, E Nepal: 2400-4000 m. Nepal, W Himalaya, E Himalaya, Tibetan Plateau, Assam-Burma and S Asia.

Use(s): Medicine (root/tuber), Poisonous (tuber).

Roscoea auriculata K.Schum. [5]

Roscoea purpurea var. *auriculata* (K.Schum.) H.Hara (1, 4)

Perennial herbs. C, E Nepal: 2400-3000 m. Nepal, E Himalaya and Tibetan Plateau.

Roscoea capitata Sm. [1, 4, 5]

Perennial herbs. C Nepal: 1000-2600 m. Nepal and Tibetan Plateau.

Roscoea cautleyoides Gagnep.

Roscoea chamaeleon Gagnep. (1, 4)

Perennial herbs. C Nepal: 2700-3000 m. Nepal and E Asia.

Roscoea ganeshensis Cowley & W.J.Baker, *Bot. Mag.* **13**: 10. 1996. [4, 28]

Perennial rhizomatous herbs. C Nepal: *ca.* 1900 m. **Endemic to Nepal**.

Central: *Gorkha/Dhading,* Buri Gandaki valley, near Abuthum Lekh, *ca.* 1900 m, B. Baker, T. Burkitt, J. Miller & R. Shrestha 34 (Holotype: K).

Roscoea nepalensis Cowley, *Kew Bull.* **34**: 811. 1980. [4, 28]

Perennial tuberous herbs. W, C Nepal: 2400-3100 m. **Endemic to Nepal**.

West: *Jumla*, valley of N Jumla, 8000-9000 ft. (2438-2743 m), O. Polunin, WR Sykes & LHJ Williams 4391 (Holotype: BM).

Roscoea purpurea Sm. [1, 4, 5]

Roscoea purpurea var. *gigantea* Wall. (1, 4, 8)

भुइँ सरो, काकोली (Bhui saro, Kaakoli)

Perennial herbs. W, C, E Nepal: 1500-2900 m. Nepal and E Himalaya.

Use(s): Medicine (tuber), Food (tuber), Forage.

Roscoea tumjensis Cowley, *Kew Bull.* **36**: 755. 1982. [4, 28]

Perennial rhizomatous herbs. C, E Nepal: 2700-3100 m. **Endemic to Nepal**.

Central: *Gorkha*, Shiar Khola, above Tumje, *ca.* 2700 m, PC Gardner 790 (Holotype: BM).

Zingiber Mill.

Zingiber capitatum Roxb. [1, 4]

Dymczewiczia capitata (Roxb.) Horan.

Perennial herbs. C Nepal: 100-200 m. Nepal and E Himalaya.

Zingiber chrysanthum Roscoe [1, 4, 5]

Golden flowered zinger

Perennial herbs. C, E Nepal: 100-1600 m. Nepal and E Himalaya.

Use(s): Medicine (rhizome), Food.

Zingiber officinale Roscoe [5]

अदुवा, सुठो (Aduwaa, Sutho); Ginger

Perennial herbs. Widely cultivated in Nepal. Native to Asia.

Use(s): Medicine (rhizome), Food (rhizome), Ritual.

Zingiber zerumbet (L.) Roscoe ex Sm. [1, 4, 8]

Amomum zerumbet L.

Perennial herbs. C Nepal: 100-200 m. Nepal, Assam-Burma, S Asia, E Asia and SE Asia.

Use: Medicine (rhizome).

Exotic/Cultivated Taxa

Curcuma amada Roxb. (**आमा हल्दी,** Aama haldi; Mango ginger). Native to E Himalaya. *Use(s)*: Medicine (rhizome), Volatile oil, Food (rhizome).

Curcuma phaeocaulis Valeton (*Curcuma zedoaria* (Christm.) Rosc.; **कच्हुर, बन हलेदो,** Kachahur, Ban haledo); Zedoary, Wild turmeric). *Use(s)*: Medicine (rhizome), Volatile oil.

Elettaria cardamomum (L.) Maton (**सुकुमेल,** Sukumel; Green cardamom). Native to S Asia. [15]

STRELITZIACEAE

Strelitzia nicolai Regel & Korn. (Giant white Bird of Paradise). Native to Africa. [14]

Strelitzia reginae Banks (Bird of Paradise). Native to Africa. [14, 54]

POALES

TYPHACEAE

Typha L.

Typha angustifolia L. [1, 4, 5]

Typha minor Curtis

पटेर, खर (Pater, Khar)

Perennial herbs. W, C, E Nepal: 100-300 m. Nepal, Asia, Europe and N America.

Use(s): Medicine (rhizome, spike, pollen, fruit), Forage, Construction materials.

Typha elephantina Roxb. [1, 4, 5]

खर, पटेर (Khar, Pater)

Perennial herbs. W, C, E Nepal: 200-300 m. Nepal, E Himalaya, S Asia, SW Asia and Africa.

Use(s): Medicine (root, fruit), Forage, Construction materials.

BROMELIACEAE

Ananas Mill.

Ananas comosus (L.) Merr. [4, 5]

Bromelia comosa L.

भुइँ कटहर (Bhui katahar); Pineapple

Perennial herbs. C Nepal: 500-1500 m. Widely cultivated for edible fruit. Native to S America.

Use(s): Food (fruit), Medicine (leaves, fruit)

XYRIDACEAE

Xyris L.

Xyris capensis Thunb.

var. **schoenoides** (Mart.) Nilsson [1, 4]

Xyris schoenoides Mart.

Perennial herbs. C, E Nepal: 1800-2700 m. Nepal, Tibetan Plateau, Assam-Burma, S Asia, SE Asia, Africa and America.

Xyris pauciflora Willd. [1, 4, 5]

Xyris pauciflora var. *oryzetorum* Miq.

Annual herbs. C, E Nepal: 200-1900 m. Nepal, E Himalaya, Tibetan Plateau, S Asia, E Asia, SE Asia and Australasia.

Use(s): Medicine (whole plant), Food (bulb).

ERIOCAULACEAE

Eriocaulon L.

Eriocaulon alpestre Hook.f. & Thomson ex Körn. [1, 4]

Annual herbs. C Nepal: *ca.* 3500 m. Nepal, E Himalaya, Tibetan Plateau, S Asia, E Asia and SE Asia.

Eriocaulon cinereum R.Br. [1, 4, 5]

Eriocaulon formosanum Hayata; *Eriocaulon heteranthum* Benth.

बुच्चे झार (Bucche jhaar); Pipewort

Annual wetland herbs. W, C, E Nepal: 1700-2400 m. Nepal, E Himalaya, Tibetan Plateau, Assam-Burma, S Asia, E Asia, SE Asia, SW Asia, Africa and Australasia.

Eriocaulon exsertum Satake, *Bull. Univ. Mus. Univ. Tokyo* **2**: 156, f. 9. 1971. [1, 4, 26]

Annual stemless herbs. E Nepal: 200-300 m. **Endemic to Nepal.**

East: *Jhapa*, Ghorwa – Sanischare, 200-300 m, H. Hara, H. Kanai, S. Kurosawa, J. Murata & M. Togashi 6305472 (Holotype: TI).

Eriocaulon hamiltonianum Mart. [8]

Annual herbs. W Nepal: 1100-1200 m. Nepal, Assam-Burma, S Asia and SW Asia.

Eriocaulon kathmanduense Satake, *Bull. Univ. Mus. Univ. Tokyo* **2**: 157. 1971. [1, 4, 26]

Annual stemless herbs. C Nepal: *ca.* 1600 m. **Endemic to Nepal.**

Central: *Kathmandu*, Sundarijal waterfall, *ca.* 1600 m, H. Hara, H. Kanai & S. Kurosawa 6305471 (Holotype: TI).

Eriocaulon luzulifolium Mart. [1, 4, 5]

Eriocaulon nepalense var. *luzulifolium* (Mart.) Praj. & J.Parn.

Short pipewort

Annual herbs. C Nepal: 300-1700 m. Nepal, S Asia, E Asia and SE Asia.

Eriocaulon nepalense Prescott ex Bong. [1, 4, 5]

Eriocaulon viride Koern. (1, 4, 5)

भूरी घाँस, टाउके फूल (Bhuri ghaans, Taauke phool); Pipe rush

Annual herbs. W, C, E Nepal: 1500-3000 m. Nepal and E Asia.

Use: Medicine (whole plant).

Eriocaulon obclavatum Satake, *Bull. Univ. Mus. Univ. Tokyo* **2**: 158. 1971. [1, 4, 26]

Annual stemless herbs. E Nepal: 200-300 m. **Endemic to Nepal.**

East: *Jhapa*, Ghorwa – Sanichare, 200-300 m, H. Hara, H. Kanai, S. Kurosawa, J. Murata & M. Togashi 6305469 (Holotype: TI).

Eriocaulon oryzetorum Mart. [1, 4]

Pipewort

Annual marshy herbs. W, C Nepal: 1000-2000 m. Nepal, S Asia, E Asia and SE Asia.

Eriocaulon quinquangulare L. [1, 4]

Annual marshy herbs. C Nepal: 100-1500 m. Nepal, Assam-Burma and S Asia.

Eriocaulon staintonii Satake, *J. Jap. Bot.* **48**: 316. 1973. [1, 4, 5, 26]

Annual stemless herbs. C, E Nepal: 700-1900 m. **Endemic to Nepal**.

East: *Sankhuwasabha*, Haitsar, N of Num, Arun valley, 6000 ft. (1828 m), JDA Stainton 1386 (Holotype: BM).

Eriocaulon trisectoides Satake, *Bull. Univ. Mus. Univ. Tokyo* **2**: 159. 1971. [1, 4, 5, 26]

Annual stemless herbs. C, E Nepal: 700-2400 m. **Endemic to Nepal**.

East: *Sankhuwasabha/Taplejung*, Mul Pokhari - Nissim - Dumhan, 700-2100 m, H. Hara, H. Kanai, S. Kurosawa, J. Murata, M. Togashi & T. Tuyama 6305468 (Holotype: TI).

Eriocaulon xeranthemum Mart. [1, 4]

Annual wetland herbs. C Nepal: 100-700 m. Nepal, S Asia and Africa.

Doubtful/Uncertain taxa

Eriocaulon sollyanum Royle [104]

JUNCACEAE

Juncus L.

Juncus allioides Franch. [1, 4, 5]

Perennial herbs. W, C, E Nepal: 2000-5000 m. Nepal, E Himalaya, Tibetan Plateau and E Asia.

Juncus amplifolius A.Camus [1, 4, 5]

Perennial herbs. E Nepal: 3300-4000 m. Nepal, E Himalaya, Tibetan Plateau, Assam-Burma and E Asia.

Juncus articulatus L. [1, 4, 5]

Juncus lampocarpus Ehrh. ex Hoffm.

Perennial herbs. W, C, E Nepal: 1200-3900 m. Nepal, E Himalaya, Tibetan Plateau, S Asia, E Asia, SE Asia, N Asia, SW Asia, Europe, Africa and N America.

Juncus benghalensis Kunth [1, 4, 5]

Juncus sphenostemon Buchenau

Perennial herbs. W, C, E Nepal: 2200-4700 m. Nepal, W Himalaya, E Himalaya, Tibetan Plateau and E Asia.

Juncus biglumoides H.Hara [1, 4, 8]

Perennial herbs. C Nepal: 3500-4200 m. Nepal and E Himalaya.

Juncus brachystigma Sam. [1, 4, 5]

Perennial herbs. W, C, E Nepal: 2300-4600 m. Nepal, E Himalaya, Tibetan Plateau and E Asia.

Juncus bryophilus W.W.Sm.

Perennial herbs. C Nepal: 3600-4500 m. Nepal and E Himalaya.

Juncus bufonius L. [1, 4, 5]

Annual or perennial wetland herbs. W, C, E Nepal: 1100-4100 m. Nepal, E Himalaya, Tibetan Plateau, S Asia, E Asia, N Asia, C Asia, SW Asia, Europe, N America and S America.

Juncus cephalostigma Sam. [1, 4, 5]

Perennial herbs. C, E Nepal: 3300-4600 m. Nepal, E Himalaya, Tibetan Plateau, Assam-Burma, S Asia and E Asia.

Juncus chrysocarpus Buchenau [1, 4, 5]

Perennial herbs. W, C, E Nepal: 2400-4500 m. Nepal, E Himalaya and Tibetan Plateau.

Juncus clarkei Buchenau [5]

Perennial herbs. W, C, E Nepal: 1800-4700 m. Nepal, E Himalaya, Tibetan Plateau, S Asia and E Asia.

Juncus concinnus D.Don [1, 4, 5]

Juncus albescens Satake (1, 4); *Juncus elegans* Royle ex D.Don; *Juncus luteocarpus* Satake (1, 4)

Perennial herbs. W, C, E Nepal: 1600-5200 m. Nepal, W Himalaya, E Himalaya, Tibetan Plateau, Assam-Burma, S Asia and E Asia.

Juncus duthiei (C.B.Clarke) Noltie [4, 5]

Microschoenus duthiei C.B.Clarke

Perennial herbs. C, E Nepal: 3000-4900 m. Nepal, E Himalaya and Tibetan Plateau.

Juncus effusus L. [1, 4, 5]

ज्वाने (Jwaane)

Perennial herbs. W, C, E Nepal: 1300-3000 m. Nepal, E Himalaya, Tibetan Plateau, E Asia, S Asia, SE Asia, Europe, Africa, America and Australasia.

Use(s): Medicine (root, stem), Construction materials (stem).

Juncus ganeshii Miyam. & H.Ohba, *J. Jap. Bot.* **70**: 245. 1995. [4, 5, 27]

Perennial rhizomatous herbs. C, E Nepal: 4400-4500 m. **Endemic to Nepal**.

Central: *Rasuwa*, Jaisuli Kund – Paldol Base Camp, *ca.* 4500 m, F. Miyamoto, KR Rajbhandari, S. Akiyama, M. Amano, H. Ikeda & H. Tsukaya 9410148 (Holotype: TI).

Juncus glaucoturgidus Noltie [4, 5]

Perennial herbs. C, E Nepal: 3700-4300 m. Nepal and E Himalaya.

Juncus gracilicaulis A.Camus [5]

Juncus concinnus var. *gracilicaulis* (A.Camus) R.C.Srivastava

Perennial herbs. W, C, E Nepal: 2500-3700 m. Nepal, E Himalaya and E Asia.

Juncus grisebachii Buchenau [1, 4, 5]

Juncus phaeocarpus A.M.Lu & Z.Y.Zhang

Perennial herbs. C, E Nepal: 2300-5200 m. Nepal, E Himalaya, Tibetan Plateau and S Asia.

Juncus harae Miyam. & H.Ohba [8]

Perennial herbs. C, E Nepal: 3500-3900 m. Nepal, E Himalaya and E Asia.

Juncus himalensis Klotzsch [1, 4, 5]

Juncus himalensis var. *schlagintweitii* (Buchenau) Buchenau

Perennial herbs. W, C, E Nepal: 1700-5200 m. Nepal, W Himalaya, E Himalaya, Tibetan Plateau, S Asia, E Asia, C Asia and SW Asia.

Juncus hydrophilus Noltie [4]

Perennial herbs. C, E Nepal: 2300-3600 m. Nepal and E Himalaya.

Juncus inflexus L. [1, 4, 5]

Juncus brachytepalus V.I.Krecz. & Gontsch

Perennial herbs. W, C, E Nepal: 1400-3500 m. Nepal, E Himalaya, Tibetan Plateau, S Asia, E Asia, SE Asia, N Asia, SW Asia, Europe, Africa and N America.

Juncus khasiensis Buchenau [5]

Perennial herbs. W, C, E Nepal: 1600-3700 m. Nepal, E Himalaya and Assam-Burma.

Juncus kingii Rendle [1, 4, 5]

Juncus longibracteatus A.M.Lu & Z.Y.Zhang

Perennial herbs. W, C, E Nepal: 3200-5000 m. Nepal, E Himalaya, Tibetan Plateau and E Asia.

Juncus leptospermus Buchenau [8]

Juncus mairei H.Lév.

Perennial caespitose herbs. C, E Nepal: 100-3600 m. Nepal, E Himalaya, S Asia and E Asia.

Juncus leucanthus Royle ex D.Don [1, 4, 5]

Perennial herbs. W, C, E Nepal: 2900-4900 m. Nepal, E Himalaya, Tibetan Plateau and E Asia.

Juncus leucomelas Royle ex D.Don [1, 4]

Juncus bhutanensis Satake (1, 4)

Perennial herbs. W, C, E Nepal: 2900-5600 m. Nepal, W Himalaya, E Himalaya, Tibetan Plateau, S Asia, E Asia and SW Asia.

Juncus membranaceus Royle [1, 4]

Juncus hoffmeisteri Klotzsch

Perennial herbs. W, C Nepal: 2700-3700 m. Nepal, W Himalaya, Tibetan Plateau, S Asia, E Asia and SW Asia.

Juncus minimus Buchenau [1, 4]

Juncus unifolius A.M.Lu & Z.Y.Zhang

Perennial herbs. C, E Nepal: 3900-5500 m. Nepal, E Himalaya, Tibetan Plateau and E Asia.

Juncus mustangensis Miyam. & H.Ohba, *J. Jap. Bot.* **78**(3): 154. 2003. [27, 117]

Perennial rhizomatous herbs. C Nepal: 3800-3900 m. **Endemic to Nepal**.

Central: *Mustang*, Lower Mustang, between Sangda La and Phalyak, 3870 m, F. Miyamoto *et al* 20210100 (Holotype: TI; Isotype: KATH).

Juncus nepalicus Miyam. & H.Ohba [4, 8]

Juncus amplifolius var. *pumilus* A.Camus

Perennial herbs. E Nepal: 3900-4400 m. Nepal, E Himalaya, Tibetan Plateau and E Asia.

Juncus ochraceus Buchenau [1, 4, 5]

Juncus tratangensis Satake

Perennial herbs. C, E Nepal: 1200-3700 m. Nepal, E Himalaya, Tibetan Plateau, S Asia and E Asia.

Juncus perpusillus Sam. [4, 5]

Perennial herbs. C, E Nepal: 3500-4600 m. Nepal, E Himalaya, Tibetan Plateau and E Asia.

Juncus prismatocarpus R.Br.

Juncus leschenaultii Gay ex Laharpe (1, 4)

Perennial or annual marshy herbs. W, C, E Nepal: 100-2100 m. Nepal, E Himalaya, Tibetan Plateau, S Asia, E Asia, SE Asia, SW Asia and Australasia.

Juncus rohtangensis Goel & Aswal [5]

Juncus sikkimensis var. *monocephalus* Hook.f. (1)

Perennial herbs. W, C, E Nepal: 3200-5000 m. Nepal, E Himalaya and Tibetan Plateau.

Juncus sherei Miyam. & H.Ohba [4, 8]

Perennial herbs. C, E Nepal: 3100-4300 m. Nepal and E Himalaya.

Juncus sikkimensis Hook.f. [1, 4, 5]

Juncus pseudocastaneus (Lingelsh.) Sam. (1); *Juncus sikkimensis* var. *pseudocastaneus* Lingelsh.

Perennial herbs. W, C, E Nepal: 3900-4600 m. Nepal, E Himalaya, Tibetan Plateau, S Asia and E Asia.

Juncus sphacelatus Decne. [1, 4, 5]

Perennial herbs. W, C, E Nepal: 3600-5200 m. Nepal, W Himalaya, E Himalaya, Tibetan Plateau, E Asia, S Asia and SW Asia.

Juncus thomsonii Buchenau [1, 4, 5]

Juncus thomsonii var. *fulvis* K.Y.Wu

Perennial herbs. W, C, E Nepal: 2700-5200 m. Nepal, E Himalaya, Tibetan Plateau, S Asia, E Asia, C Asia and SW Asia.

Juncus trachyphyllus Miyam. & H.Ohba [5]

Perennial herbs. C Nepal: *ca.* 4300 m. Nepal and E Asia.

Juncus trichophyllus W.W.Sm. [4]

Perennial herbs. C, E Nepal: 3400-4200 m. Nepal and E Himalaya.

Juncus triglumis L. [1, 4, 5]

Perennial herbs. W, C, E Nepal: 2800-4900 m. Nepal, W Himalaya, E Himalaya, Tibetan Plateau, S Asia, E Asia, N Asia and Europe.

Juncus uniflorus W.W.Sm. [1, 4, 5]

Perennial herbs. C, E Nepal: 3500-4700 m. Nepal, E Himalaya and Tibetan Plateau.

Juncus wallichianus J.Gay ex Laharpe [1, 4, 5]

Juncus prismatocarpus subvar. *unitubulosus* Buchenau; *Juncus pseudokrameri* Satake

ज्वाने (Jwaane)

Perennial herbs. W, C, E Nepal: 200-3100 m. Nepal, E Himalaya, S Asia, E Asia and N Asia.

Luzula DC.

Luzula effusa Buchenau [1, 4, 5]

Perennial herbs. E Nepal: 2200-3400 m. Nepal, E Himalaya, Tibetan Plateau, Assam-Burma, E Asia and SE Asia.

Luzula multiflora (Ehrh.) Lej. [1, 4, 5]

Juncus multiflorus Ehrh.

Perennial herbs. W, C, E Nepal: 3000-4100 m. Nepal, E Himalaya, Tibetan Plateau, S Asia, E Asia, N Asia and Australasia.

Use: Medicine (rhizome).

Luzula oligantha Sam. [1, 4, 5]

Perennial tufted herbs. C, E Nepal: 1900-4000 m. Nepal, E Himalaya, Tibetan Plateau, E Asia and N Asia.

Luzula plumosa E.Mey. [1, 4, 5]

Luzula formosana Ohwi; *Luzula pilosa* var. *plumosa* (Wall. ex E.Mey.) Franch.

Perennial tufted herbs. C, E Nepal: 2400-3400 m. Nepal, E Himalaya, Tibetan Plateau, S Asia, E Asia and SE Asia.

CYPERACEAE

Actinoscirpus (Ohwi) R.W.Haines & Lye

Actinoscirpus grossus (L.f.) Goetgh. & D.A.Simpson

Scirpus grossus L.f.; *Schoenoplectus grossus* (L.f.) Palla (1, 4)

var. **grossus**

मोथे, कसुर (Mothe, Kasur); Giant bur-rush

Perennial stoloniferous herbs. C, E Nepal: 100-900 m. Nepal, Assam-Burma, S Asia, E Asia, SE Asia, SW Asia and Australasia.

Use(s): Medicine, Food (root/tuber), Construction materials (culm).

var. **kysoor** (Roxb.) Noltie [79]

Scirpus kysoor Roxb.

कसुर (Kasur); Bull rush

Perennial stoloniferous herbs. W, C, E Nepal: 100-900 m. Nepal, E Himalaya, Assam-Burma and S Asia.

Use(s): Medicine (tuber), Fodder, Construction materials (culm).

Blysmus Panz. ex Schult.

Blysmus compressus (L.) Panz. ex Link [1, 4]

Schoenus compressus L.; *Scirpus compressus* (L.) Pers.

Flat sedge

Perennial herbs. W, C, E Nepal: 2600-4100 m. Nepal, Tibetan Plateau, S Asia, E Asia, SW Asia and Europe.

Bulbostylis Kunth

Bulbostylis barbata (Rottb.) C.B.Clarke [1, 4, 5]

Scirpus barbatus Rottb.

झुसे झार (Jhuse jhaar); Water grass

Annual cespitose herbs. W, C, E Nepal: 100-1300 m. Nepal, Asia, Africa and Australasia.

Use: Medicine (whole plant).

Bulbostylis densa (Wall.) Hand.-Mazz. [1, 4, 5]

Scirpus densus Wall.

झुसे झार (Jhuse jhaar)

Annual wetland herbs. W, C, E Nepal: 1300-3500 m. Nepal, Asia, Africa and Australasia.

Use: Forage.

Carex L.

Carex alopecuroides D.Don ex Tilloch & Taylor [1, 4, 5]

Perennial herbs. C, E Nepal: 1200-3200 m. Nepal, E Himalaya, S Asia, E Asia and SE Asia.

Carex anomoea Hand.-Mazz. [5]

Carex inaequalis Boott ex C.B.Clarke, nom. Illeg. (1, 4)

Perennial herbs. E Nepal: 2500-3400 m. Nepal, E Himalaya and E Asia.

Carex atrata L.

subsp. **atrata** [1, 4]

Black alpine sedge

Perennial herbs. E Nepal: 3400-5000 m. Nepal, E Himalaya, S Asia, E Asia, N Asia and Europe.

subsp. **pullata** (Boott) Kük. [1, 4]

Carex atrata var. *pullata* Boott; *Carex duthiei* C.B.Clarke (5)

Perennial herbs. W, C, E Nepal: 3400-4400 m. Nepal, E Himalaya, Tibetan Plateau and E Asia.

Carex atrofusca Schkuhr

subsp. **atrofusca**

Perennial herbs. W Nepal: 4600-4800 m. Nepal, Asia, Europe and N America.

Note: Dolpa, Kagmara Base Camp, 4650 m, M. Minaki *et al* 9106110 (TI)

subsp. **minor** (Boott) T.Koyama [1, 4, 5]

Carex ustulata var. *minor* Boott; *Carex atrofusca* var. *minor* (Boott) Kük.

Perennial herbs. W, C, E Nepal: 3600-5600 m. Nepal, E Himalaya, Tibetan Plateau, S Asia, N Asia and SW Asia.

Carex baccans Nees [1, 4, 5]

Perennial herbs. C, E Nepal: 1300-2000 m. Nepal, S Asia, E Asia and SE Asia.

Carex bhutanensis S.R.Zhang [51]

Kobresia prainii Kük (5); *Kobresia prainii* Kük. var. *elliptica* Y.C.Yang; *Kobresia utriculata* C.B.Clarke

Perennial herbs. E Nepal: *ca.* 4100 m. Nepal, E Himalaya and Tibetan Plateau.

Carex bigelowii Torr. ex Schwein. [1]

Carex rigida Gooden.

Perennial herbs. C Nepal: *ca.* 3800 m. Nepal, Arctic and alpine N Hemisphere.

Carex bonatiana (Kük.) N.A.Ivanova [51]

Kobresia bonatiana Kük.; *Kobresia fragilis* C.B.Clarke (1, 4, 5); *Kobresia clarkeana* (Kük.) Kük.; *Kobresia curvata* C.B.Clarke (5)

Perennial herbs. W, C, E Nepal: 2300-4200 m. Nepal, E Himalaya, Tibetan Plateau and E Asia.

Carex breviculmis R.Br. [1, 4]

Carex royleana Nees

Perennial herbs. C Nepal: *ca.* 3200 m. Nepal, Assam-Burma, S Asia, E Asia and N Asia.

Carex cardiolepis Nees [5]

Perennial herbs. C Nepal: 3700-3800 m. Nepal, W Himalaya, Tibetan Plateau, S Asia, E Asia and SW Asia.

Carex composita Boott [5]

Perennial herbs. E Nepal: 2100-2200 m. Nepal, E Himalaya, S Asia and E Asia.

Carex condensata Nees [5]

Perennial herbs. W, C, E Nepal: 900-1800 m. Nepal, E Himalaya, S Asia and SE Asia.

Carex continua C.B.Clarke [8]

Carex rafflesiana var. *continua* (C.B.Clarke) Kük.

Perennial herbs. C Nepal: 800-1900 m. Nepal, Assam-Burma, S Asia, E Asia and SE Asia.

Carex crebra V.I.Krecz. [5]

Perennial herbs. C Nepal: *ca.* 4000 m. Nepal, Tibetan Plateau and E Asia.

Carex cruciata Wahlenb. [1, 4]

Carex bengalensis Roxb.

लामो हात कटुवा, तिन सिका घाँस (Laamo haat katuwaa, Tinsikaa ghaans)

Perennial herbs. W, C, E Nepal: 800-3400 m. Nepal, E Himalaya, Tibetan Plateau, S Asia, E Asia, SE Asia and Africa.

Use: Forage.

Carex cruenta Nees [1, 4, 5]

Perennial herbs. W, C, E Nepal: 3700-5600 m. Nepal, W Himalaya, Tibetan Plateau, S Asia, E Asia and SW Asia.

Carex daltonii Boott [1, 4, 5]

Perennial herbs. C Nepal: 1300-2200 m. Nepal and E Himalaya.

Carex decora Boott [1, 4, 5]

Perennial herbs. C, E Nepal: 2400-4000 m. Nepal and E Himalaya.

Carex delicata C.B.Clarke [8]

Perennial herbs. C Nepal: 3500-3600 m. Nepal, S Asia, N Asia, C Asia and SW Asia.

Carex diluta M.Bieb. [1, 4, 5]

Perennial herbs. C Nepal: 2700-3200 m. Nepal, S Asia, N Asia, C Asia and SW Asia.

Carex dimorpholepis Steud. [1, 4]

Carex cernua Boott

Perennial herbs. W, C Nepal: 800-2300 m. Nepal, Assam-Burma, S Asia, E Asia and SE Asia.

Carex dispalata Boott

subsp. **laxiflorens** T.Koyama [1, 4]

Perennial herbs. E Nepal: *ca.* 2100 m. Nepal, Tibetan Plateau and E Asia.

Carex doniana Spreng.

Carex alopecuroides var. *chlorostachys* C.B.Clarke (4); *Carex japonica* subsp. *chlorostachys* (C.B.Clarke) T.Koyama

Perennial rhizomatous herbs. C, E Nepal: 1200-3200 m. Nepal, E Asia and SE Asia.

Carex eleusinoides Turcz. ex Kunth [5]

Perennial herbs. E Nepal: 3500-4000 m. Nepal, E Asia, N Asia and N America.

Carex esbirajbhandarii (Rajbh. & H.Ohba) O.Yano, *J. Linn. Soc.* **179**: 21. 2015. [51]

Kobresia esbirajbhandarii Rajbh. & H.Ohba, *J. Jap. Bot.* **62**: 272 (1987). (4, 5, 26)

Perennial rhizomatous herbs. C, E Nepal: 3900-4700 m. **Endemic to Nepal**.

East: *Solukhumbu*, Dudh Khola – Dukhkunda glacier, *ca.* 4600 m, KR Rajbhandari 61776 (Holotype: TI).

Carex fastigiata Franch. [1, 4]

Perennial herbs. C Nepal: *ca.* 3700 m. Nepal, E Himalaya and E Asia.

Carex fedia Nees [5]

Carex wallichiana Nees

Perennial herbs. C Nepal: 200-300 m. Nepal, Assam-Burma, S Asia, E Asia, SE Asia and SW Asia.

Carex filicina Nees [1, 4, 5]

Carex pseudofilicina Hayata; *Cyperus caricinus* D.Don

सिन्दुरपांग (Sindurpaang-Tamang)

Perennial herbs. W, C, E Nepal: 1200-4000 m. Nepal, Tibetan Plateau, Assam-Burma, S Asia, E Asia, SE Asia and SW Asia.

Use: Forage.

Carex finitima Boott [1, 8]

Carex fusiformis subsp. *finitima* (Boott) Noltie (4)

Perennial herbs. C, E Nepal: 2700-3900 m. Nepal, E Himalaya, E Asia and SE Asia.

Carex fissiglumis (C.B.Clarke) S.R.Zhang & O.Yano [51]

Kobresia fissiglumis C.B.Clarke (4); *Kobresia esenbeckii* var. *fissiglumis* (C.B.Clarke) Noltie

Perennial herbs. W, C Nepal: 3600-5600 m. Nepal, E Himalaya and Tibetan Plateau.

Carex fucata Boott ex C.B.Clarke [1, 4]

Carex sikkimensis C.B.Clarke

Perennial herbs. W, C, E Nepal: 2400-3500 m. Nepal and E Himalaya.

Carex fusiformis Nees [5]

Perennial herbs. C Nepal: 2800-2900 m. Nepal, E Himalaya, Tibetan Plateau and E Asia.

Carex gammiei (C.B.Clarke) S.R.Zhang & O.Yano [51]

Kobresia gammiei C.B.Clarke (4, 5); *Kobresia williamsii* T.Koyama

Perennial herbs. W, C, E Nepal: 3600-4000 m. Nepal, E Himalaya and Tibetan Plateau.

Carex gandakiensis Katsuy., *Fl. Mustang*: 444 (2008). [11, 26]

Perennial stoloniferous herbs. W, C, E Nepal: 3300-3500 m. **Endemic to Nepal**.

Central: *Mustang*, Chim Gaon, N of Tukucha, Kali Gandaki, 11000 ft. (3350 m), JDA Stainton *et al* 921 (Holotype: BM; Isotypes: E, TI).

Carex gracilenta Boott ex Boeckler [5]

Carex infuscata var. *gracilienta* (Boott ex Boeckler) P.C.Li

Perennial herbs. W, C, E Nepal: 3400-4300 m. Nepal, E Himalaya, Tibetan Plateau, S Asia and E Asia.

Carex griersonii Noltie [1, 4, 5]

Perennial herbs. W, C Nepal: 2200-3400 m. Nepal and E Himalaya.

Carex haematostoma Nees [1, 4, 5]

Carex bhutanica T.Koyama

Perennial herbs. W, C, E Nepal: 3200-4400 m. Nepal, E Himalaya,Tibetan Plateau and E Asia.

Carex handel-mazzettii (N.A.Ivanova) S.R.Zhang [51]

Kobresia handel-mazzettii N.A.Ivanova; *Kobresia royleana* var. *himalaica* Rajbh. & H.Ohba (4, 8).

Perennial herbs. W, C, E Nepal: 3500-4600 m. Nepal, Tibetan Plateau and E Asia.

Carex harae (Rajbh. & H.Ohba) O.Yano [51]

Kobresia harae Rajbh. & H.Ohba (4, 5, 26)

Perennial rhizomatous herbs. C, E Nepal: 3300-4000 m. Nepal and E Himalaya.

Carex hebecarpa C.A.Mey. [1, 4]

Perennial herbs. C, E Nepal: 1300-3500 m. Nepal, E Himalaya and S Asia.

Carex hemineuros T.Koyama [1, 4, 5]

Perennial herbs. W, C Nepal: 3000-3200 m. Nepal and Tibetan Plateau.

Carex henryi (C.B.Clarke) L.K.Dai [57]

Carex longicruris var. *henryi* C.B.Clarke

Perennial herbs. C Nepal: 2200-2300 m. Nepal and E Asia.

Carex himalaica T.Koyama, *Bot. Mag. Tokyo* **86**: 278. 1973. [1, 4, 5, 26]

Perennial rhizomatous herbs. E Nepal: 2800-4200 m. **Endemic to Nepal**.

East: *Sankhuwasabha*, Kasuwa Khola, N of Num, Arun Valley, 12000 ft. (3657 m), JDA Stainton 522 (Holotype: BM).

Carex inanis Kunth [1, 4, 5]

Perennial herbs. W, C, E Nepal: 1400-4100 m. Nepal, W Himalaya, E Himalaya, Tibetan Plateau and E Asia.

Carex inclinis Boott ex C.B.Clarke [1, 4, 5]

Perennial herbs. C, E Nepal: 2700-3600 m. Nepal and E Himalaya.

Carex infuscata Nees [1, 4, 5]

Perennial herbs. C, E Nepal: 2600-4300 m. Nepal, W Himalaya, E Himalaya, S Asia and SW Asia.

Carex insignis Boott [1, 4]

Perennial herbs. E Nepal: 1600-1900 m. Nepal, E Himalaya, Tibetan Plateau, S Asia and SE Asia.

Carex jackiana Boott [5]

Perennial herbs. E Nepal: 2800-2900 m. Nepal, Assam-Burma, SE Asia and Australasia.

Carex kanaii (Rajbh. & H.Ohba) S.R.Zhang & O.Yano [51]

Kobresia kanaii Rajbh. & H.Ohba (4, 5, 26)

Perennial rhizomatous herbs. C, E Nepal: 4100-4800 m. Nepal, Tibetan Plateau and S Asia.

Carex kumaonensis Kük. [1, 4, 8]

Perennial herbs. W, C Nepal: 1800-3700 m. Nepal and E Himalaya.

Carex laeta Boott [1, 4, 5]

Carex pisanensis T.Koyama

Perennial herbs. W, C, E Nepal: 2200-5000 m. Nepal, E Himalaya and E Asia.

Carex lehmannii Drejer [1, 4, 5]

Carex alpina var. *lehmannii* (Drejer) R.C.Srivast.

Perennial herbs. W, C, E Nepal: 3400-4100 m. Nepal, E Himalaya, Tibetan Plateau, S Asia and E Asia.

Carex lenta D.Don [1, 4, 5]

Perennial herbs. C Nepal: 1400-1900 m. Nepal, E Himalaya, Tibetan Plateau and E Asia.

Carex ligulata Nees

Carex hebecarpa subsp. *ligulata* (Nees) T.Koyama (1, 4); *Carex hebecarpa* var. *ligulata* (Nees) Kük.

Perennial herbs. C Nepal: 600-2000 m. Nepal, E Himalaya, S Asia and E Asia.

Carex longicruris Nees [1, 4, 5]

Perennial herbs. W, C Nepal: 1800-2500 m. Nepal, E Himalaya, S Asia and SW Asia.

Carex longipes D.Don ex Tilloch & Taylor [1, 4, 5]

Carex longipes var. *nepalensis* (Spreng.) Boott

Perennial herbs. W, C, E Nepal: 1400-2700 m. Nepal, W Himalaya, E Himalaya, S Asia, E Asia and SE Asia.

Carex mallae (Rajbh. & H.Ohba) O.Yano, *J. Linn. Soc.* **179**: 23. 2015. [51]

Kobresia mallae Rajbh. & H.Ohba (4, 26)

Perennial rhizomatous herbs. W, C Nepal: 3500-4600 m. **Endemic to Nepal**.

Central: *Mustang*, Thorang phedi, *ca.* 4000 m, H. Ohba *et al* 8330827 (Holotype: TI).

Carex maritima Gunnerus [1, 4]

Perennial herbs. C, E Nepal: 4100-4300 m. Nepal, N Asia, C Asia, Europe, N America and S America.

Carex microglochin Wahlenb. [1, 4, 5]

Uncinia microglochin (Wahlenb.) Spreng.

Perennial herbs. W, C, E Nepal: 3600-4600 m. Nepal, W Himalaya, E Himalaya, Tibetan Plateau, S Asia, E Asia, N Asia, Europe and N America.

Carex montis-everestii Kük. [4, 8]

Perennial herbs. C Nepal: 4500-4800 m. Nepal, Tibetan Plateau, S Asia and SW Asia.

Carex moorcroftii Falc. ex Boott [5]

Carex melanantha var. *moorcroftii* (Falc. ex Boott) Kük.

Perennial herbs. W, C, E Nepal: 4700-5700 m. Nepal, E Himalaya, Tibetan Plateau and E Asia.

Carex munda Boott [1, 4, 5]

Perennial herbs. C, E Nepal: 2400-4100 m. Nepal, E Himalaya and Tibetan Plateau.

Carex myosurus Nees [1, 4, 5]

Perennial herbs. W, C, E Nepal: 800-3000 m. Nepal, Tibetan Plateau, Assam-Burma, S Asia, E Asia and SE Asia.

Carex nakaoana T.Koyama [1, 4, 5]

Perennial herbs. W, C, E Nepal: 2900-5000 m. Nepal, E Himalaya and Tibetan Plateau.

Carex nivalis Boott [1, 4, 5]

गौरी धान (Gauri dhaan)

Perennial herbs. W, C, E Nepal: 3800-4900 m. Nepal, W Himalaya, Tibetan Plateau, E Asia, C Asia and SW Asia.

Use: Ritual.

Carex notha Kunth [5]

Perennial herbs. W, C, E Nepal: 2400-3500 m. Nepal and E Himalaya.

Carex nubigena D.Don ex Tilloch & Taylor [1, 4, 5]

Perennial herbs. W, C, E Nepal: 1500-4000 m. Nepal, E Himalaya, Tibetan Plateau, S Asia, E Asia, SE Asia, N Asia and SW Asia.

Carex obscura Nees [1, 4, 5]

var. **obscura**

Perennial herbs. W, C, E Nepal: 3000-4300 m. Nepal, W Himalaya, S Asia and SW Asia.

var. **brachycarpa** C.B.Clarke [1, 4]

Perennial herbs. C Nepal: *ca.* 4000 m. Nepal, W Himalaya, E Himalaya, Tibetan Plateau and E Asia.

Carex orbicularis Boott [1, 4, 5]

Perennial herbs. W, C, E Nepal: 2300-4800 m. Nepal, W Himalaya, Tibetan Plateau, S Asia, E Asia, N Asia and SW Asia.

Carex parva Nees [1, 4, 5]

Carex unifoliata Kük. & Hand.-Mazz.

Perennial herbs. W, C, E Nepal: 2300-4800 m. Nepal, E Himalaya, Tibetan Plateau, E Asia and C Asia.

Carex phacota Spreng. [1, 4]

Carex lenticularis D.Don ex Tilloch & Taylor

Lakeshore sedge

Perennial swampy or marshy herbs. C Nepal: *ca.* 1300 m. Nepal, Assam-Burma, S Asia, E Asia and SE Asia.

Carex plectobasis V.I.Krecz. [5]

Carex hirtella Drejer (1, 4)

Perennial herbs. W, C Nepal: 2900-4900 m. Nepal, W Himalaya, Tibetan Plateau, E Asia and SW Asia.

Carex polycephala Boott [1, 4, 5]

Perennial herbs. C, E Nepal: 500-3200 m. Nepal and E Himalaya.

Carex pomiensis Y.C.Yang [8]

Perennial herbs. W, C, E Nepal: 2400-2500 m. Nepal and Tibetan Plateau.

Carex prainii Kük. [51]

Kobresia sikkimensis Kük. (1, 4, 5)

Perennial herbs. C, E Nepal: 3400-4800 m. Nepal, E Himalaya and Tibetan Plateau.

Carex pseudofoetida Kük.

subsp. **pseudofoetida** (1, 4)

Perennial herbs. C Nepal: *ca.* 4400 m. Nepal, W Himalaya, E Himalaya, Tibetan Plateau, S Asia, C Asia, N Asia and SW Asia.

subsp. **afghanica** Kük. [5]

Perennial herbs. W, C, E Nepal: 3600-4500 m. Nepal, E Himalaya, Tibetan Plateau, S Asia and SW Asia.

Carex pseuduncinoides (Noltie) O.Yano & S.R.Zhang [51]

Kobresia pseuduncinoides Noltie; *Kobresia kansuensis* Kük. (5)

Perennial herbs. W, C, E Nepal: 3800-4800 m. Nepal, E Himalaya and Tibetan Plateau.

Carex psychrophila Nees [1, 4]

Carex celsa Boott

Perennial herbs. W, C Nepal: 3000-3700 m. Nepal, W Himalaya, E Himalaya, Tibetan Plateau, S Asia, E Asia and SW Asia.

Carex pulchra Boott [1, 4, 5]

Perennial herbs. E Nepal: 2100-3500 m. Nepal and E Himalaya.

Carex rara Boott [1, 4, 5]

Carex rara var. *biwensis* (Franch.) Kük. ex Matsum.

Perennial herbs. W, C, E Nepal: 2400-3800 m. Nepal, E Himalaya, Tibetan Plateau and E Asia.

Carex remota L.

subsp. **rochebrunii** (Franch. & Sav.) Kük. [5]

Carex rochebrunii Franch. & Sav. (1, 4)

Perennial herbs. W, C, E Nepal: 1700-3600 m. Nepal, E Himalaya, Tibetan Plateau and E Asia.

subsp. **stewartii** Kük. [5]

Perennial herbs. W, C, E Nepal: 2300-3500 m. Nepal, E Himalaya, S Asia and SW Asia.

Carex rhombifructus Ohwi, *J. Jap. Bot.* **44**: 60. 1969. [1, 4, 26]

Perennial rhizomatous herbs. W, C Nepal: 1600-3100 m. **Endemic to Nepal**.

Central: *Lalitpur*, Godawari – Phulchowki, 1600-2500 m, H. Kanai, G. Murata, H. Ohashi & T. Yamazaki 1138 (Holotype: TI).

Carex rufulistolon T.Koyama, *Acta Phytotax. Geobot.* **15** : 111.1954. [1, 4, 5, 26]

Perennial rhizomatous herbs. W, C Nepal: 2400-3900 m. **Endemic to Nepal**.

Central: *Manang*, Pisan – Tatpani, *ca.* 3100 m, Nakao s.n. (Holotype: KYO).

Carex setigera D.Don [1, 4]

var. **setigera**

Perennial herbs. W, C, E Nepal: 2400-2800 m. Nepal, E Himalaya, Tibetan Plateau, S Asia and E Asia.

var. **schlagintweitiana** (Boeckeler) Kük.

Carex schlagintweitiana Boeckeler; *Carex schlagintweitiana* subsp. *deformis* Noltie (4, 5)

Perennial herbs. C, E Nepal: 3000-4000 m. Nepal, W Himalaya, Tibetan Plateau, S Asia, E Asia and SW Asia.

Carex setosa Boott [1, 4]

Carex pachyrrhiza Franch. (8)

Perennial herbs. W, C, E Nepal: 3400-4500 m. Nepal, W Himalaya, E Himalaya, Tibetan Plateau, S Asia, E Asia and SW Asia.

Carex speciosa Kunth [126]

subsp. **speciosa**

Perennial herbs. C Nepal: 2300-2500 m. Nepal, S Asia, E Asia and SE Asia.

subsp. **dilatata** Noltie [4]

Perennial herbs. C, E Nepal: 1200-2200 m. Nepal, E Himalaya and E Asia.

Carex staintonii X.F.Jin, H.Ikeda & O.Yano, *J. Jap. Bot.* **90**(2): 111. 2015. [50]

Perennial herbs. C Nepal: 2400-2700 m. **Endemic to Nepal.**

Central: *Rasuwa/Sindhupalchowk*, Gosainkunda, Melamchi, 8000 ft. (*ca.* 2400 m), JDA Stainton 3772 (Holotype: TI; Isotype: BM).

Carex stracheyi Boott ex C.B.Clarke [1, 4, 8]

Perennial herbs. W, C Nepal: 2100-3200 m. Nepal and E Himalaya.

Carex stramentitia Boott ex Boeckeler [1, 4, 5]

Perennial herbs. C, E Nepal: 1500-2600 m. Nepal, E Himalaya, Assam-Burma, S Asia, E Asia and SE Asia.

Carex supina Willd. ex Wahlenb. [1, 4, 5]

Perennial herbs. W, C Nepal: 4000-5700 m. Nepal, W Himalaya, C Himalaya, Tibetan Plateau, C Asia, SW Asia and Europe.

Carex teres Boott [1, 4]

Perennial herbs. C, E Nepal: *ca.* 3100 m. Nepal, E Himalaya, Tibetan Plateau and S Asia.

Carex uncinoides Boott [51]

Kobresia uncinoides (Boott) C.B.Clarke (1, 4, 5)

Perennial herbs. W, C, E Nepal: 3500-4600 m. Nepal, E Himalaya, Tibetan Plateau, Assam-Burma and E Asia.

Carex vaginosa (C.B.Clarke) S.R.Zhang [51]

Kobresia vaginosa C.B.Clarke (4), *Kobresia nepalensis* var. *vaginosa* (C.B.Clarke) R.C.Srivast.

Perennial herbs. W, C, E Nepal: 2900-4600 m. Nepal, W Himalaya, E Himalaya, Tibetan Plateau and E Asia.

Carex vesiculosa Boott [1, 4, 8]

Carex diffusa Boot ex C.B.Clarke

Perennial herbs. W, C, E Nepal: 900-2300 m. Nepal, E Himalaya and SE Asia.

Carex vidua Boott ex C.B.Clarke [51]

Kobresia vidua (Boott ex C.B.Clarke) Kük. (1, 4, 5)

Perennial herbs. W, C, E Nepal: 2800-4400 m. Nepal, E Himalaya, Tibetan Plateau and E Asia.

Carex viridula Michx. [5]

Carex serotina Merat (1, 4)

Yellow sedge

Perennial herbs. W Nepal: 2400-2600 m. Nepal, Asia and Europe.

Carex wallichiana Spreng.

Carex foliosa D.Don ex Tilloch & Taylor (1, 4)

Perennial herbs. C, E Nepal: 1800-3000 m. Nepal, Himalaya and S Asia.

Carex winterbottomii C.B.Clarke [1, 4]

Carex setigera var. *fasciculata* Strachey

Perennial herbs. C Nepal: *ca.* 3200 m. Nepal, W Himalaya and S Asia.

Cladium P.Browne

Cladium jamaicense Crantz

subsp. **chinense** (Nees) T.Koyama [1, 4, 5]

Cladium chinense Nees; *Cladium mariscus* subsp. *chinense* (Nees) Govaerts

Perennial herbs. C Nepal: 2600-2800 m. Nepal, S Asia and E Asia.

Use: Paper pulp (stem).

Courtoisina Soják

Courtoisina cyperoides (Roxb.) Soják

Kyllinga cyperoides Roxb.; *Mariscus cyperoides* (Roxb.) A.Dietr. (1, 4)

Annual wetland weedy herbs. W, C, E Nepal: 200-1200 m. Nepal, Asia, Africa and Australasia.

Cyperus L.

Cyperus alulatus J.Kern [1, 4, 8]

Annual marshy herbs. W Nepal: 700-1200 m. Nepal, E Himalaya and SW Asia.

Cyperus castaneus Willd. [1, 4]

Annual herbs. E Nepal: *ca.* 300 m. Nepal, E Himalaya, Assam-Burma, S Asia, E Asia, SE Asia, Africa and Australasia.

Cyperus compactus Retz. [5]

Mariscus compactus (Retz.) Bold. (1, 4)

Perennial aquatic herbs. W, C, E Nepal: 100-300 m. Nepal, Tibetan Plateau, Assam-Burma, S Asia, E Asia, SE Asia, SW Asia and Australasia.

Use: Mats (culm).

Cyperus compressus L. [1, 4, 5]

मोथे झार, झुसुमा (Mothe jhaar; Jhusumaa: Tharu); Papyrus plant

Annual herbs. W, C, E Nepal: 200-1400 m. Nepal, E Himalaya, Tibetan Plateau, Assam-Burma, S Asia, E Asia, SE Asia, SW Asia, Africa, America and Australasia.

Use(s): Medicine (whole plant), Forage.

Cyperus corymbosus Rottb. [1, 4, 5]

Cyperus articulatus L.; *Cyperus diphyllus* Retz.

Perennial aquatic herbs. W, C Nepal: 100-1400 m. Nepal, Pantropical and Subtropical.

Use: Mats (culm).

Cyperus cuspidatus Kunth [1, 4, 5]

Cyperus uncinatus C.B.Clarke

Annual tufted wetland herbs. W, C, E Nepal: 400-2500 m. Nepal, Asia, Africa, America and Australasia.

Cyperus cyperinus (Retz.) Suringar [5]

Kyllinga cyperina Retz.; *Mariscus cyperinus* (Retz.) Vahl (1, 4)

Perennial herbs. C, E Nepal: 600-2100 m. Nepal, E Himalaya, Tibetan Plateau, Assam-Burma, S Asia, E Asia, SE Asia, SW Asia and Australasia.

Cyperus cyperoides (L.) Kuntze [5]

Scirpus cyperoides L.; *Mariscus sumatrensis* (Retz.) J.Raynal (1, 4)

फिरफिरे घाँस, करौंते (Phirphire ghaans, Karaunte)

Perennial herbs. W, C, E Nepal: 100-2400 m. Nepal, Asia, Africa and Australasia.

Use(s): Forage, Medicine (root), Construction materials.

Cyperus difformis L. [1, 4, 5]

Annual aquatic herbs. W, C, E Nepal: 100-2700 m. Nepal, Asia, Europe, Africa and Australasia.

Cyperus diffusus Vahl [1, 4]

Dwarf umbrella grass

Perennial herbs. C, E Nepal: 200-900 m. Nepal, E Himalaya, Tibetan Plateau, Assam-Burma, S Asia, E Asia, SE Asia and Australasia.

Cyperus digitatus Roxb. [1, 4, 5]

Cyperus digitatus var. *khasiana* (C.B.Clarke) J.Kern

Perennial wetland herbs. W, C Nepal: 200-1100 m. Nepal, Tibetan Plateau, Assam-Burma, S Asia, E Asia, SE Asia, SW Asia, Africa, America and Australasia.

Cyperus distans L.f. [1, 4, 5]

मोथे (Mothe)

Perennial wetland herbs. W, C, E Nepal: 100-1100 m. Nepal, W Himalaya, E Himalaya, Assam-Burma, S Asia, E Asia, SE Asia, Africa, America and Australasia.

Use: Medicine (leaves, root).

Cyperus dubius Rottb.

Kyllinga triceps Rottb. (1, 4)

Perennial rhizomatous herbs. E Nepal. Nepal and old world tropics.

Use: Medicine (root oil).

Cyperus esculentus L. [1, 4, 5]

कसुर (Kasur); Yellow nut sedge

Perennial wetland herbs. W, C, E Nepal: 100-2300 m. Nepal, S Asia, Europe, Africa and America.

Use(s): Medicine (tuber), Food (tuber).

Cyperus exaltatus Retz. [4, 5]

Perennial aquatic herbs. W, C, E Nepal: 100-500 m. Nepal, W Himalaya, Assam-Burma, S Asia, E Asia, SE Asia, SW Asia, Africa and Australasia.

Use: Fiber (paper making).

Cyperus fuscus L. [8]

Cyperus virescens Hoffm.

Annual herbs. C Nepal: 100-2000 m. Nepal, W Himalaya, Tibetan Plateau, S Asia, E Asia, SE Asia, N Asia, C Asia, SW Asia, Europe and Africa.

Cyperus haspan L. [1, 4, 5]

Cyperus haspan var. *flaccidissimus* Kük.; *Cyperus haspan* var. *juncoides* Kük.

Sharp-edge sedge

Perennial wetland herbs. C, E Nepal: 100-1400 m. Nepal, Asia, Africa, America and Australasia.

Use: Forage.

Cyperus imbricatus Retz. [1, 4, 5]

Cyperus imbricatus var. *densespicatus* (Hayata) Ohwi; *Cyperus imbricatus* subsp. *elongates* (Boeckeler) T.Koyama

Perennial wetland herbs. W, C Nepal: 100-200 m. Nepal, Assam-Burma, S Asia, E Asia, SW Asia, Africa, America and Australasia.

Cyperus involucratus Rottb.

Cyperus alternifolius subsp. *flabelliformis* (Rottb.) Kük. (4); *Cyperus flabelliformis* Rottb.

छाता मोथे (Chhata mothe)

Perennial herbs. E Nepal: 100-200 m. Nepal, SW Asia and Africa.

Cyperus iria L. [1, 4, 5]

छाता मोथे (Chhaate mothe); Umbrella sedge

Annual or occasionally perennial herbs. W, C, E Nepal: 100-2000 m. Nepal, Asia, Africa and Australasia.

Use(s): Medicine (whole plant), Forage, Mats (culm).

Cyperus laxus Lam. [5]

Perennial herbs. C, E Nepal: 100-1600 m. Nepal, native to Africa and W Hemisphere.

Cyperus longus L. [1, 4]

मोथे (Mothe)

Perennial wetland herbs. W, C Nepal: *ca.* 800 m. Nepal, S Asia, SW Asia, Europe and Africa.

Use: Medicine (rhizome).

Cyperus malaccensis Lam. [1, 4]

Cyperus fortunei Steud.

Perennial wetland herbs. E Nepal: 100-700 m. Nepal, Assam-Burma, S Asia, E Asia, SE Asia, SW Asia and Australasia.

Use(s): Mats, baskets (culm).

Cyperus michelianus (L.) Delile [1, 4]

Scirpus michelianus L.

Annual herbs. E Nepal: *ca.* 200 m. Nepal, Assam-Burma, Asia, Africa, Europe and Australasia.

Cyperus microiria Steud. [8]

Annual herbs. W Nepal: 100-3800 m. Nepal, S Asia, E Asia and SE Asia.

Cyperus niveus Retz. [1, 4, 5]

सानो मोथे, सेतो मोथे (Saano mothe, Seto mothe)

Perennial herbs. W, C Nepal: 400-2900 m. Nepal, W Himalaya, E Himalaya, Tibetan Plateau, Assam-Burma, E Asia, S Asia, SE Asia, SW Asia and Africa.

Cyperus nutans Vahl [1, 4, 5]

var. **nutans**

Perennial wetland herbs. W, C, E Nepal: 300-2600 m. Nepal, E Himalaya, S Asia, E Asia, SW Asia, Africa and Australasia.

var. **eleusinoides** (Kunth) Haines [5]

Cyperus eleusinoides Kunth (1, 4)

Perennial wetland herbs. W, C, E Nepal: 300-1800 m. Nepal, Asia, Africa and Australasia.

Cyperus pangorei Rottb. [1, 4, 8]

Perennial wetland herbs. C, E Nepal: 200-1300 m. Nepal, Assam-Burma, S Asia, E Asia and SW Asia.

Cyperus paniceus (Rottb.) Boeckeler

Kyllinga panicea Rottb.; *Mariscus paniceus* (Rottb.) Vahl (1, 4)

Perennial herbs. C Nepal: *ca.* 1300 m. Nepal, E Himalaya, Assam-Burma, S Asia, SE Asia, SW Asia and Australasia.

Cyperus pilosus Vahl [1, 4, 5]

Cyperus paniculatus D.Don; *Cyperus obliquus* Nees

Perennial wetland herbs. W, C, E Nepal: 100-1100 m. Nepal, E Himalaya, Tibetan Plateau, Assam-Burma, S Asia, E Asia, SE Asia and Australasia.

Use: Forage.

Cyperus platystylis R.Br. [1, 4, 5]

Perennial aquatic herbs. C, E Nepal: 100-200 m. Nepal, E Himalaya, Tibetan Plateau, Assam-Burma, S Asia, E Asia, SE Asia and Australasia.

Cyperus procerus Rottb. [1, 4, 5]

Perennial aquatic herbs. E Nepal: 100-200 m. Nepal, S Asia, E Asia, SE Asia, Africa and Australasia.

Cyperus rotundus L. [1, 4, 5]

Cyperus comosus L.; *Cyperus tuberosus* Rottb. (1, 4)

नगर मोथे (Nagar mothe); Brown nutsedge

Perennial herbs. W, C, E Nepal: 100-2400 m. Nepal, S Asia, SE Asia, Africa and Australasia.

Use(s): Medicine (tuber), Food (tuber), Forage.

Cyperus squarrosus L. [5]

Cyperus aristatus Rottb.; *Mariscus squarrosus* (L.) C.B.Clarke; *Mariscus aristatus* (Rottb.) Cherm. (1, 4)

Annual wetland herbs. W, C Nepal: 1100-2700 m. Nepal, Asia, Europe, Africa, America and Australasia.

Use: Medicine (root/stolon).

Cyperus tenuiculmis Boeckeler [1, 4, 5]

Perennial aquatic herbs. W, C, E Nepal: 200-2000 m. Nepal, E Himalaya, Assam-Burma, S Asia, E Asia, SE Asia, Africa and Australasia.

Cyperus tenuispica Steud. [1, 4, 5]

Annual aquatic herbs. W, C, E Nepal: 100-1400 m. Nepal, Asia, Africa and Australasia.

Cyperus trisulcus D.Don [1, 4]

Perennial herbs. C Nepal: *ca.* 1300 m. Endemic to Nepal (?)

Eleocharis R.Br.

Eleocharis acutangula (Roxb.) Schult. [1, 4, 5]

Scirpus acutangulus Roxb.

Perennial wetland herbs. C, E Nepal: 100-1000 m. Nepal, S Asia, E Asia, SE Asia, Assam-Burma, Africa, America and Australasia.

Eleocharis atropurpurea (Retz.) J.Presl. & C.Presl. [4, 5]

Scirpus atropurpureus Retz.

Purple spike rush

Annual marshy herbs. W, C, E Nepal: 100-1200 m. Nepal, E Himalaya, S Asia, E Asia, SE Asia, SW Asia, Europe, Africa, America and Australasia.

Eleocharis congesta D.Don [1, 4, 5]

Scirpus congestus (D.Don) Buch.-Ham. ex D.Don

Perennial wetland herbs. W, C, E Nepal: 200-3000 m. Nepal, Asia and Australasia.

Eleocharis dulcis (Burm.f.) Trin. ex Hensch. [1, 4]

Andropogon dulcis Burm.f.; *Scirpus plantagineus* Retz.

कसुर (Kasur); Chinese water chestnut

Perennial wetland herbs. E Nepal: 200-2000 m. Nepal, Tibetan Plateau, Assam-Burma, S Asia, E Asia, SE Asia, SW Asia, Africa and Australasia.

Use: Food (tuber).

Eleocharis palustris (L.) Roem. & Schult. [1, 4, 5]

Scirpus palustris L.

Perennial herbs. W, C Nepal: 2000-3800 m. Nepal, E Asia, N Asia, C Asia, SW Asia, Europe, Africa and N America.

Eleocharis quinqueflora (Hartmann) O.Schwarz [1, 4, 5]

Scirpus quinqueflorus Hartmann; *Eleocharis obscura* T.Koyama

Perennial herbs. W, C Nepal: 2000-4300 m. Nepal, Asia, Europe, Africa, N America and S America.

Eleocharis retroflexa (Poir.) Urban

Scirpus retroflexus Poir.

subsp. **chaetaria** (Roem. & Schult.) T.Koyama [1, 4]

Eleocharis chaetaria Roem. & Schult.

Annual wetland herbs. E Nepal: 200-300 m. Nepal, Assam-Burma, S Asia, E Asia, SE Asia, Africa, America and Australasia.

Eleocharis tetraquetra Nees [1, 4, 5]

Perennial wetland herbs. C, E Nepal: 1400-3500 m. Nepal, E Himalaya, S Asia, E Asia, SE Asia, N Asia, SW Asia and Australasia.

Eleocharis uniglumis (Link) Schult. [56]

Scirpus uniglumis Link

Perennial herbs. W Nepal: *ca.* 3400 m. Nepal, S Asia, E Asia, N Asia, C Asia, SW Asia, Europe, Africa and N America.

Eriophorum L.

Eriophorum comosum (Wall.) Nees [1, 4, 5]

Scirpus comosus Wall.; *Erioscirpus comosus* (Wall.) Palla

फुर्के झार (Phurke jhaar); Hairy cotton grass

Perennial herbs. W, C, E Nepal: 300-2600 m. Nepal and Asia.

Use: Forage.

Eriophorum microstachyum Boeckeler [1, 4]

Erioscirpus microstachyus (Boeckeler) Palla; *Scirpus microstachyus* (Boeckeler) T.Koyama

Perennial herbs. C, E Nepal: 500-2900 m. Nepal and E Himalaya.

Fimbristylis Vahl

Fimbristylis aestivalis (Retz.) Vahl [1, 4, 5]

Scirpus aestivalis Retz.

Annual wetland herbs. C, E Nepal: 200-2000 m. Nepal, E Himalaya, S Asia, E Asia, SE Asia, N Asia and Australasia.

Fimbristylis bisumbellata (Forssk.) Bubani [1, 4, 5]

Scirpus bisumbellatus Forssk.

पानि मोथे, पानि छतेरा (Paani mothe, Paani chhatera)

Annual wetland herbs. W, C Nepal: 100-1100 m. Nepal, Assam-Burma, S Asia, E Asia, SE Asia, N Asia, SW Asia, Europe, Africa and Australasia.

Fimbristylis complanata (Retz.) Link [1, 4, 5]

Scirpus complanatus Retz.

Perennial wetland herbs. W, C, E Nepal: 300-3100 m. Nepal, E Himalaya, Tibetan Plateau, S Asia, E Asia, SE Asia, SW Asia, Africa, S America and Australasia.

Fimbristylis cymosa R.Br. [4]

जहर मोथे (Jahar mothe); Hurricane grass

Perennial wetland herbs. E Nepal: 100-400 m. Nepal, S Asia, E Asia, SE Asia, Africa and Australasia.

Fimbristylis dichotoma (L.) Vahl [1, 4]

Scirpus dichomotus L.

subsp. **dichotoma**

पानि मोथे (Paani mothe); Forked fimbry

Perennial or annual wetland herbs. W, C, E Nepal: 100-2100 m. Nepal, Tibetan Plateau, S Asia, E Asia, N Asia, C Asia, SW Asia, Africa, America and Australasia.

Use(s): Medicine (tuber), Forage.

subsp. **podocarpa** (Nees) T.Koyama [4]

Fimbristylis podocarpa Nees

Woolly fimbry

Annual herbs. E Nepal: *ca.* 200 m. Nepal, E Himalaya, S Asia, E Asia, SE Asia, Africa and Australasia.

Fimbristylis falcata (Vahl) Kunth [1, 4, 5]

Scirpus falcatus Vahl

मोथे (Mothe)

Perennial herbs. W, C Nepal: 200-1100 m. Nepal, S Asia, SE Asia, SW Asia and Africa.

Use: Medicine (root).

Fimbristylis fimbristyloides (F.Muell.) Druce [1, 4, 5]

Abildgaardia fimbristyloides F.Muell.; *Fimbristylis nanofusca* Tang & F.T.Wang

Annual tufted herbs. C, E Nepal: 500-1700 m. Nepal, E Himalaya, Assam-Burma, S Asia, E Asia, SE Asia and Australasia.

Fimbristylis fusca (Nees) Benth. ex C.B.Clarke [1, 4, 5]

Abildgaardia fusca Nees; *Fimbristylis fusca* var. *cantoniensis* C.B.Clarke

Perennial herbs. W, C, E Nepal: 100-2000 m. Nepal, Assam-Burma, S Asia, E Asia, SE Asia and Australasia.

Fimbristylis fuscinux C.B.Clarke [5]

Tufted herbs. C Nepal: 500-600 m. Nepal and E Himalaya.

Fimbristylis littoralis Gaudich. [5]

Fimbristylis miliacea (L.) Vahl (1, 4)

ज्वाने झार (Jwaane jhaar)

Annual or short-lived perennial wetland herbs. W, C, E Nepal: 100-2000 m. Nepal, E Himalaya, S Asia, E Asia, SE Asia, SW Asia, Africa, America and Australasia.

Use: Forage.

Fimbristylis ovata (Burm.f.) J.Kern [1, 4, 5]

Carex ovata Burm.f.

भुलना (Bhulnaa)

Perennial wetland herbs. W, C, E Nepal: 200-1100 m. Nepal, E Himalaya, S Asia, E Asia, SE Asia, SW Asia, Africa, S America and N America.

Use: Forage.

Fimbristylis pierotii Miq. [1, 4, 5]

Fimbristylis pinetorum Merr.

Perennial herbs. C Nepal: 1600-2400 m. Nepal, W Himalaya, E Asia and SE Asia.

Fimbristylis quinquangularis (Vahl) Kunth [1, 4, 8]

Scirpus quinquangularis Vahl

ज्वाने झार (Jwaane jhaar)

Annual wetland herbs. W, C, E Nepal: 100-2000 m. Nepal, Tibetan Plateau, S Asia, E Asia, SE Asia, C Asia, SW Asia, Africa and Australasia.

Use: Forage.

Fimbristylis rigidula Nees [1, 4]

Fimbristylis hanceana Boeckeler

Perennial herbs. C Nepal: 700-1300 m. Nepal, W Himalaya, Assam-Burma, S Asia, E Asia and SE Asia.

Fimbristylis schoenoides (Retz.) Vahl [1, 4, 5]

Scirpus schoenoides Retz.

Perennial wetland herbs. W, C, E Nepal: 200-1200 m. Nepal, E Himalaya, S Asia, E Asia, SE Asia, SW Asia, Africa and Australasia.

Use: Forage.

Fimbristylis squarrosa Vahl [1, 4, 5]

जिरे झार (Jire jhaar)

Annual wetland herbs. W, C, E Nepal: 500-1500 m. Nepal, Tibetan Plateau, Assam-Burma, S Asia, E Asia, SE Asia, Europe, Africa and Australasia.

Use: Medicine (whole plant).

Fimbristylis stolonifera C.B.Clarke [1, 4]

Perennial herbs. W, C Nepal: 2000-2900 m. Nepal, E Himalaya and E Asia.

Fimbristylis umbellaris (Lam.) Vahl [1, 4, 5]

Scirpus umbellaris Lam.; *Scirpus globulosus* Retz.; *Fimbristylis globulosa* (Retz.) Kunth

गुदा मोथे (Gudaa mothe)

Perennial wetland herbs. C Nepal: 1500-1600 m. Nepal, Assam-Burma, S Asia, E Asia, SE Asia and Australasia.

Fuirena Rottb.

Fuirena ciliaris (L.) Roxb. [4, 5]

Scirpus ciliaris L.

Annual wetland herbs. E Nepal: *ca.* 1000 m. Nepal, Assam-Burma, S Asia, E Asia, SE Asia, Africa and Australasia.

Isolepis R.Br.

Isolepis setacea (L.) R.Br. [1, 4, 5]

Scirpus setaceus L.

Bristle club rush

Annual herbs. W, C, E Nepal: 2200-4000 m. Nepal, Asia, Europe, Africa, N America and Australasia.

Kobresia Willd.

Kobresia capillifolia (Decne.) C.B.Clarke [1, 4, 8]

Elyna capillifolia Decne.

Perennial herbs. C Nepal: 1200-4800 m. Nepal, W Himalaya, E Himalaya, Tibetan Plateau, S Asia, E Asia, N Asia, C Asia and SW Asia.

Kobresia cercostachys (Franch.) C.B.Clarke [5]

Carex cercostachys Franch.

Perennial herbs. W, C, E Nepal: 4000-4600 m. Nepal, E Himalaya, Tibetan Plateau and E Asia.

Kobresia curticeps (C.B.Clarke) Kük. [1, 4, 5]

Carex curticeps C.B.Clarke

Perennial herbs. C, E Nepal: 3400-3700 m. Nepal, E Himalaya, Tibetan Plateau and S Asia.

Kobresia curvirostris (C.B.Clarke.) C.B.Clarke [5]

Hemicarex curvirostris C.B.Clarke

Perennial herbs. E Nepal: 2900-3000 m. Nepal, E Himalaya, S Asia and SE Asia.

Kobresia duthiei C.B.Clarke [1, 4, 5]

Carex clavispica S.R.Zhang (51)

Perennial herbs. W, C, E Nepal: 3700-5700 m. Nepal, E Himalaya, Tibetan Plateau and E Asia.

Kobresia esenbeckii (Kunth) Noltie [4, 5]

Carex esenbeckii Kunth.; *Kobresia seticulmis* Boeckeler

Perennial herbs. W, C, E Nepal: 2900-4600 m. Nepal, E Himalaya, Tibetan Plateau, Assam-Burma, S Asia and E Asia.

Kobresia filicina (C.B.Clarke) C.B.Clarke [5]

Hemicarex filicina C.B.Clarke

Perennial herbs. C, E Nepal: 3700-3800 m. Nepal, E Himalaya, Tibetan Plateau, S Asia and E Asia.

Kobresia gandakiensis Rajbh. & H.Ohba [4, 26]

Carex breviprophylla O.Yano (51)

Perennial rhizomatous herbs. C Nepal: 1200-2000 m. Nepal and E Himalaya.

Kobresia laxa Nees [1, 4]

Elyna laxa (Nees) Kunth; *Hemicarex laxa* (Nees) Benth.; *Schoenoxiphium laxum* (Nees) Ivan.

Perennial herbs. W, C, E Nepal: 3300-3800 m. Nepal, W Himalaya, Tibetan Plateau, S Asia, C Asia and SW Asia.

Kobresia macrantha Boeckeler [5]

Kobresia macrantha var. *nudicarpa* (Y.C.Yang) P.C.Li; *Kobresia nudicarpa* (Y.C.Yang) S.R.Zhang

Perennial herbs. C, E Nepal: 3700-6700 m. Nepal, Tibetan Plateau, E Asia and N Asia.

Kobresia nepalensis (Nees) Kük. [1, 4, 5]

Uncinia nepalensis Nees; *Carex neesii* S.R.Zhang (51); *Kobresia nepalensis* var. *elachista* (C.B.Clarke) Kük.

Perennial herbs. W, C, E Nepal: 2900-5700 m. Nepal, W Himalaya, E Himalaya, Tibetan Plateau, S Asia, E Asia and SW Asia.

Kobresia nitens C.B.Clarke [1, 4]

Carex ovoidispica O.Yano (51)

Perennial herbs. W Nepal: 5000-5600 m. Nepal, E Himalaya, S Asia and SW Asia.

Kobresia pygmaea (C.B.Clarke) C.B.Clarke [1, 4, 5]

Hemicarex pygmaea C.B.Clarke; *Kobresia microstachya* K.V.Ivanova; *Kobresia pygmaea* var. *filiculmis* Kük.

Perennial herbs. W, C, E Nepal: 1100-5600 m. Nepal, W Himalaya, E Himalaya, Tibetan Plateau, Assam-Burma, E Asia and N Asia.

Kobresia royleana (Nees) Boeckeler [4]

Trilepis royleana Nees; *Kobresia stenocarpa* var. *royleana* (Nees) C.B.Clarke

Perennial herbs. W, C Nepal: 3500-4600 m. Nepal, W Himalaya, Tibetan Plateau, E Asia, C Asia and SW Asia.

Kobresia schoenoides (C.A.Mey.) Steud. [1, 4, 5]

Elyna schoenoides C.A.Mey.; *Kobresia deasyi* C.B.Clarke (1, 4); *Kobresia pamiroalaica* N.A.Ivanova

Perennial herbs. W, C, E Nepal: 3500-5600 m. Nepal, W Himalaya, E Himalaya, Tibetan Plateau, S Asia (Pakistan), E Asia, N Asia, C Asia and SW Asia.

Kobresia simpliciuscula (Wahlenb.) Mack.

Carex simpliciuscula Wahlenb.; *Kobresia caricina* Willd. (1, 4)

Perennial herbs. W, C Nepal: 4000-5600 m. Nepal, Europe and N America.

Kyllinga Rottb.

Kyllinga brevifolia Rottb. [1, 4, 5]

Cyperus brevifolius (Rottb.) Hassk.

subsp. **brevifolia**

Perennial weedy marshy herbs. W, C, E Nepal: 100-2300 m. Nepal, Asia, Africa, America and Australasia.

Use(s): Medicine (root/rhizome, leaves), Forage.

subsp. **leiolepis** (Franch. & Sav.) T.Koyama [1, 4, 5]

Kyllinga monocephala var. *leiolepis* Franch. & Sav.; *Kyllinga gracillima* Miq.

Perennial marshy herbs. W, C, E Nepal: 1000-2900 m. Nepal, Himalaya, Tibetan Plateau, E Asia and N Asia.

Kyllinga cylindrica Nees

Kyllinga odorata subsp. *cylindrica* (Nees) T.Koyama (1, 4, 5)

Perennial herbs. W, C, E Nepal: 1100-2000 m. Nepal, E Himalaya, S Asia, E Asia, SE Asia and Africa.

Kyllinga nemoralis (J.R.Forst. & G.Forst.) Dandy ex Hutch. & Dalziel [1, 4, 5]

Thryocephalon nemorale J.R.Forst. & G.Forst.

Perennial herbs. W, C, E Nepal: 300-1200 m. Nepal, E Himalaya, Assam-Burma, S Asia, E Asia, SE Asia, Africa and Australasia.

Use: Medicine (root).

Kyllinga squamulata Vahl [1, 4, 5]

Kyllinga cristata Afzel. ex A.Rich

Annual tufted herbs. W, C, E Nepal: 1100-2600 m. Nepal, E Himalaya, S Asia, E Asia, SE Asia, SW Asia, Africa and Australasia.

Lipocarpha R.Br.

Lipocarpha chinensis (Osbeck) J.Kern [1, 4, 5]

Scirpus chinensis Osbeck

भकुन्डे घाँस (Bhakunde ghaans)

Perennial wetland herbs. W, C, E Nepal: 300-2100 m. Nepal, Asia, Africa and Australasia.

Lipocarpha gracilis (Rich.) Nees

Hypolytrum gracile Rich.; *Hypaelyptum sphacelatum* Vahl; *Lipocarpha sphacelata* (Vahl) Kunth (1, 4)

Annual or perennial tufted wetland herbs. W, C, E Nepal. Nepal and Pantropical.

Use: Forage.

Lipocarpha squarrosa (L.) Goetgh. [5]

Scirpus squarrosus L.; *Rikliella squarrosa* (L.) Raynal (4)

Annual tufted wetland herbs. C, E Nepal: 100-1400 m. Nepal, W Himalaya, Assam-Burma, S Asia, E Asia, SE Asia and SW Asia.

Pycreus P.Beauv.

Pycreus diaphanus (Schrad. ex Roem. & Schult.) S.S.Hooper & T.Koyama [1, 4, 5]

Cyperus diaphanus Schrad. ex Roem. & Schult.

Annual tufted herbs. C, E Nepal: 1100-2100 m. Nepal, W Himalaya, E Himalaya, Tibetan Plateau, S Asia, E Asia, SE Asia and N Asia.

Pycreus flavidus (Retz.) T.Koyama [1, 4, 5]

Cyperus flavidus Retz.

Perennial aquatic herbs. W, C, E Nepal: 100-3800 m. Nepal, Asia, Europe, Africa and Australasia.

Pycreus pumilus (L.) Nees [1, 4, 5]

Cyperus pumilus L.

Annual wetland herbs. W, C, E Nepal: 100-1100 m. Nepal, Asia, Africa and Australasia.

Pycreus sanguinolentus (Vahl) Nees [1, 4]

Cyperus sanguinolentus Vahl

झुसे मोथे, रगते मोथे (Jhuse mothe, Ragate mothe); Purple glume flat sedge

Annual aquatic herbs. W, C, E Nepal: 100-2900 m. Nepal, Asia, Africa and Australasia.

Use: Forage.

Pycreus stramineus C.B.Clarke [1, 4]

Cyperus substramineus Kuk.

Annual wetland or wasteland herbs. E Nepal: *ca.* 200 m. Nepal, S Asia and SE Asia.

Pycreus sulcinux (C.B.Clarke) C.B.Clarke [5]

Cyperus sulcinux C.B.Clarke

Annual herbs. E Nepal: *ca.* 1500 m. Nepal, Asia and Australasia.

Pycreus unioloides (R.Br.) Urb. [1, 4, 5]

Cyperus unioloides R.Br.

Perennial wetland herbs. C, E Nepal: 900-1900 m. Nepal, Asia, Africa, America and Australasia.

Rhynchospora Vahl

Rhynchospora rubra (Lour.) Makino [1, 4, 5]

Schoenus ruber Lour.; *Rhynchospora wallichii* (Nees) Kunth

Red star sedge

Annual or perennial herbs. C Nepal: 1300-1400 m. Nepal, S Asia, E Asia, SE Asia, Africa and Australasia.

Rhynchospora rugosa (Vahl) Gale

subsp. **brownii** (Roem. & Schult.) T.Koyama [1, 4, 5]

Rhynchospora brownii Roem. & Schult.

Perennial wetland herbs. C, E Nepal: 1800-2300 m. Nepal, S Asia, E Asia, SE Asia, Europe, Africa and Australasia.

Schoenoplectus (Rchb.) Palla

Schoenoplectus articulatus (L.) Palla [1, 4]

Scirpus articulatus L.; *Schoenoplectiella articulata* (L.) Lye

Annual aquatic herbs. W Nepal. Nepal and Asia.

Use(s): Medicine (tubers), Thatching (dried plants).

Schoenoplectus fuscorubens (T.Koyama) T.Koyama [1, 4]

Scirpus fuscorubens T.Koyama; *Schoenoplectiella fuscorubens* (T.Koyama) Hayasaka (8)

Annual herbs. W, C Nepal: 1100-3200 m. Nepal, E Himalaya and Tibetan Plateau.

Schoenoplectus juncoides (Roxb.) Palla [1, 4, 5]

Scirpus juncoides Roxb.; *Schoenoplectiella juncoides* (Roxb.) Lye; *Scirpus quadrangulus* D.Don (1, 4)

डाडु झार, अरबाजे झार (Dadu jhaar, Arbaje jhaar); Rock bulrush

Annual wetland herbs. W, C, E Nepal: 1100-2700 m. Nepal, S Asia, E Asia and SE Asia.

Use(s): Forage, Green manure, Construction materials (stem).

Schoenoplectus mucronatus (L.) Palla [1, 4, 5]

Scirpus mucronatus L.

ठुलो मोथे (Thulo mothe)

subsp. **mucronatus**

Perennial aquatic herbs. C, E Nepal: 80-2000 m. Nepal, Asia and Europe.

Use(s): Forage, Construction materials (stem).

subsp. **robustus** (Miq.) T.Koyama

Scirpus mucronatus var. *robustus* Miq. (1, 4)

Perennial herbs. C, E Nepal: 700-1500 m. Nepal, S Asia, E Asia and SE Asia.

Schoenoplectus supinus (L.) Palla

subsp. **lateriflorus** (J.F.Gmel.) T.Koyama [1, 4]

Scirpus lateriflorus J.F.Gmel.; *Schoenoplectiella lateriflora* (J.F.Gmel.) Lye

पानि मोथे (Paani mothe)

Annual wetland herbs. E Nepal: 700-1200 m. Nepal, S Asia, E Asia, SE Asia and Australasia.

Schoenoplectus tabernaemontani (C.C.Gmel.) Palla [1, 4]

Scirpus tabernaemontani C.C.Gmel.

Perennial herbs. W, C Nepal: 2500-3700 m. Nepal, S Asia, C Asia, SW Asia and Europe.

Scirpus L.

Scirpus rosthornii Diels [1, 4]

Scirpus chinensis Munro

Annual herbs. E Nepal: 300-2600 m. Nepal, E Himalaya and E Asia.

Scleria P.J.Bergius

Scleria benthamii C.B.Clarke

Scleria khasiana C.B.Clarke

Annual or perennial herbs. W, C, E Nepal. Nepal, Asia and Australasia.

Scleria biflora Roxb. [1, 4, 5]

Scleria biflora var. *ferruginea* J.Kern

करौंते झार (Karaunte jhaar)

Annual herbs. C, E Nepal: 600-1800 m. Nepal, W Himalaya, Assam-Burma, S Asia, E Asia and SE Asia.

Use(s): Food (young plant), Forage.

Scleria levis Retz. [1, 4, 5]

Scleria alta Boeckler (5); *Scleria hebecarpa* Nees

करौंते (Karaunte)

Perennial herbs. W, C, E Nepal: 100-1900 m. Nepal, Assam-Burma, S Asia, E Asia, SE Asia and Australasia.

Use: Medicine (rhizome, fruit).

Scleria parvula Steud. [1, 4, 5]

Scleria fenestrata Franch. & Sav.

Annual herbs. C, E Nepal: 1300-1600 m. Nepal, E Himalaya, Tibetan Plateau, S Asia, E Asia, SE Asia and Africa.

Scleria pergracilis (Nees) Kunth [1, 4, 5]

Hypoporum pergracile Nees

Annual herbs. C Nepal: 1300-1800 m. Nepal, W Himalaya, Tibetan Plateau, Assam-Burma, E Asia, S Asia, SE Asia, Africa to Australasia.

Use(s): Medicine (stem), Mosquito repellent (leaves).

Scleria terrestris (L.) Fassett [1, 4]

Zizania terrestris L.

Perennial herbs. C Nepal: *ca.* 2000 m. Nepal, W Himalaya, E Himalaya, Tibetan Plateau, Assam-Burma, S Asia, E Asia, SE Asia and Australasia.

Trichophorum Pers.

Trichophorum pumilum (Vahl) Schinz & Thell. [8]

Scirpus pumilus Vahl; *Baeothryon pumilum* (Vahl) T.Koyama

Dwarf clubrush

Perennial herbs. W, C Nepal: 1200-4000 m. Nepal, Asia, Europe and N America.

Doubtful/Uncertain taxa

Bulboschoenus maritimus (L.) Palla [*Schoenoplectus maritimus* (L.) Lye] [124]

Cyperus scariosus R.Br. [124]

Cyperus wallichianus Spreng. [1, 4]

Schoenoplectiella roylei (Nees) Lye [*Schoenoplectus roylei* (Nees) Ovcz. & Czukav.] [124]

POACEAE (GRAMINEAE)

Achnatherum P.Beauv.

Achnatherum sibiricum (L.) Keng ex Tzvelev

Avena sibirica L.; *Stipa sibirica* (L.) Lam. (1, 4, 8)

Needle grass

Perennial herbs. W, C Nepal: 2600-3600 m. Nepal, Himalaya, Tibetan Plateau, S Asia, E Asia, N Asia, C Asia and SW Asia.

Agropyron Gaertn.

Agropyron thomsonii Hook.f.

Elymus thomsonii (Hook.f.) Melderis (1, 4)

Perennial herbs. W, C Nepal: 2700-3800 m. Nepal and S Asia.

Agrostis L.

Agrostis gigantea Roth [1, 4]

Agrostis alba L.; *Agrostis stolonifera* var. *gigantea* (Roth) Koch; *Agrostis stolonifera* var. *ramosa* (Gray) Veldkamp (5)

Perennial herbs. W, C Nepal: 2900-4000 m. Nepal, Asia, Europe and Africa.

Agrostis hookeriana C.B.Clarke ex Hook.f. [1, 4, 5]

Agrostis poluninii Bor

Annual herbs. C, E Nepal: 3500-5000 m. Nepal, E Himalaya, Tibetan Plateau, S Asia and E Asia.

Agrostis inaequiglumis Griseb. [1, 4, 5]

Annual herbs. C, E Nepal: 3500-4500 m. Nepal, E Himalaya and S Asia.

Use: Forage.

Agrostis micrantha Steud. [1, 4, 5]

Agrostis myriantha Hook.f. (1, 4)

Perennial herbs. W, C, E Nepal: 1400-3500 m. Nepal, E Himalaya, Tibetan Plateau, Assam-Burma, E Asia and SE Asia.

Use: Forage.

Agrostis munroana Aitch. & Hemsl. [1, 4, 5]

Calamagrostis munroana (Aitch. & Hemsl.) Boiss.

Annual herbs. W, C, E Nepal: 2500-4900 m. Nepal, W Himalaya, S Asia, E Asia and SW Asia.

Agrostis nervosa Nees ex Trin. [1, 4, 5]

Agrostis clarkei Hook.f.; *Agrostis sikkimensis* Bor (1,4)

Annual herbs. W, C, E Nepal: 2300-4300 m. Nepal, E Himalaya, Assam-Burma, S Asia and E Asia.

Agrostis pendryi Paszko [8]

Annual herbs. E Nepal: 3100-4200 m. Nepal and W Himalaya.

Agrostis pilosula Trin. [1, 4, 5]

Agrostis ciliata Trin.; *Agrostis wallichiana* Steud.; *Calamagrostis jacquemontii* Hook.f.

Annual herbs. W, C, E Nepal: 1550-4600 m. Nepal, E Himalaya, Tibetan Plateau, S Asia and SW Asia.

Agrostis stolonifera L. [1, 4, 5]

Agrostis palustris Hudson

Creeping bent

Perennial herbs. W, C Nepal: 1800-3500 m. Nepal, Asia and Europe.

Use: Forage.

Alloteropsis J.Presl.

Alloteropsis cimicina (L.) Stapf [5]

Milium cimicinum L.; *Agrostis cimicina* (L.) Poir.; *Panicum cimicinum* (L.) Retz.

Annual herbs. W, C, E Nepal: 700-800 m. Nepal, Assam-Burma, S Asia, E Asia, SE Asia, Africa and Australasia.

Alloteropsis semialata (R.Br.) Hitchc. [5]

Panicum semialatum R.Br.; *Urochloa semialata* (R.Br.) Kunth

Perennial herbs. W Nepal: *ca.* 1000 m. Nepal, S Asia, E Asia, SE Asia, Africa and Australasia.

Alopecurus L.

Alopecurus aequalis Sobol. [1, 4, 5]

Alopecurus aristulatus Michaux

Orange foxtail

Annual or short-lived perennial herbs. W, C, E Nepal: 800-2900 m. Nepal, Asia, Europe and N America.

Alopecurus geniculatus L. [1, 4]

Alopecurus australis Nees; *Alopecurus paludosus* Crantz

Marsh foxtail

Perennial herbs. W Nepal: *ca.* 1700 m. Nepal and temperate regions.

Use: Forage.

Alopecurus nepalensis Trin. ex Steud. [5]

Alopecurus borii Tzvelev

Annual swampy herbs. W Nepal: 200-300 m. Nepal, E Himalaya, S Asia, N Asia and SW Asia.

Ampelocalamus S.L.Chen, T.H.Wen & G.Y.Sheng

Ampelocalamus patellaris (Gamble) Stapleton [4, 17]

Dendrocalamus patellaris Gamble; *Patellocalamus patellaris* (Gamble) W.T.Lin

गोपी बाँस, लेवस बाँस (Gopi baans, Lewas baans)

Perennial woody thornless bamboos. C, E Nepal: 1200-1800 m. Nepal, E Himalaya, Assam-Burma, S Asia, E Asia and SE Asia.

Andropogon L.

Andropogon munroi C.B.Clarke [1, 4]

Andropogon tristis Nees ex Hack.; *Cymbopogon hookeri* (Munro ex Hack.) Stapf ex Bor; *Cymbopogon munroi* (C.B.Clarke) Noltie (5)

ढल्के खर (Dhalke khar)

Perennial herbs. W, C, E Nepal: 1800-4000 m. Nepal, W Himalaya, E Himalaya, Assam-Burma and E Asia.

Use: Forage.

Aniselytron Merr.

Aniselytron treutleri (Kuntze) Soják [8]

Milium treutleri Kuntze; *Calamagrostis treutleri* (Kuntze) U.Shukla.

Perennial herbs. C Nepal: *ca.* 2700 m. Nepal, E Himalaya, Assam-Burma, S Asia, E Asia and SE Asia.

Anthoxanthum L.

Anthoxanthum hookeri (Griseb.) Rendle [1, 4, 5]

Ataxia hookeri Griseb.; *Hierochloe hookeri* (Griseb.) Maxim.

Perennial herbs. W, C, E Nepal: 2300-3400 m. Nepal, E Himalaya, Tibetan Plateau, Assam-Burma, S Asia and E Asia.

Anthoxanthum odoratum L. [5]

Sweet vernal grass

Perennial herbs. C Nepal: *ca.* 1500 m. Nepal, Tibetan Plateau, E Asia, N Asia and Europe.

Use: Perfume (essential oil).

Anthoxanthum sikkimense (Maxim.) Ohwi [1, 4]

Hierochloe sikkimensis Maxim.; *Hierochloe gracillima* Hook.f.

Annual herbs. C Nepal: *ca.* 3800 m. Nepal, E Himalaya, S Asia and E Asia.

Apluda L.

Apluda mutica L. [1, 4, 5]

Apluda aristata L.

डाक्ले खर, डाक्ले झार (Daakle khar, Dakle jhaar)

Perennial herbs. W, C, E Nepal: 100-2500 m. Nepal, Himalaya, Tibetan Plateau, Assam-Burma, S Asia, SE Asia, SW Asia and Australasia.

Use(s): Forage, Construction materials.

Apocopis Nees

Apocopis paleaceus (Trin.) Hochr. [1, 4, 5]

Ischaemum paleaceum Trin.; *Andropogon himalayensis* Steud.; *Andropogon paleaceus* (Trin.) Steud.

Perennial herbs. W, C, E Nepal: 900-2000 m. Nepal, E Himalaya, Assam-Burma, S Asia, E Asia and SE Asia.

Aristida L.

Aristida adscensionis L. [1, 4, 5]

Aristida vulgaris Trin. & Rupr.

Annual or perennial herbs. W, C, E Nepal: 200-2100 m. Nepal, E Asia, N Asia, Pantropical and Warm Temperate.

Use: Forage.

Aristida cumingiana Trin. & Rupr. [1, 4]

Aristida delicatula Hochst. Ex A.Rich.

Annual herbs. C Nepal: *ca.* 1400 m. Nepal, Assam-Burma, S Asia, E Asia, SE Asia, Africa and Australasia.

Aristida cyanantha Nees ex Steud. [1, 4]

Chaetaria cyanantha (Steud.) Nees

Perennial herbs. W Nepal: 900-1500 m. Nepal, W Himalaya, S Asia and SW Asia.

Use: Forage.

Arthraxon P.Beauv.

Arthraxon hispidus (Thunb.) Makino [5]

Phalaris hispida Thunb.; *Artharxon micans* (Nees) Hochst. (1, 4); *Arthraxon quartinianus* (A.Rich.) Nash (1, 4, 5)

झ्यामटे (Jhyamte); Small carpet grass

Annual herbs. W, C, E Nepal: 300-2900 m. Nepal, Asia, Africa and Australasia.

Use: Forage.

Arthraxon lanceolatus (Roxb.) Hochst. [1, 4]

Andropogon lanceolatus Roxb.

Annual herbs. W, C, E Nepal: 1000-2800 m. Nepal, S Asia, E Asia, SE Asia and Africa.

Use: Forage.

Arthraxon lancifolius (Trin.) Hochst. [1, 4, 5]

Andropogon lancifolius Trin.; *Andropogon multicaulis* Steud.

चित्रे घाँस, चित्रे बाँस (Chitre ghaans, Chitre baans)

Annual herbs. W, C, E Nepal: 900-1900 m. Nepal, Asia and Africa.

Use: Forage.

Arthraxon microphyllus (Trin.) Hochst. [5]

Andropogon microphyllus Trin.; *Arthraxon sikkimensis* Bor (1, 4)

Annual slender herbs. C, E Nepal: 1700-2700 m. Nepal, E Himalaya, E Asia and SE Asia.

Use: Forage.

Arthraxon prionodes (Steud.) Dandy [1, 4, 5]

Andropogon prionodes Steud.

Perennial herbs. W, C Nepal: 1100-2200 m. Nepal, E Himalaya, Tibetan Plateau, Assam-Burma, S Asia, SW Asia and Africa.

Arthraxon submuticus (Nees ex Steud.) Hochst. [1, 4]

Andropogon submuticus Nees ex Steud.; *Batratherum submuticum* (Nees ex Steud.) W.Watson

Annual herbs. E Nepal: *ca.* 1800 m. Nepal, W Himalaya, S Asia, E Asia and SE Asia.

Arthraxon typicus (Buse) Koord.

Lucaea typica Buse; *Arthraxon breviaristatus* Hack. [1, 4]

Perennial herbs. W Nepal: *ca.* 2900 m. Nepal, E Himalaya, Assam-Burma, E Asia and SE Asia.

Arundinaria Michx.

Arundinaria racemosa Munro [17]

Sarocalamus racemosus (Munro) Stapleton (122); *Yushania racemosa* (Munro) R.B.Majumdar

Spreading woody thornless bamboo. E Nepal: 2900-3600 m. Nepal and E Himalaya.

Use(s): Construction materials (culm), Fodder.

Arundinella Raddi

Arundinella bengalensis (Spreng.) Druce [1, 4, 5]

Panicum bengalense Spreng.; *Arundinella wallichii* Nees ex Steud.

Perennial herbs. W, C, E Nepal: 100-1600 m. Nepal, E Himalaya, Tibetan Plateau, Assam-Burma, E Asia and SE Asia.

Use: Forage.

Arundinella birmanica Hook.f. [1, 4, 5]

Arundinella setosa Trin.

Annual herbs. C, E Nepal: 1300-2600 m. Nepal and Assam-Burma.

Use: Forage.

Arundinella hookeri Munro ex Keng [1, 4, 5]

Arundinella villosa var. *himalaica* Hook.f.

फुर्के खर (Phurke khar)

Perennial herbs. W, C, E Nepal: 2300-3500 m. Nepal, E Himalaya, Tibetan Plateau, Assam-Burma, E Asia and SE Asia.

Use(s): Forage, Construction materials.

Arundinella nepalensis Trin. [1, 4, 5]

Arundinella glabra Nees ex Hook. & Arn.

फुर्के खर, मुसे खरी (Phurke khar, Muse khari); River grass

Perennial herbs. W, C, E Nepal: 100-2500 m. Nepal, Asia, Africa and Australasia.

Use(s): Fence, Construction materials, Forage, Medicine (leaves).

Arundinella pumila (Hochst. ex A.Rich.) Steud. [1, 4, 5]

Acratherum pumilum Hochst. ex A.Rich.; *Arundinella tenella* Nees ex Steud.

Annual herbs. W, C, E Nepal: 600-1600 m. Nepal, Assam-Burma, S Asia, SE Asia and Africa.

Use: Forage.

Arundinella setosa Trin.

var. **esetosa** Bor ex S.M.Phillips & S.L.Chen [1, 4]

Arundinella bidentata Keng

Perennial herbs. W, C, E Nepal: 500-2600 m. Nepal, E Himalaya, Assam-Burma, S Asia, E Asia, SE Asia and Australasia.

Arundo L.

Arundo donax L. [1, 4, 5]

Donax arundinaceus P.Beauv.

ठुलो नर्कट (Thulo narkat); Reed bamboo

Perennial herbs. W, C, E Nepal: 1500-2500 m. Nepal, Asia, Europe and Africa.

Use(s): Medicine (root, leaves), Forage, Utensils and handicrafts, Fence.

Avena L.

Avena barbata Pott ex Link [1, 4]

Avena hirsuta Moench

Annual herbs. C Nepal: *ca.* 3200 m. Nepal, W Himalaya, S Asia, N Asia, Europe and Africa.

Use: Forage.

Avena fatua L. [1, 4, 5]

जंगली जौ (Jangali jau); Wild oat

Annual herbs. W, C, E Nepal: 100-3800 m. Nepal, Asia, Africa and Europe.

Use(s): Forage, Construction materials, Medicine (root, seeds).

Avena sativa L. [5]

जौ (Jau); Oats

Annual herbs. W, C, E Nepal: 1500-2200 m. Widely cultivated in Nepal. Native to SW Asia and Europe.

Use: Medicine (seeds, straw).

Axonopus P.Beauv.

Axonopus compressus (Sw.) P.Beauv. [1, 4, 5]

Milium compressum Sw.

Cow grass

Perennial herbs. C, E Nepal: 400-1800 m (Naturalized). Native to America.

Bambusa Schreb.

Bambusa balcooa Roxb. [4, 5, 17]

Dendrocalamus balcooa (Roxb.) Voigt

धनु बाँस, बन बाँस (Dhanu baans, Ban baans); Bamboo

Perennial herbs. C Nepal: 900-1600 m. Nepal, Assam-Burma and S Asia.

Use(s): Medicine (siliceous secretion of the culm), Utensils and handicrafts, Construction materials, Ritual.

Bambusa jaintiana Majumdar

Bambusa alamii Stapleton (4, 17)

मुगी बाँस (Mugi baans)

Perennial herbs. C, E Nepal: 100-1000 m. Nepal, E Himalaya and S Asia.

Use(s): Forage, Utensils and handicrafts, Construction materials, Ritual.

Bambusa multiplex (Lour.) Raeusch. ex Schult. & Schult.f. [4, 17]

Arundo multiplex Lour.; *Bambusa nana* Roxb.

ठुलो निंगालो (Thulo nigalo); Chinese dwarf bamboo

Perennial herbs. C Nepal: 200-2000 m. Nepal, E Himalaya, S Asia, E Asia and SE Asia.

Use(s): Fodder, Utensils and handicrafts, Construction materials, Ritual.

Bambusa nepalensis Stapleton [4, 17]

तामा बाँस, फुस्रे बाँस (Tama baans, Phusre baans)

Perennial herbs. C Nepal: 1600-1700 m. Nepal, widely planted.

Use(s): Food (young shoot), Forage, Construction materials, Utensils and handicrafts, Ritual.

Bambusa nutans Wall. ex Munro [1, 4, 17]

Bambusa falconeri Munro

थारु बाँस, साते बाँस (Thaaru baans, Saate baans)

Perennial herbs. W, C Nepal: 100-1700 m. Nepal, W Himalaya, E Himalaya and S Asia.

Use(s): Forage, Utensils and handicrafts, Construction materials, Ritual.

Bambusa teres Buch.-Ham. ex Munro

Bambusa nutans subsp. *cupulata* Stapleton (4, 8, 17)

माल बाँस (Maal baans)

Perennial herbs. E Nepal: 100-1600 m. Nepal, E Himalaya, Tibetan Plateau, Assam-Burma, S Asia and E Asia.

Use(s): Forage, Construction materials, Utensils and handicrafts.

Bambusa tulda Roxb. [4, 17]

Dendrocalamus tulda (Roxb.) Voigt

काँडा बाँस, तारु बाँस (Kandaa baans, Taaru baans)

Perennial herbs. C Nepal: 100-1400 m. Nepal, E Himalaya, S Asia, E Asia and SE Asia.

Use(s): Medicine (siliceous secretion of the culm), Food, Utensils and handicrafts, Construction materials, Ritual.

Bothriochloa Kuntze

Bothriochloa bladhii (Retz.) S.T.Blake [5]

Andropogon bladhii Retz.; *Bothriochloa glabra* (Roxb.) A.Camus (1, 4); *Bothriochloa intermedia* (R.Br.) A.Camus (1, 4); *Bothriochloa intermedia* var. *punctata* (Roxb.) Keng

Perennial herbs. W, C, E Nepal: 200-2600 m. Nepal, Asia, Africa and Australasia.

Use: Fodder.

Bothriochloa ischaemum (L.) Keng [1, 4, 5]

Andropogon ischaemum L.

Perennial herbs. W, C, E Nepal: 100-3600 m. Nepal, Asia, Europe and Africa.

Bothriochloa kuntzeana (Hack.) Henrard [1, 4, 5]

Andropogon kuntzeanus Hack.

Perennial herbs. W, C Nepal: 1800-2900 m. Nepal and S Asia.

Bothriochloa pertusa (L.) A.Camus [1, 4, 5]

Holcus pertusus L.; *Andropogon pertusus* (L.) Willd.

Sour grass

Perennial herbs. W, C, E Nepal: 100-3800 m. Nepal, S Asia, E Asia, SE Asia, SW Asia and Africa.

Use: Fodder.

Brachiaria (Trin.) Griseb.

Brachiaria distachya (L.) Stapf

Panicum distachyon L.; *Urochloa distachya* (L.) T.Q.Nguyen (5)

Annual herbs. C Nepal: 200-1100 m. Nepal, S Asia, SE Asia, Africa and Australasia.

Use: Forage.

Brachiaria kurzii (Hook.f.) A.Camus [1, 4]

Panicum kurzii Hook.f.; *Urochloa kurzii* (Hook.f.) T.Q.Nguyen (5)

Annual herbs. C, E Nepal: 100-1600 m. Nepal, S Asia, E Asia, SE Asia and Australasia.

Brachiaria mutica (Forssk.) Stapf.

Panicum muticum Forssk.; *Urochloa mutica* (Forssk.) T.Q.Nguen

Buffalo grass

Perennial aquatic herbs. W, C, E Nepal: *ca.* 200 m. Nepal, E Asia, Africa, America and Pantropical.

Use: Forage.

Brachiaria ramosa (L.) Stapf [1, 4]

Panicum ramosum L.; *Urochloa ramosa* (L.) T.Q.Nguyen (5);

बाँस पाते, लिखे बन्सो (Baans paate, Likhe banso); Signal grass

Annual wetland herbs. W, C, E Nepal: 100-1800 m. Nepal, Asia, Africa and Old World tropics.

Use(s): Food (grains), Forage.

Brachiaria reptans (L.) C.A.Gardner & C.E.Hubb. [1, 4]

Panicum reptans L.; *Urochloa reptans* (L.) Stapf

Running grass

Annual weedy herbs. C Nepal: 100-1200 m. Nepal, Australasia and Pantropical.

Use: Fodder.

Brachiaria subquadripara (Trin.) Hitchc. [1, 4]

Panicum subqudriparum Trin.; *Urochloa subquadripara* (Trin.) R.D.Webster (5)

Annual mat forming herbs. W, C, E Nepal: 200-1000 m. Nepal, E Asia, Australasia and Pantropical.

Brachiaria villosa (Lam.) A.Camus [1, 4]

Panicum villosum Lam.; *Brachiaria villosa* var. *barbata* Bor (1, 4); *Urochloa villosa* (Lam.) T.Q.Nguen; *Urochloa villosa* var. *barbata* (Bor) Noltie (5)

Annual herbs. W, C, E Nepal: 100-2400 m. Nepal, E Himalaya, Tibetan Plateau, Assam-Burma, S Asia, E Asia, SE Asia and Africa.

Brachypodium P.Beauv.

Brachypodium pinnatum (L.) P.Beauv. [1, 4, 5]

Bromus pinnatus L.; *Agropyron pinnatum* (L.) Chevall.

Tor grass

Perennial herbs. W, C, E Nepal: 1800-3800 m. Nepal, Tibetan Plateau, E Asia, N Asia, C Asia, SW Asia and Europe.

Brachypodium sylvaticum (Huds.) P.Beauv. [1, 4, 5]

Festuca sylvatica Huds.; *Brachypodium sylvaticum* var. *luzoniense* (Hack.) H.Hara (1, 4)

Perennial herbs. W, C, E Nepal: 1800-3500 m. Nepal, Asia, Europe and Africa.

Briza L.

Briza media L. [1, 4, 5]

Poa media (L.) Cav.

Perennial herbs. W, C, E Nepal: 800-3800 m. Nepal, Asia and Europe.

Bromus L.

Bromus catharticus Vahl [8]

Bromus unioloides Kunth; *Bromus willdenowii* Kunth (1, 4)

Perennial herbs. C Nepal: 1400-2000 m. Nepal, E Asia, native to S America.

Use: Fodder.

Bromus himalaicus Stapf [1, 4]

Bromopsis himalaica (Stapf) Holub

Perennial herbs. W, C, E Nepal: 3300-3800 m. Nepal, E Himalaya, Tibetan Plateau and E Asia.

Bromus japonicus Houtt. [1, 4, 5]

Annual herbs. W, C Nepal: 3600-4200 m. Nepal, Tibetan Plateau, E Asia, N Asia, C Asia, SW Asia, Europe and Africa.

Bromus nepalensis Melderis [4, 27]

Perennial herbs. W, C Nepal: *ca.* 3000 m. Nepal and Tibetan Plateau.

Bromus porphyranthos Cope [1, 4]

Bromus grandis (Stapf) Melderis; *Bromus himalaicus* var. *grandis* Stapf (5)

Perennial herbs. W, C, E Nepal: 2800-4200 m. Nepal, E Himalaya, Tibetan Plateau, S Asia, E Asia and SW Asia.

Bromus staintonii Melderis [1, 4, 5]

Bromus staintonii var. *pilosiusculus* Melderis (1, 4)

Perennial herbs. W, C, E Nepal: 2900-3400 m. Nepal, W Himalaya, E Himalaya and Tibetan Plateau.

Bromus tectorum L. [1, 4, 5]

Festuca tectorum (L.) Jess.

जुके (Juke-Tibetan); Downy broome

Annual herbs. W, C Nepal: 3100-4100 m. Nepal, Asia, Europe and Africa.

Use(s): Medicine (grains), Forage.

Calamagrostis Adans.

Calamagrostis emodensis Griseb. [1, 4, 5]

Calamagrostis garhwalensis C.E.Hubb. & Bor (1, 4)

Garhwal reed grass

Perennial herbs. W, C, E Nepal: 2300-3700 m. Nepal, W Himalaya, E Himalaya, Tibetan Plateau, S Asia and E Asia.

Calamagrostis epigeios (L.) Roth

Arundo epigeios L.; *Calamagrostis epigejos* (L.) Roth (1, 4)

Perennial herbs. C, E Nepal: 3000-3200 m. Nepal, Asia, Europe and Africa.

Calamagrostis filiformis Griseb.

Agrostis triaristata (Hook.f.) Bor (1, 4, 5); *Deyeuxia filiformis* (Griseb.) Hook.f.; *Deyeuxia triaristata* Hook.f.

Perennial rhizomatous herbs. C, E Nepal: 3100-4200 m. Nepal and E Himalaya.

Calamagrostis hedinii Pilg.

Calamagrostis pseudophragmites subsp. *tartarica* (Hook.f.) Tzvelev (8)

Perennial herbs. C Nepal: 2700-3800 m. Nepal, W Himalaya and Tibetan Plateau.

Calamagrostis pseudophragmites (Haller f.) Koeler [1, 4, 5]

Arundo pseudophragmites Haller f.; *Calamagrostis nepalensis* Nees ex Steud.

Perennial herbs. W, C, E Nepal: 1500-4600 m. Nepal, Asia and Europe.

Capillipedium Stapf

Capillipedium assimile (Steud.) A.Camus [1, 4, 5]

Andropogon assimilis Steud.; *Bothriochloa assimilis* (Steud. ex Zoll.) Ohwi

मुसे खरी, खरु (Muse khari, Kharu)

Perennial herbs. W, C, E Nepal: 200-2100 m. Nepal, E Himalaya, Tibetan Plateau, Assam-Burma, S Asia, E Asia and SE Asia.

Use: Forage.

Capillipedium parviflorum (R.Br.) Stapf [1, 4, 5]

Holcus parviflorus R.Br.; *Bothriochloa parviflora* (R.Br.) Ohwi

Golden beard

Perennial herbs. W, C, E Nepal: 200-2700 m. Nepal, Asia, Africa, Australasia and Pantropical.

Use: Forage.

Centotheca Desv.

Centotheca lappacea (L.) Desv. [1, 4, 5]

Cenchrus lappaceus L.

Perennial herbs. C, E Nepal: 200-500 m. Nepal, E Himalaya, Assam-Burma, S Asia, E Asia, SE Asia, Africa and Australasia.

Use: Forage.

Cephalostachyum Munro

Cephalostachyum capitatum Munro [4]

गोहिया बाँस, गोबिया (Gohiya baans, Gobiya); Tufted bamboo

Perennial herbs. W, C, E Nepal: 1900-3000 m. Nepal, E Himalaya and Assam-Burma.

Use(s): Medicine (leaves), Forage, Construction materials.

Cephalostachyum latifolium Munro [4, 17]

Schizostachyum latifolium (Munro) R.B.Majumdar

गोपी बाँस, मुरली बाँस (Gopi baans, Murali baans)

Perennial herbs. E Nepal: 100-3000 m. Nepal, E Himalaya, Assam-Burma and E Asia.

Use(s): Forage, Utensils and handicrafts, Construction materials, Ornamental, Ritual.

Chionachne R.Br.

Chionachne gigantea (J.Koenig) Veldcamp

Coix gigantea J.Koenig; *Chionachne koenigii* (Spreng.) Thwaites (1, 4); *Coix barbata* Roxb.

काँस (Kans)

Perennial herbs. W Nepal: *ca.* 600 m. Nepal, Tibetan Plateau, Assam-Burma, S Asia, E Asia, SE Asia and SW Asia.

Use: Medicine (whole plant).

Chloris Sw.

Chloris barbata Sw. [5]

Andropogon barbatus L.; *Chloris inflata* Link

Giant finger grass

Annual or perennial herbs. W Nepal: *ca.* 100 m. Nepal, Asia, Africa, America and Australasia.

Chloris virgata Sw. [1, 4, 5]

Chloris caudata Trin. ex Bunge

Feather finger grass

Annual herbs. W Nepal: 2000-2900 m. Nepal, Asia, Africa, America and Australasia.

Use(s): Medicine (stem, leaves), Forage.

Chrysopogon Trin.

Chrysopogon aciculatus (Retz.) Trin. [1, 4, 5]

Andropogon aciculatus Retz.

घोडे दुबो, कुरे घाँस (Ghode dubo, Kure ghaans); Love grass

Perennial herbs. W, C, E Nepal: 100-1700 m. Nepal, Asia and Australasia.

Use(s): Medicine (whole plant), Forage.

Chrysopogon fulvus (Spreng.) Chiov. [1, 4]

Pollinia fulva Spreng.; *Andropogon monticola* Roem. & Schult.

घोरिया (Ghoriyaa); Red beard grass

Perennial herbs. C Nepal: 300-900 m. Nepal, Asia and Africa.

Use: Forage.

Chrysopogon gryllus (L.) Trin. [1, 4, 5]

Andropogon gryllus L.

धाप्ले घाँस, पटपटे खर (Dhaaple ghaans, Patapate khar); Scented grass

Perennial herbs. W, C, E Nepal: 800-2900 m. Nepal, Asia, Europe and Africa.

Use(s): Medicine (root), Forage.

Chrysopogon lawsonii (Hook.f.) Veldkamp

Andropogon lawsonii Hook.f.; *Vetiveria lawsonii* (Hook.f.) Blatt. & McCann (1, 4)

Perennial herbs. E Nepal: *ca.* 200 m. Nepal and S Asia.

Use: Forage.

Chrysopogon serrulatus Trin. [1, 4, 5]

Andropogon trinii Steud.

Golden beard grass

Perennial herbs. W, C, E Nepal: 200-1400 m. Nepal and W Himalaya.

Use: Forage.

Chrysopogon zizanioides (L.) Roberty [5]

Phalaris zizanioides L.; *Vetiveria zizanioides* (L.) Nash

खस खस (Khas khas); Vetiver grass

Perennial herbs. W, C, E Nepal: 100-1100 m. Nepal, Tibetan Plateau, Assam-Burma, S Asia, E Asia, SE Asia and Africa.

Use(s): Forage, Medicine (root), Ritual.

Coelachne R.Br.

Coelachne simpliciuscula (Wight & Arn. ex Steud.) Munro ex Benth. [1, 4, 5]

Panicum simpliciusculum Wight & Arn. ex Steud.

Annual aquatic herbs. C, E Nepal: 1000-2500 m. Nepal, E Himalaya, Assam-Burma, S Asia, E Asia, SE Asia and Africa.

Use(s): Food, Forage.

Coix L.

Coix lacryma-jobi L. [1, 4, 5]

Coix arundinacea Lam.

भिरकौंलो, जाबे (Bhirkaulo, Jaabe); Job's tear millet

Perennial herbs. W, C, E Nepal: 100-2100 m. Nepal, E Himalaya, Tibetan Plateau, Assam-Burma and Asia.

Use(s): Medicine (root, seeds), Food, Forage, Construction materials.

Colpodium Trin.

Colpodium wallichii (Stapf) Bor [1, 4, 5]

Catabrosa wallichii Stapf; *Paracolpodium wallichii* (Stapf) E.B.Alexeev

Perennial herbs. W, C, E Nepal: 4100-5000 m. Nepal, E Himalaya, Tibetan Plateau and S Asia.

Cymbopogon Spreng.

Cymbopogon distans (Nees ex Steud.) W.Watson [1, 4, 5]

Andropogon distans Nees ex Steud.

पिरे घाँस (Pire ghaans); Lemon grass

Perennial herbs. W, C Nepal: 1400-2900 m. Nepal, W Himalaya, Tibetan Plateau, S Asia, E Asia and SW Asia.

Use(s): Medicine (whole plant), Forage, Construction materials.

Cymbopogon flexuosus (Nees ex Steud.) W.Watson [1, 4]

Andropogon flexuosus Nees ex Steud.; *Cymbopogon flexuosus* var. *sikkimensis* Bor (5)

कागती घाँस (Kaagati ghaans); East Indian lemon grass

Perennial herbs. C Nepal: 900-1100 m. Nepal, Assam-Burma, S Asia, E Asia and SE Asia.

Use(s): Medicine (oil), Food, Forage.

Cymbopogon jwarancusa (Jones) Schult. [1, 4, 5]

Andropogon jwarancusa Jones

कागती घाँस, पहेंलो उसिर (Kaagati ghaans, Panhelo usir); Elephant grass

Perennial herbs. W, C, E Nepal: 200-2400 m. Nepal, E Himalaya, Tibetan Plateau, S Asia, E Asia, SE Asia and SW Asia.

Use(s): Medicine (whole plant), Construction materials.

Cymbopogon martini (Roxb.) W.Watson [1, 4, 5]

Andropogon martini Roxb.

पाम रोजा (Paam rojaa); Palma rose

Perennial herbs. W, C Nepal: 900-2000 m. Nepal, Assam-Burma, SE Asia and E Asia.

Use(s): Medicine (whole plant), Forage, Ritual.

Cymbopogon microthecus (Hook.f.) A.Camus [1, 4]

Andropogon microtheca Hook.f.

Perennial herbs. C, E Nepal: 400-1300 m. Nepal, E Himalaya, Assam-Burma and S Asia.

Cymbopogon osmastonii R.Parker [5]

Perennial herbs. W, C, E Nepal: 100-200 m. Nepal and S Asia.

Cymbopogon pendulus (Nees ex Steud.) W.Watson [1, 4, 5]

Andropogon pendulus Nees ex Steud.

पिरे, कागती घाँस (Pire, Kaagati ghaans); Lemon grass

Perennial herbs. W, C, E Nepal: 100-2600 m. Nepal, E Himalaya, Assam-Burma, E Asia and SE Asia.

Use: Forage.

Cymbopogon pospischilii (K.Schum.) C.E.Hubb. [5]

Andropogon pospischilii K.Schum.; *Cymbopogon stracheyi* (Hook.f.) Raizada & S.K.Jain (1, 4)

Bitter turpentine grass

Perennial herbs. W, C, E Nepal: 900-3300 m. Nepal, W Himalaya, Tibetan Plateau, S Asia, E Asia, SW Asia and Africa.

Cymbopogon schoenanthus (L.) Spreng. [1, 4]

Andropogon schoenanthus L.

रूसा घाँस, पिरे (Rusaa ghaans, Pire); Camel grass

Perennial herbs. C Nepal: 2400-3000 m. Nepal, W Himalaya, S Asia, SW Asia and Africa.

Use: Medicine (whole plant).

Cynodon Rich.

Cynodon dactylon (L.) Pers. [1, 4, 5]

Panicum dactylon L.

दुबो (Dubo); Bermuda grass

Perennial herbs. W, C, E Nepal: 100-3000 m. Nepal, Tibetan Plateau, E Asia, Pantropical and Temperate region.

Use(s): Medicine (whole plant), Forage, Utensils and handicrafts, Ornamental, Ritual.

Cynodon plectostachyus (K.Schum.) Pilg. [1, 4]

Leptochloa plectostachya K.Schum.

Star grass

Perennial herbs. E Nepal: 100-200 m. Nepal, S Asia and Africa.

Use: Forage.

Cynodon radiatus Roth ex Roem. & Schultes [5]

Cynodon arcuatus J.Presl (1, 4, 5)

Annual herbs. W, C, E Nepal: 100-3500 m. Nepal, Asia, Africa and Australasia.

Use: Forage.

Cyrtococcum Stapf

Cyrtococcum patens (L.) A.Camus [1, 4]

Panicum patens L.

var. **patens**

Perennial creeping herbs. C, E Nepal: 100-1500 m. Nepal, E Himalaya, Tibetan Plateau, Assam-Burma, S Asia, E Asia, SE Asia and Australasia.

Use: Forage.

var. **latifolium** (Honda) Ohwi

Panicum patens forma *latifolium* Honda; *Cyrtococcum accrescens* (Trin.) Stapf (1, 4)

Bow grass

Perennial mat forming herbs. W, C, E Nepal: 700-2100 m. Nepal and SE Asia.

Dactylis L.

Dactylis glomerata L. [1, 4, 5]

Dactylis glomerata subsp. *himalayensis* Domin

Orchard grass

Perennial herbs. W, C Nepal: 2500-3700 m. Nepal, Asia, Europe and Africa.

Use: Forage.

Dactyloctenium Willd.

Dactyloctenium aegyptium (L.) Willd. [1, 4, 5]

Cynosurus aegypticus L.; *Eleusine aegyptia* (L.) Desf.

माकुरे घाँस, दत्यौरे झार (Maakure ghaans, Datyaure jhaar); Egyptian finger grass

Annual herbs. W, C, E Nepal: 100-1000 m. Nepal, E Asia, Tropical and warm temperate regions of Old World.

Use: Medicine (grains).

Danthonia DC.

Danthonia cachemyriana Jaub. & Spach. [1, 4]

Danthonia exilis Hook.f.

Perennial herbs. C, E Nepal: 3300-4400 m. Nepal, Tibetan Plateau, S Asia, N Asia and SW Asia.

Use: Forage.

Danthonia cumminsii Hook.f. [1, 4, 5]

Danthonia cachemyriana var. *minor* Hook.f.

Perennial herbs. W, C, E Nepal: 2200-4800 m. Nepal, W Himalaya, E Himalaya, Tibetan Plateau, S Asia, E Asia and SW Asia.

Dendrocalamus Nees

Dendrocalamus hamiltonii Nees & Arn. ex Munro [1, 4, 5, 17]

चोया बाँस, तामा बाँस (Choya baans, Taama baans); Tufted bamboo

Perennial bamboo. C, E Nepal: 900-2000 m. Nepal, E Himalaya, Assam-Burma, S Asia, E Asia and SE Asia.

Use(s): Medicine (tender shoot, leaves), Food (young shoot), Fence, Construction materials, Utensils and handicrafts, Ritual.

Dendrocalamus hookeri Munro [4, 5, 17]

Bambusa hookeri (Munro) Rivière & C. Rivière

भालु बाँस, कालो बाँस (Bhaalu baans, Kaalo baans)

Perennial bamboo. E Nepal: 1500-1700 m. Nepal, E Himalaya and SE Asia.

Use(s): Food (young shoot), Construction materials, Utensils and handicrafts.

Dendrocalamus strictus (Roxb.) Nees [4, 5, 17]

Bambos stricta Roxb.; *Bambusa stricta* (Roxb.) Roxb.

तारु बाँस, माल बाँस (Taaru baans, Maal baans)

Perennial bamboo. C, E Nepal: 100-1800 m. Nepal, Tibetan Plateau, E Asia, S Asia, SE Asia and Africa.

Use(s): Medicine (leaves), Food (young shoot), Fodder, Utensils and handicrafts, Construction materials.

Deschampsia P.Beauv.

Deschampsia cespitosa (L.) P.Beauv. [1, 4, 5]

Aira cespitosa L.

Perennial herbs. W, C, E Nepal: 2900-4600 m. Nepal, Asia, Europe, Africa and N America.

Use(s): Medicine (whole plant), Forage, Utensils and handicrafts.

Deschampsia flexuosa (L.) Trin. [1, 4]

Aira flexuosa L.

Perennial herbs. W Nepal: *ca.* 3000 m. Nepal, E Asia, SE Asia, N Asia, SW Asia, Europe, Africa, N America and S America.

Desmostachya Stapf

Desmostachya bipinnata (L.) Stapf [1, 4, 5]

Briza bipinnata L.; *Eragrostis cynosuroides* (Retz.) P.Beauv.

कुश (Kush); Sacrificial grass

Perennial herbs. W, C, E Nepal: 100-700 m. Nepal, Assam-Burma, S Asia, E Asia, SE Asia, SW Asia, Africa and Australasia.

Use(s): Medicine (root), Forage, Utensils and handicrafts, Ritual.

Deyeuxia Clarion ex P.Beauv.

Deyeuxia diffusa Keng [8]

Calamagrostis diffusa (Keng) Keng.f.

Perennial herbs. C Nepal: *ca.* 3100 m. Nepal, Tibetan Plateau and E Asia.

Deyeuxia holciformis (Jaub. & Spach) Bor [1, 4]

Calamagrostis holciformis Jaub. & Spach (5); *Deyeuxia compacta* Munro ex Hook.f. (8)

Perennial rhizomatous herbs. W, C Nepal: 3600-4300 m. Nepal, W Himalaya, E Himalaya, Tibetan Plateau and SW Asia.

Deyeuxia nepalensis Bor [1, 4]

Calamagrostis staintonii G.Singh (5)

Perennial herbs. W, C Nepal: 3500-4600 m. Nepal and E Asia.

Deyeuxia nivicola Hook.f. [1, 4]

Calamagrostis nivicola (Hook.f.) Hand.-Mazz. (5)

Perennial tufted herbs. W, C, E Nepal: 3900-5700 m. Nepal, E Himalaya, Tibetan Plateau and SW Asia.

Deyeuxia petelotii (Hitchc.) S.M.Phillips & Wen L.Chen

Aulacolepis petelotii Hitchc.; *Agrostis petelotii* (Hitchc.) Noltie (5); *Calamagrostis petelotii* (Hitchc.) Govaerts

Perennial herbs. W, C, E Nepal: 2000-2800 m. Nepal, E Himalaya, E Asia and SE Asia.

Deyeuxia pulchella Hook.f. [1, 4]

Calamagrostis lahulensis G.Singh; *Calamagrostis pulchella* Griseb. (5); *Deyeuxia pulchella* var. *laxa* P.C.Kuo & S.L.Liu (8)

Perennial herbs. W, C, E Nepal: 2200-5000 m. Nepal, W Himalaya, E Himalaya, Tibetan Plateau and E Asia.

Deyeuxia scabrescens (Griseb.) Munro ex Duthie [1, 4]

Calamagrostis scabrescens Griseb. (5)

Perennial tufted herbs. W, C, E Nepal: 3000-4600 m. Nepal, W Himalaya, E Himalaya, Tibetan Plateau, Assam-Burma and E Asia.

Deyeuxia zenkeri (Trin.) Veldkamp

Agrostis zenkeri Trin. (1, 4, 5); *Calamagrostis zenkeri* (Trin.) Davidse

Annual herbs. C, E Nepal: 900-3100 m. Nepal and S Asia.

Dichanthium Willemet

Dichanthium annulatum (Forssk.) Stapf [1, 4, 5]

Andropogon annulatus Forssk.

अपन्गास घाँस, सेतो रांकी (Apangas ghaans, Seto raanki)

Perennial herbs. W, C, E Nepal: 100-1100 m. Nepal, Assam-Burma, S Asia, E Asia, SE Asia, SW Asia, Africa and Australasia.

Use: Forage.

Dichanthium caricosum (L.) A.Camus [5]

Andropogon caricosus L.

Hay grass

Perennial herbs. E Nepal: 300-1000 m. Nepal, Asia, Africa, S America and Australasia.

Use: Forage.

Digitaria Heist. ex Fabr.

Digitaria abludens (Roem. & Schult.) Veldkamp [1, 4, 5]

Panicum abluden Roem. & Schult.

Annual herbs. W, C, E Nepal: 300-2000 m. Nepal, E Himalaya, Assam-Burma, S Asia, E Asia, SE Asia and SW Asia.

Use(s): Forage, Ritual.

Digitaria ciliaris (Retz.) Koeler [1, 4, 5]

Panicum ciliare Retz.; *Panicum adscendens* Kunth

चित्रे बन्सो (Chitre banso); Summer grass

Annual herbs. W, C, E Nepal: 300-2000 m. Nepal, Tibetan Plateau, E Asia, N Asia and Tropics of old world.

Use: Forage.

Digitaria cruciata (Nees ex Steud.) A.Camus [1, 4, 5]

Panicum cruciatum Nees ex Steud.

Crab grass

Annual herbs. W, C, E Nepal: 1000-3500 m. Nepal, E Himalaya, Tibetan Plateau, Assam-Burma and E Asia.

Use: Forage.

Digitaria longiflora (Retz.) Pers. [1, 4, 5]

Paspalum longiflorum Retz.

Annual or perennial herbs. W, C, E Nepal: 300-1800 m. Nepal, Asia, America and Tropics of the Old World.

Use: Forage.

Digitaria radicosa (J.Presl) Miq. [1, 4, 5]

Panicum radicosum J.Presl

Trailing crab grass

Annual herbs. C, E Nepal: 700-2800 m. Nepal, Assam-Burma, S Asia, E Asia, SE Asia and Australasia.

Use: Forage.

Digitaria setigera Roth [1, 4, 5]

बन्सो (Banso), Bristly crab grass

Annual herbs. W, C, E Nepal: 700-2200 m. Nepal, E Himalaya, Assam-Burma, S Asia, E Asia SE Asia, Africa and Australasia.

Use: Forage.

Digitaria stricta Roth [1, 4, 5]

Digitaria royleana (Nees ex Hook.f.) Prain; *Setaria stricta* (Roth ex Roem. & Schult.) Kunth

var. **stricta**

Annual herbs. W, C, E Nepal: 1000-2400 m. Nepal, E Himalaya, Assam-Burma, S Asia and E Asia.

var. **denudata** (Link) Henrard [1, 4]

Digitaria denudata Link

Annual herbs. C Nepal: *ca.* 1300 m. Nepal, E Himalaya, Tibetan Plateau, Assam-Burma, S Asia and E Asia.

var. **glabrescens** Bor [1, 4]

Annual herbs. W Nepal: *ca.* 2400 m. Nepal, W Himalaya and E Asia.

Use: Forage.

Digitaria ternata (A.Rich.) Stapf [1, 4, 5]

Cynodon ternatus A.Rich.; *Paspalum ternatum* (A.Rich.) Hook.f.

Annual herbs. W, C, E Nepal: 1600-2500 m. Nepal, E Himalaya, Tibetan Plateau, S Asia, E Asia, SE Asia and Africa.

Use: Forage.

Digitaria violascens Link [1, 4, 5]

Violet crab grass

Annual herbs. W, C, E Nepal: 800-2300 m. Nepal, Asia, S America, Australasia and Pantropical.

Use: Forage.

Dimeria R.Br.

Dimeria fuscescens Trin. [1, 4]

Perennial herbs. C Nepal: 1000-2000 m. Nepal, S Asia and SE Asia.

Dimeria ornithopoda Trin. [1, 4, 5]

Dimeria tenera Trin.

Annual erect herbs. C, E Nepal: 1100-1900 m. Nepal, Tibetan Plateau, Assam-Burma, Asia and Australasia.

Duthiea Hack.

Duthiea brachypodium (P.Candargy) Keng & Keng f. [5]

Triavenopsis brachypodium P.Candargy; *Duthiea nepalensis* Bor (1, 4)

Perennial herbs. W, C Nepal: 3800-4500 m. Nepal, E Himalaya, Tibetan Plateau and E Asia.

Echinochloa P.Beauv.

Echinochloa colona (L.) Link [1, 4, 5]

Panicum colonum L.

समा घाँस (Sama ghaans); Barnyard grass

Annual wetland herbs. W, C, E Nepal: 100-2400 m. Nepal, Tibetan Plateau, E Asia, N Asia, Africa and Old world.

Use(s): Medicine (whole plant), Food (grains), Forage.

Echinochloa crusgalli (L.) P.Beauv. [1, 4, 5]

Panicum crusgalli L.

var. **crusgalli**

समा घाँस (Samaa ghaans); Barnyard grass

Annual wetland herbs. W, C, E Nepal: 300-2400 m. Nepal, Assam-Burma, Asia, Africa and America.

Use(s): Medicine (whole plant), Food (grains), Forage.

var. **breviseta** (Doll) Neilr. [8]

Panicum crus-galli var. *brevisetum* Doll

Annual wetland herbs. W, C, E Nepal: 1800-2400 m. Nepal, Assam-Burma, S Asia, SE Asia and Africa.

Echinochloa frumentacea Link [5]

Panicum frumentaceum Roxb.

Barnyard millet

Annual wetland weedy herbs. W, C, E Nepal: 1600-1700 m. Nepal and E Asia.

Use(s): Medicine (stem, leaves), Forage.

Echinochloa glabrescens Munro ex Hook.f. [1, 4, 8]

Echinochloa crus-galli var. *formosensis* Ohwi

Annual wetland herbs. W, C Nepal: 900-1200 m. Nepal, W Himalaya, E Himalaya, Assam-Burma and SW Asia.

Echinochloa picta (J.Koenig) P.W.Michael. [5]

Panicum pictum J.Koenig

Variegated cockspur grass

Perennial herbs. W, C, E Nepal: 100-300 m. Nepal, SE Asia, Africa and Australasia.

Echinochloa pyramidalis (Lam.) Hitchc. & Chase [1, 4, 5]

Panicum pyramidale Lam.

Antelope grass

Perennial wetland herbs. E Nepal: 200-2000 m. Nepal, S Asia, Africa and America.

Use: Forage.

Eleusine Gaertn.

Eleusine coracana (L.) Gaertn. [1, 4, 5]

Cynosurus coracanus L.

कोदो (Kodo); Finger millet

Annual herbs. W, C, E Nepal: 1100-2000 m. Widely cultivated in Nepal. Cultivated in tropics of Old World, introduced into the New World.

Use(s): Medicine (seeds), Food (seeds).

Eleusine indica (L.) Gaertn. [1, 4]

Cynosurus indicus L.

कोदो घाँस, बन काँडे (Kodo ghaans, Ban kaande); Goose grass

Annual herbs. W, C, E Nepal: 600-2600 m. Nepal, Tibetan Plateau, E Asia, Pantropical and Subtropical.

Use(s): Forage, Medicine (whole plant).

Elymus L.

Elymus antiquus (Nevski) Tzevelev

Agropyron antiquum Nevski; *Elymus microlepis* (Melderis) Melderis (4)

Wild rye grass

Perennial herbs. W Nepal: *ca.* 4000 m. Nepal, Tibetan Plateau, S Asia and E Asia.

Elymus canaliculatus (Nevski) Tzvelev [1]

Agropyron canaliculatum Nevski; *Elymus longearistatus* (Boiss.) Tzvelev

Perennial herbs. Nepal: 3400-4300 m. Nepal, Tibetan Plateau, S Asia, C Asia, N Asia and SW Asia.

Elymus dahuricus Griseb. [1, 4]

Clinelymus dahuricus (Turcz. ex Griseb.) Nevski

Perennial herbs. W, C Nepal: 2600-3000 m. Nepal, E Himalaya, Tibetan Plateau, S Asia, E Asia, N Asia, C Asia, SE Asia and SW Asia.

Elymus nepalensis (Melderis) Melderis, *Enum. Fl. Pl. Nepal* 1: 131. 1978. [4, 5, 27, 29]

Agropyron nepalense Melderis

Perennial herbs. W, C Nepal: 2000-4000 m. **Endemic to Nepal**.

West: *Jumla*, Lumsa, NW Jumla, 9500 ft. (2895 m), O. Polunin, WR Sykes, LHJ Williamms 5111 (Holotype: BM).

Elymus nutans Griseb. [1, 4, 5]

Clinelymus nutans (Grisebach) Nevski.

Perennial herbs. W, C, E Nepal: 2700-4600 m. Nepal, E Himalaya, Tibetan Plateau, S Asia, E Asia, SE Asia, C Asia, N Asia and SW Asia.

Elymus schrenkianus (Fisch. & C.A.Mey. ex Schrenk) Tzvelev [1, 4, 5]

Triticum schrenkianum Fisch. & C.A.Mey. ex Schrenk; *Agropyron schrenkianum* (Fisch. & C.A.Mey.) Drobov

Perennial herbs. W, C, E Nepal: 2700-4400 m. Nepal, E Himalaya, Tibetan Plateau, E Asia, N Asia and C Asia.

Elymus semicostatus (Nees ex Steud.) Melderis [1, 4, 5]

Triticum semicostatum Nees ex Steud., *Agropyron semicostatum* Nees ex Steud.

Drooping wild rye

Perennial herbs. W, C Nepal: 1700-4000 m. Nepal, W Himalaya, E Himalaya and SW Asia.

Elymus sibiricus L. [1, 4, 5]

Clinelymus sibiricus (L.) Nevski

Siberian wild rye

Perennial herbs. W, C Nepal: 3300-4100 m. Nepal, Tibetan Plateau, S Asia, E Asia, N Asia, SE Asia, C Asia and Europe.

Elymus sikkimensis (Melderis) Melderis [1, 4, 5]

Agropyron sikkimensis Melderis

Perennial herbs. W, C, E Nepal: 2000-3200 m. Nepal, E Himalaya and S Asia.

Elymus tangutorum (Nevski) Hand.-Mazz. [1, 4]

Clinelymus tangutorum Nevski

Perennial herbs. W Nepal: 2700-3000 m. Nepal, E Himalaya, Tibetan Plateau, E Asia and N Asia.

Elytrophorus P.Beauv.

Elytrophorus spicatus (Willd.) A.Camus [1, 4, 5]

Dactylis spicata Willd.; *Elytrophorus articulatus* P.Beauv.

Silver spike grass

Annual seasonally submerged herbs. C, E Nepal: 200-1400 m. Nepal and old world.

Enteropogon Nees

Enteropogon dolichostachyus (Lag.) Keng

Chloris dolichostachya Lag. (1, 4, 5); *Melica digitata* Roxb.

Perennial herbs. W, C, E Nepal: 100-500 m. Nepal, E Himalaya, Assam-Burma, S Asia, E Asia, SE Asia, SW Asia and Australasia.

Eragrostiella Bor

Eragrostiella nardoides (Trin.) Bor [1, 4, 5]

Eragrostis nardoides Trin.

चुपी (Chupi)

Perennial herbs. W, C, E Nepal: 100-1800 m. Nepal, Himalaya and S Asia.

Use: Forage.

Eragrostris L.

Eragrostis atrovirens (Desf.) Trin. ex Steud. [1, 4, 5]

Poa atrovirens Desf.

Perennial herbs. W, C, E Nepal: 200-3500 m. Nepal, Asia and Africa.

Use: Forage.

Eragrostis cilianensis (All.) Janch. [48]

Poa cilianensis All.

Annual herbs. W Nepal: *ca.* 2000 m. Nepal, Tibetan Plateau, E Asia, N Asia, Pantropical and Subtropical.

Use: Forage.

Eragrostis coarctata Stapf [1, 4, 8]

Perennial herbs. C, E Nepal: 100-1500 m. Nepal, Assam-Burma, S Asia and SE Asia.

Use: Forage.

Eragrostis cylindrica (Roxb.) Arn. [1, 4, 5]

Poa cylindrica Roxb.

Perennial herbs. W, C, E Nepal: 300-2200 m. Nepal and E Asia.

Use: Forage.

Eragrostis ferruginea (Thunb.) P.Beauv. [1, 4, 5]

Poa ferruginea Thunb.

Perennial herbs. C Nepal: 3500-3700 m. Nepal, E Himalaya, Tibetan Plateau, S Asia, E Asia and SE Asia.

Use: Forage.

Eragrostis gangetica (Roxb.) Steud. [1, 4, 5]

Poa gangetica Roxb.

बन्सो, चरी दाना (Banso, Chari danaa)

Annual herbs. C, E Nepal: 1700-1800 m. Nepal, Assam-Burma, S Asia, SE Asia, Africa and S America.

Use: Forage.

Eragrostis japonica (Thunb.) Trin. [1, 4, 5]

Poa japonica Thunb.

रांकी (Ranki)

Annual herbs. W, C, E Nepal: 100-1700 m. Nepal, E Himalaya, Assam-Burma, S Asia, E Asia and SE Asia.

Use(s): Forage, Construction materials.

Eragrostis minor Host [1, 4, 8]

Poa eragrostis L.

Annual herbs. W, C Nepal: 1400-3600 m. Nepal, Asia, Europe, Africa and Mediterranean region.

Use: Forage.

Eragrostis nigra Nees ex Steud. [1, 4, 5]

Eragrostis atropurpurea Hochst. ex Steud.; *Eragrostis deqenensis* B.S.Sun & S.Wang

फुर्के खर (Phurke khar)

Perennial herbs. W, C, E Nepal: 900-3000 m. Nepal, E Himalaya, Tibetan Plateau, Assam-Burma, S Asia, E Asia and SE Asia.

Use: Forage.

Eragrostis nutans (Retz.) Nees ex Steud. [1, 4]

Poa nutans Retz.; *Poa elegans* Roxb.

Perennial herbs. W, C, E Nepal: 700-2000 m. Nepal, S Asia, E Asia, SE Asia and N America.

Use: Forage.

Eragrostis papposa (Roem. & Schult.) Steud. [1, 4, 8]

Megastachya papposa Roem. & Schult.

Perennial herbs. W, C Nepal: 2700-2800 m. Nepal, W Himalaya and Europe.

Use: Forage.

Eragrostis pilosa (L.) P.Beauv. [1, 4]

Poa pilosa L.

कि (Ki-Gurung)

Annual herbs. W, C, E Nepal: 200-2800 m. Nepal, Tibetan Plateau, E Asia, SE Asia, Europe, Africa and Australasia.

Use: Forage.

Eragrostis tenella (L.) P.Beauv. ex Roem. & Schult. [1, 4]

Poa tenella L.; *Eragrostis amabilis* (L.) Wight & Arn (5)

जुंगे बन्सो (Junge banso)

Annual herbs. W, C, E Nepal: 100-1400 m. Nepal, Tibetan Plateau, E Asia and Old world.

Use: Forage.

Eragrostis tenuifolia (A.Rich.) Hochst. ex Steud. [5]

Poa tenuifolia A.Rich.

Perennial herbs. C Nepal: 2400-3200 m. Nepal, native of S America, N America and Australasia.

Eragrostis tremula Hochst. ex Steud. [1, 4, 5]

Poa tremula Lam.

Annual herbs. C, E Nepal: 100-900 m. Nepal, Assam-Burma, S Asia, SE Asia, SW Asia and Africa.

Use: Forage.

Eragrostis unioloides (Retz.) Nees ex Steud. [1, 4, 5]

Poa unioloides Retz.

तिलके झार (Tilke jhaar)

Annual or rare perennial wetland herbs. W, C, E Nepal: 100-2200 m. Nepal, Asia and Africa.

Use: Forage.

Eragrostis viscosa (Retz.) Trin. [1, 4]

Poa viscosa Retz.; *Eragrostis tenella* var. *viscosa* (Retz.) Stapf

Sticky love grass

Annual herbs. E Nepal: *ca.* 800 m. Nepal, Himalaya, S Asia and Africa.

Use: Forage.

Eriochloa Kunth.

Eriochloa procera (Retz.) C.E.Hubb. [5]

Agrostis procera Retz.

Spring grass

Annual or short-lived perennial semi-aquatic herbs. W, C, E Nepal: 100-200 m. Nepal, Assam-Burma, S Asia, E Asia, SE Asia, Africa, America and Australasia.

Use: Forage.

Eulalia Kunth

Eulalia fastigiata (Nees ex Steud.) Stapf ex Bor [1, 4]

Saccharum fastigiatum Nees ex Steud.

Perennial herbs. C Nepal: *ca.* 900 m. Nepal, S Asia and SE Asia.

Eulalia leschenaultiana (Decne.) Ohwi [1, 4]

Andropogon leschenaultianus Decne.

Perennial herbs. C, E Nepal: 300-700 m. Nepal, Tibetan Plateau, S Asia, E Asia, SE Asia and Australasia.

Eulalia mollis (Griseb.) Kuntze [1, 4, 5]

Erianthus mollis Griseb.

Perennial herbs. W, C, E Nepal: 1200-3700 m. Nepal, E Himalaya, Tibetan Plateau and E Asia.

Eulalia quadrinervis (Hack.) Kuntze [1, 4, 5]

Pollinia quadrinervis Hack.

Perennial herbs. W, C, E Nepal: 1300-2700 m. Nepal, E Himalaya, Tibetan Plateau, Assam-Burma, E Asia and SE Asia.

Eulalia staintonii Bor [1, 4, 5]

Perennial herbs. W, C Nepal: 1900-2700 m. Nepal and S Asia.

Eulalia trispicata (Schult.) Henrard [1, 4, 5]

Andropogon trispicatus Schult.

Perennial herbs. W, C, E Nepal: 1100-2900 m. Nepal, E Himalaya, Assam-Burma, S Asia, E Asia, SE Asia and Australasia.

Use(s): Thatching (dried plant), Forage.

Eulaliopsis Honda

Eulaliopsis binata (Retz.) C.E.Hubb. [1, 4, 5]

Andropogon binatus Retz.

बाबियो (Baabiyo); Sabai grass

Perennial herbs. W, C, E Nepal: 100-2600 m. Nepal, E Himalaya, Assam-Burma, S Asia, E Asia, SE Asia and SW Asia.

Use(s): Medicine (leaves), Forage, Utensils and handicrafts, Construction materials.

Eulaliopsis sykesii Bor, *Kew Bull.* **12**(3): 412. 1957. [1, 4, 27]

Perennial herbs. W Nepal: 2500-2600 m. **Endemic to Nepal**.

West: Hunta, Nagdung Khola, 8500 ft. (2590 m), O. Polunin, WR Sykes, LHJ Williams 3170 (Holotype: K).

Use(s): Forage, Construction materials.

Festuca L.

Festuca chumbiensis E.B.Alexeev [4]

Perennial herbs. E Nepal: *ca.* 3600 m. Nepal and Tibetan Plateau.

Festuca cumminsii Stapf [5]

Perennial herbs. W, C, E Nepal: 2800-5400 m. Nepal, W Himalaya, E Himalaya, Tibetan Plateau, E Asia, N Asia, C Asia and SW Asia.

Festuca eriobasis H.Scholz, *Willdenowia* **18**(2): 431. 1989. [4, 27]

Perennial herbs. C Nepal: 4100-4600 m. **Endemic to Nepal**.

Central: *Rasuwa*, Langschisa Kharka, 4530 m, G. Miehe & S. Miehe 11830 (Holotype: BM).

Festuca forrestii St.-Yves

Festuca forrestii subsp. *nepalensis* H.Scholz (4)

Perennial herbs. W, C, E Nepal: 4100-4700 m. Nepal, Tibetan Plateau and E Asia.

Festuca gigantea (L.) Vill. [1, 4, 5]

Bromus giganteus L.

Tall bearded fescue

Perennial herbs. W, C Nepal: 2200-3900 m. Nepal, Asia and Europe.

Use: Forage.

Festuca leptopogon Stapf [1, 4, 5]

Festuca subulata Trinius var. *leptopogon* (Stapf) St.-Yves

धिम्सी घाँस (Dhimsi ghaans)

Perennial herbs. W, C, E Nepal: 1800-3800 m. Nepal, E Himalaya, Tibetan Plateau, Assam-Burma, E Asia and SE Asia.

Festuca modesta Steud. [1, 4]

Perennial herbs. W, C Nepal: 2600-4100 m. Nepal, W Himalaya, Tibetan Plateau, S Asia and E Asia.

Festuca nepalica E.B.Alexeev, *Bjull. Moskovsk. Obač. Isp. Prir., Otd. Biol.* **82**: 99. 1977. [4, 27]

Perennial herbs. W, C Nepal: 5000-5100 m. **Endemic to Nepal**.

West: *Dolpa*, Tingjegaon, 16600 ft. (5060 m), O. Polunin, WR Sykes & LHJ Williams 1162 (Holotype: K).

Festuca nitidula Stapf [1, 4]

Perennial herbs. C Nepal: 3500-4300 m Nepal, W Himalaya, Tibetan Plateau and E Asia.

Festuca ovina L. [1, 4, 5]

Perennial herbs. W, C, E Nepal: 3200-5600 m. Nepal and temperate regions.

Festuca parvigluma Steud. [1, 4, 5]

Perennial herbs. W, C, E Nepal: 2400-3700 m. Nepal, Tibetan Plateau, Assam-Burma and E Asia.

Festuca poluninii E.B. Alexeev, *Bjull. Moskovsk. Obač. Isp. Prir., Otd. Biol.* **83**(4): 111. 1978. [4, 27]

Perennial herbs. W Nepal: 4100-4200 m. **Endemic to Nepal**.

West: *Jumla*, Maharigaon, 13500 ft, (4114 m), O. Polunin, WR Sykes & LHJ Williams 2144 (Holotype: BM).

Festuca polycolea Stapf [1, 4, 5]

Festuca polycolea var. *brevis* Stapf (1, 4); *Festuca ovina* subsp. *polycolea* (Stapf) St.-Yves

Perennial herbs. W, C, E Nepal: 3000-4700 m. Nepal, E Himalaya and S Asia.

Festuca rubra L. [1, 4, 5]

Festuca rubra var. *nankataizanensis* Ohwi; *Festuca rubra* var. *niitakensis* Ohwi.

Red fescue

Perennial herbs. W, C, E Nepal: 1700-4200 m. Nepal, Tibetan Plateau, S Asia, E Asia, N Asia, SW Asia, Europe and N America.

Festuca sibirica Hack. ex Boiss.

Leucopoa albida (Turcz. ex Trin.) V.I.Krecz. & Bobrov (1, 4)

Perennial herbs. W Nepal: *ca.* 5000 m. Nepal, Tibetan Plateau, S Asia, E Asia and N Asia.

Festuca stapfii E.B.Alexeev [5]

Festuca undata var. *aristata* Stapf (1, 4)

Perennial herbs. W, C Nepal: 3300-4100 m. Nepal and E Himalaya.

Festuca undata Stapf [1, 4]

Perennial herbs. C, E Nepal: 3200-4000 m. Nepal and E Himalaya.

Festuca wallichiana E.B.Alexeev [4]

Perennial herbs. W, C Nepal: 2400-5400 m. Nepal, E Himalaya, Tibetan Plateau and S Asia.

Garnotia Brongn.

Garnotia acutigluma (Steud.) Ohwi [8]

Urachne acutigluma Steud.

Perennial herbs. E Nepal: *ca.* 1600 m. Nepal, E Himalaya, Assam-Burma, E Asia and SE Asia.

Garnotia emodi (Endl.) Janowski [1, 4, 5]

Berghausia emodi Endl.; *Garnotia polypogonoides* Munro ex Oliv.

Perennial herbs. C, E Nepal: 900-2500 m. Nepal, E Himalaya and Assam-Burma.

Garnotia tenella (Arn. ex Miq.) Janowski [1, 4, 5]

Berghausia tenella Arn. ex Miq.

Annual herbs. C, E Nepal: 1000-2400 m. Nepal, E Himalaya, Assam-Burma, S Asia, E Asia and SE Asia.

Glyceria R.Br.

Glyceria tonglensis C.B.Clarke [1, 4, 5]

Perennial herbs. W, C, E Nepal: 2100-3800 m. Nepal, W Himalaya, E Himalaya, Tibetan Plateau, Assam-Burma, S Asia and E Asia.

Hackelochloa Kuntze

Hackelochloa granularis (L.) Kuntze [1, 4, 5]

Cenchrus granularis L.

Annual herbs. W, C, E Nepal: 100-1400 m. Nepal, E Asia and Tropics.

Use: Medicine (whole plant).

Helictotrichon Besser ex Roem. & Schult.

Helictotrichon junghuhnii (Buse) Henrard [5]

Avena junghuhnii Buse; *Helictotrichon virescens* (Nees ex Steud.) Henrard (1, 4); *Helictotrichon asperum* (Munro ex Thwaites) Bor (1, 4); *Helictotrichon parviflorum* (Hook.f.) Bor (1, 4)

Perennial herbs. W, C, E Nepal: 2100-4400 m. Nepal, W Himalaya, E Himalaya, Tibetan Plateau, Assam-Burma, S Asia, E Asia, SE Asia and SW Asia.

Hemarthria R.Br.

Hemarthria compressa (L.f.) R.Br. [1, 4, 5]

Rottboellia compressa L.f.; *Rottboellia glabra* Roxb.

घोडे दुबो, मकैया घाँस (Ghode dubo, Makaiya ghaans); Whip grass

Perennial floating herbs. W, C, E Nepal: 100-1500 m. Nepal, E Himalaya, Assam-Burma, S Asia, E Asia, SE Asia and SW Asia.

Use(s): Medicine (plant ash), Forage, Ritual.

Hemarthria vaginata Buse [5]

Hemarthria protensa Nees ex Steud.

Perennial herbs. E Nepal: 200-300 m. Nepal, E Himalaya, Tibetan Plateau, Assam-Burma, S Asia, E Asia and SE Asia.

Heteropogon Pers.

Heteropogon contortus (L.) P.Beauv. ex Roem. & Schult. [1, 4, 5]

Andropogon contortus L.

अर्थुंगे, दाप्सु खर (Arthunge, Daapsu khar); Black spear grass

Perennial herbs. W, C, E Nepal: 300-2600 m. Nepal, Tibetan Plateau, E Asia, Pantropical, Subtropical and Temperate.

Use(s): Medicine (root), Forage, Utensils and handicrafts, Construction materials.

Heteropogon melanocarpus (Elliott) Benth. [5]

Andropogon melanocarpus Elliott

Annual erect herbs. C Nepal: 200-300 m. Nepal, S Asia, E Asia, SW Asia, Africa and America.

Hierochloe R.Br.

Hierochloe flexuosa Hook.f. [1, 4]

Anthoxanthum flexuosum (Hook.f.) Veldkamp

Perennial herbs. W, C, E Nepal: 400-4000 m. Nepal and E Himalaya.

Hierochloe laxa Hook.f. [1, 4]

Perennial herbs. W, C Nepal: 3500-4100 m. Nepal, W Himalaya, S Asia and SW Asia.

Himalayacalamus Keng f.

Himalayacalamus asper Stapleton, *Edinburgh J. Bot.* **51**(3): 310. 1994. [4, 17, 27]

घोडे निगालो, मालिंगे निंगालो (Ghode nigaalo, Maalinge nigaalo); Tibetan princess bamboo

Perennial arborescent herbs. C Nepal: 1200-2000 m. **Endemic to Nepal**.

Central: *Kaski*, Pipar, *ca.* 2000 m, CMA Stapleton 314 (Holotype: E).

Himalayacalamus brevinodus Stapleton [4, 8, 17]

मालिंगे निंगालो (Maalinge nigaalo)

Perennial arborescent herbs. E Nepal: *ca.* 1500 m. Nepal and E Himalaya.

Use(s): Food (tender shoot), Forage, Construction materials.

Himalayacalamus cupreus Stapleton, *Edinburgh J. Bot.* **51**(3): 314. 1994. [4, 17, 27]

मालिंगे निंगालो (Maalinge nigaalo)

Perennial arborescent herbs. C Nepal: *ca.* 2500 m. **Endemic to Nepal**.

Central: *Kaski*, Pipar, *ca.* 2500 m, CMA Stapleton 306 (Holotype: E).

Himalayacalamus falconeri (Hook.f. ex Munro) Keng f. [4, 5, 17]

Thamnocalamus falconeri Hook. ex Munro; *Arundinaria falconeri* (Hook.f. ex Munro) Gamble

सिंगाने, ठुड्डी निंगालो (Singaane, Thudi nigaalo); Candy cane bamboo

Perennial arborescent herbs. C, E Nepal: 1200-2300 m. Nepal and E Himalaya.

Use(s): Food (tender shoot), Construction materials.

Himalayacalamus fimbriatus Stapleton, *Edinburgh J. Bot.* **51**(3): 316. 1994. [4, 17, 27]

तिते निंगालो (Tite Nigaalo)

Perennial arborescent herbs. W, C Nepal: *ca.* 1200 m. **Endemic to Nepal**.

Central: *Kathmandu*, *ca.* 1200 m, CMA Stapleton 910 (Holotype: E).

Use(s): Food (tender shoot), Forage, Utensils and handicrafts, Construction materials.

Himalayacalamus hookerianus (Munro) Stapleton [4, 17]

Arundinaria hookeriana Munro; *Drepanostachyum hookerianum* (Munro) Keng f.

पदंग (Padang-Gurung); Himalayan blue bamboo

Perennial arborescent herbs. E Nepal: 1800-2200 m. Nepal, E Himalaya and Assam-Burma.

Use(s): Forage, Utensils and handicrafts, Construction materials.

Himalayacalamus planatus Stapleton, *J. Bot. Res. Inst. Texas* **1**(1): 137. 2007. [8]

मालिंगे निंगालो (Maalinge nigaalo)

Perennial herbs. C Nepal: *ca.* 2440 m. **Endemic to Nepal**.

Central: *Rasuwa*, Syabru, 8000 ft. (2438 m), CMA Stapleton 328 (Holotype: K).

Uses(s): Forage, Household equipments (baskets, trays and mats).

Himalayacalamus porcatus Stapleton, *Edinburgh J. Bot.* **51**(3): 318. 1994. [4, 8, 27]

सेतो निंगालो (Seto nigaalo)

Perennial arborescent herbs. C Nepal: 2200-2600 m. **Endemic to Nepal**.

Central: *Rasuwa*, Syabru, 7500 ft. (2286 m), CMA Stapleton 332 (Holotype: E).

Hordeum L.

Hordeum brevisubulatum (Trin.) Link

Hordeum secalinum var. *brevisubulatum* Trin.

subsp. **turkestanicum** Tzvelev

Hordeum turkestanicum Nevski (1, 4)

Perennial herbs. W, C, E Nepal: 2600-4000 m. Nepal, Asia and Europe.

Use: Forage.

Hordeum distichon L.

Hordeum nudum (L.) Ard.

जौ, दुई हरे (Jau, Dui hare); Two rowed barley

Annual herbs. W, C, E Nepal: 200-300 m. Widely cultivated in temperate regions.

Use: Forage.

Hordeum vulgare L. [1, 4, 5]

जौ (Jau); Barley

Annual herbs. W, C, E Nepal: 2200-3500 m. Widely cultivated in Nepal and S Asia. Native to N Asia.

Use(s): Medicine (seeds), Food (seeds).

Hygroryza Nees

Hygroryza aristata (Retz.) Nees ex Wight & Arn. [1, 4, 5]

Pharus aristatus Retz.; *Leersia aristata* (Retz.) Roxb.

नाभो धान (Naabho dhaan); Wild rice

Perennial aquatic herbs. W, C, E Nepal: 100-1300 m. Nepal, Tibetan Plateau, Assam-Burma, S Asia, E Asia and SE Asia.

Use(s): Medicine (root, seeds), Forage.

Hymenachne P.Beauv.

Hymenachne amplexicaulis (Rudge) Nees [5]

Panicum amplexicaule Rudge; *Hymenachne pseudointerrupta* Mull. Hal. (1,4)

Perennial aquatic floating herbs. C, E Nepal: 100-1400 m. Nepal, Assam-Burma, S Asia, E Asia and SE Asia.

Ichnanthus P.Beauv.

Ichnanthus pallens (Sw.) Munro ex Benth.

Panicum pallens Sw.

var. **major** (Nees) Stieber

Panicum pallens var. *majus* Nees; *Ichnanthus vicinus* (F.M.Bailey) Merr. (1, 4)

Perennial rambling herbs. C Nepal: *ca.* 900 m. Nepal, Assam-Burma, S Asia, E Asia, SE Asia, Africa, S America and Australasia.

Imperata Cirillo

Imperata cylindrica (L.) P.Beauv. [1, 4, 5]

Lagurus cylindricus L.; *Imperata arundinacea* Cirillo

var. **cylindrica**

सिरु (Siru); Cogon grass

Perennial herbs. W, C, E Nepal: 100-2400 m. Nepal, Asia, Europe and Africa.

Use(s): Medicine (root), Forage, Utensils and handicrafts, Construction materials.

var. **major** (Nees) C.E.Hubb. [1,4]

Imperata koenigii var. *major* Nees

Perennial herbs. E Nepal: 1500-2000 m. Nepal, Asia, Africa and Australasia.

Use(s): Medicine (root/rhizome, seeds), Forage, Utensils and handicrafts, Construction materials.

Isachne R.Br.

Isachne albens Trin. [1, 4, 5]

Panicum albens (Trin.) Steud.

Perennial wetland herbs. W, C, E Nepal: 1400-2300 m. Nepal, E Himalaya, Tibetan Plateau, Assam-Burma, S Asia, E Asia, SE Asia and Africa.

Isachne globosa (Thunb.) Kuntze [1, 4, 5]

Milium globosum Thunb.; *Isachne miliacea* Roth (1, 4)

Swamp millet

Perennial wetland herbs. C, E Nepal: 500-1500 m. Nepal, E Himalaya, S Asia, E Asia, SE Asia and Australasia.

Isachne scabrosa Hook.f. [1, 4]

Perennial ascending herbs. C Nepal: *ca.* 2400 m. Nepal, E Himalaya, Tibetan Plateau and Assam-Burma.

Isachne sikkimensis Bor [1, 4, 5]

Annual decumbent herbs. C, E Nepal: 1600-3400 m. Nepal, E Himalaya, Tibetan Plateau and S Asia.

Ischaemum L.

Ischaemum rugosum Salisb. [1, 4, 5]

Ischaemum rugosum var. *segetum* (Trin.) Hack. (1, 4)

बन्सो, मल्लिडो (Banso, Mallido)

Annual wetland herbs. W, C, E Nepal: 100-1600 m. Nepal, E Himalaya, Tibetan Plateau, Assam-Burma, S Asia, E Asia, SE Asia and Australasia.

Use: Forage.

Koeleria Pers.

Koeleria litvinowii Domin

subsp. **argentea** (Griseb.) S.M.Philips & Z.L.Wu

Koeleria argentea Griseb. (5); *Koeleria argentea* var. *nepalensis* Domin

Perennial tufted herbs. W, C Nepal: 2500-4100 m. Nepal and Asia.

Koeleria pyramidata (Lam.) P.Beauv.

Poa pyramidata Lam.; *Koeleria cristata* Pers. (1, 4)

Crested hair grass

Perennial herbs. W, C Nepal: 2300-4200 m. Nepal, E Himalaya, S Asia, E Asia, N Asia, SW Asia and Europe.

Leersia Sw.

Leersia hexandra Sw. [1, 4, 5]

करौंते झार, नाभो धान (Karaunte jhaar, Naabho dhaan); Southern cutgrass

Perennial aquatic floating herbs. W, C, E Nepal: 200-1600 m. Invasive species in Nepal. E Himalaya, Assam-Burma, S Asia, E Asia, SE Asia, Africa, America, Australasia and Pantropical.

Use(s): Medicine (whole plant), Forage.

Leptochloa Pal.

Leptochloa chinensis (L.) Nees [5]

Poa chinensis L.

Annual or rarely perennial wetland herbs. E Nepal:100-200 m. Nepal, E Himalaya, Assam-Burma, S Asia, E Asia, SE Asia, Africa and Australasia.

Leptochloa panicea (Retz.) Ohwi [5]

Poa panicea Retz.

Annual herbs. E Nepal: 100-200 m. Nepal, S Asia, E Asia, SE Asia, Africa and America.

Littledalea Hemsl.

Littledalea tibetica Hemsl. [1, 4, 5]

Perennial herbs. W, C Nepal: 2900-4200 m. Nepal, Tibetan Plateau and E Asia.

Lolium L.

Lolium multiflorum Lam. [5]

Italian rye grass

Annual to short-lived perennial herbs. E Nepal: *ca.* 2800 m. Nepal, Tibetan Plateau, E Asia, N Asia, SW Asia, Europe and Africa.

Lolium perenne L. [1, 4, 5]

Perennial herbs. C Nepal: *ca.* 2200 m. Commonly cultivated in Asia, Europe and Africa.

Lolium temulentum L. [1, 4, 5]

Annual herbs. W Nepal: 1300-1400 m. Nepal, Tibetan Plateau, E Asia, N Asia, SW Asia, Europe and Africa.

Lophatherum Brongn.

Lophatherum gracile Brongn. [5]

Acroelytrum japonicum Steudel

Perennial herbs. C Nepal: 400-500 m. Nepal, Tibetan Plateau, Assam-Burma, S Asia, E Asia, SE Asia and Australasia.

Melica L.

Melica onoei Franch. & Sav. [5]

Perennial tufted herbs. W, C, E Nepal: 2500-3300 m. Nepal, W Himalaya, Tibetan Plateau, S Asia, E Asia, N Asia and SW Asia.

Melica persica Kunth

Melica jacquemontii Decne. (1, 4, 8)

Perennial herbs. W, C Nepal: *ca.* 4000 m. Nepal, W Himalaya, Tibetan Plateau, S Asia, E Asia, C Asia, SW Asia and Africa.

Melica scaberrima (Nees ex Steud.) Hook.f. [1, 4, 5]

Glyceria scaberrima Nees ex Steud.; *Festuca scaberrima* Nees ex Steud.

Perennial herbs. W, C Nepal: 2700-4000 m. Nepal, W Himalaya, Tibetan Plateau, S Asia and E Asia.

Melocanna Trin.

Melocanna baccifera (Roxb.) Kurz [4, 17]

Bambusa baccifera Roxb.; *Melocanna bambusoides* Trin.

फिलिम बाँस, लाहुरे बाँस (Philim baans, Laahure bans); Muli bamboo

Perennial herbs. W, C, E Nepal: 400-2700 m. Nepal, S Asia and SE Asia.

Use: Construction materials.

Microstegium Nees

Microstegium ciliatum (Trin.) A.Camus [1, 4, 5]

Pollinia ciliata Trin.

खुनखुने (Khunkhune)

Perennial rambling herbs. W, C, E Nepal: 500-2100 m. Nepal, E Himalaya, Assam-Burma, S Asia, E Asia and SE Asia.

Use: Forage.

Microstegium fasciculatum (L.) Henrard

Andropogon fasciculatus L.; *Microstegium vagans* (Nees ex Steud.) A.Camus (1, 4)

Perennial mat forming prostrate herbs. C, E Nepal: 1300-1500 m. Nepal, E Himalaya, Assam-Burma, S Asia, E Asia, SE Asia and Africa.

Microstegium nudum (Trin.) A.Camus [1, 4, 5]

Pollinia nuda Trin.

Annual rambling herbs. W, C, E Nepal: 800-3000 m. Nepal, E Himalaya, Tibetan Plateau, S Asia, E Asia, SE Asia, SW Asia, Africa and Australasia.

Microstegium petiolare (Trin.) Bor [1, 4, 5]

Spodiopogon petiolaris Trin.; *Andropogon petiolaris* (Trin.) Steud.

Perennial rambling or scandent herbs. W, C Nepal: 1700-2300 m. Nepal, E Himalaya, Assam-Burma and E Asia.

Microstegium vimineum (Trin.) A.Camus [1, 4, 5]

Andropogon vimineus Trin.; *Microstegium willdenovianum* Nees

Annual mat forming decumbent herbs. C, E Nepal: 700-1800 m. Nepal, Asia, Africa and N America.

Miscanthus Andersson

Miscanthus nepalensis (Trin.) Hack. [1, 4, 5]

Eulalia nepalensis Trin.

Himalayan fairy grass

Perennial herbs. W, C, E Nepal: 1100-3000 m. Nepal, E Himalaya, Tibetan Plateau, Assam-Burma, S Asia, E Asia, SE Asia and Australasia.

Miscanthus nudipes (Griseb.) Hack. [1, 4]

Erianthus nudipes Griseb.

Pink feather grass

Perennial erect herbs. E Nepal: *ca.* 3200 m. Nepal, E Himalaya, Tibetan Plateau, S Asia and E Asia.

Mnesithea Kunth

Mnesithea laevis (Retz.) Kunth [5]

Rottboellia laevis Retz.

Perennial tufted erect herbs. W, C, E Nepal: 100-700 m. Nepal, S Asia, E Asia, SE Asia, SW Asia and Australasia.

Mnesithea striata (Nees ex Steud.) de Koning & Sosef

Rottboellia striata Nees ex Steud.; *Coelorachis striata* (Nees ex Steud.) A.Camus (1, 4)

Perennial herbs. E Nepal: 500-1400 m. Nepal, E Himalaya, Tibetan Plateau, Assam-Burma, E Asia and SE Asia.

Muhlenbergia Schreb.

Muhlenbergia duthieana Hack. [1, 4, 5]

Perennial herbs. W, C Nepal: 2000-3400 m. Nepal, W Himalaya and S Asia.

Muhlenbergia himalayensis Hack. ex Hook.f. [1, 4, 5]

Perennial herbs. W, C, E Nepal: 1800-2900 m. Nepal, W Himalaya, E Himalaya, Tibetan Plateau, S Asia, E Asia and SW Asia.

Muhlenbergia huegelii Trin. [1, 4, 5]

Muhlenbergia longistolon Ohwi; *Muhlenbergia viridissima* Nees ex Steud.

Perennial decumbent or prostrate herbs. W, C, E Nepal: 2000-2500 m. Nepal and Asia.

Neyraudia Hook.f.

Neyraudia arundinacea (L.) Henrard [1, 4]

Aristida arundinacea L.; *Arundo madagascariensis* Kunth

लेसे घाँस (Lese ghaans); Madagascar grass

Perennial herbs. W, C, E Nepal: 100-2900 m. Nepal, W Himalaya, S Asia, E Asia, SE Asia, SW Asia and Africa.

Use(s): Ritual, Forage.

Neyraudia reynaudiana (Kunth) Keng ex Hitchc. [1, 4]

Arundo reynaudiana Kunth; *Neyraudia arundinacea* var. *zollingeri* (Buse) Henrard (5)

लेसे घाँस (Lese ghaans); Burma reed

Perennial robust herbs. W, C, E Nepal: 200-2100 m. Nepal, E Himalaya, Tibetan Plateau, Assam-Burma, E Asia and SE Asia.

Ophiuros C.F.Gaertn.

Ophiuros exaltatus (L.) Kuntze [5]

Aegilops exaltata L.

Cane grass

Perennial robust herbs. C Nepal: *ca.* 200 m. Nepal, S Asia, E Asia, SE Asia and Australasia.

Oplismenus P.Beauv.

Oplismenus burmannii (Retz.) P.Beauv. [1, 4, 5]

Panicum burmannii Retz.

ओते घाँस (Ote ghaans)

Annual prostrate herbs. W, C, E Nepal: 100-1800 m. Nepal and Pantropical.

Use: Forage.

Oplismenus compositus (L.) P.Beauv. [1, 4, 5]

Panicum compositum L.

Perennial prostrate herbs. W, C, E Nepal: 300-2800 m. Nepal, Tibetan Plateau, E Asia, SE Asia, Africa, Australasia and Pantropical.

Oplismenus undulatifolius (Ard.) Roem. & Schult. [1, 4]

Panicum undulatifolium Ard.; *Oplismenus hirtellus* subsp. *undulatifolius* (Ard.) U.Scholz (5)

Perennial prostrate herbs. W, C, E Nepal: 2100-2300 m. Nepal, S Asia, E Asia, Africa and N Hemisphere.

Orinus Hitchc.

Orinus thoroldii (Stapf ex Hemsl.) Bor [5]

Diplachne thoroldii Stapf ex Hemsl.

Perennial herbs. C Nepal: 3700-3800 m. Nepal, W Himalaya, Tibetan Plateau and N Asia.

Orthoraphium Nees

Orthoraphium roylei Nees

Stipa roylei (Nees) Duthie (1, 4, 5)

Perennial herbs. W, C, E Nepal: 2000-4000 m. Nepal, Himalaya, Tibetan Plateau and Assam-Burma.

Oryza L.

Oryza meyeriana (Zoll. & Moritzi) Baill.

Padia meyeriana Zoll. & Moritzi

subsp. **meyeriana**

जंगली धान (Jangali dhaan); Wild rice

Perennial herbs. C Nepal: 300-400 m. Nepal, Tibetan Plateau, Assam-Burma, S Asia, E Asia and SE Asia.

Use: Forage.

subsp. **granulata** (Nees & Arn. ex G.Watt) Tateoka [5]

Oryza granulata Nees & Arn. ex G.Watt; *Oryza meyeriana* var. *granulata* (Nees & Arn. ex G.Watt) Duistermaat

Perennial herbs. C Nepal: 300-400 m. Nepal, S Asia, E Asia and SE Asia.

Oryza minuta J.Presl [1, 4]

Perennial herbs. C Nepal: 100-400 m. Nepal, E Himalaya, Assam-Burma and SE Asia.

Use: Forage.

Oryza officinalis Wall. ex Watt [5]

Oryza latifolia var. *silvatica* A.Camus; *Oryza minuta* var. *silvatica* (A.Camus) Veldkamp

Perennial herbs. C Nepal: 100-1000 m. Nepal, E Himalaya, Assam-Burma, S Asia, E Asia, SE Asia and Australasia.

Oryza rufipogon Griff. [1, 4, 5]

Oryza nivara S.D.Sharma & Shastry (5)

जंगली धान (Jangali dhaan); Red rice

Annual aquatic herbs. W, C Nepal: 100-600 m. Nepal, Assam-Burma, S Asia, E Asia, SE Asia and Australasia.

Use: Forage.

Oryza sativa L. [1, 4, 5]

धान (Dhaan); Rice

Annual marshy herbs. W, C, E Nepal: 200-1800 m. Widely cultivated as cereal crop.

Panicum L.

Panicum antidotale Retz. [1, 4, 5]

Panicum subalbidum Kunth

Blue panic grass

Perennial herbs. W, C Nepal: 100-200 m. Nepal, E Himalaya, S Asia and SW Asia.

Use: Medicine (whole plant).

Panicum curviflorum Hornem. [5]

Panicum trypheron Schult. (1, 4); *Panicum tenellum* Roxb.

Annual herbs. C, E Nepal: 100-1200 m. Nepal, S Asia, E Asia, SE Asia and SW Asia.

Panicum dichotomiflorum Michx. [5]

Panicum paludosum Roxb. (1, 4, 5)

Annual aquatic herbs. W, C, E Nepal: 100-1200 m. Nepal, S Asia, E Asia, SE Asia and tropics of the new world.

Use: Forage.

Panicum humidorum Buch.-Ham. ex Hook.f. [1, 4]

Panicum humidorum Buch.-Ham. ex Wall.

Perennial herbs. Nepal, E Himalaya and SE Asia.

Panicum humile Nees ex Steud.

Panicum walense Mez. (1, 4, 5)

Annual herbs. C, E Nepal: 500-1200 m. Nepal, Tibetan Plateau, S Asia, E Asia, SE Asia and Africa.

Use: Forage.

Panicum miliaceum L. [1, 4, 5]

चिनो, ठुलो कागुनो (Chino, Thulo kaaguno); Broomcorn millet

Annual herbs. W, C Nepal: 1500-2400 m. Widely cultivated in Nepal and W Himalaya.

Panicum notatum Retz. [1, 4, 5]

Panicum montanum Roxb.

रक्ते (Rakte)

Perennial decumbent or rambling herbs. W, C, E Nepal: 500-1200 m. Nepal, E Himalaya, Tibetan Plateau, Assam-Burma, S Asia, E Asia and SE Asia.

Use: Forage.

Panicum repens L. [1, 4]

Torpedo grass

Perennial wetland weedy herbs. C, E Nepal: 100-700 m. Nepal, E Asia, Pantropical and Subtropical.

Use: Forage.

Panicum sumatrense Roth [1, 4]

Panicum psilopodium Trin. (1, 4)

श्यामक (Shyamak) Barefoot panic grass

Annual erect herbs. W, C, E Nepal: 100-2300 m. Nepal, Tibetan Plateau, Assam-Burma, S Asia, E Asia and Australasia.

Use(s): Medicine (grain), Forage.

Paspalidium Stapf

Paspalidium flavidum (Retz.) A.Camus [1, 4, 5]

Panicum flavidum Retz.

Perennial or annual herbs. W, C, E Nepal: 100-1700 m. Nepal, E Himalaya, S Asia, E Asia, SE Asia, Africa and Australasia.

Use: Forage.

Paspalidium geminatum (Forssk.) Stapf [49]

Panicum geminatum Forssk.

Egyptian panic grass

Perennial wetland herbs. C Nepal: 500-600 m. Nepal, Asia and Africa.

Paspalidium punctatum (Burm.f.) A.Camus [1, 4, 5]

Panicum punctatum Burm.f.

Perennial marshy herbs. E Nepal: 100-200 m. Nepal, Assam-Burma, S Asia, E Asia, SE Asia and Africa.

Use: Forage.

Paspalum L.

Paspalum conjugatum P.J.Bergius [1, 4, 5]

जनै घाँस (Janai ghaans); Buffalo grass

Perennial wetland herbs. E Nepal: 200-700 m. Nepal, E Asia, Pantropical and Subtropical.

Use: Forage.

Paspalum dilatatum Poir. [5]

Digitaria dilatata (Poir.) H.J.Coste

Perennial herbs. W, C, E Nepal: 800-1800 m. Nepal, E Asia, native to S America.

Paspalum distichum L. [1, 4, 5]

Digitaria paspalodes Michx.; *Paspalum paspalodes* (Michx.) Scribn.

जनै घाँस (Janai ghaans); Knotgrass

Perennial or wetland herbs. W, C, E Nepal: 100-2000 m. Nepal, E Asia, Pantropical and Warm Temperate.

Use: Forage.

Paspalum notatum Flüggé [8]

Perennial mat forming herbs. W, C, E Nepal: 1000-1700 m. Nepal, native to America.

Paspalum scrobiculatum L. [1, 4, 5]

कोदे बन्सो, जनै घाँस (Kode banso, Janai ghaans)

Perennial wetland herbs. W, C, E Nepal: 100-2200 m. Nepal, E Asia, tropics and subtropics of old world.

Use(s): Medicine (whole plant), Forage.

Pennisetum (Hochst.) Hook.f.

Pennisetum clandestinum Hochst. ex Chiov. [5]

Cenchrus clandestinus (Hochst. ex Chiov.) Morrone

Perennial mat forming herbs. C Nepal: *ca.* 1500 m. Nepal, E Asia, native to Africa.

Pennisetum flaccidum Griseb. [1, 4, 5]

Pennisetum centrasiaticum Tzvelev

पांग (Pang-Tamang); Himalayan fountain grass

Perennial herbs. W, C Nepal: 1600-4300 m. Nepal and Asia.

Use(s): Medicine (whole plant), Forage.

Pennisetum glaucum (L.) R.Br.

Panicum glaucum L.; *Pennisetum typhoides* (Burm.) Stapf & C.E.Hubb.; *Setaria glauca* (L.) P.Beauv. (1, 4)

बज्र, बले बन्सो (Bajra, Bale banso); Pearl millet

Annual erect herbs. W, C, E Nepal: 900-2300 m. Nepal, E Asia, native to Africa.

Use(s): Forage, Food (grains).

Pennisetum hohenackeri Hochst. ex Steud. [1, 4]

Pennisetum alopecuros Steud.

Perennial herbs. C Nepal: 2400-3000 m. Nepal, S Asia, SW Asia and Africa.

Pennisetum hordeoides (Lam.) Steud. [1, 4]

Panicum hordeoides Lam.; *Cenchrus hordeoides* (Lam.) Morrone

Annual ascending herbs. C Nepal: 300-2100 m. Nepal, S Asia and Africa.

Pennisetum orientale Rich. [1, 4, 5]

Perennial herbs. W, C Nepal: 700-2000 m. Nepal, Asia, Africa and America.

Pennisetum pedicellatum Trin. [5]

दिनानाथ घाँस (Dinanath ghaans); Fountain grass

Annual ascending herbs. C Nepal: 1000-1100 m. Nepal, S Asia, SE Asia, Africa, S America and Australasia.

Use: Forage.

Pennisetum polystachion (L.) Schult. [1, 4]

Panicum polystachion L.

यम झार (Yam jhaar); Mission grass

Perennial herbs. C, E Nepal: 300-1600 m. Nepal and old world.

Use: Forage.

Pennisetum purpureum Schumach. [5]

Gymnotrix nitens Andersson; *Pennisetum benthamii* Steud.

नेपिएर घाँस (Napier ghaans); Elephant grass, Napier grass

Perennial herbs. C Nepal: *ca.* 2100 m. Nepal, S Asia, E Asia and native to Africa.

Use: Forage.

Perotis Aiton

Perotis hordeiformis Nees ex Hook. & Arn. [1, 4]

Perotis chinensis Gand.

Annual ascending herbs. W, C, E Nepal: 300-1200 m. Nepal, Assam-Burma, S Asia, E Asia, SE Asia and SW Asia.

Perotis indica (L.) Kuntze [1, 4, 5]

Anthoxanthum indicum L.; *Perotis latifolia* Aiton

Annual ascending herbs. W, C, E Nepal: 100-1200 m. Nepal, E Himalaya, Assam-Burma, S Asia, E Asia, SE Asia, Africa and Australasia.

Phacelurus Griseb.

Phacelurus speciosus (Steud.) C.E.Hubb. [5]

Andropogon speciosus Steud.

Perennial herbs. W Nepal: 2500-2700 m. Nepal, S Asia and SW Asia.

Phalaris L.

Phalaris arundinacea L. [1, 4]

Typhoides arundinacea (L.) Moench

सेतो दुबो (Seto dubo); Reed canary grass

Perennial herbs. E Nepal: *ca.* 2100 m. Nepal, Tibetan Plateau, E Asia, N Asia and N Hemisphere.

Phalaris minor Retz. [5]

Phalaris minor var. *nepalensis* (Trin.) Bor (1, 4)

रगते झार, रक्ते (Ragate jhaar, Rakte)

Annual herbs. W, C Nepal: 200-1900 m. Nepal, Asia, Europe and Africa; native to the Mediterranean region.

Use: Forage.

Phleum L.

Phleum alpinum L. [1, 4, 5]

Phleum commutatum Gaudin

Perennial herbs. W, C, E Nepal: 2400-4700 m. Nepal, Asia, Europe, N America and S America.

Phragmites Adans.

Phragmites australis (Cav.) Trin. ex Steud. [1, 4, 5]

Arundo australis Cav.; *Arundo phragmites* L.; *Phragmites communis* Trin

Common reed

Perennial herbs. C Nepal: 1300-3600 m. Nepal, W Himalaya, Tibetan Plateau, E Asia and temperate regions.

Use(s): Medicine (root, leaves), Forage, Utensils and handicrafts, Food (young shoots).

Phragmites karka (Retz.) Trin. ex Steud. [1, 4, 5]

Arundo karka Retz.; *Phragmites vallatoria* (L.) Veldkamp

नर्कट (Narkat)

Perennial herbs. W, C, E Nepal: 100-2800 m. Nepal, Assam-Burma, S Asia, E Asia, SE Asia, Africa and Australasia.

Use(s): Medicine (root), Forage, Construction materials, Utensils and handicrafts.

Phyllostachys Sieb. & Zucc.

Phyllostachys nigra (Lodd. ex Lindl.) Munro [5]

Bambusa nigra Lodd. ex Lindl.

Black bamboo

Perennial herbs. C Nepal: *ca.* 1300 m. Nepal, Tibetan Plateau and E Asia.

Piptatherum P.Beauv.

Piptatherum aequiglume (Duthie ex Hook.f) Roshev.

Oryzopsis aequiglumis Duthie ex Hook.f. (5)

Perennial herbs. W, C, E Nepal: 2300-4300 m. Nepal, W Himalaya, E Himalaya, Tibetan Plateau, S Asia, E Asia and SW Asia.

Piptatherum gracile Mez

Oryzopsis gracilis (Mez) Pilg. (1, 4, 5)

Perennial herbs. C Nepal: 3000-3800 m. Nepal, W Himalaya, Tibetan Plateau, S Asia, E Asia and SW Asia.

Piptatherum laterale (Regel) Nevski

Milium laterale Regel; *Oryzopsis lateralis* (Munro ex Regel) Stapf ex Hook.f. (1, 4, 5)

तुची (Tuchi)

Perennial herbs. W, C Nepal: 2600-4000 m. Nepal, W Himalaya, Tibetan Plateau, S Asia, E Asia, C Asia, N Asia and SW Asia.

Piptatherum munroi (Stapf ex Hook.f.) Mez

Oryzopsis munroi Stapf ex Hook.f. (1, 4, 5)

Perennial herbs. W, C, E Nepal: 2100-4900 m. Nepal and Asia.

Poa L.

Poa albertii Regel

subsp. **arnoldii** (Melderis) Olonova & G.H.Zhu

Poa arnoldii Melderis (1, 4); *Poa mustangensis* Rajbh. (4, 5)

Annual herbs. W Nepal: 4000-5600 m. Nepal, E Himalaya, Tibetan Plateau and E Asia.

Use: Forage.

subsp. **poophagorum** (Bor) Olonova & G.H.Zhu

Poa poophagorum Bor (1, 4, 5)

Perennial herbs. W, C, E Nepal: 3100-5800 m. Nepal, E Himalaya, Tibetan Plateau, S Asia, E Asia and N Asia.

Use: Forage.

Poa alpina L. [1, 4]

Perennial herbs. W Nepal: *ca.* 3800 m. Nepal, Tibetan Plateau, S Asia, E Asia, N Asia, C Asia, SW Asia, Europe and N America.

Use: Forage.

Poa annua L. [1, 4, 5]

पोडे घाँस (Pode ghaans)

Annual herbs. W, C, E Nepal: 1300-4500 m. Nepal, Asia, Europe, Africa, N America, S America and Australasia.

Use: Forage.

Poa bulbosa L. [1, 4]

Perennial herbs. C, E Nepal: *ca.* 4200 m. Nepal, W Himalaya, Tibetan Plateau, S Asia, N Asia, C Asia, SW Asia, Europe and Africa.

Use: Forage.

Poa calliopsis Litv. ex Ovcz. [1, 4, 5]

Poa phariana Bor.

Perennial herbs. W, C Nepal: 4300-5600 m. Nepal and Asia.

Use: Forage.

Poa eleanorae Bor [1, 4, 8]

Perennial herbs. C, E Nepal: 3600-4200 m. Nepal, E Himalaya, Tibetan Plateau, S Asia and E Asia.

Use: Forage.

Poa falconeri Hook.f. [1, 4]

Perennial herbs. W, C Nepal: 3300-4100 m. Nepal, W Himalaya, Tibetan Plateau and S Asia.

Use: Forage.

Poa harae Rajb. [4, 5]

Perennial tufted herbs. E Nepal: 4400-4600 m. Nepal, E Himalaya and Assam-Burma.

Poa hideaki-ohbae Rajbh., *J. Jap. Bot.* **88**(2): 77. 2013. [8, 29]

Perennial herbs. C Nepal: *ca.* 3400 m. **Endemic to Nepal**.

Central: *Dhading*, Pati Kharka near Pabil Kharka, *ca.* 3400 m, KR Rajbhandari 17562 (Holotype: KATH; Isotype: TI).

Poa himalayana Nees ex Steud. [1, 4]

Poa gracilior Keng ex L.Liu; *Poa stewartiana* Bor.

Annual or short-lived perennial herbs. W, C, E Nepal: 2900-4000 m. Nepal, W Himalaya, Tibetan Plateau, S Asia, E Asia and SW Asia.

Use: Forage.

Poa hirtiglumis Hook.f. [1, 4]

Perennial herbs. C, E Nepal: 400-4900 m. Nepal, E Himalaya, Tibetan Plateau and E Asia.

Use: Forage.

Poa hylobates Bor [1, 4]

Poa elanata Keng ex Tzvelev

Annual herbs. W, C Nepal: 3000-3800 m. Nepal, Tibetan Plateau, E Asia and N Asia.

Use: Forage.

Poa imperialis Bor, *Kew Bull.* 1957: 414. 1958. [1, 4, 27]

Perennial stoloniferous herbs. E Nepal: 4400-4500 m. **Endemic to Nepal**.

East: *Taplejung*, Tamur valley, Yangma Khola, NE of Walungchung Gola, 14500 ft. (4420 m), JDA Stainton 1105 (Holotype: BM).

Use: Forage.

Poa kanaii Rajbh. [4, 5, 27]

Perennial tufted herbs. C Nepal: *ca.* 4900 m. Nepal and W Himalaya.

Poa khasiana Stapf [1, 4]

Poa formosae Ohwi

Perennial herbs. C, E Nepal: 2900-3600 m. Nepal, Tibetan Plateau, S Asia and E Asia.

Use: Forage.

Poa langtangensis Melderis, *Enum. Fl. Pl. Nepal* **1**: 143. 1978. [1, 4, 27]

Perennial rhizomatous herbs. C Nepal: *ca.* 4000 m. **Endemic to Nepal.**

Central: *Rasuwa,* Langtang valley, 13500 ft. (4114 m), AD Schilling, DC Sayers & MS Bista 461 (Holotype: K).

Use: Forage.

Poa lhasaensis Bor

Poa jaunsarensis Bor (1, 4)

Perennial herbs. W, C, E Nepal: 3600-3700 m. Nepal, W Himalaya, Tibetan Plateau, S Asia and E Asia.

Use: Forage.

Poa mairei Hack. [8]

Poa ludens R.R.Stewart (1, 4); *Poa pseudopratensis* Hook.f.

Perennial tufted herbs. C, E Nepal: 3600-4300 m. Nepal, E Himalaya, Tibetan Plateau and E Asia.

Use: Forage.

Poa muktinathensis Rajbh., *J. Jpn. Bot.* **91** Suppl. 175-177. 2016. [75]

Annual herbs. C Nepal: *ca.* 5200 m. **Endemic to Nepal.**

Central: *Mustang*, Thorung La, above Muktinath, *ca.* 5200 m, KR Rajbhandari 8350A (Holotype: KATH).

Poa nemoralis L. [1, 4]

Perennial herbs. C Nepal: 2600-4100 m. Nepal, E Himalaya, Tibetan Plateau, S Asia, E Asia, N Asia, C Asia, SW Asia and Europe.

Use: Forage.

Poa nepalensis (Wall. ex Griseb.) Duthie [1, 4, 5]

Poa annua var. *nepalensis* Wall. ex Griseb.

Perennial herbs. W, C, E Nepal: 1900-3000 m. Nepal and Asia.

Use: Forage.

Poa nitidespiculata Bor

Perennial herbs. C Nepal: *ca.* 3700 m. Nepal, E Himalaya and Tibetan Plateau.

Use: Forage.

Poa pagophila Bor [1, 4, 5]

Poa levipes (Keng) Keng ex L.Liu; *Poa nigropurpurea* C.Ling

Perennial herbs. W, C, E Nepal: 3500-5200 m. Nepal, W Himalaya, E Himalaya, Tibetan Plateau, S Asia and E Asia.

Use: Forage.

Poa polycolea Stapf [1, 4, 5]

Poa chalarantha Keng; *Poa nubigena* Keng

Perennial herbs. E Nepal: 3500-5600 m. Nepal, E Himalaya, Tibetan Plateau, S Asia, E Asia, N Asia and SW Asia.

Use: Forage.

Poa pratensis L. [1, 4, 8]

subsp. **pratensis**

Perennial herbs. C, E Nepal: 2700-4400 m. Nepal, Asia, Europe, Africa, N America, S America and Australasia.

Use: Forage.

subsp. **alpigena** (Lindm.) Hiitonen [4, 5]

Poa alpigena Lindm.

Perennial herbs. W, C Nepal: 3800-4600 m. Nepal, E Asia, N Asia, Europe, native to N America and S America.

Use: Forage.

subsp. **angustifolia** (L.) Dumort. [4, 5]

Poa angustifolia L.; Poa *pratensis* var. *angustifolia* (L.) Gaudin

Perennial herbs. W, C Nepal: 2900-3800 m. Nepal, E Himalaya, Tibetan Plateau, S Asia, E Asia, N Asia, C Asia, SW Asia and Europe.

Use: Forage.

Poa rajbhandarii Noltie [5]

Annual or short-lived perennial herbs. C, E Nepal: 2800-4100 m. Nepal, E Himalaya, Tibetan Plateau and E Asia.

Use: Forage.

Poa sikkimensis (Stapf) Bor [1, 4, 5]

Poa annua var. *sikkimensis* Stapf

Annual or short-lived perennial herbs. W, C, E Nepal: 3200-4400 m. Nepal, E Himalaya, Tibetan Plateau and E Asia.

Use: Forage.

Poa stapfiana Bor [1, 4, 5]

Poa tremula Stapf

Perennial herbs. W, C, E Nepal: 2600-4200 m. Nepal, W Himalaya, E Himalaya, Tibetan Plateau, S Asia, E Asia and SW Asia.

Use: Forage.

Poa supina Schrad. [1, 4]

Perennial herbs. C Nepal: 1900-3600 m. Nepal, Asia, Europe and N America.

Use: Forage.

Poa szechuensis Rendle

Poa tibeticola Bor (1, 4)

Perennial herbs. W, C Nepal: 2700-4400 m. Nepal, E Himalaya and Tibetan Plateau.

Use: Forage.

Poa tibetica Munro ex Stapf [1, 4]

Poa chushualana Rajeshwari, R.R.Rao & Arti

Perennial herbs. C Nepal: 3000-4500 m. Nepal and Asia.

Use: Forage.

Pogonatherum P.Beauv.

Pogonatherum crinitum (Thunb.) Kunth [1, 4, 5]

Andropogon crinitus Thunb.

मुसे खरी, खरुकी (Muse khari, Kharuki)

Perennial mat forming herbs. W, C, E Nepal: 700-1700 m. Nepal, Asia, Africa and Australasia.

Use(s): Medicine (whole plant), Forage.

Pogonatherum paniceum (Lam.) Hack. [1, 4, 5]

Saccharum paniceum Lam.

पाब्गा (Pabg-Tamang); Miniature bamboo

Perennial mat forming herbs. W, C, E Nepal: 900-3200 m. Nepal, Asia, Africa and Australasia.

Use: Forage.

Polypogon Desf.

Polypogon fugax Nees ex Steud. [1, 4, 5]

Polypogon littoralis var. *higegaweri* (Steud.) Hook.f.

झ्याप्ले घाँस, खरायो पाईले (Jhyaple ghaans, Kharaayo paile)

Annual herbs. W, C, E Nepal: 1300-3600 m. Nepal, Asia, Europe and Africa.

Use: Forage.

Polypogon monspeliensis (L.) Desf. [1, 4, 5]

Alopecurus monspeliensis L.

Annual herbs. W, C, E Nepal: 100-2000 m. Nepal, Asia, Europe and Africa.

Pseudechinolaena Stapf

Pseudechinolaena polystachya (Kunth) Stapf [1, 4, 5]

Echinolaena polystachya Kunth

Annual mat forming herbs. C, E Nepal: 900-2000 m. Nepal, Tibetan Plateau, E Asia and Pantropical.

Pseudopogonatherum A.Camus

Pseudopogonatherum contortum (Brongn.) A.Camus [1, 4]

Pogonatherum contortum Brongn.; *Eulalia contorta* (Brongn.) Kuntze (5)

Annual herbs. W, C, E Nepal: 100-1100 m. Nepal, Tibetan Plateau, Assam-Burma, S Asia, E Asia, SE Asia and Australasia.

Pseudosorghum A.Camus

Pseudosorghum fasciculare (Roxb.) A.Camus [1, 4, 5]

Andropogon fascicularis Roxb.

Annual decumbent herbs. C, E Nepal: 100-1000 m. Nepal, Assam-Burma, S Asia, E Asia and SE Asia.

Ptilagrostis Griseb.

Ptilagrostis mongholica (Turcz. ex Trin.) Griseb.

Stipa mongholica Turcz. ex Trin. (1, 4, 5)

Perennial herbs. W, C, E Nepal: 3800-4900 m. Nepal, W Himalaya, E Himalaya, Tibetan Plateau, E Asia and N Asia.

Puccinellia Parl.

Puccinellia himalaica Tzvelev [55]

Perennial tufted herbs. C Nepal: *ca.* 2700 m. Nepal, Tibetan Plateau, S Asia, E Asia, N Asia and SW Asia.

Puccinellia ladakhensis (H.Hartmann) Dickoré

Poa ladakhensis H.Hartmann

Perennial herbs. C Nepal: 4200-4300 m. Nepal, W Himalaya, Tibetan Plateau and S Asia.

Puccinellia tenuiflora (Griseb.) Scribn. & Merr.

Atropis tenuiflora Griseb.

Perennial herbs. W, C, E Nepal: 400-4000 m. Nepal, Tibetan Plateau, S Asia, E Asia, N Asia, C Asia and SW Asia.

Rottboellia L.f.

Rottboellia cochinchinensis (Lour.) Clayton [5]

Stegosia cochinchinensis Lour.; *Rottboellia exaltata* (L.) L.f. (1, 4)

Annual herbs. W, C Nepal: 100-1200 m. Nepal, Asia, Africa and Australasia.

Use: Fodder.

Saccharum L.

Saccharum arundinaceum Retz. [1, 4, 5]

Hardy sugarcane

Perennial herbs. C Nepal: 200-300 m. Nepal, E Himalaya, Tibetan Plateau, Assam-Burma, S Asia, E Asia and SE Asia.

Use(s): Medicine (root), Food, Forage, Construction materials.

Saccharum bengalense Retz. [5]

Saccharum munja Roxb.

बरुवा (Baruwaa); Baruwa grass

Perennial herbs. W, C Nepal: 100-400 m. Nepal, W Himalaya, S Asia and SW Asia.

Use(s): Medicine (root), Forage, Utensils and handicrafts, Ornamental.

Saccharum filifolium Steud.

Erianthus filifolius Nees ex Steud. (1, 4)

Perennial woody herbs. W, C, E Nepal: 2100-2500 m. Nepal, W Himalaya, S Asia and SW Asia.

Saccharum longesetosum (Andersson) V.Naray. ex Bor [5]

Erianthus longesetosus Andersson (1, 4); *Erianthus longesetosus* var. *hookeri* (Hack.) Bor (1, 4); *Erianthus hookeri* Hack.

Perennial herbs. W, C, E Nepal: 200-2200 m. Nepal, E Himalaya, Tibetan Plateau, Assam-Burma, SE Asia and E Asia.

Saccharum narenga (Nees ex Steud.) Wall. ex Hack. [5]

Eriochrysis narenga Nees ex Steud. (1); *Narenga porphyrocoma* (Hance) Bor (1, 4)

खडाइ (Khadai)

Perennial erect herbs. W, C, E Nepal: 100-700 m. Nepal, Tibetan Plateau, Assam-Burma, S Asia, E Asia, SE Asia and SW Asia.

Use(s): Handicraft, Construction materials.

Saccharum officinarum L. [5]

उखु (Ukhu); Sugarcane

Perennial robust herbs. C Nepal: 400-500 m. Widely cultivated as cash crop for sugar. Native to SE Asia.

Use: Food (stem juice).

Saccharum procerum Roxb. [1, 4, 5]

Erianthus procerus (Roxb.) Raizada

Devil sugarcane

Perennial herbs. W, C Nepal: 100-700 m. Nepal and Asia.

Use: Forage.

Saccharum ravennae (L.) L. [4]

Andropogon ravennae L.; *Erianthus elephantinus* Hook.f.; *Erianthus ravennae* (L.) P.Beauv.

Ravenna grass

Perennial herbs. E Nepal: 500-1500 m. Nepal, W Himalaya, S Asia, N Asia, C Asia, SW Asia and Europe.

Saccharum rufipilum Steud. [5]

Erianthus fulvus Nees ex Hack.; *Erianthus rufipilus* (Steud.) Griseb. (1, 4)

Perennial herbs. W, C, E Nepal: 1300-3200 m. Nepal, E Himalaya, Tibetan Plateau, Assam-Burma, S Asia, E Asia and SW Asia.

Saccharum sikkimense (Hook.f.) V.Naray. ex Bor [5]

Erianthus sikkimensis Hook.f. (1, 4)

Perennial herbs. C, E Nepal: 100-300 m. Nepal, S Asia and E Asia.

Saccharum spontaneum L. [1, 4, 5]

काँस (Kaans); Thatch grass

Perennial herbs. W, C, E Nepal: 100-1900 m. Nepal, Asia, Africa and Australasia.

Use(s): Medicine (whole plant), Forage, Utensils and handicrafts, Construction materials, Ritual.

Saccharum williamsii (Bor) Bor ex Cope, *Grass, Burma, Ceylon, India & Pakistan* 214. 1960. [1, 4, 27, 29]

Erianthus williamsii Bor

Perennial rhizomatous herbs. C Nepal: 2700-2900 m. **Endemic to Nepal**.

Central: *Myagdi*, NW Gurjakhani, 9500 ft. (2895 m), JDA Stainton, WR Sykes & LHJ Williams 3676 (Holotype: BM).

Use: Forage.

Sacciolepis Nash

Sacciolepis indica (L.) Chase [1, 4, 5]

Aira indica L.

कागुने बन्सो (Kaagune banso)

Annual marshy herbs. W, C, E Nepal: 200-2400 m. Nepal, E Himalaya, Assam-Burma, S Asia, E Asia, SE Asia, Africa and Australasia.

Use: Forage.

Sacciolepis interrupta (Willd.) Stapf [1, 4, 5]

Panicum interruptum Willd.

Perennial partly submerged or floating herbs. E Nepal: *ca.* 200 m. Nepal, Assam-Burma, S Asia, E Asia, SE Asia and Africa.

Sacciolepis myosuroides (R.Br.) A.Camus [1, 4, 5]

Panicum myosuroides R.Br.

Annual marshy herbs. C, E Nepal: 100-2100 m. Nepal, Asia and Australasia.

Schizachyrium Nees

Schizachyrium brevifolium (Sw.) Büse [1, 4, 5]

Andropogon brefolius Sw.

Annual rambling or prostrate herbs. W, C, E Nepal: 100-1800 m. Nepal, E Himalaya, Tibetan Plateau, Assam-Burma, S Asia, E Asia, SE Asia, SW Asia, Africa and America.

Schizachyrium delavayi (Hack.) Bor [1, 4, 5]

Andropogon delavayi Hack.; *Andropogon bootanensis* Hook.f.

Perennial herbs. C Nepal: 1900-2600 m. Nepal, E Himalaya and Tibetan Plateau, Assam-Burma and E Asia.

Sclerostachya (Andersson ex Hack.) A.Camus

Sclerostachya fusca (Roxb.) A.Camus [5]

Saccharum fuscum Roxb.; *Miscanthus fuscus* (Roxb.) Benth.

Perennial erect herbs. W, C, E Nepal: *ca.* 100 m. Nepal, S Asia and SE Asia.

Secale L.

Secale cereale L. [5]

Triticum cereale (L.) Salisb.

Rye

Annual herbs. C Nepal: *ca.* 2800 m. Nepal, Tibetan Plateau, E Asia and N Asia.

Setaria P.Beauv.

Setaria forbesiana (Nees ex Steud.) Hook.f. [1, 4, 5]

Panicum forbesianum Nees ex Steud.

Perennial erect herbs. C Nepal: 1200-2000 m. Nepal, E Himalaya, Assam-Burma, E Asia and SE Asia.

Setaria homonyma (Steud.) Chiov. [5]

Panicum homonymum Steud.

Annual herbs. W Nepal: *ca.* 1000 m. Nepal, S Asia and Africa.

Use: Forage.

Setaria intermedia Roem. & Schult. [5]

Panicum tomentosum Roxb.; *Setaria tomentosa* (Roxb.) Kunth (1, 4)

Annual ascending or decumbent herbs. W, C, E Nepal: 400-1200 m. Nepal, Asia and Africa.

Setaria italica (L.) P.Beauv. [1, 4, 5]

Panicum italicum L.

कागुनो (Kaaguno); Fox tail millet

Annual herbs. W, C, E Nepal: 1100-2700 m. Widely cultivated.

Use(s): Food/Medicine (grains), Forage.

Setaria palmifolia (J.Köenig) Stapf [1, 4, 5]

Panicum palmifolium J.Köenig

बाँसपाते कागुनो (Baanspate kaaguno)

Perennial herbs. W, C, E Nepal: 300-1800 m. Nepal, Tibetan Plateau, old world and new world.

Setaria parviflora (Poir.) Kerguélen

Cenchrus parviflorus Poir.; *Setaria geniculata* P.Beauv (1, 4)

कनिके कागुनो, कागुने झार (Kanike kaaguno, Kaagune jhaar)

Annual or short-lived perennial herbs. C, E Nepal: 700-2200 m. Nepal, E Asia, Pantropical and Subtropical.

Setaria plicata (Lam.) T.Cooke [1, 4, 8]

Panicum plicatum Lam.

Perennial herbs. W, C, E Nepal: 900-2000 m. Nepal, Tibetan Plateau, S Asia, E Asia and SE Asia.

Use(s): Medicine (leaves, grains), Food (tender shoots).

Setaria pumila (Poir.) Roem. & Schult. [5]

Panicum pumilum Poir.; *Setaria pallide-fusca* (Schumach.) Stapf & C.E.Hubb. (1, 4)

कागुने झार, कनिके कागुनो (Kaagune jhaar, Kanike kaaguno); Yellow fox tail

Annual herbs. W, C, E Nepal: 100-2500 m. Nepal, Tibetan Plateau, Asia and Europe.

Use: Forage.

Setaria sphacelata (Schumach.) Stapf & C.E.Hubb. ex Moss [5]

Panicum sphacelatum Schumach.

Perennial herbs. E Nepal: 1700-1800 m. Nepal, Asia, Africa, N America, S America and Australasia.

Use: Forage.

Setaria verticillata (L.) P.Beauv. [1, 4, 5]

Panicum verticillatum L.

झुसे कागुने घाँस (Jhuse kaagune ghaans); Bristle grass

Annual herbs. W, C, E Nepal: 300-900 m. Nepal, E Asia and old world.

Use(s): Forage, Food (grains).

Setaria viridis (L.) P.Beauv. [1, 4, 5]

Panicum viride L.

Green bristle grass

Annual herbs. W, C Nepal: 1000-3900 m. Nepal, Tibetan Plateau, E Asia and N Asia.

Use: Medicine (whole plant).

Sinarundinaria Nakai

Sinarundinaria falcata (Nees) C.S.Chao & Renvoize

Arundinaria falcata Nees; *Drepanostachyum falcatum* (Nees) Keng f. (4, 5, 17); *Drepanostachyum khasianum* (Munro) Keng f. (8, 17)

घोडे निगालो, तिते निंगालो (Ghode nigaalo, Tite nigaalo); Himalayan weeping bamboo

Perennial herbs. W, C, E Nepal: 1500-2200 m. Nepal, E Himalaya, Assam-Burma, S Asia and SE Asia.

Use(s): Food (young shoot), Forage, Utensils and handicrafts, Construction materials.

Sinarundinaria intermedia (Munro) C.S.Chao & Renvoize

Arundinaria intermedia Munro; *Drepanostachyum intermedium* (Munro) Keng f. (4, 5, 17)

तिते निंगालो (Tite nigaalo)

Perennial herbs. C, E Nepal: 200-2000 m. Nepal, E Himalaya and S Asia.

Use(s): Forage, Construction materials, Utensils and handicrafts.

Sinarundinaria langtangensis R.Manandhar & Bajracharya, *Pleione* **6**(2): 381. 2012. [119]

Perennial herbs. C Nepal: 3000-3100 m. **Endemic to Nepal**.

Central: *Rasuwa*, Ghoda Tabela, 3010 m, R. Manandhar 216 (Holotype: KATH; Isotype: TUCH).

Sinochasea Keng

Sinochasea trigyna Keng

Trikeraia oreophila Cope (5)

Perennial herbs. E Nepal: 4200-4300 m. Nepal, E Himalaya, Tibetan Plateau and N Asia.

Sorghum Moench

Sorghum bicolor (L.) Moench [5]

Holcus bicolor L.; *Sorghum roxburghii* Stapf (1, 4)

जुनेलो (Junelo); Indian broomcorn millet

Annual herbs. W, C Nepal: 900-2000 m. Cultivated in Nepal and S Asia. Native to Africa.

Sorghum halepense (L.) Pers. [1, 4]

Holcus halepensis L.; *Andropogon halepensis* (L.) Brot.

Perennial herbs. W, C, E Nepal: 100-2000 m. Nepal, S Asia, E Asia, C Asia, SW Asia and Europe.

Use(s): Forage, Medicine (seeds).

Sorghum miliaceum (Roxb.) Snowden [1, 4]

Andropogon miliaceus Roxb.

Perennial herbs. W, C, E Nepal: 700-800 m. Nepal, S Asia, N America and Australasia.

Use: Forage.

Sorghum nitidum (Vahl) Pers. [5]

Holcus nitidus Vahl

Perennial herbs. W Nepal: 100-200 m. Nepal, E Himalaya, Assam-Burma, S Asia, E Asia, SE Asia and Australasia.

Spodiopogon Trin.

Spodiopogon dubius Hack. [5]

Perennial herbs. W, C Nepal: 1500-2100 m. Nepal, W Himalaya and Tibetan Plateau.

Sporobolus R.Br.

Sporobolus diandrus (Retz.) P.Beauv. [1, 4, 5]

Agrostis diandra Retz.; *Sporobolus indicus* var. *flaccidus* (Roem. & Schult.) Veldkamp

कोदे घाँस (Kode ghaans)

Perennial herbs. W, C, E Nepal: 100-1400 m. Nepal, E Himalaya, Assam-Burma, S Asia, E Asia, SE Asia, SW Asia and Australasia.

Use: Forage.

Sporobolus fertilis (Steud.) Clayton [1, 4, 5]

Agrostis fertilis Steud.; *Sporobolus indicus* var. *major* (Buse) Baaijens

बन्धन झार, घोबे (Bandhan jhaar, Ghobe)

Perennial herbs. W, C, E Nepal: 400-2400 m. Nepal, E Himalaya, Tibetan Plateau, Assam-Burma, S Asia, E Asia, SE Asia and Australasia.

Use(s): Forage, Construction materials.

Sporobolus helvolus (Trin.) T.Durand & Schinz [1, 4]

Vilfa helvola Trin.

Perennial herbs. C Nepal: 500-1500 m. Nepal, S Asia, SE Asia, SW Asia and Africa.

Sporobolus pilifer (Trin.) Kunth [1, 4, 5]

Vilfa pilifera Trin.

Annual herbs. W, C, E Nepal: 900-2700 m. Nepal, E Himalaya, S Asia, E Asia, SE Asia and Africa.

Stipa L.

Stipa breviflora Griseb. [1, 4]

Perennial herbs. C Nepal: 3000-3800 m. Nepal, W Himalaya, Tibetan Plateau, S Asia, E Asia, C Asia and N Asia.

Stipa capensis Thunb. [5]

Annual herbs. C Nepal: *ca.* 3800 m. Nepal, Asia, Europe and Africa.

Stipa capillacea Keng

Stipa koelzii R.R.Stewart (5)

Perennial herbs. W, C, E Nepal: 3300-4100 m. Nepal, W Himalaya, E Himalaya, Tibetan Plateau and E Asia.

Stipa capillata L. [1, 4]

Needle grass

Perennial herbs. C Nepal: *ca.* 4000 m. Nepal, W Himalaya, S Asia, E Asia, N Asia, C Asia, SW Asia and Europe.

Use: Forage.

Stipa consanguinea Trin. & Rupr. [1, 4]

Perennial herbs. W, C Nepal: 2600-4000 m. Nepal and N Asia.

Stipa duthiei Hook.f. [1, 4]

Perennial herbs. C Nepal: *ca.* 3000 m. Nepal, E Himalaya, Tibetan Plateau and S Asia.

Stipa hookeri Stapf

Perennial herbs. W, C, E Nepal. Nepal, W Himalaya, E Himalaya and Tibetan Plateau.

Stipa milleri Noltie [5]

Perennial herbs. C, E Nepal: 4400-4600 m. Nepal, E Himalaya and S Asia.

Stipa przewalskyi Roshev. [8]

Perennial herbs. C Nepal: 3800-3900 m. Nepal, Tibetan Plateau and E Asia.

Stipa roborowskyi Roshev. [8]

Stipa basiplumosa var. *longearistata* Munro ex Hook.f.

Perennial herbs. W, C, E Nepal. Nepal, W Himalaya, Tibetan Plateau and N Asia.

Stipa sareptana Roshev.

var. **krylovii** (Roshev.) P.C.Kuo & Y.H.Sun

Stipa krylovii Roshev. (76)

Perennial tufted herbs. C Nepal: *ca.* 3600 m. Nepal, W Himalaya, Tibetan Plateau, E Asia, N Asia and C Asia.

Stipa staintonii Bor, *Bull. Bot. Surv. India* **7**: 133. 1965. [1, 4, 5, 27]

Perennial herbs. W, C Nepal: 2600-4000 m. **Endemic to Nepal**.

West: *Rukum*, Seng Khola, 12500 ft. (3810 m), JDA Stainton, WR Sykes & LHJ Williams 4677 (Holotype: K).

Thamnocalamus Munro

Thamnocalamus chigar (Stapleton) Stapleton

Borinda chigar Stapleton, *Edinb. J. Bot.* **51**(2): 275-295. 1994. (4, 27, 118)

चिगर (Chigar)

Perennial bamboo. W, C Nepal: *ca.* 3000 m. **Endemic to Nepal**.

Central: *Kaski*, Karuwa to Pipar, *ca.* 3000 m, CMA Stapleton 315 (Holotype: E).

Thamnocalamus nepalensis (Stapleton) Stapleton [5]

Thamnocalamus spathiflorus subsp. *nepalensis* Stapleton (4, 8, 17)

मालिङ्गो, रातो निंगालो (Maalingo, Raato nigaalo)

Perennial bamboo. C Nepal: 3200-3500 m. Nepal and Tibetan Plateau.

Use(s): Forage, Utensils and handicrafts, Construction materials.

Thamnocalamus spathiflorus (Trin.) Munro [1, 4, 17]

var. **spathiflorus**

Arundinaria spathiflora Trin.; *Thamnocalamus aristatus* (Gamble) E.G.Camus (1)

Perennial bamboo. C, E Nepal: 2200-3200 m. Nepal, W Himalaya, E Himalaya and Tibetan Plateau.

var. **crassinodus** (T.P.Yi) Stapleton [17]

Fargesia crassinoda T.P.Yi

Perennial bamboo. C Nepal: 2400-3500 m. Nepal and Tibetan Plateau.

Themeda Forssk.

Themeda anathera (Nees ex Steud.) Hack. [1, 4, 5]

Anthistiria anathera Nees ex Steud.

धड्दी (Dhaddi)

Perennial herbs. W, C Nepal: 700-2300 m. Nepal, E Himalaya, Tibetan Plateau, S Asia, E Asia and SW Asia.

Use(s): Forage, Construction materials.

Themeda arundinacea (Roxb.) A.Camus [1, 4, 5]

Anthistiria arundinacea Roxb.; *Themeda subsericans* (Nees ex Steud.) Ridl. (5)

उरेली, धड्दी (Ureli, Dhaddi)

Perennial erect herbs. W, C Nepal: 100-1700 m. Nepal, E Himalaya, Assam-Burma, S Asia, E Asia and SE Asia.

Use(s): Medicine (root), Food, Forage, Utensils and handicrafts.

Themeda caudata (Nees ex Hook. & Arn.) A.Camus [1, 4, 5]

Anthistiria caudata Nees ex Hook. & Arn.; *Themeda gigantea* subsp. *caudata* (Nees ex Hook. & Arn.) Hack.

उरेली, उल्ला (Ureli, Ullaa)

Perennial herbs. C, E Nepal: 100-2100 m. Nepal, E Himalaya, Tibetan Plateau, Assam-Burma, S Asia, E Asia and SE Asia.

Use(s): Forage, Construction materials.

Themeda hookeri (Griseb.) A.Camus [1, 4, 5]

Anthistiria hookeri Griseb.

उरेली (Ureli)

Perennial ascending herbs. W, C, E Nepal: 1500-3000 m. Nepal, E Himalaya, Tibetan Plateau and E Asia.

Use(s): Forage, Construction materials.

Themeda quadrivalvis (L.) Kuntze [1, 4]

Andropogon quadrivalvis L.

Annual erect or ascending herbs. W, C, E Nepal: 600-3000 m. Nepal, Assam-Burma, S Asia, E Asia, SE Asia, Africa, N America, S America and Australasia.

Use(s): Forage, Construction materials.

Themeda triandra Forssk. [1, 4, 5]

खर (Khar); Kangaroo grass

Perennial herbs. W, C, E Nepal: 1700-4100 m. Nepal, Asia, Africa and Australasia.

Use(s): Forage, Construction materials.

Themeda villosa (Lam.) A.Camus [1, 4, 5]

Anthistiria villosa Lam.; *Anthistiria gigantea* subsp. *villosa* (Lam.) Hook.f.

उरेली (Ureli)

Perennial herbs. W, C, E Nepal: 100-2000 m. Nepal, E Himalaya, Tibetan Plateau, S Asia, E Asia, SE Asia and Africa.

Use(s): Forage, Food (young shoots), Construction materials.

Thysanolaena Nees

Thysanolaena latifolia (Roxb. ex Hornem.) Honda [5]

Melica latifolia Roxb. ex Hornem.; *Agrostis maxima* Roxb.; *Thysanolaena maxima* (Roxb.) Kuntze (1, 4)

अमरिसो (Amriso); Broom grass

Perennial herbs. W, C, E Nepal: 100-2000 m. Nepal, E Himalaya, Tibetan Plateau, Assam-Burma, S Asia, E Asia, SE Asia and Africa.

Use(s): Medicine (roots), Forage, Utensils and handicrafts.

Tragus Haller

Tragus berteronianus Schult.

Cenchrus racemosus L.; *Tragus racemosus* (L.) All. (8)

Annual herbs. E Nepal: 300-400 m. Nepal, Asia, Europe, Africa, N America and S America.

Tripogon Roem. & Schult.

Tripogon bromoides Roth [1, 4]

Perennial herbs. C Nepal: 1200-3300 m. Nepal, S Asia and SE Asia.

Tripogon capillatus Jaub. & Spach [8]

Annual herbs. C, E Nepal: 1700-2100 m. Nepal, S Asia and SW Asia.

Tripogon filiformis Nees ex Steud. [1, 4, 5]

Tripogon filiformis var. *tenuispicus* Hook.f.

Perennial herbs. W, C, E Nepal: 300-3500 m. Nepal and Asia.

Tripogon purpurascens Duthie [1, 4]

Tripogon hookerianus Bor; *Tripogon jacquemontii* var. *submuticus* Hook.f.

Perennial herbs. W Nepal: *ca.* 2300 m. Nepal, W Himalaya, E Himalaya, Tibetan Plateau, S Asia, E Asia, N Asia and SW Asia.

Tripogon trifidus Munro ex Stapf [1, 4, 5]

Perennial herbs. C, E Nepal: 1200-3200 m. Nepal, E Himalaya, Tibetan Plateau, Assam-Burma, S Asia, E Asia and SE Asia.

Trisetum Pers.

Trisetum aeneum (Hook.f.) R.R.Stewart [1, 4]

Avena aenea Hook.f.

Perennial herbs. W, C Nepal: 4000-4300 m. Nepal, W Himalaya and S Asia.

Use: Forage.

Trisetum clarkei (Hook.f.) R.R.Stewart [1, 4]

Avena clarkei Hook.f.

Perennial herbs. C, E Nepal: 3500-4500 m. Nepal, W Himalaya, Tibetan Plateau, S Asia E Asia, N Asia and SW Asia.

Trisetum scitulum Bor [1, 4, 5]

Perennial herbs. C, E Nepal: 3700-4400 m. Nepal, E Himalaya, Tibetan Plateau, S Asia and E Asia.

Trisetum spicatum (L.) K.Richt.

subsp. **virescens** (Regel) Tzvelev

Avena flavescens var. *virescens* Regel; *Trisetum spicatum* subsp. *himalaicum* Hultén ex Veldcamp (1, 4)

Perennial herbs. W, C, E Nepal: 2500-5600 m. Nepal, Asia, Europe, N America, S America and Australasia.

Triticum L.

Triticum aestivum L. [1, 4, 5]

गहुँ (Gahun); Wheat

Annual herbs. W, C, E Nepal: 1800-3500 m (Widely cultivated). Native to Mediterranean region and SW Asia.

Urochloa P.Beauv.

Urochloa panicoides P.Beauv. [1, 4]

Annual marshy herbs. C Nepal: *ca.* 700 m. Nepal, E Himalaya, S Asia, E Asia and Africa.

Urochloa setigera (Retz.) Stapf

Panicum setigerum Retz.; *Brachiaria setigera* (Retz.) C.E.Hubb. (1, 4)

Signal grass

Annual herbs. E Nepal: *ca.* 1500 m. Nepal, Assam-Burma, S Asia, E Asia, SE Asia and Africa.

Vulpia C.C.Gmel.

Vulpia myuros (L.) C.C.Gmel. [5]

Festuca myuros L.

Annual herbs. C Nepal: *ca.* 1500 m. Nepal, Asia, Europe and Africa.

Yushania Keng f.

Yushania emeryi (Stapleton) Demoly

Borinda emeryi Stapleton, *Edinburgh J. Bot.* **51**(2): 286. 1994. (4, 8, 17, 27)

कालो मालिङ्गो, कालो निंगालो (Kaalo maalingo, Kaalo nigaalo)

Evergreen perennial bamboo. E Nepal: 2000-4100 m. **Endemic to Nepal**.

East: *Sankhuwasabha*, Younglay, N of Num, Barun Khola, 3750 m, HB Emery 54 (Holotype: K).

Yushania maling (Gamble) R.B.Majumdar & Karthik. [4, 5, 17]

Arundinaria maling Gamble; *Sinarundinaria maling* (Gamble) C.S.Chao & Renvoize

खोस्रे मालिङ्गो, धनु बाँस (Khosre maalingo, Dhanu baans)

Evergreen perennial bamboo. E Nepal: 2400-3000 m. Nepal and E Himalaya.

Use(s): Food, Forage, Fence, Utensils and handicrafts.

Yushania microphylla (Munro) R.B.Majumdar [4, 17]

Arundinaria microphylla Munro

मालिङ्गो (Maalingo)

Evergreen perennial bamboo. E Nepal: 2300-3500 m. Nepal, E Himalaya and S Asia.

Yushania pantlingii (Gamble) R.B.Majumdar [1, 4]

Arundinaria pantlingii Gamble; *Semiarundinaria pantlingii* (Gamble) Nakai

Evergreen perennial bamboo. E Nepal: *ca.* 2300 m. Nepal, E Himalaya and S Asia.

Zea L.

Zea mays L. [1, 4, 5]

मकै (Makai); Maize, Corn

Annual herbs. W, C, E Nepal: 200-2000 m. Widely cultivated as cereal crop. Native to America.

Use(s): Food (grains), Medicine (cornsilk, maize meal).

Doubtful/Uncertain Taxa

Achnatherum brandisii (Mez) Z.L.Wu (*Stipa brandisii* Mez); W, C Nepal: 2600-3200 m. [103]

Agrostis brachiata Munro ex Hook.f. [103]

Bambusa bambos (L.) Voss [*Bambusa arundinacea* (Retz.) Willd.] **तारु बाँस, कट बाँस, बंस लोचन** (Taaru baans, Kat baans; Bansa lochan); Spiny bamboo. *Use(s)*: Medicine (root, leaves, sprouts), Food (sprouts).

Bambusa vulgaris Schrader ex Wendland [17]

Cymbopogon citratus (DC.) Stapf (Lemon grass), Native to S Asia. *Use(s):* Medicine (leaves), Essential oil.

Dendrocalamus giganteus Wall. ex Munro, **ढुंग्रे बास** (Dhungre baans); E Nepal: below 1000 m. Cultivated. Native to Malaysia [17]

Dendrocalamus hamiltonii var. **undulatus** Stapleton [17]

Themeda laxa (Andersson) A.Camus (Kew, K000245947; Wallich, N. 8775) [107]

Trisetum flavescens (L.) P.Beauv. (*Avena flavescens* L.); E Nepal: 3300-3900 m.

EUDICOTS

RANUNCULALES

PAPAVERACEAE (incl. FUMARIACEAE)

Argemone L.

Argemone mexicana L. [2, 4, 6, 10]

थाकल, सुंगुरे काँडा (Thaakal, Sungure kaanda); Mexican poppy

Prickly annual herbs. W, C, E Nepal: 100-1400 m. Invasive species in Nepal. Native to Mexico.

Use(s): Medicine (aerial part, latex, seeds), Ritual, Seed oil is used for lighting.

Argemone ochroleuca Sweet [4, 10]

Annual herbs. E Nepal: 100-200 m. Nepal, W Himalaya, S Asia, N America, S America and Australasia.

Use: Medicine (whole plant).

Cathcartia Hook.f. ex Hook.

Cathcartia villosa Hook.f. ex Hook. [10]

Meconopsis villosa (Hook.f. ex Hook.) G.Taylor (2, 4, 6)

Perennial herbs. E Nepal: 2600-3700 m. Nepal and E Himalaya.

Corydalis Vent.

Corydalis calycina Lidén, *Bull. Brit. Mus. (Nat. Hist.) Bot.* **18**(6): 532. 1989. [4, 8, 10, 12, 27, 29]

Annual or short-lived perennial herbs. W, C Nepal: 3000-5500 m. **Endemic to Nepal**.

West: *Jumla*, Maharigaon, 10500 ft. (3200 m), O. Polunin, WR Sykes & LHJ Williams 158 (Holotype: BM).

Corydalis cashmeriana Royle [2, 4, 6, 10]

subsp. **cashmeriana**

टोंग्रे-सिल्बा (Tongre-silbaa – Tibetan)

Perennial herbs. W, C, E Nepal: 2400-5500 m. Nepal, W Himalaya and Tibetan Plateau.

Use: Medicine (whole plant).

subsp. **longicalcarata** (D.G.Long) Lidén [8, 10]

Corydalis ecristata var. *longicalcarata* D.G.Long; *Corydalis cashmeriana* var. *longicalcarata* (D.G.Long) R.C.Srivastava; *Corydalis ecristata* subsp. *longicalcarata* (D.G.Long) C.Y.Wu

Perennial herbs. C, E Nepal: 3200-4600 m. Nepal, E Himalaya and Tibetan Plateau.

Corydalis casimiriana Duthie & Prain [2, 4, 6, 10]

subsp. **casimiriana**

Annual or perennial herbs. W, C, E Nepal: 2300-3800 m. Nepal and W Himalaya.

supsp. **brachycarpa** Lidén [10]

Annual or perennial herbs. C, E Nepal: 2800-4200 m. Nepal, E Himalaya and Tibetan Plateau.

Corydalis cavei D.G.Long [4, 8, 10]

Perennial herbs. C, E Nepal: 2700-4500 m. Nepal, E Himalaya and Tibetan Plateau.

Corydalis chaerophylla DC. [2, 4, 6, 10]

ओखरे घाँस, कर्मे घाँस (Okhare ghaans, Karme ghaans)

Perennial herbs. W, C, E Nepal: 1800-5500 m. Nepal, W Himalaya, E Himalaya and Assam-Burma.

Use: Medicine (root).

Corydalis clavibracteata Ludlow, *Bull. Brit. Mus. (Nat. Hist.) Bot.* **5**: 53. 1975. [2, 4, 8, 10, 12, 27, 29]

Perennial dwarf herbs. W Nepal: 3600-4700 m. **Endemic to Nepal**.

West: *Dolpa*, Ringmigaon, slopes of Kanjiroba, 15500 ft. (4720 m), O. Polunin, WR Sykes & LHJ Williams 3540 (Holotype: BM; Isotype: E).

Corydalis conspersa Maxim. [4, 10]

Corydalis zambuii C.E.C.Fisch. & Kaul.

Perennial herbs. C Nepal: *ca.* 1500 m. Nepal, Tibetan Plateau and E Asia.

Use: Medicine (whole plant).

Corydalis cornuta Royle [4, 6, 10]

Annual or biennial herbs. W, C Nepal: 2200-3500 m. Nepal, W Himalaya, E Himalaya and Africa.

Corydalis diphylla Wall. [10]

Perennial herbs. W Nepal: *ca.* 2500 m. Nepal and W Himalaya.

Corydalis ecristata (Prain) D.G.Long [4, 8, 10]

Corydalis cashmeriana var. *ecristata* Prain

Perennial herbs. E Nepal: 3900-5000 m. Nepal, E Himalaya and Tibetan Plateau.

Corydalis elegans Wall. ex Hook.f. & Thomson [2, 4, 6, 10]

subsp. **elegans**

Perennial herbs. W Nepal: 3900-4200 m. Nepal, W Himalaya and Tibetan Plateau.

subsp. **robusta** Lidén [4, 10]

Perennial herbs. W Nepal: 3900-4700 m. **Endemic to Nepal**.

West: *Jumla*, Bhurchula Lekh, S Jumla, 4120 m, O. Polunin, WR Sykes & LHJ Williams 4633 (Holotype: BM; Isotypes: E, KATH).

Corydalis filicina Prain [4, 10]

Perennial herbs. W, C Nepal: 3000-4500 m. Nepal and E Himalaya.

Corydalis filiformis Royle [4, 6, 10]

Annual or perennial herbs. W, C Nepal: 2500-4100 m. Nepal and W Himalaya.

Corydalis flabellata Edgew. [2, 4, 6, 10]

Perennial herbs. C Nepal: 2800-3500 m. Nepal, W Himalaya and Tibetan Plateau.

Corydalis flaccida Hook.f. & Thomson [2, 4, 6, 10]

Perennial herbs. W, C, E Nepal: 3000-4300 m. Nepal, E Himalaya, Tibetan Plateau, Assam-Burma and E Asia.

Corydalis geraniifolia Hook.f. & Thomson [8, 10]

Corydalis chaerophylla var. *geraniifolia* (Hook.f. & Thomson) H.Hara (2, 4)

Perennial herbs. E Nepal: 3000-4800 m. Nepal and E Himalaya.

Corydalis govaniana Wall. [2, 4, 6, 10]

भुतकेश (Bhutkesh)

Perennial herbs. W, C, E Nepal: 1900-5000 m. Nepal, W Himalaya and Tibetan Plateau.

Use: Medicine (root).

Corydalis hendersonii Hemsl. [2, 4, 6, 10]

Corydalis nepalensis Kitam.

Perennial herbs. W, C Nepal: 4000-6100 m. Nepal, W Himalaya, Tibetan Plateau and E Asia.

Use: Medicine (whole plant).

Corydalis hookeri Prain [2, 4, 6, 10]

Corydalis denticulatobracteata Fedde; *Corydalis paniculata* C.Y.Wu & H.Chuang

Perennial herbs W, C, E Nepal: 2700-5500 m. Nepal, E Himalaya and Tibetan Plateau.

Corydalis jigmei C.E.C.Fisch. & Kaul [10]

Perennial herbs. E Nepal: 4100-5300 m. Nepal, E Himalaya and Tibetan Plateau.

Corydalis juncea Wall. [2, 4, 6, 10]

Perennial herbs. W, C, E Nepal: 2500-5100 m. Nepal, E Himalaya, Tibetan Plateau and Assam-Burma.

Corydalis lathyroides Prain [2, 4, 6, 10, 29]

Corydalis brevicalcarata Ludlow (2)

Perennial herbs. W, C, E Nepal: 3000-5500 m. Nepal, W Himalaya and E Himalaya.

Corydalis latiflora Hook.f. & Thomson [2, 4, 6, 10, 29]

Corydalis alburyi Ludlow & Stearn (2); *Corydalis gerdae* Fedde (2, 8); *Corydalis latiflora* subsp. *gerdae* (Fedde) Liden ex C.Y.Wu; *Corydalis mitae* Kitam.

Perennial herbs. W, C Nepal: 3000-5700 m. Nepal, E Himalaya and Tibetan Plateau.

Corydalis leptocarpa Hook.f. & Thomson [2, 4, 6, 10]

Perennial herbs. C, E Nepal: 1800-3000 m. Nepal, E Himalaya, Assam-Burma, E Asia and SE Asia.

Corydalis longipes DC. [2, 4, 6, 10]

Perennial (rarely annual) herbs. W, C, E Nepal: 2000-5300 m. Nepal and Tibetan Plateau.

Corydalis lowndesii Liden [4, 6, 10]

Perennial herbs. C, E Nepal: 2800-5000 m. Nepal and Tibetan Plateau.

Corydalis magni Pusalkar [10]

Perennial herbs. W Nepal: 4500-6100 m. Nepal, W Himalaya and Tibetan Plateau.

Corydalis megacalyx Ludlow, *Bull. Brit. Mus. (Nat. Hist.) Bot.* **5**: 58. 1975. [2, 4, 6, 10, 12, 27, 29]

भुतकेश, तोंगजिल-सेर्पो (Bhutkesh; Tongzil-serpo: Amchi)

Perennial cushion forming herbs. W, C, E Nepal: 3600-5500 m. **Endemic to Nepal**.

West: *Rukum*, E. Chalikhe Pahad, 15000 ft. (4570 m), JDA Stainton, WR Sykes & LHJ Williams 3697 (Holotype: BM; Isotype: KATH).

Use: Medicine (whole plant).

Corydalis meifolia Wall. [2, 4, 6, 10]

Corydalis meifolia var. *sikkimensis* Wall. (2, 4)

घो-ठुल-सर्बो (Gho-thul-serbo – Tibetan)

Perennial herbs. W, C, E Nepal: 3900-5200 m. Nepal, W Himalaya, E Himalaya and Tibetan Plateau.

Use: Medicine (whole plant).

Corydalis polygalina Hook.f. & Thomson [2, 4, 6, 10]

Perennial herbs. E Nepal: 4000-4800 m. Nepal, E Himalaya and Tibetan Plateau.

Corydalis pseudojuncea Ludlow [2, 4, 6, 10]

Perennial herbs. W, C Nepal: 2600-5500 m. Nepal, W Himalaya and Tibetan Plateau.

Corydalis pseudolongipes Lidén [4, 6, 10]

Perennial (rarely annual) herbs. C, E Nepal: 2800-3500 m. Nepal, E Himalaya, Tibetan Plateau and Africa.

Corydalis shakyae Lidén [4, 8, 10]

Perennial(?) herbs. W Nepal: 3600-3900 m. Nepal and W Himalaya.

Corydalis sikkimensis (Prain) Fedde [4, 8, 10]

Corydalis duthie var. *sikkimensis* Prain

Perennial herbs. E Nepal: 3900-4300 m. Nepal and E Himalaya.

Corydalis simplex Lidén, *Bull. Brit. Mus. (Nat. Hist.), Bot.* **18**(6): 514. 1989. [4, 10, 29]

रोग-पो-जोम्स-स्काईज (Rog-po-joms-skies – Tibetan)

Perennial herbs. W Nepal: 4000-5500 m. **Endemic to Nepal**.

West: *Dolpa*, Jangla Bhanjyang, *ca.* 4300 m, S. Einarrson, L. Skarby & B. Wetterhall 1970 (Holotype: BM).

Use: Medicine (leaves, seeds).

Corydalis spicata Lidén, *Bull. Brit. Mus. (Nat. Hist.) Bot.* **18**(6): 528. 1989. [4, 8, 10, 12, 27, 29]

Perennial(?) herbs. W Nepal: 2500-2600 m. **Endemic to Nepal**.

West: *Jumla*, Kabra, 2580 m, PR Shakya, MN Subedi & RK Uprety 8802 (Holotype: GB; Isotype: KATH).

Corydalis staintonii Ludlow [2, 4, 6, 10, 29]

Corydalis chasmophila Ludlow (2); *Corydalis sykesii* Ludlow (2)

Perennial herbs. W, C, E Nepal: 3600-5500 m. Nepal and E Himalaya.

Corydalis stipulata Lidén, *Bull. Brit. Mus. (Nat. Hist.) Bot.* **18**(6): 522. 1989. [4, 8, 10, 12, 27, 29]

Perennial herbs. C Nepal: 2800-4000 m. **Endemic to Nepal**.

Central: *Rasuwa*, Mailung Khola, S. Ganesh Himal, 11000 ft. (3350 m), JDA Stainton 5127 (Holotype: BM; Isotypes: E, KATH).

Corydalis stracheyi Prain [4, 6, 10]

Perennial herbs. W, C, E Nepal: 3600-5200 m. Nepal, W Himalaya, E Himalaya, Tibetan Plateau, Assam-Burma and E Asia.

Corydalis stricta Steph. ex Fisch. [2, 4, 6, 10]

Corydalis astragalina Hook.f. & Thomson; *Corydalis schlagenweitii* Fedde

Perennial herbs. W, C Nepal: 3500-5500 m. Nepal, W Himalaya, Tibetan Plateau, E Asia, N Asia and SW Asia.

Corydalis terracina Lidén, *Bull. Brit. Mus. (Nat. Hist.) Bot.* **18**(6): 522. 1989. [4, 8, 10, 12, 27, 29]

Perennial rhizomatous herbs. W, C Nepal: 1300-2700 m. **Endemic to Nepal**.

Central: *Kaski*, Chipli, N. Pokhara, 8000 ft. (2440 m), JDA Stainton, WR Sykes & LHJ Williams 4880 (Holotype: BM; Isotype: KATH).

Corydalis trifoliata Franch. [2, 4, 8, 10]

Perennial herbs. C, E Nepal: 3000-4300 m. Nepal, E Himalaya, Tibetan Plateau, Assam-Burma and E Asia.

Corydalis uncinata Lidén, *Bull. Brit. Mus. (Nat. Hist.) Bot.* **18**(6): 520. 1989. [4, 10, 12, 27, 29]

Perennial fleshy herbs. W, C Nepal: 3600-5500 m. **Endemic to Nepal**.

West: *Dolpa*, Phoksumdo Taal, 12000 ft. (3660 m), JDA Stainton 5508 (Holotype: BM; Isotypes: E, KATH).

Corydalis uncinatella Lidén, *Bull. Brit. Mus. (Nat. Hist.) Bot.* **18**(6): 520. 1989. [4, 6, 10, 12, 27, 29]

Perennial rhizomatous herbs. W, E Nepal: 3600-3900 m. **Endemic to Nepal**.

West: *Rukum*, Toridwari Bhanjyang, 12500 ft. (3810 m), JDA Stainton 5481 (Holotype: BM; Isotype: E).

Corydalis vaginans Royle [4, 6, 10]

Corydalis ramosa Wall. ex Hook.f. & Thomson (2); *Corydalis ramosa* var. *vaginans* (Royle) Hook.f. & Thomson

Annual or biennial herbs. W Nepal: *ca.* 3000 m. Nepal and W Himalaya.

Use: Medicine (sap).

Dactylicapnos Wall.

Dactylicapnos cordata Lidén [10]

Perennial climbing herbs. E Nepal: 1300-1900 m. Nepal and E Himalaya.

Dactylicapnos grandifoliolata Merr. (10)

Dactylicapnos ventii (Khanh) Lidén

Perennial climbing herbs. E Nepal: 1500-2800 m. Nepal, E Himalaya, Tibetan Plateau and Assam-Burma.

Dactylicapnos macrocapnos (Prain) Hutch. [6, 10]

Dicentra macrocapnos Prain (2, 4)

subsp. **macrocapnos**

Perennial climbing herbs. W, C Nepal: 900-2600 m. Nepal and W Himalaya.

subsp. **echinosperma** Lidén [10]

Perennial climbing herbs. C Nepal: 900-2700 m. Nepal and Tibetan Plateau.

Dactylicapnos odontocarpa Lidén [10]

Perennial climbing herbs. C, E Nepal: 1600-2600 m. Nepal and E Himalaya.

Dactylicapnos roylei (Hook.f. & Thomson) Hutch. [6, 10]

Dicentra roylei Hook.f. & Thomson (2, 4)

Annual climbing herbs. C, E Nepal: 2000-3000 m. Nepal, W Himalaya, E Himalaya, Tibetan Plateau, Assam-Burma and E Asia.

Dactylicapnos scandens (D.Don) Hutch. [6, 10]

Diclytra scandens D.Don; *Dactylicapnos thalictrifolia* Wall.; *Dicentra scandens* (D.Don) Walp. (2, 4); *Dicentra thalictrifolia* (Wall.) Hook.f. & Thomson

जोगी लहरा, सानो पहेली (Jogi laharaa, Saano paheli)

Perennial climbing herbs. C, E Nepal: 1500-2800 m. Nepal, E Himalaya, Tibetan Plateau, Assam-Burma, S Asia, E Asia and SE Asia.

Use(s): Medicine (whole plant), Fodder.

Dicranostigma Hook.f. & Thomson

Dicranostigma lactucoides Hook.f. & Thomson [2, 4, 6, 10]

Stylophorum lactucoides (Hook.f. & Thomson) Benth. & Hook.f.

दुधे साग (Dudhe saag), **क्येराप** (Kyeraap: Sherpa)

Perennial herbs. W, C Nepal: 2400-4400 m. Nepal, W Himalaya, Tibetan Plateau and E Asia.

Use(s): Food, Medicine (whole plant).

Fumaria L.

Fumaria indica (Hausskn.) Pugsley [2, 4, 6, 10]

Fumaria vaillantii var. *indica* Hausskn.

ढुकुरे झार, कैरुवा (Dhukure jhaar, Kairuwaa)

Annual herbs. W, C, E Nepal: 100-2500 m. Nepal, W Himalaya, E Himalaya, S Asia and C Asia.

Use(s): Medicine (whole plant), Fodder.

Hypecoum L.

Hypecoum leptocarpum Hook.f. & Thomson [2, 4, 6, 10]

परवट (Parwat)

Annual herbs. W, C Nepal: 1700-4200 m. Nepal, E Himalaya, Tibetan Plateau, E Asia, N Asia, C Asia and SW Asia.

Use: Medicine (whole plant).

Meconopsis Prain

Meconopsis autumnalis P.Egan, *Phytotaxa* **20:** 48. 2011. [10, 29, 31]

Perennial herbs. C Nepal: 3300-4200 m. **Endemic to Nepal**.

Central: *Rasuwa*, Thulo Bhedi Kharka – Jaisuli Kund, 4160 m, F. Miyamoto *et al* 9440053 (Holotype: E; Isotypes: KATH, TI).

Meconopsis bella Prain [2, 4, 6, 10]

Perennial herbs. C, E Nepal: 3600-5300 m. Nepal and E Himalaya.

Meconopsis chankheliensis Grey-Wilson, *Curtis's Bot. Mag.* **23**(2): 203. 2006. [8, 10, 27, 29]

Perennial herbs. W Nepal: 3100-4600 m. **Endemic to Nepal**.

West: *Jumla*, Maharigaon, 15000 ft. (4570 m), O. Polunin, WR Sykes & LHJ Williams 224 (Holotype: BM).

Meconopsis dhwojii G.Tayl. ex Hay, *New Fl. & Silva* **4**: 225. 932. [2, 4, 6, 10, 12, 27, 29, 34]

Biennial herbs. C, E Nepal: 3600-4800 m. **Endemic to Nepal**.

East: Sangmo, Lal Dhwoj 297 (Holotype: BM).

Meconopsis discigera Prain [2, 4, 6, 10]

Perennial herbs. E Nepal: 3100-4600 m. Nepal and E Himalaya.

Meconopsis ganeshensis Grey-Wilson, *Curtis's Bot. Mag.* **23**(2): 188. 2006. [10, 27, 29, 34]

Perennial herbs. C Nepal: 3600-4600 m. **Endemic to Nepal**.

Central: *Dhading*, Ganesh Himal, Aankhu Khola, 13500 ft. (4110 m), JDA Stainton 3990 (Holotype: BM; Isotype: E).

Meconopsis gracilipes G.Tayl., *Gen. Meconopsis*: 38.1934. [2, 4, 10, 12, 27, 29]

Perennial herbs. C Nepal: 2400-4900 m. **Endemic to Nepal**.

Central: *Gorkha*, Khorlak, 12000-16000 ft. (3660-4880 m), Lal Dhwoj 17 (Holotype: BM).

Meconopsis grandis Prain [2, 4, 6, 10]

subsp. **grandis**

चिल्दर, उपल नोन्गोंग्पो (Childar, Upal ngongpo: Amchi)

Perennial herbs. C, E Nepal: 3100-4500 m. Nepal and E Himalaya.

Use(s): Medicine (leaves, flowers), Food (seeds), Fodder.

subsp. **jumlaensis** C.Grey-Wilson [10]

Perennial herbs. W Nepal: 3300-4400 m. **Endemic to Nepal**.

Western: *Jumla*, 3505 m, O. Polunin, WR Sykes & LHJ Williams 4371 (Holotype: BM).

Meconopsis horridula Hook.f. & Thomson [2, 4, 6, 10]

Meconopsis horridula var. *rudis* Hook.f. & Thomson; *Meconopsis rudi* (Prain) Prain

दनगिन-दा (Dangin-daa: Sherpa), **छार-गुन, क्यासर** (Chhar-gun, Kyaashar: Tibetan)

Perennial prickly herbs. W, C, E Nepal: 3700-5300 m. Nepal, E Himalaya, Tibetan Plateau, Assam-Burma and E Asia.

Use(s): Medicine (whole plant), Food.

Meconopsis lamjungensis T.Yoshida, H.Sun & Grey-Wilson, *Curti's Botanical Magazine* **29**(2): 194. 2012. [29, 37]

Perennial herbs. C Nepal: 4000-4100 m. **Endemic to Nepal**.

Central: *Lamjung*, 4050 m, JDA Stainton *et al* 6254 (Holotype: BM).

Meconopsis lyrata (H.A.Cummins & Prain ex Prain) Fedde ex Prain [2, 4, 6, 10]

Cathcartia lyrata H.A.Cummins & Prain ex Prain

Perennial herbs. C, E Nepal: 3200-4600 m. Nepal, E Himalaya, Tibetan Plateau and E Asia.

Meconopsis manasluensis P.A.Egan, *Phytotaxa* **20:** 50. 2011. [8, 10, 29, 31]

Perennial herbs. C Nepal: *ca.* 4000 m. **Endemic to Nepal**.

Central: *Gorkha*, E. Samdo – S. Sanam Khola River Valley, *ca.* 4000 m, H. Ikeda *et al* 20815156 (Holotype: E; Isotype: E).

Meconopsis napaulensis DC. [2, 4, 6, 10, 12, 27, 29, 34]

क्यासर (Kyaasar: Sherpa)

Perennial herbs. C, E Nepal: 3200-4500 m. **Endemic to Nepal**.

Central: Nepal, N. Wallich s.n. (Holotype: G-DC).

Use(s): Food, Medicine (root).

Meconopsis paniculata (D.Don) Prain [2, 4, 6, 10, 34]

Papaver paniculatum D.Don; *Meconopsis longipetiolata* G.Taylor ex Hay

ग्यासुर, उपल सेर्पो (Gyashur: Sherpa; Upal-serpo: Amchi)

Perennial herbs. W, C, E Nepal: 2700-5000 m. Nepal, E Himalaya and Tibetan Plateau.

Use(s): Medicine (flowers, leaves), Food, Fodder.

Meconopsis pinnatifolia C.Y.Wu & H.Chuang ex L.H.Zhou [10]

Perennial herbs. C Nepal: 3600-4900 m. Nepal and Tibetan Plateau.

Meconopsis regia G.Tayl., *Journ. Bot.* **67**: 258.1929. [2, 4, 6, 10, 12, 27, 29, 34]

Perennial herbs. W, C Nepal: 3000-4300 m. **Endemic to Nepal**.

Central: *Gorkha*, Barpak, 14000-15000 ft. (4270-4370 m), Lal Dhwoj 18 (Holotype: BM; Isotype: BM).

CITES Appendix: III.

Meconopsis robusta Hook.f. & Thomson [10]

Perennial herbs. W Nepal: 2400-4100 m. Nepal and W Himalaya.

Meconopsis simikotensis Grey-Wilson, *Alpine Gardener* **74** (2): 220. 2006. [10, 29]

Perennial herbs. W Nepal: 3300-4000 m. **Endemic to Nepal**.

West: *Humla*, JDA Stainton, WR Sykes & LHJ Williams 4270 (Holotype: BM).

Meconopsis simplicifolia (D.Don) Walp. [2, 4, 6, 10]

Papaver simplicifolium D.Don; *Stylophorum simplicifolium* (D.Don) Spreng.

Perennial herbs. C, E Nepal: 3300-4900 m. Nepal and E Himalaya.

Meconopsis sinuata Prain [2, 4, 6, 10]

Perennial herbs. C, E Nepal: 3600-4300 m. Nepal and E Himalaya.

Meconopsis staintonii Grey-Wilson, *Curtis's Bot. Mag.* **23**(2): 190. 2006. [8, 10, 27, 29]

Perennial herbs. C Nepal: 2500-4300 m. **Endemic to Nepal**.

Central: *Mustang*, Lete, Kali Gandaki Valley, 12000 ft. (3660 m), JDA Stainton, WR Sykes & LHJ Williams 5583 (Holotype: BM).

Meconopsis taylorii L.H.J.Williams, *Trans. Roy. Soc. Edinb.* **41**: 347. 1972. [2, 4, 10, 12, 27, 29, 34]

Perennial herbs. C Nepal: 3600-4600 m. **Endemic to Nepal**.

Central: *Kaski*, Seti Khola, Annapurna Himal, 15000 ft. (4570 m), JDA Stainton, WR Sykes & LHJ Williams 6593 (Holotype: BM; Isotypes: BM, E).

Meconopsis wallichii Hook. [10, 34]

Perennial herbs. E Nepal: 3100-4300 m. Nepal and E Himalaya.

Papaver L.

Papaver dubium L.

subsp. **glabrum** (Royle) Kadereit [10]

Papaver glabrum Royle; *Papver dubium* var. *glabrum* Koch (2, 4)

Annual herbs. W, C Nepal: 2500-2900 m. Nepal, W Himalaya, S Asia, SW Asia, Europe and N America.

Use: Medicine (petals, latex-immature capsule).

Papaver rhoeas L. [2, 4, 10]

सेती बिराउली (Seti birauli)

Annual herbs. C Nepal: 2500-3100 m (Cultivated). Native to Europe, W Asia, W Africa and Mediterranean region.

Use: Medicine (petals, capsules).

Papaver somniferum L. [2, 4, 10]

अफिम (Aphim); Opium poppy

Annual herbs. W Nepal: 1800-2000 m (Cultivated). Native to S Europe, C Asia and SW Asia.

Use(s): Medicine (whole plant), Food (seeds).

Exotic/Cultivated Taxa

Eschscholzia californica Cham. (**पिताम्बर,** Pitambar; Californian poppy). Native to N America. [14, 54]

Papaver orientale L. (**अफिम फूल,** Aphim phul; Oriental poppy). Native to Iran, Mediterranean region. [14]

Doubtful/Uncertain Taxa

Corydalis hamata Franch. (misapplied name for *Corydalis conspersa* Maxim.) [2, 4]

Corydalis nana Royle (misapplied name for *Corydalis magni* Pusalkar) [2, 4]

Corydalis pachypoda (Franch.) Hand.-Mazz. [2, 4]

Corydalis rutiflora (Sm.) DC. (misapplied name for *Corydalis diphylla* Wall.) [2]

Fumaria parviflora Lam. (misapplied name for *Fumaria indica* (Hausskn.) Pugsley) [2, 4, 6]

Fumaria vaillantii Loisel. (misapplied name for *Fumaria indica* (Hausskn.) Pugsley) [2, 4]

Hypecoum parviflorum Kar. & Kir. (misapplied name for *Hypecoum leptocarpum* Hook.f. & Thomson) [2, 4]

Meconopsis rebeccae H.S.Debnath & Nayar [4]

CIRCAEASTERACEAE

Circaeaster Maxim.

Circaeaster agrestis Maxim. [2, 4, 6, 11]

Annual dwarf herbs. W, C, E Nepal: 2700-4100 m. Nepal, W Himalaya, E Himalaya, Tibetan Plateau and E Asia.

LARDIZABALACEAE

Decaisnea Hook.f. & Thomson

Decaisnea insignis (Griff.) Hook.f. & Thomson [2, 4, 6]

Slackia insignis Griff.

Deciduous shrubs. C, E Nepal: 2400-2700 m. Nepal, E Himalaya, Tibetan Plateau, Assam-Burma and E Asia.

Holboellia Wall.

Holboellia angustifolia Wall.

Holboellia acuminata Lindl.; *Holboellia latifolia* var. *acuminata* (Lindl.) Gagnep.; *Holboellia latifolia* var. *angustifolia* (Wall.) Hook.f. & Thomson. (2, 4, 6)

गुफला (Gufalaa)

Evergreen climbing shrubs. C, E Nepal: 1600-3000 m. Nepal, E Himalaya, Tibetan Plateau, Assam-Burma, S Asia and E Asia.

Use: Food (fruit).

Holboellia latifolia Wall. [2, 4, 6]

Stauntonia latifolia Wall.

गुफला, मलकाटी (Gufalaa, Malkati)

Evergreen climbing shrubs. W, C, E Nepal: 1800-4000 m. Nepal, E Himalaya, Tibetan Plateau, Assam-Burma, S Asia and E Asia.

Use(s): Food (fruit), Medicine (root), Fodder.

MENISPERMACEAE

Cissampelos L.

Cissampelos nepalensis D.G.Rhodes, *Phytologia* **30**(6): 475. 1975. [127]

Climbing shrubs. W, C, E Nepal. Endemic to Nepal (?)

East: *Ilam*, D. Nicolson 3121 (Holotype: US).

Cissampelos pareira L.

Cissampelos hirsuta Buch.-Ham. ex DC.; *Cissampelos pareira* var. *hirsuta* (Buch.-Ham. ex DC.) Forman (2, 4, 6)

चिल्लो बाटुल पाते, गुदर गानो (Chillo baatul paate, Gudar gaano)

Climbing shrubs. W, C, E Nepal: 100-2200 m. Nepal and Pantropical.

Use: Medicine (whole plant).

Cocculus DC.

Cocculus hirsutus (L.) Diels [2, 4]

Menispermum hirsutum L.; *Cocculus villosus* DC.

काँसे लहरो (Kaanse laharo)

Climbing shrubs. C Nepal. Nepal, S Asia and Tropical Africa.

Use: Medicine (root, leaves).

Cocculus laurifolius DC. [2, 4, 6]

Menispermum laurifolium (DC.) Roxb.

तिलफोरा (Tilphoraa)

Shrubs or small trees. W, C Nepal: 900-1900 m. Nepal, W Himalaya, Tibetan Plateau, Assam-Burma, S Asia, E Asia and SE Asia.

Use: Medicine (whole plant).

Cocculus orbiculatus (L.) DC. [2, 4, 6]

Menispermum orbiculatum L.; *Cocculus mollis* Hook.f. & Thomson

Woody climbers. C Nepal: 1500-2000 m. Nepal, E Asia and SE Asia. Introduced in Indian Ocean islands and Pacific islands.

Parabaena Miers.

Parabaena sagittata Miers ex Hook.f. & Thomson [2, 4]

Climbing shrubs. Nepal, E Himalaya, Tibetan Plateau, Assam-Burma, S Asia, E Asia and SE Asia.

Pericampylus Miers.

Pericampylus glaucus (Lam.) Merr. [2, 4, 6]

Menispermum glaucum Lam.

दुधे लहरो, पिपल पाती (Dudhe laharo, Pipal paati)

Woody climbers. W, E Nepal: 300-1800 m. Nepal, E Himalaya, Assam-Burma, E Asia and SE Asia.

Use(s): Medicine (whole plant), Fodder.

Stephania Lour.

Stephania elegans Hook.f. & Thomson [2, 4, 6]

तारो लहरा, बाटुलपाते (Taaro laharaa: Chepang; Baatulpaate)

Herbaceous climbing shrubs. W, C, E Nepal: 1000-1800 m. Nepal, E Himalaya, Assam-Burma and E Asia.

Use(s): Medicine (root), Ritual, Construction materials.

Stephania glabra (Roxb.) Miers [2, 4, 6]

Cissampelos glabra Roxb.

निम लहरा, बर्कुली लहरा (Nim laharaa, Barkuli laharaa)

Herbaceous climbing shrubs. W, C, E Nepal: 400-2500 m. Nepal, W Himalaya, E Himalaya, Tibetan Plateau, Assam-Burma, S Asia and SE Asia.

Use(s): Medicine (whole plant), Construction materials.

Stephania glandulifera Miers [2, 4, 6]

गुजर गानो, बाटुलपाते, बाटुले लहरा (Gujar gaano, Baatulpaate, Baatule laharaa)

Deciduous climbers. C, E Nepal: 1300-2100 m. Nepal and E Himalaya.

Use(s): Medicine (whole plant), Construction materials.

Stephania gracilenta Miers [2, 4, 6]

Climbing shrubs or herbaceous climbers. W, C, E Nepal: 600-2600 m. Nepal, Tibetan Plateau and S Asia.

Use(s): Medicine (whole plant), Construction materials.

Stephania japonica (Thunb.) Miers [2, 4, 6]

Menispermum japonicum Thunb.

var. **japonica**

पाठा, बाटुले पात (Paatha, Baatule paat)

Climbing shrubs. C, E Nepal: 1400-2600 m. Nepal, S Asia, E Asia, SE Asia and Australasia.

Use(s): Medicine (root), Construction materials.

var. **discolor** (Blume) Forman [2, 4, 6]

Clypea discolor Blume; *Cissampelos hernandiifolia* Willd.

Climbing shrubs. C, E Nepal: 200-1500 m. Nepal, Assam-Burma, S Asia, E Asia, SE Asia and Australasia.

Use: Medicine (root, leaves).

Tiliacora Colebr.

Tiliacora acuminata (Lam.) Hook.f. & Thomson [2, 4, 6]

Menispermum acuminatum Lam.; *Tiliacora racemosa* Colebr.

रुख काने (Rukh kaane)

Woody climbing shrubs. W Nepal: *ca.* 200 m. Nepal, Assam-Burma and S Asia.

Use: Medicine (stem).

Tinospora Miers.

Tinospora cordifolia (Willd.) Hook.f. & Thomson [6]

Menispermum cordifolium Willd.

गुर्जो लहरो (Gurjo laharo)

Deciduous climbing shrubs. C Nepal: 100-200 m. Nepal, Assam-Burma and S Asia.

Tinospora sinensis (Lour.) Merr. [2, 4, 6]

Campylus sinensis Lour.; *Menispermum malabaricum* Lam.

गुर्जो को लहरा, गुर्जो (Gurjo ko laharaa, Gurjo)

Climbing shrubs. W, C, E Nepal: 200-500 m. Nepal, E Himalaya, Assam-Burma, S Asia, E Asia and SE Asia.

Use(s): Medicine (whole plant), Ornamental, Ritual.

BERBERIDACEAE

Berberis L. (चुत्रो, Chutro; Barberry)

Berberis angulosa Wall. ex Hook.f. & Thomson [2, 4, 6, 18]

var. **angulosa**

Berberis parisepala Ahrendt (2, 4)

चुत्रे काँडा (Chutre kaanda)

Deciduous shrubs. C, E Nepal: 3000-4500 m. Nepal, E Himalaya and Tibetan Plateau.

Use(s): Medicine (root), Food (fruit), Fuel.

var. **fasciculata** Ahrendt [18]

Deciduous shrubs. C Nepal: 3600-3700 m. Nepal and E Himalaya.

Berberis aristata DC. [2, 4, 6, 18]

Berberis ceratophylla G.Don (2, 4); *Berberis chitria* Buch.-Ham. ex Lindl. (2, 4); *Berberis chitria* var. *sikkimensis* C.K.Schneid.; *Berberis sikkimensis* Ahrendt (2, 4); *Berberis sikkimensis* var. *baileyi* Ahrendt (2, 4, 11)

बाह्रमासे चुत्रों, क्येर्वा (Baarhamaase chutro; Kyerwa); Barberry

Erect spiny shrubs. W, C, E Nepal: 1300-3400 m. Nepal, E Himalaya and W Himalaya.

Use(s): Medicine (root, bark, leaves, flowers, fruit), Food (fruit), Fence, Dye.

Berberis asiatica Roxb. ex DC. [2, 4, 6, 18]

चुत्रों, रसाजन (Chutro, Rasajan); Barberry

Evergreen shrubs. W, C, E Nepal: 1000-2700 m. Nepal, W Himalaya and Tibetan Plateau.

Use(s): Medicine (bark, fruit), Food (fruit), Fence, Dye (bark).

Berberis concinna Hook.f. [2, 4, 6, 18]

Berberis concinna var. *brevior* Ahrendt

var. **concinna**

पुचु (Puchu)

Deciduous shrubs. W, C, E Nepal: 2600-4600 m. Nepal, E Himalaya and Tibetan Plateau.

var. **extensiflora** Ahrendt [8, 18]

Deciduous shrubs. C Nepal: 3000-3400 m.

Berberis everestiana Ahrendt

var. **everestiana** [11]

कनिके चुत्रो (Kanike chutro)

Deciduous dwarf shrubs. C Nepal: 2800-3550 m. Nepal, E Himalaya, Tibetan Plateau.

var. **ventosa** Ahrendt [2, 4, 6, 18]

Deciduous shrubs. W, C, E Nepal: 2700-4600 m. Nepal and E Himalaya.

Use: Food (fruit).

Berberis glaucocarpa Stapf [2, 4, 6, 18]

Semi evergreen shrubs. W, C Nepal: 2300-3500 m. Nepal and W Himalaya.

Berberis hamiltoniana Ahrendt [2, 4, 18]

Deciduous shrubs. W, C Nepal: 2900-4300 m. Nepal and W Himalaya.

Berberis hookeri Lem. [2, 4, 6, 18]

Evergreen shrubs. C, E Nepal: 2000-3400 m. Nepal, E Himalaya, Tibetan Plateau and Assam-Burma.

Berberis insignis Hook.f. & Thomson [2, 4, 6, 18]

Evergreen shrubs. E Nepal: 2000-3400 m. Nepal, E Himalaya and Assam-Burma.

Use: Medicine (root, stem, bark, fruit).

Berberis jaeschkeana C.K.Schneid.

var. **usteriana** C.K.Schneid. [2, 4, 6, 18]

Deciduous shrubs. W, C Nepal: 2600-3800 m. Nepal and W Himalaya.

Berberis karnaliensis Bh.Adhikari [8, 18]

Deciduous shrubs. W Nepal: 1800-3400 m. Nepal and W Himalaya.

Berberis koehneana C.K.Schneid. [2, 4, 6, 18]

Berberis koehneana var. *auramea* Ahrendt (2, 4, 11)

Deciduous shrubs. W, C Nepal: 2400-4000 m. Nepal and W Himalaya.

Berberis kumaonensis C.K.Schneid. [2, 4, 18]

Deciduous semi-prostrate shrubs. W Nepal: 3000-4000 m. Nepal and W Himalaya.

Berberis mucrifolia Ahrendt, *J. Roy. Hort. Soc. Lond.* **81**: 135. 1956. [2, 4, 6, 12, 18, 26, 29]

चुत्रो (Chutro)

Small deciduous shrubs. W, C, E Nepal: 2000-4500 m. **Endemic to Nepal**.

Central: *Mustang*, Tegar, 14500 ft. (4420 m), JDA Stainton, WR Sykes & LHJ Williams 8111 (Holotype: BM).

Use: Food (fruit).

Berberis orthobotrys Bienert ex Aitch.

var. **rubicunda** Ahrendt [18]

Deciduous shrubs. C, E Nepal: 2800-4200 m.

Berberis pendryi Bh.Adhikari, *Edinburgh J. Bot.* **69**(3): 477. 2012. [8, 18, 29]

Deciduous shrubs. C Nepal: *ca.* 3600 m. **Endemic to Nepal**.

Central: *Mustang*, below Muktinath, *ca.* 3600 m, C. Pendry *et al* EA_34 (Holotype: E; Isotype: KATH).

Berberis petiolaris Wall. ex G.Don [2, 4, 6, 18]

var. **petiolaris**

Berberis petiolaris var. *extensa* Ahrendt ex Rao (4)

Deciduous shrubs or small trees. W Nepal: 2700-3500 m. Nepal and W Himalaya.

var. **garhwalana** Ahrendt [18]

Deciduous shrubs or small trees. W Nepal: *ca.* 2700 m. Nepal and W Himalaya.

Berberis thomsoniana C.K.Schneid. [6, 18]

Deciduous shrubs. W, C, E Nepal: 3000-3650 m. Nepal and E Himalaya.

Berberis tsarica Ahrendt [2, 4, 18]

Deciduous shrubs. C, E Nepal: 4000-4700 m. Nepal, E Himalaya and Tibetan Plateau.

Use: Medicine (bark, flowers, fruit).

Berberis virescens Hook.f. [2, 4, 6, 18]

Deciduous shrubs. E Nepal: 3500-4000 m. Nepal and E Himalaya.

Berberis wallichiana DC. [2, 4, 6, 18]

Berberis poluninii Ahrendt (2, 4); *Berberis wallichiana* var. *atroviridis* Hook.f. & Thomson

बाह्रमासे चुत्रो (Baarhamaase chutro)

Evergreen shrubs. C, E Nepal: 2300-3300 m. Nepal, E Himalaya and Assam-Burma.

Use(s): Medicine (root, leaves, young twig), Food (fruit).

Mahonia Nutt.

Mahonia acanthifolia G.Don [2, 4, 6, 32]

Mahonia borealis Takeda (2, 4)

जमाने मान्द्रो (Jamaane maandro)

Evergreen shrubs or small trees. W, C, E Nepal: 2100-2900 m. Nepal and E Himalaya.

Use(s): Medicine (bark), Food (fruit).

Mahonia napaulensis DC. [2, 4, 6, 11, 32]

जमाने मान्द्रो (Jamaane maandro)

Evergreen shrubs or small trees. C Nepal: 1200-1900 m. Nepal, E Himalaya, Tibetan Plateau, Assam-Burma, cultivated elsewhere.

Use(s): Medicine (bark), Food (fuel), Fuel, Dye (bark, root).

Podophyllum L.

Podophyllum hexandrum Royle [2, 4, 6]

Podophyllum emodi Wall. ex Hook.f. & Thomson; *Sinopodophyllum hexandrum* (Royle) Ying

लघु पत्र; ओल-मोसे (Laghu patra; Wol-mose: Amchi); Himalayan Mayapple

Perennial rhizomatous herbs. W, C, E Nepal: 2300-4500 m. Nepal, W Himalaya, E Himalaya, Tibetan Plateau and SW Asia.

Use(s): Medicine (rhizome, fruit), Food, Commercial.

CITES Appendix: II.

Doubtful/Excluded Taxa

Berberis aristata var. **floribunda** (G.Don) Hook.f. & Thomson [2, 4]
Berberis bhutanensis Ahrendt [2, 4]
Berberis brandisiana Ahrendt [98]
Berberis coriaria Royle ex Lindl. [2, 4]
Berberis edgeworthiana C.K.Schneid. [2, 4]
Berberis erythroclada Ahrendt (Lekh chutro) [2, 4, 6]
Berberis everestiana Ahrendt var. **everestiana** [2, 4]
Berberis hobsoni Ahrendt [2, 4]
Berberis jaeschkeana C.K.Schneid. var. **jaeschkeana** [2, 4]
Berberis jaeschkeana var. **bimbilaica** Ahrendt [2, 4]
Berberis khasiana Ahrendt [2, 4]
Berberis lycium Royle [2, 4]
Berberis macrosepala Hook.f. & Thomson [98]

Berberis micrantha (Hook.f. & Thomson) Ahrendt. [2, 4]

Berberis orthobotrys var. **canescens** Ahrendt [98]

Berberis praecipua C.K.Schneid. [4]

Berberis tsarica var. **ritangensis** Ahrendt [11]

Berberis ulcina Hook.f. & Thomson [98]

Berberis umbellata Wall. ex G.Don [2, 4]

RANUNCULACEAE

Aconitum L.

Aconitum alpinonepalense Tamura [2, 4, 29]

Perennial or pseudo-annual herbs. E Nepal: 4300-4400 m. Nepal and Tibetan Plateau.

Aconitum amplexicaule Lauener, *Notes Roy. Bot. Gard. Edinb.* **26**: 7. 1964. [2, 4, 6, 12, 28, 29]

Perennial tuberous herbs. W, C Nepal: 3800-4200 m. **Endemic to Nepal**.

West: *Rukum*, Dogadi Khola, 12500 ft. (3810 m), JDA Stainton, WR Sykes & LHJ Williams 3780 (Holotype: BM; Isotype: E).

Aconitum angulatum Tamura, *Acta Phytoatax. Geobot.* **23**: 100. 1968. [2, 4, 12, 28, 29]

Perennial tuberous herbs. C, E Nepal: 3800-4200 m. **Endemic to Nepal**.

East: Deoma, *ca.* 3900 m, K. Nishioka 1049 (Holotype: KYO).

Aconitum balangrense Lauener [2, 4, 6]

Perennial herbs. W, C, E Nepal: 3000-3900 m. Endemic to Nepal?

Use: Medicine (root).

Aconitum balfourii Stapf [2, 4]

Caltha codua Buch.-Ham.

गोबरी (Gobaree)

Perennial herbs. W, C, E Nepal: 2700-3800 m. Nepal and S Asia (India).

Use: Medicine (whole plant).

Aconitum bhedingense Lauener, *Notes Roy. Bot. Gard. Edinb.* **25**: 26. 1963. [2, 4, 8, 12, 28, 29]

Perennial herbs. C, E Nepal: 3000-3900 m. **Endemic to Nepal**.

Central: *Dolakha*, Bheding, 12000-13000 ft. (3660-3960 m), Lal Dhwoj 325 (Holotype: E; Isotype: BM).

Aconitum deltoideum Lauener, *Notes Roy. Bot. Gard. Edinb.* **32**: 411. 1973. [2, 4, 29]

Perennial herbs. E Nepal: *ca.* 2900 m. **Endemic to Nepal**.

East: *Sankhuwasabha*, Milke daanda, *ca.* 2900 m, LHJ Williams & JDA Stainton 8414 (Holotype: BM).

Aconitum dhwojii Lauener, *Notes Roy. Bot. Gard. Edinb.* **25**: 25. 1963. [2, 4, 12, 28, 29]

Erect tuberous herbs. C Nepal: 4500-4800 m. **Endemic to Nepal**.

Central: *Gorkha*, Khorlak, 15000-16000 ft. (4570-4880 m), Lal Dhwoj 29 (Holotype: BM; Isotype: E).

Aconitum elwesii Stapf [2, 4]

Herbs. E Nepal: *ca.* 3000 m. Nepal, E Himalaya and Tibetan Plateau.

Use: Poisonous plant (root).

Aconitum falconeri Stapf [6]

Perennial herbs. C Nepal: 4300-4400 m. Nepal and S Asia (India).

Aconitum ferox Wall. ex Ser. [2, 4]

Aconitum virosum D.Don

सेतो बिख, निलो बिख (Seto bikh, Nilo bikh)

Perennial herbs. W, C, E Nepal: 2100-4700 m. Nepal and E Himalaya.

Use(s): Medicine (root), To poison arrows for hunting (root).

Aconitum funiculare Stapf [6]

Perennial herbs. E Nepal: *ca.* 3000 m. Nepal and E Himalaya.

Aconitum gammiei Stapf [2, 4, 6]

Aconitum dissectum D.Don

मधु बिख, बिख (Madhu bikh, Bikh), **पोंगर** (Pongar: Sherpa)

Perennial herbs. C, E Nepal: 3300-4300 m. Nepal, E Himalaya, Tibetan Plateau and E Asia.

Use: Medicine (root).

Aconitum heterophylloides (Brühl) Stapf [4, 6, 29]

Aconitum ferox var. *heterophylloides* Brühl; *Aconitum deinorrhizum* Stapf (2, 4); *Aconitum heterophylloides* var. *leucanthum* (Brühl) Lauener (2, 4); *Aconitum nepalense* Lauener (2, 4); *Caltha nirbisia* Buch.-Ham.

अत्तिस (Attis)

Perennial herbs. W, C, E Nepal: 2800-4600 m. Nepal and E Himalaya.

Use: Medicine (root).

Aconitum heterophyllum Wall. ex Royle [2, 4, 6, 11]

Aconitum atees Royle

अत्तिस (Atis); Aconite

Perennial herbs. W, C Nepal: 2800-3700 m. Nepal and W Himalaya (Kashmir).

Use(s): Medicine (tuber), Poisonous.

IUCN Category: **Endangered** A2cd (ver 3.1).

Aconitum hookeri Stapf [2, 4, 6]

Perennial herbs. C, E Nepal: 4000-4900 m. Nepal and E Himalaya.

Aconitum laciniatum (Brühl) Stapf [2, 4]

Aconitum ferox var. *laciniata* Brühl

var. **laciniatum**

कालो बिख, मुडुला (Kaalo bikh, Mudula)

Biennial herbs. C, E Nepal: 3800-4600 m. Nepal, E Himalaya and Tibetan Plateau.

Use: Poisonous (tuber/rhizome).

var. **fractiflexum** Lauener [2, 4]

Herbs. E Nepal: *ca.* 2800 m.

Aconitum laeve Royle [2, 4, 6]

Perennial robust herbs. W Nepal: *ca.* 2900 m. Nepal, W Himalaya and S Asia.

Use: Medicine (root/tuber).

Aconitum naviculare (Brühl) Stapf [11]

Aconitum ferox var. *naviculare* Brühl; *Aconitum naviculare* var. *leiocarpum* Tamura (2, 4)

बिखमा; बोंगकर (Bikhmaa; Bongkar: Amchi)

Perennial herbs. W, C Nepal: *ca.* 4900 m. Nepal, E Himalaya and Tibetan Plateau.

Use: Medicine (whole plant).

Aconitum novoluridum Munz [2, 4, 6]

Aconitum luridum Hook.f. & Thomson

अतिसिंगी बिख, बिख (Atisingi bikh, Bikh)

Perennial herbs. E Nepal: 3600-5000 m. Nepal, E Himalaya and Tibetan Plateau.

Use: Medicine (whole plant).

Aconitum orochryseum Stapf [2, 4, 8]

निर्मसी (Nirmasi)

Perennial herbs. C, E Nepal: 3500-4900 m. Nepal and E Himalaya.

Use(s): Medicine (root), Commercial.

Aconitum palmatum D.Don

Aconitum bisma (Buch.-Ham.) Rapaics (2, 4)

बिखमा (Bikhmaa)

Perennial herbs. C Nepal. Nepal, E Himalaya and Tibetan Plateau.

Use(s): Medicine (root), Commercial.

Aconitum poluninii Lauener, *Notes Roy. Bot. Gard. Edinb.* **26**: 7. 1964. [2, 4, 6, 12, 28, 29]

Erect herbs. W, E Nepal: 2600-3800 m. **Endemic to Nepal**.

West: *Mugu*, Mugu – Purana Mugu, Mugu Khola, 12500 ft. (3810 m), O. Polunin, WR Sykes & LHJ Williams 5338 (Holotype: BM; Isotype: E).

Aconitum pulchellum Hand.-Mazz. [2, 4, 6]

Perennial herbs. E Nepal: *ca.* 4200 m. Nepal, E Himalaya, Tibetan Plateau, Assam-Burma and E Asia.

Aconitum rotundifolium Kar. & Kir. [2, 4, 11]

Aconitum napellus var. *rotundifolium* (Kar. & Kir.) Hook.f. & Thomson

मधु बिख (Madhu bikh)

Perennial herbs. C, E Nepal: 2700-4400 m. Nepal, W Himalaya, Tibetan Plateau, N Asia and SW Asia.

Use: Medicine (root).

Aconitum scaposum Franch.

Aconitum jucundum Diels (2, 4); *Aconitum scaposum* var. *hupehanum* Rapaics

Perennial herbs. E Nepal: 2500-3200 m. Nepal, E Himalaya, Assam-Burma and E Asia.

Aconitum soongaricum Stapf [4, 8, 29]

Perennial herbs. W Nepal: 3100-3900 m. Nepal, S Asia, E Asia and C Asia.

Aconitum spicatum (Brühl) Stapf [2, 4, 6, 11]

Aconitum ferox var. *spicata* Brühl

बिख, अतिबिष; बनाग्ना-नाक्पो (Bikh, Atibish; Bangna-nakpo: Amchi)

Perennial herbs. W, C, E Nepal: 1800-4700 m. Nepal, E Himalaya and Tibetan Plateau.

Use(s): Medicine (root/tuber), Commercial, To poison arrows for hunting.

Aconitum staintonii Lauener, *Notes Roy. Bot. Gard. Edinb.* **26**: 4. 1964. [2, 4, 12, 28, 29]

Erect herbs. E Nepal: 3500-4100 m. **Endemic to Nepal**.

East: *Taplejung*, Wolangchung Gola, 11500 ft. (3500 m), JDA Stainton 1034 (Holotype: BM; Isotype: E).

Aconitum tabatae Tamura, *Acta Phytotax. Geobot.* **37**: 152. 1986. [4, 6, 12, 28, 29]

Erect herbs. W Nepal: 3800-3900 m. **Endemic to Nepal**.

West: *Rukum*, Purbang, 3850 m, H. Tabata *et al* 3638 (Holotype: KYO; Isotype: KATH).

Aconitum tamuranum Lauener, *Notes Roy. Bot. Gard. Edinb.* **26**: 3. 1964. [2, 4, 29]

Herbs. W Nepal: 3900-4800 m. **Endemic to Nepal**.

West: *Jumla*, Thakurji Lekh, S of Jumla, 3960 m, O. Polunin, WR Sykes & LHJ Williams 423 (Holotype: BM).

Aconitum violaceum Jacquem. ex Stapf [2, 4, 6]

Aconitum multifidum Royle

मधु बिख, बिख (Madhu bikh, Bikh)

Perennial herbs. W, C Nepal: *ca.* 4400 m. Nepal, W Himalaya and S Asia.

Use: Medicine (root).

IUCN Category: **Vulnerable** A2cd (ver 3.1).

Aconitum williamsii Lauener, *Notes Roy. Bot. Gard. Edinb.* **26**: 6. 1964. [2, 4, 6, 12, 28, 29]

Erect tuberous herbs. W, C Nepal: 3300-3700 m. **Endemic to Nepal**.

Central: *Myagdi*, Gurjakhani, 10750 ft. (3280 m), JDA Stainton, WR Sykes & LHJ Williams 4483 (Holotype: BM; Isotype: E).

Actaea L.

Actaea acuminata Wall. ex Royle [6]

Actaea spicata var. *acuminata* (Wall. ex Royle) H.Hara (2, 4); *Actaea spicata* forma *acuminata* (Wall.) Huth

Perennial herbs. W Nepal: 2500-3900 m. Nepal, W Himalaya, E Himalaya, Tibetan Plateau and SW Asia.

Use: Medicine (tuber).

Adonis L.

Adonis aestivalis L. [2, 4, 6]

Adonis inglisii Royle

Pheasant's eye

Annual herbs. W Nepal: 3000-3300 m. Nepal, W Himalaya, Tibetan Plateau, N Asia, SW Asia and Europe.

Use: Medicine (whole plant).

Adonis chrysocyathus Hook.f. & Thomson [2, 4, 6]

Chrysocyathus falconeri Chrtek & Slovikova

Perennial herbs. W Nepal: 3600-4100 m. Nepal, W Himalaya, S Asia and N Asia.

Use: Medicine (whole plant).

Adonis nepalensis Simonovicz [2, 4, 6, 11]

Chrysocyathus nepalensis (Simonovicz) Chrtek & Slovikova

Perennial herbs. C, E Nepal: 3800-4500 m. Nepal and E Himalaya.

Anemone L.

Anemone demissa Hook.f. & Thomson [2, 4, 6]

var. **demissa**

Anemone demissa var. *connectens* Brühl

Perennial herbs. W, C, E Nepal: 2700-5600 m. Nepal, E Himalaya, Tibetan Plateau, Assam-Burma, S Asia and E Asia.

Use: Medicine (whole plant).

var. **villosissima** Brühl [2, 4]

Anemone demissa subsp. *villosissima* (Brühl) R.P.Chaudhary

Perennial herbs. E Nepal: 3000-4000 m. Nepal, E Himalaya, Tibetan Plateau and E Asia.

Anemone elongata D.Don [2, 4, 6]

Anemonastrum elongatum (D.Don) Holub

रतन जोत, धनेरो (Ratan jot, Dhanero)

Herbs. W, C, E Nepal: 1800-3700 m. Nepal, E Himalaya, Tibetan Plateau, Assam-Burma and S Asia.

Use: Medicine (whole plant).

Anemone fuscopurpurea H.Hara, *J. Jap. Bot.* **48**: 353. 1973. [2, 4, 12, 28, 29]

Perennial rhizomatous herbs. E Nepal: 3600-4400 m. **Endemic to Nepal**.

East: *Sankhuwasabha*, Banduke Pokhari – Saju Pokhari, 4000-4200 m, H. Kanai *et al* 721252 (Holotype: TI; Isotype: TI).

Anemone geum H.Lév. [2, 4, 6]

subsp. **geum**

Herbs. W, C, E Nepal: 2400-5100 m. Nepal, E Himalaya, Tibetan Plateau, S Asia and N Asia.

subsp. **ovalifolia** (Brühl) R.P.Chaudhary [4, 8, 11]

Anemone obtusiloba subsp. *ovalifolia* Brühl; *Anemone ovalifolia* (Brühl) Hand.-Mazz.

Perennial herbs. W, C, E Nepal: 3600-5100 m. Nepal, E Himalaya, Tibetan Plateau, Assam-Burma and E Asia.

Anemone obtusiloba D.Don [2, 4, 6, 11]

Anemone discolor Royle; *Anemone obtusiloba* var. *leiocarpa* Tamura (4, 8)

subsp. **obtusiloba**

रतन जोत, काँग्रे झार (Ratan jot, Kangre jhaar)

Perennial herbs. W, C, E Nepal: 2300-4200 m. Nepal, W Himalaya, E Himalaya, Assam-Burma, E Asia and N Asia.

Use: Medicine (root, seeds).

subsp. **nepalensis** R.P.Chaudhary [4]

Herbs. C, E Nepal: 2900-3500 m.

subsp. **potentilloides** (Lauener) R.P.Chaudhary [4]

Anemone obtusiloba var. *potentilloides* Lauener

Herbs. Nepal, W Himalaya and SW Asia.

Anemone polyanthes D.Don [2, 4, 6]

Anemone villosa Royle

सुब्का-मार्पो (Subka-marpo: Tibetan)

Perennial herbs. W, C, E Nepal: 2300-4400 m. Nepal, W Himalaya and E Himalaya.

Use: Medicine (root).

Anemone polycarpa W.E.Evans [2, 8, 29]

Anemone obtusiloba subsp. *saxicola* Brühl; *Anemone rupestris* subsp. *polycarpa* (W.E.Evans) W.T.Wang (4)

Herbs. C, E Nepal: 3000-5000 m. Nepal, E Himalaya, Tibetan Plateau and E Asia.

Anemone rivularis Buch.-Ham. ex DC. [2, 4, 6, 11]

Anemone dubia Wall. ex Wight & Arn.

कंग्राते, हात्ती पाईले, सुब्का-कार्पो (Kangarate, Haatti paaile, Subka-karpo: Amchi); Common wild Anemone

Perennial herbs. W, C, E Nepal: 1600-4000 m. Nepal, E Himalaya, Tibetan Plateau, Assam-Burma, S Asia, E Asia, SE Asia and N Asia.

Use(s): Medicine (root, leaves, seeds), Food (seeds).

Anemone rupestris Wall. ex Hook.f. & Thomson [2, 4, 6]

subsp. **rupestris**

Anemone obtusiloba subsp. *omalocarpella* Brühl (2, 4)

Perennial herbs. C, E Nepal: 2700-4800 m. Nepal, E Himalaya, Tibetan Plateau and E Asia.

subsp. **gelida** (Maxim.) Lauener [4, 11]

Anemone gelida Maxim.

Perennial herbs. C, E Nepal: 3500-4600 m. Nepal, E Himalaya, Tibetan Plateau and E Asia.

Anemone rupicola Cambess. [2, 4, 6, 11]

सुब्का-कार्पो (Subkaa-kaarpo: Tibetan); Rock Anemone, Wood Anemone

Perennial herbs. W, C, E Nepal: 2700-4800 m. Nepal, W Himalaya, E Himalaya, Tibetan Plateau, E Asia and SW Asia.

Use: Medicine (root, fruit).

Anemone smithiana Lauener & Panigrahi [2, 4, 6, 11]

Perennial herbs. W, C, E Nepal: 3300-4600 m. Nepal, E Himalaya and Tibetan Plateau.

Anemone trullifolia Hook.f. & Thomson [2, 4, 6]

Anemone trullifolia var. *luxurians* Tamura

झारे जरा, जिब्रे झार (Jhaare jaraa, Jibre jhaar)

Herbs. E Nepal: 3000-4600 m. Nepal, E Himalaya, Tibetan Plateau and E Asia.

Use: Medicine (root).

Anemone vitifolia Buch.-Ham. ex DC. [2, 4, 6, 11]

कपतासे, माउरे मुलो (Kaptaase, Maaure mulo); Grape leaved Anemone

Perennial herbs. W, C, E Nepal: 1300-3300 m. Nepal, W Himalaya, E Himalaya, Tibetan Plateau, Assam-Burma, S Asia and E Asia.

Use: Medicine (root, leaves, fruit).

Aquilegia L.

Aquilegia moorcroftiana Wall. ex Royle [2, 4, 6]

Aquilegia kanaoriensis Jacquem. ex Cambess.

Perennial herbs. W, C Nepal: 3800-4200 m. Nepal, W Himalaya, Tibetan Plateau, S Asia and N Asia.

Aquilegia nivalis (Baker) Falc. ex B.D.Jacks. [11]

Aquilegia glauca var. *nivalis* Baker

Perennial herbs. C Nepal: 3700-3800 m. Nepal and W Himalaya.

Aquilegia pubiflora Wall. ex Royle [2, 4, 6]

Aquilegia vulgaris subsp. *pubiflora* (Wall.) Hook.f. & Thomson

Perennial herbs. W Nepal: 2400-3200 m. Nepal, W Himalaya and S Asia.

Use: Medicine (root, seeds).

Calathodes Hook.f. & Thomson

Calathodes palmata Hook.f. & Thomson [2, 4, 6]

Perennial herbs. E Nepal: *ca.* 3200 m. Nepal, E Himalaya and Tibetan Plateau.

Callianthemum C.A.Mey.

Callianthemum pimpinelloides (D.Don) Hook.f. & Thomson [2, 4, 6]

Ranunculus pimpinelloides D.Don

Perennial herbs. W, C, E Nepal: 2800-4000 m. Nepal, W Himalaya, E Himalaya, Tibetan Plateau, S Asia, E Asia and SW Asia.

Caltha L.

Caltha govaniana Wall. ex Royle [2, 4]

Herbs. C Nepal: *ca.* 4100 m. Nepal, W Himalaya, E Himalaya and S Asia.

Caltha palustris L. [6]

var. **palustris**

Herbs. W, C, E Nepal: 3300-3600 m. Nepal, E Himalaya and S Asia.

var. **himalaica** Tamura [4]

Herbs. W, E Nepal: 2900-3600 m. Nepal, E Himalaya and Tibetan Plateau.

var. **himalensis** (D.Don) Mukerjee [2, 4, 8]

Caltha himalensis D.Don

एक आंख्ले फूल (Ek aankhle phool)

Herbs. W, C, E Nepal: 2400-4200 m. Nepal, E Himalaya and S Asia.

Use: Medicine (root, leaves).

Caltha scaposa Hook.f. & Thomson [2, 4, 6]

Perennial herbs. W, C, E Nepal: 2700-5500 m. Nepal, E Himalaya, Tibetan Plateau and E Asia.

Cimicifuga L.

Cimicifuga foetida L. [2, 4]

Actaea cimicifuga L.; *Actaea frigida* (Wall. ex Royle) Prantl (6); *Cimicifuga frigida* Royle (4, 11)

Perennial herbs. W, C, E Nepal: 2500-4000 m. Nepal, W Himalaya, E Himalaya, Tibetan Plateau, Assam-Burma, E Asia, N Asia and C Asia.

Use: Medicine (root).

Clematis L.

Clematis acuminata DC. [2, 4, 6]

Clematis acuminata var. *wallichii* Hook.f. & Thomson

रनसाग (Ransaag)

Woody climbers. W, C, E Nepal: 600-3000 m. Nepal, E Himalaya and S Asia.

Use: Food (tender shoot).

Clematis alternata Kitam. & Tamura [2, 4, 8, 29]

Archiclematis alternata (Kitam. & Tamura) Tamura

रंगेली गुरु बुटी (Rangeli guru buti)

Woody climbers. C Nepal: 2000-3000 m. Nepal and Tibetan Plateau.

Use: Medicine (leaves).

Clematis barbellata Edgew. [2, 4, 6]

var. **barbellata**

जुंगे लहरा (Junge laharaa)

Woody climbers. W, C Nepal: 2500-3200 m. Nepal, W Himalaya and S Asia.

Use: Medicine (leaves, flowers).

var. **obtusa** Kitam. & Tamura [11]

जुंगे लहरा (Junge laharaa)

Woody climbers. W, C Nepal: 2500-3700 m. Nepal and Tibetan Plateau.

Clematis bracteolata Tamura, *Acta Phytotax. Geobot.* **23**: 29. 1968. [2, 4, 11, 12, 28, 29]

Scandent shrubs. W Nepal: *ca.* 3700 m. **Endemic to Nepal**.

West: *Dolpa*, Sangda, *ca.* 3700 m, K. Nishioka 558 (Holotype: KYO).

Clematis brevicaudata DC. [6]

Woody climbers. W, C Nepal: 2100-2500 m. Nepal, Tibetan Plateau, E Asia and N Asia.

Clematis buchananiana DC. [2, 4, 6, 11]

Clematis buchananiana var. *rugosa* Hook.f. & Thomson

जुंगे लहरा, पिनासे लहरा, भेडा लहरा (Junge laharaa, Pinaase laharaa, Bhedaa laharaa)

Woody climbers. W, C, E Nepal: 1300-3300 m. Nepal, W Himalaya, E Himalaya, Tibetan Plateau, Assam-Burma and E Asia.

Use(s): Medicine (root, leaves, stem, flowers), Food (tender leaves), Fermenter.

Clematis connata DC. [2, 4, 6]

Clematis venosa Royle

var. **connata**

भैंसे लहरो (Bhainse laharo)

Woody climbers. W, C, E Nepal: 2200-3300 m. Nepal, W Himalaya, E Himalaya, Tibetan Plateau and E Asia.

Use: Medicine (whole plant).

var. **confusa** (Grey-Wilson) W.T.Wang

Clematis confusa Grey-Wilson (4)

Woody climbers. E Nepal: 2600-2700 m. Nepal, E Himalaya and E Asia.

Clematis gouriana Roxb. ex DC. [2, 4, 6]

Clematis vitalba subsp. *gouriana* (Roxb.) Kuntze

जुंगे लहरा (Junge laharaa)

Woody climbers. W, C, E Nepal: 1400-1600 m. Nepal, Assam-Burma, S Asia, E Asia and Australasia.

Use: Medicine (leaves).

Clematis grata Wall. [2, 4, 6, 11]

Clematis cordata Royle

Woody climbers. W, C Nepal: 1500-3200 m. Nepal, W Himalaya, E Himalaya, Tibetan Plateau and SW Asia.

Clematis graveolens Lindl. [2, 4]

Clematis orientalis subsp. *graveolens* (Lindl.) Kuntze

इमाऊ फंजन, रेको-पसेरेन (Emaau phanjan: Tibetan, Reko-paseren: Gurung)

Woody climbers. W, C Nepal: 900-3000 m. Nepal, W Himalaya and SW Asia.

Use: Medicine (stem, seeds).

Clematis grewiiflora DC. [2, 4, 6]

Clematis loasaefolia DC.

भेंडे कुरो (Bhende kuro)

Woody climbers. W, C, E Nepal: 900-1600 m. Nepal, E Himalaya, Tibetan Plateau, Assam-Burma and S Asia.

Use: Medicine (achene).

Clematis montana Buch.-Ham. ex DC. [2, 4, 6]

Clematis anemoniflora D.Don

var. **montana**

जुंगे लहरो, अंगुरी झार (Junge laharo, Anguri jhaar)

Woody climbers. W, C, E Nepal: 1600-4000 m. Nepal, W Himalaya, E Himalaya, Tibetan Plateau, S Asia, E Asia and SW Asia.

Use(s): Medicine (root, stem, leaves, flowers), Forage.

var. **brevifolia** Kuntze [87]

Woody climbers. C, E Nepal: 1000-4200 m. Nepal, W Himalaya, E Himalaya, Tibetan Plateau and Assam-Burma.

Clematis napaulensis DC. [2, 4, 6]

Clematis cirrhosa var. *nepalensis* (DC.) Kuntze

भैंसी लहरो (Bhainsi laharo); **इम्गो नाक्पो** (Imgo nak po: Am)

Woody climbers. W, C, E Nepal: 1200-3400 m. Nepal, E Himalaya, Tibetan Plateau, Assam-Burma and E Asia.

Use: Medicine (whole plant).

Clematis orientalis L. [4, 8, 29]

var. **orientalis**

Suffruticose climbers. W, C Nepal: *ca.* 2000 m. Nepal and Asia.

var. **uniflora** Tamura [4, 8]

Suffruticose climbers. C Nepal: 2600-3300 m.

Clematis phlebantha L.H.J.Williams, *J. Roy. Hort. Soc., London* **93**: 345. 1968. [2, 4, 6, 12, 28, 29]

Trailing shrubs. W, C Nepal: 2600-3700 m. **Endemic to Nepal.**

West: *Dolpa*, Suligad, 9500 ft. (3000 m), JDA Stainton 4281 (Holotype: BM; Isotypes: E, K).

Use: Medicine (leaves).

Clematis puberula Hook.f. & Thomson [2, 4, 6, 11]

Clematis parviloba var. *puberula* (Hook.f. & Thomson) Kuntze

सेतो जुही (Seto juhi)

Herbaceous climbers. W, C Nepal: 200-2700 m. Nepal, W Himalaya, E Himalaya, Tibetan Plateau, Assam-Burma and E Asia.

Clematis rehderiana Craib [2, 4, 6, 11]

Woody climbers. W, C Nepal: 2700-3000(-4200) m. Nepal, Tibetan Plateau and E Asia.

Clematis roylei Rehder [2, 4, 8]

Clematis nutans Royle (6)

Woody climbers. W, C Nepal: 450-3300 m. Nepal and W Himalaya.

Clematis sikkimensis (Hook.f. & Thomson) Drumm. ex Burkill [2, 4]

Clematis acuminata var. *sikkimensis* Hook.f. & Thomson

Woody climbers. C Nepal. Nepal, E Himalaya and Assam-Burma.

Clematis smilacifolia Wall. [2, 4, 6]

Clematis subpeltata Wall.

Woody climbers. W, C Nepal: 300-1800 m. Nepal, E Himalaya, Tibetan Plateau, Assam-Burma, S Asia, E Asia and SE Asia.

Use: Medicine (root).

Clematis staintonii W.T.Wang, *Acta Phytotax. Sin.* **38**(4): 333. 2000. [29]

Climbers. C Nepal: *ca.* 3000 m. **Endemic to Nepal.**

Central: *Mustang*, *ca.* 3000 m, JDA Stainton, WR Sykes & LHJ Williams 1659 (Holotype: P).

Clematis terniflora DC. [2, 4]

Clematis dioscoreifolia Levl. & Vaniot (14)

जुंगे लहरा (Junge laharaa); Sweet autumn Clematis

Climbing shrubs. C Nepal: 1400-1600 m. Planted as hedge plant. Native to E Asia.

Clematis tibetana Kuntze [2, 4, 6]

var. **tibetana**

Clematis tibetana subsp. *brevipes* (Tamura) Yonek. (8, 11)

इमोंग-नाग्पो (Imong-naagpo: Amchi)

Climbing shrubs. W, C Nepal: 1500-4000 m. Nepal, Tibetan Plateau and S Asia.

Use: Medicine (stem, leaves, flowers).

var. **laciniifolia** Grey-Wilson [4]

Woody climbers. C Nepal: *ca.* 3900 m.

var. **vernayi** (C.E.C.Fisch.) W.T.Wang [8]

Clematis vernayi C.E.C.Fisch.; *Clematis tibetana* subsp. *vernayi* (C.E.C.Fisch.) Grey-Wilson

Woody climbers. W, C Nepal: 1700-4200 m. Nepal, Tibetan Plateau and S Asia.

Clematis tongluensis (Brühl) Tamura [2, 4, 6]

Clematis montana var. *tongluensis* Brühl

Woody climbers. E Nepal: 2600-3000 m. Nepal and E Himalaya.

Clematis tortuosa Wall. ex C.E.C.Fisch. [2, 4, 6, 11]

Clematis buchananiana var. *tortuosa* Hook.f. & Thomson

Woody climbers. W, C, E Nepal: 2000-3900 m. Nepal, W Himalaya, E Himalaya, Assam-Burma and S Asia.

Clematis zemuensis W.W.Sm. [2, 4, 6]

Woody climbers. C, E Nepal: 2700-3700 m. Nepal and E Himalaya.

Delphinium L.

Delphinium altissimum Wall. [4, 6, 29]

Herbs. W, C Nepal: 2000-2300 m. Nepal, E Himalaya and Tibetan Plateau.

Delphinium brunonianum Royle [2, 4, 6, 11]

Delphinium jacquemontianum Cambess.

मौरो मुलो; जगो-पोए (Mauro mulo; Jago-poe: Amchi); Musk Larkspur

Perennial herbs. W, C Nepal: 3500-6000 m. Nepal, W Himalaya, Tibetan Plateau, S Asia and SW Asia.

Use(s): Medicine (stem, leaves, flowers), Non-vertebrate poison.

Delphinium caeruleum Jacquem. ex Cambess. [2, 4, 8, 11]

Delphinium grandiflorum var. *kunawarensis* Brühl

तिमुसा (Timusaa: Amchi)

Perennial herbs. W, C Nepal: 3000-5800 m. Nepal, E Himalaya, Tibetan Plateau and E Asia.

Use(s): Medicine (roots, leaves, flowers), Non-vertebrate poison.

Delphinium cooperi Munz [2, 4, 6]

Perennial herbs. W, E Nepal: 2500-4700 m. Nepal and E Himalaya.

Delphinium densiflorum Duthie ex Huth [2, 4, 6]

Perennial herbs. W, C Nepal: 4000-6000 m. Nepal, Tibetan Plateau and S Asia.

Delphinium dictyocarpum DC.

subsp. **tibeticum** Brühl [4, 6, 98]

Herbs. C Nepal: *ca.* 4000 m.

Delphinium glaciale Hook.f. & Thomson [2, 4, 6]

Perennial herbs. C, E Nepal: 5000-6500 m. Nepal, E Himalaya and Tibetan Plateau.

Delphinium grandiflorum L. [2, 4, 6]

मुडुलो, रुखवा मिन्दु (Mudulo, Rukhvaa mindu: Tibetan); Siberian larkspur

Perennial herbs. W, C Nepal: 3700-4000 m. Nepal, Tibetan Plateau, E Asia and N Asia.

Delphinium himalayae Munz, *J. Arn. Arb.* **49**: 145. 1968. [2, 4, 6, 11, 12, 28, 29]

अतिस (Atis)

Perennial tuberous herbs. W, C Nepal: 3000-4500 m. **Endemic to Nepal**.

Central: C. Wigram 118 (Holotype: E).

Use(s): Medicine (root/tuber), Commercial.

Delphinium incisum (Hook.f. & Thomson) Wall. ex Munz [2, 4]

Delphinium ranunculifolium var. *incisum* Hook.f. & Thomson

Herbs. C Nepal and W Himalaya.

Delphinium kamaonense Huth [2, 4, 6, 11]

Delphinium grandiflorum var. *kumaonenss* Huth ex Brühl

छारकांग (Chharkang: Tibetan)

Perennial herbs. W, C Nepal: 3000-4500 m. Nepal, W Himalaya and Tibetan Plateau.

Use: Medicine (whole plant).

Delphinium ludlowii Munz [2, 4, 6]

Perennial herbs. E Nepal: 1400-3700 m. Nepal and E Himalaya.

Delphinium nepalense Kitam. & Tamura [2, 4, 11]

Perennial herbs. W, C, E Nepal: 4000-5500 m. Nepal and E Himalaya.

Delphinium nortonii Dunn [2, 4, 8]

var. **nortonii**

Herbs. W, E Nepal: 3500-5000 m. Nepal, E Himalaya and Tibetan Plateau.

var. **curvicalcaratum** Tamura [2, 4]

Herbs. W, E Nepal: 3500-4500 m.

Delphinium pyramidale Royle [2, 4]

Delphinium ranunculifolium Wall. ex Hook.f. & Thomson

Perennial herbs. W Nepal: 2000-3500 m. Nepal and W Himalaya.

Use: Medicine (whole plant).

Delphinium scabriflorum D.Don [2, 4, 8, 11]

Delphinium carela Buch.-Ham. ex D.Don

बिख, विषादी घाँस (Bikh, Bishaadi ghaans)

Perennial herbs. W, C, E Nepal: 1500-2500 m. Nepal and E Himalaya.

Use: Medicine (leaves).

Delphinium stapeliosum Brühl ex Huth [2, 4, 6]

Herbs. W, C, E Nepal: 1200-3500 m. Nepal, E Himalaya and Assam-Burma.

Delphinium tabatae Tamura [4]

Herbs. W Nepal: 2700-2800 m. Nepal and Tibetan Plateau.

Delphinium umbrosum Hand.-Mazz.

var. **drepanocentrum** (Brühl ex Huth) W.T.Wang & M.J.Warnock

Delphinium altissimum var. *drepanocentrum* Brühl ex Huth; *Delphinium drepanocentrum* (Brühl) Munz (2, 4, 6)

Herbs. C, E Nepal: 2300-3800(-5500) m. Nepal, E Himalaya and Tibetan Plateau.

Delphinium unifolium Tamura, *J. Phytogeogr. Taxon.* **27**: 35. 1979. [4, 29]

Delphinium viscosum var. *sordidum* Kitam. & Tamura

Herbs. C Nepal: *ca.* 3500 m. **Endemic to Nepal**.

Central: Banga Khola, *ca.* 3500 m, S. Nakao 25279 (Holotype: KYO).

Delphinium vestitum Wall. ex Royle [2, 4, 6, 11]

Delphinium rectivenium Royle

मौरे मुलो (Maure mulo)

Perennial herbs. W, C, E Nepal: 2700-4700 m. Nepal, W Himalaya, E Himalaya and Tibetan Plateau.

Use: Medicine (root).

Delphinium viscosum Hook.f. & Thomson [2, 4, 6, 11]

Delphinium trilobatum Huth

Perennial herbs. C, E Nepal: 3000-5200 m. Nepal and E Himalaya.

Delphinium williamsii Munz, *J. Arn. Arb.* **48**: 541. 1967. [2, 4, 12, 28, 29]

Perennial herbs. W Nepal: 1500-2600 m. **Endemic to Nepal**.

West: *Kalikot*, Gilam, Tila Valley, 6000 ft. (1830 m), O. Polunin, WR Sykes & LHJ Williams 1907 (Holotype: BM; Isotype: E).

Dichocarpum W.T.Wang & P.G.Xiao

Dichocarpum adiantifolium (Hook.f. & Thomson) W.T.Wang & P.K.Hsiao

Isopyrum adiantifolium Hook.f. & Thomson (2, 4, 6)

Perennial herbs. W, C, E Nepal: 1700-2600 m. Nepal, E Himalaya and Assam-Burma.

Halerpestes Greene

Halerpestes cymbalaria (Pursh) Greene [4]

Ranunculus cymbalaria Pursh

Perennial herbs. Nepal, E Himalaya, E Asia, SW Asia and N Asia.

Use: Medicine (whole plant).

Halerpestes lancifolia (Bertol.) Hand.-Mazz.

Ranunculus lancifolius Bertol.; *Ranunculus tricuspis* var. *lancifolius* (Bertol.) H.Hara (2, 4)

Perennial herbs. C Nepal: 2600-5100 m. Nepal, W Himalaya and Tibetan Plateau.

Halerpestes tricuspis (Maxim.) Hand.-Mazz. [4]

Ranunculus tricuspis Maxim. (2, 4); *Halerpestes tricuspis* var. *variifolia* (Tamura) W.T.Wang (4, 11)

Perennial herbs. W, C, E Nepal: 2600-4400 m. Nepal, W Himalaya, E Himalaya, Tibetan Plateau, E Asia and N Asia.

Use: Medicine (whole plant).

Naravelia DC.

Naravelia zeylanica (L.) DC. [2, 4, 6]

Atragene zeylanica L.

रसगर्गी (Rasgargi)

Climbing shrubs. E Nepal: *ca.* 400 m. Nepal, E Himalaya, S Asia and E Asia.

Use: Medicine (whole plant).

Oxygraphis Bunge

Oxygraphis endlicheri (Walp.) Bennet & S.Chandra [6, 11]

Callianthemum endlicheri Walp.; *Oxygraphis polypetala* (Royle) Hook.f. & Thomson (2, 4)

Perennial herbs. W, C, E Nepal: 2200-5000 m. Nepal, W Himalaya, E Himalaya and Tibetan Plateau.

Oxygraphis nepalensis Tamura, *J. Geobot.* **26**(3): 68. 1978. [4, 29]

Herbs. C Nepal: 3000-3500 m. **Endemic to Nepal**.

Central: *Mustang*, Tukucha, Kali Gandaki, 13500 ft. (4114 m), JDA Stainton, WR Sykes & LHJ Williams 1135 (Holotype: BM).

Paraquilegia J.R.Drumm. & Hutch.

Paraquilegia microphylla (Royle) J.R.Drumm. & Hutch. [2, 4, 8, 11]

Isopyrum microphyllum Royle

युमो देउ जिन (Yumo deu jin: Amchi)

Perennial herbs. W, C Nepal: 3400-4900 m. Nepal, W Himalaya, E Himalaya, Tibetan Plateau, E Asia, N Asia and C Asia.

Use: Medicine (stem, leaves, flowers).

Paroxygraphis W.W.Sm.

Paroxygraphis sikkimensis W.W.Sm. [2, 4, 6]

Perennial herbs. C, E Nepal: 3200-4300 m. Nepal and E Himalaya.

Ranunculus L.

Ranunculus adoxifolius Hand.-Mazz. [2, 4, 6]

Ranunculus caespitosus Wall. ex Hook.f. & Thomson

Perennial herbs. W, C, E Nepal: 3600-4300 m. Nepal, E Himalaya and Tibetan Plateau.

Ranunculus albertii Regel & Schmalh. [4, 29]

Ranunculus sulphureus var. *albertii* (Regel & Schmalh.) Maxim.

Perennial herbs. Nepal: 4100-4400 m. Nepal, N Asia and C Asia.

Ranunculus arvensis L. [6]

Annual herbs. W Nepal: *ca.* 1500 m (Naturalized). Native to W Asia and Europe.

Ranunculus brotherusii Freyn [2, 4, 6]

चु-रुक्पा (Chu-rukpa: Amchi)

Perennial herbs. W, C, E Nepal: 3000-5000 m. Nepal, W Himalaya, E Himalaya, Tibetan Plateau, S Asia, E Asia, N Asia and C Asia.

Use: Medicine (stem, leaves, flowers).

Ranunculus cantoniensis DC. [2, 4, 6, 29]

Ranunculus napaulensis DC. (4, 6); *Ranunculus trilobatus* D.Don

Perennial herbs. W, C, E Nepal: 1300-2000 m. Nepal, W Himalaya, E Himalaya and E Asia.

Use: Medicine (whole plant).

Ranunculus diffusus DC. [2, 4, 6]

Ranunculus mollis Wall. ex D.Don

var. **diffusus**

ब्रिमोरमेंदो, नक्कोरे झार (Brimornendo, Nakkore jhaar)

Perennial herbs. W, C, E Nepal: 1500-4200 m. Nepal, E Himalaya, Tibetan Plateau, Assam-Burma, S Asia, E Asia and SW Asia.

Use(s): Medicine (root), Food (tender leaves, shoot).

var. **incisus** Tamura [4, 11]

Perennial herbs. W, C Nepal: 1500-3000 m.

Ranunculus ficariifolius H.Lév. & Vaniot [2, 4, 6]

Ranunculus flaccidus Hook.f. & Thomson

Perennial herbs. W, C, E Nepal: 3200-4100 m. Nepal, E Himalaya, E Asia and SE Asia.

Ranunculus flavidus (Hand.-Mazz.) C.D.K.Cook [4, 11]

Batrachium flavidum Hand.-Mazz.

Aquatic annual herbs. W, C Nepal: 2700-4000 m. Nepal and W Himalaya.

Ranunculus glabricaulis (Hand.-Mazz.) L.Liou [4, 29]

Ranunculus hirtellus var. *glabricaulis* Hand.-Mazz.

Perennial herbs. C Nepal: 3700-4200 m. Nepal, Tibetan Plateau and E Asia.

Ranunculus himalaicus Tamura, *J. Geobot.* **26**(3): 68. 1978. [4, 29]

Herbs. W Nepal: 4600-5600 m. **Endemic to Nepal**.

Western: *Humla*, Sisne Himal, 15000 ft. (4572 m), O. Polunin, WR Sykes & LHJ Williams 257 (Holotype: BM).

Ranunculus hirtellus Royle [2, 4, 6, 11]

Ranunculus attenuatus Royle ex D.Don

var. **hirtellus**

Perennial herbs. W, C, E Nepal: 2800-5500 m. Nepal, W Himalaya, E Himalaya, Tibetan Plateau and SW Asia.

Use: Medicine (whole plant).

var. **leiocarpus** Tamura (4)

Herbs. W, C, E Nepal: 3100-4200 m.

Ranunculus laetus Wall. ex Hook.f. & Thomson [2, 4, 6]

Ranunculus distans Wall. ex Royle

Perennial herbs. W Nepal: 1700-3800 m. Nepal, W Himalaya, E Himalaya, S Asia, C Asia and SW Asia.

Use(s): Medicine (leaves), Fodder.

Ranunculus longicaulis C.A.Mey [33]

Ranunculus pulchellus var. *longicaulis* Trautv.

Herbs. W, C Nepal: 3000-4500 m. Nepal, W Himalaya, E Himalaya, N Asia and C Asia.

Ranunculus makaluensis Kadota, *Himal. Pl.* 2: 106. 1991. [4, 12, 28, 29]

Perennial dwarf herbs. E Nepal: 4300-4400 m. **Endemic to Nepal**.

East: *Sankhuwasabha*, Merek, 4340 m, M. Suzuki *et al* 8850675 (Holotype: TI; Isotype: TNS).

Ranunculus membranaceus Royle [8, 11]

Ranunculus pulchellus var. *sericeus* Hook.f. & Thomson (2, 4, 6)

var. **membranaceus**

Perennial herbs. W, C, E Nepal: 800-4500 m. Nepal, Tibetan Plateau, S Asia, E Asia, N Asia, Europe and N America.

var. **stracheyanus** (Maxim.) Yonek. [8, 11]

Ranunculus affinis var. *stracheyanus* Maxim.; *Ranunculus pulchellus* var. *stracheyanus* (Maxim.) Hand.-Mazz. (2)

Perennial herbs. W, C, E Nepal: 3000-5600 m. Nepal, E Himalaya, Tibetan Plateau, E Asia, N Asia and SW Asia.

Ranunculus munroanus J.R.Drumm. ex Dunn [2, 4, 6, 11]

Perennial herbs. W, C Nepal: 3800-4200 m. Nepal, W Himalaya and Tibetan Plateau.

Ranunculus muricatus L. [4, 6, 98]

Annual herbs. C Nepal: 1300-1400 m. Nepal, Native to W Asia and Europe.

Ranunculus natans C.A.Mey. [2, 4, 6]

Ranunculus hyperboreus var. *natans* (C.A.Mey.) Regal

Perennial herbs. W, C Nepal: 2700-3800 m. Nepal, W Himalaya and N Asia.

Ranunculus oreionannos C.Marquand & Airy Shaw [2, 4]

Perennial herbs. W, C, E Nepal: 4500-5900 m. Nepal and Tibetan Plateau.

Ranunculus pegaeus Hand.-Mazz. [2, 4, 6]

var. **pegaeus**

Perennial herbs. W, C, E Nepal: 3600-4500 m. Nepal, E Himalaya, Tibetan Plateau, S Asia and E Asia.

var. **curvistylis** Y.Kadota [4]

Perennial herbs. C Nepal: 3900-4000 m.

Ranunculus potaninii Kom.

Ranunculus pulchellus var. *potaninii* (Kom.) Hand.-Mazz. (2)

Perennial herbs. W Nepal: *ca.* 4300 m. Nepal, Tibetan Plateau and E Asia.

Ranunculus pseudohirculus Schrenk [33]

Ranunculus pulchellus var. *pseudohirculus* (Schrenk) Trautv.; *Ranunculus longicaulis* var. *pseudohirculus* (Schrenk) Gubanov

Herbs. W, C, E Nepal: 4200-5700 m. Nepal, E Himalaya, S Asia, N Asia and C Asia.

Ranunculus pseudopygmaeus Hand.-Mazz. [2, 4, 6, 11]

Perennial herbs. C, E Nepal: 3200-4200 m. Nepal, Tibetan Plateau and E Asia.

Ranunculus pulchellus C.A.Mey. [2, 4, 6, 33]

Ranunculus longicaulis subsp. *pulchellus* (C.A.Mey.) Gubanov

var. **pulchellus**

Perennial herbs. W, C, E Nepal: 2700-4000 m. Nepal, W Himalaya, E Himalaya, N Asia and C Asia.

var. **tibeticus** (Maxim.) Yonek. [8, 11]

Ranunculus affinis var. *tibeticus* Maxim.

Perennial herbs. W, C, E Nepal: 3600-4900 m. Nepal, W Himalaya, E Himalaya, Tibetan Plateau, E Asia and N Asia.

Ranunculus sceleratus L. [2, 4, 6, 11]

Ranunculus umbellatus Roxb. ex Willd.

नक्कोरे (Nakkore); Celery leaved buttercup

Annual herbs. W, C, E Nepal: 800-1700 m. Nepal, W Himalaya, S Asia, E Asia, SE Asia, N Asia, C Asia, SW Asia, Europe and N America.

Use(s): Medicine (whole plant), Food.

Ranunculus siamensis M.Tamura [4]

Herbs. Nepal, Assam-Burma, S Asia and SE Asia.

Ranunculus sphaerospermus Boiss. & Blanche [2, 4]

Ranunculus peltatus subsp. *sphaerospermus* (Boiss. & Blanche) Meikle

Herbs. Nepal, W Himalaya, Tibetan Plateau and SW Asia.

Ranunculus tanguticus (Maxim.) Ovcz. [2]

Ranunculus affinis var. *tanguticus* Maxim.; *Ranunculus brotherusii* var. *tanguticus* (Maxim.) Tamura (11)

Perennial herbs. W, C, E Nepal: 3500-4900 m. Nepal, Tibetan Plateau and E Asia.

Ranunculus trichophyllus Chaix ex Vill. [2, 4, 6, 11]

var. **trichophyllus**

पानि खर (Paani khar)

Aquatic annual herbs. W, C, E Nepal: 3000-4800 m. Nepal, Temperate Eurasia, Africa and N America.

Use: Medicine (whole plant).

var. **glaber** (Tamura) Tamura

Batrachium trichophyllum var. *glaber* Tamura; *Ranunculus trichophyllus* var. *glaber* (Tamura) H.Hara (2, 4)

Herbs. C, E Nepal: 2800-4500 m. Nepal and E Himalaya.

Thalictrum L.

Thalictrum alpinum L. [2, 4, 6, 11]

var. **alpinum**

Perennial herbs. W, C, E Nepal: 2800-5000 m. Nepal, W Himalaya, E Himalaya, Tibetan Plateau, SE Asia, N Asia, Europe and N America.

Use: Medicine (root).

var. **elatum** O.E.Ulbr.

Thalictrum alpinum var. *stipitatum* Yabe (4); *Thalictrum setulosinerve* H.Hara (2, 4)

Perennial herbs. C Nepal: 2400-4600 m. Nepal, E Himalaya, Tibetan Plateau and E Asia.

var. **minutissimum** H.Hara [2, 4]

Perennial herbs. C Nepal: 3800-5000 m.

Thalictrum chelidonii DC. [2, 4, 6]

Thalictrum cysticarpum Wall.

दाम्पाते (Daampaate)

Perennial herbs. W, C, E Nepal: 1800-3500 m. Nepal, W Himalaya, E Himalaya, Tibetan Plateau and Assam-Burma.

Thalictrum cultratum Wall. [2, 4, 6, 11]

Thalictrum chelidonii var. *cultratum* (Wall.) Hook.f. & Thomson

नगोतिन चागक्यु (Ngotin chaag kyu: Tibetan)

Perennial herbs. W, C, E Nepal: 2200-4500 m. Nepal, W Himalaya, E Himalaya, Tibetan Plateau, S Asia and E Asia.

Use: Medicine (root, leaves, flowers).

Thalictrum dalzellii Hook. [4, 6, 29]

Perennial herbs. W, E Nepal: 1300-2000 m. Nepal and S Asia (India).

Thalictrum elegans Wall. ex Royle [2, 4, 6, 11]

च्याकसेप (Chyaaksep: Tibetan)

Perennial herbs. W, C, E Nepal: 3100-4100 m. Nepal, W Himalaya, E Himalaya, Tibetan Plateau, S Asia and E Asia.

Use: Medicine (whole plant).

Thalictrum foetidum L. [2, 4, 8, 11]

Thalictrum vaginatum Royle

नगोतिन (Ngotin: Tibetan)

Perennial herbs. W, C, E Nepal: 2300-4700 m. Nepal, widely distributed in Asia and Europe.

Use: Medicine (aerial part).

Thalictrum foliolosum DC. [2, 4, 6, 11]

Thalictrum falconeri Lecoy.

दाम्पाते; चाग-क्यु (Dampaate; Chag-kyu: Amchi)

Perennial herbs. W, C, E Nepal: 1300-3400 m. Nepal, W Himalaya, E Himalaya, Tibetan Plateau, Assam-Burma, E Asia and SE Asia.

Use: Medicine (root, leaves, flowers).

Thalictrum javanicum Blume [2, 4, 6]

Thalictrum glyphocarpum Wight & Arn.

गुह्ये पाती (Guhya paati)

Perennial herbs. W, C, E Nepal: 2100-3000 m. Nepal, W Himalaya, E Himalaya, Tibetan Plateau, S Asia, E Asia and SE Asia.

Use(s): Medicine (whole plant), Food (young plant).

Thalictrum leuconotum Franch. [2, 4]

Perennial herbs. C Nepal: 3600-3800 m. Nepal, E Himalaya and E Asia.

Thalictrum pauciflorum Royle [2, 4, 6]

Thalictrum macrostigma Edgew.

Perennial herbs. W, C Nepal: 3000-3200 m. Nepal and W Himalaya.

Thalictrum platycarpum Hook.f. & Thomson [2, 4, 6, 11]

Thalictrum cultratum subsp. *platycarpum* (Hook.f. & Thomson) Brühl

Perennial herbs. W, C, E Nepal: 2800-4700 m. Nepal, W Himalaya, Tibetan Plateau, S Asia and SW Asia.

Thalictrum punduanum Wall. [2, 4, 6]

Thalictrum punduanum var. *glaucum* Hook.f. & Thomson

Perennial herbs. W, C, E Nepal: 1200-2500 m. Nepal, W Himalaya, E Himalaya, Assam-Burma and S Asia.

Thalictrum reniforme Wall. [2, 4, 6, 11]

Thalictrum chelidonii var. *reniforme* (Wall.) Hook.f. & Thomson; *Thalictrum neurocarpum* Royle (2, 4, 8)

Perennial herbs. W, C, E Nepal: 2600-3500 m. Nepal, W Himalaya, E Himalaya and Tibetan Plateau.

Use: Medicine (root, aerial part).

Thalictrum rostellatum Hook.f. & Thomson [2, 4, 6, 11]

Perennial herbs. W, C, E Nepal: 2300-3200 m. Nepal, W Himalaya, E Himalaya, Tibetan Plateau, S Asia and E Asia.

Thalictrum rotundifolium DC. [2, 4, 6, 29]

Perennial herbs. C, E Nepal: 1200-2700 m. Nepal and Tibetan Plateau.

Thalictrum rutifolium Hook.f. & Thomson [2, 4]

Perennial herbs. W Nepal: *ca.* 4300 m. Nepal, W Himalaya, E Himalaya, Tibetan Plateau and E Asia.

Thalictrum saniculiforme DC. [2, 4, 6]

Thalictrum radiatum Royle

Perennial herbs. W, C, E Nepal: 1900-2800 m. Nepal, W Himalaya, E Himalaya, Tibetan Plateau and E Asia.

Thalictrum squamiferum Lecoy. [4, 8, 11, 29]

Thalictrum glareosum Hand.-Mazz. (2, 4)

Perennial herbs. W, C Nepal: 3500-5600 m. Nepal, E Himalaya and Tibetan Plateau.

Use: Medicine (root, flowers).

Thalictrum virgatum Hook.f. & Thomson [2, 4, 6, 11]

Perennial herbs. W, C, E Nepal: 2000-4500 m. Nepal, E Himalaya, Tibetan Plateau and E Asia.

Trollius L.

Trollius acaulis Lindl. [2, 4]

Perennial herbs. W Nepal: 3600-3700 m. Nepal and W Himalaya.

Trollius pumilus D.Don [2, 4, 6]

Trollius pumilus forma *sikkimensis* (Brühl) H.Hara (2, 4, 8, 11)

Perennial herbs. W, C, E Nepal: 3500-5700 m. Nepal, E Himalaya, Tibetan Plateau, Assam-Burma and S Asia.

Exotic/Cultivated Taxa

Consolida ajacis (L.) Schur., *Delphinium ajacis* L.; *Delphinium pauciflorum* D.Don (2, 4) (**अज फूली स्वाँ,** Ajaphuli swaan: Newari; Larkspur). Native to Europe (Switzerland). [14]

Nigella damascena L. (Fennel flower). Native to S Europe, N Africa and SW Asia. [14]

Nigella sativa L. (**काल जिरा,** Kaal jira; Black cumin). Native to S Asia and SW Asia.

Ranunculus asiaticus L. (**रानी कमल,** Raani kamal; Persian buttercup). Native to mediterranean region, SW Asia and SE Europe. [14]

Doubtful/Uncertain Taxa

Aconitum wallichianum Lauener [Notes R.B.G. Edinb. **25**: 30. 1963].

Anemone coelestina var. **holophylla** (Diels) Ziman & B.E.Dutton [101]

Anemone griffithii Hook.f. & Thomson [101]

Anemone rockii Ulbr. [101]

Clematis palmata Hook. & Thomson [6]

Clematis siamensis J.R.Drummond & Craib [101]

Clematis tripartita W.T.Wang [101]

Clematis wallichii W.T.Wang [109]

Delphinium cashmerianum Royle [4, 98]

Delphinium denudatum Wall. ex Hook.f. & Thomson [4, 98]

Delphinium viscosum var. **chrysotrichum** Brühl ex Huth [101]

Oxygraphis glacialis (Fisch. ex DC.) Bunge [101]

Ranunculus himalaicus Kadota, nom. Illeg. [4]

PROTEALES

SABIACEAE

Meliosma Blume

Meliosma dilleniifolia (Wall. ex Wight & Arn.) Walp. [2, 4]

Millingtonia dilleniifolia Wall. ex Wight & Arn.

कोदे खबदे (Kode khabde)

Deciduous trees. W, C, E Nepal: 1800-2900 m. Nepal, E Himalaya, Tibetan Plateau, Assam-Burma, S Asia and E Asia.

Use: Fodder.

Meliosma pinnata (Roxb.) Maxim. [2, 4]

Millingtonia pinnata Roxb.; *Meliosma wallichii* Planch. ex Hook.f.

Semi-evergreen trees. C, E Nepal: 300-400 m. Nepal, E Himalaya, Tibetan Plateau, Assam-Burma and S Asia.

Meliosma simplicifolia (Roxb.) Walp. [2, 4]

Millingtonia simplicifolia Roxb.

कोदे खबदे (Kode khavade)

Evergreen trees. C, E Nepal: 200-2500 m. Nepal, E Himalaya, Tibetan Plateau, Assam-Burma, S Asia, E Asia and SE Asia.

Meliosma thomsonii King ex Brandis

Meliosma simplicifolia subsp. *thomsonii* (King ex Brandis) Beusekom (2, 4, 6)

Evergreen trees. C Nepal: *ca.* 1800 m. Nepal, W Himalaya, E Himalaya, Tibetan Plateau, Assam-Burma, E Asia and SE Asia.

Meliosma yunnanensis Franch.

Meliosma simplicifolia subsp. *yunnanensis* (Franch.) Beusekom (2, 4)

Evergreen trees. W, C, E Nepal: 1800-2400 m. Nepal, E Himalaya, Tibetan Plateau, Assam-Burma and E Asia.

Sabia Colebr.

Sabia campanulata Wall. ex Roxb. [2, 4]

Woody climbers. W, C, E Nepal: 1800-3400 m. Nepal, W Himalaya, E Himalaya, Tibetan Plateau, Assam-Burma and E Asia.

Sabia paniculata Edgew. ex Hook.f. & Thomson [2, 4]

Woody climbers. W, C, E Nepal: 300-1700 m. Nepal, E Himalaya, Assam-Burma, S Asia and SE Asia.

Sabia parviflora Wall. ex Roxb. [2, 4]

Woody climbers. Nepal, W Himalaya, E Himalaya, Assam-Burma, S Asia, E Asia and SE Asia.

Sabia purpurea Hook.f. & Thomson [2, 4]

Woody climbers. W, C Nepal: 800-2200 m. Nepal, W Himalaya, E Himalaya, Assam-Burma and SE Asia.

NELUMBONACEAE (incl. NYMPHAEACEAE, p.p.)

Nelumbo Adans.

Nelumbo nucifera Gaertn. [2, 4, 6]

Nelumbium speciosum Willd.; *Nymphaea nelumbo* L.

कमल, रातो कमल (Kamal, Raato kamal); Sacred water lotus, Indian lotus

Perennial aquatic herbs. W, C, E Nepal: 500-2100 m (Naturalized). Native to Tropical Asia and Australia.

Use(s): Ornamental, Ritual, Medicine (root, flowers, seeds), Food (rhizome, young leaves, seeds), Plates (leaves).

PROTEACEAE

Grevillea R.Br. ex Knight

Grevillea robusta A.Cunn. ex R.Br. [3, 4, 7, 14]

काँगियो फूल (Kaangiyo phul); Silk Oak

Evergreen trees. C, E Nepal: 700-1500 m. Widely planted as avenue tree. Native to Australia.

Helicia Lour.

Helicia nilagirica Bedd. [3, 4, 7]

Helicia erratica Hook.f.

बाँदरे (Baandare)

Evergreen trees. E Nepal: 400-1800 m. Nepal, E Himalaya, Assam-Burma, S Asia, E Asia and SE Asia.

Use: Medicine.

Exotic/Cultivated Taxa

PLATANACEAE

Platanus orientalis L. (**चिनार,** Chinar; Oriental plane). Planted as avenue tree. Native to SE Europe and SW Asia.

TROCHODENDRALES

TROCHODENDRACEAE (incl. TETRACENTRACEAE)

Tetracentron Oliv.

Tetracentron sinense Oliv. [2, 4, 6, 10]

Tetracentron sinense var. *himalense* H.Hara & Kanai

Deciduous trees. E Nepal: 1900-3200 m. Nepal, E Himalaya, Assam-Burma, E Asia and SE Asia.

CITES Appendix: III.

BUXALES

BUXACEAE

Buxus L.

Buxus rugulosa Hatus.

var. **prostrata** (W.W.Sm.) M.Cheng

Buxus microphylla var. *prostrata* W.W.Sm; *Buxus rugulosa* var. *intermedia* Hatus (3, 4, 7); *Buxus sempervirens* var. *microphylla* Hook.f.

Evergreen shrubs. W Nepal: 3000-3500 m. Nepal, W Himalaya, Tibetan Plateau and E Asia.

Buxus wallichiana Baill. [3, 4, 7]

Himalayan boxwood

Evergreen shrubs or small trees. W, C Nepal: 2000-2900 m. Nepal, W Himalaya, E Himalaya, Assam-Burma and SW Asia.

Use: Medicine (bark, leaves).

Sarcococca Lindl.

Sarcococca coriacea (Hook.) Sweet [3, 4, 7]

Pachysandra coriacea Hook.

फितिफिया (Phitiphiyaa), **पतापे** (Patape: Tamang)

Evergreen shrubs. C, E Nepal: 600-2500 m. Nepal, W Himalaya, E Himalaya, Assam-Burma and S Asia.

Use(s): Medicine (root), Fodder.

Sarcococca hookeriana Baill. [3, 4, 7]

Sarcococca pruniformis var. *hookeriana* (Baill.) Hook.f.

चिले काठ, तेल पारो (Chilekaath, Telparo); Himalayan sweet box

Evergreen shrubs. W, C, E Nepal: 1800-3500 m. Nepal, E Himalaya, Tibetan Plateau and SW Asia.

Use: Medicine (root).

Sarcococca saligna (D.Don) Müll. Arg. [3, 4, 7]

Buxus saligna D.Don; *Sarcococca pruniformis* var. *angustifolia* Lindl.

फितफिया (Phitphiya)

Evergreen shrubs. W, C Nepal: 1900-2300 m. Nepal, W Himalaya, Tibetan Plateau, SE Asia and SW Asia.

Use(s): Medicine (root), Fodder.

Sarcococca wallichii Stapf [3, 4, 7]

Evergreen shrubs. W, C, E Nepal: 1600-2900 m. Nepal, E Himalaya, Tibetan Plateau, Assam-Burma and E Asia.

CORE EUDICOTS

DILLENIALES

DILLENIACEAE

Dillenia L.

Dillenia aurea Sm. [2, 4]

Deciduous trees. C Nepal. Nepal, Assam-Burma, S Asia and SE Asia.

Use(s): Medicine (bark), Ritual, Construction material.

Dillenia indica L. [2, 4, 6]

पांच फल, राम फल (Paanch phal, Ramphal); Elephant apple

Evergreen trees. E Nepal: 100-300 m. Nepal, E Himalaya, Assam-Burma, S India, E Asia and SE Asia.

Use(s): Medicine (bark, leaves, fruit), Food.

Dillenia pentagyna Roxb. [2, 4, 6]

Dillenia floribunda Hook.f. & Thomson

टाटरि, राम फल (Taatari, Raam phal)

Deciduous trees. W, C, E Nepal: 100-1500 m. Nepal, E Himalaya, Assam-Burma, S Asia, E Asia and SE Asia.

Use(s): Medicine (bark, seeds), Food (fruit), Timber, Cordage.

Dillenia scabrella Roxb. ex Wall. [2, 4]

Deciduous trees. C Nepal: 1300-1400 m. Nepal, Assam-Burma and SE Asia.

Use: Food (fruit).

SUPERROSIDS

SAXIFRAGALES

PAEONIACEAE

Paeonia L.

Paeonia emodi Wall. ex Royle [2, 4, 6]

Paeonia anomala var. *emodi* L.; *Paeonia emodi* forma *glabrata* (Hook.f. & Thomsom) H.Hara (2, 4); *Paeonia emodi* var. *glabrata* Hook.f. & Thomson

चन्द्र, सुंगुरे औंले (Chandra, Sungure aunle)

Perennial herbs. W Nepal: 2100-2800 m. Nepal, W Himalaya and Tibetan Plateau.

Use(s): Ornamental, Medicine (tuber).

Exotic/Cultivated Taxa

Paeonia lactiflora Pall., (*Paeonia albiflora* Pall.); (Chinese peony). Native to E Asia, N Asia and C Asia. Cultivated. [14] *Use(s)*: Ornamental, Medicine (root).

Paeonia officinalis L., (Garden peony). Native to Europe. [14]

HAMAMELIDACEAE

Exbucklandia R.W.Br.

Exbucklandia populnea (R.Br. ex Griff.) R.W.Br. [2, 4, 6]

Bucklandia populnea R.Br. ex Griff.; *Bucklandia populifolia* Hook.f. & Thomson

पिपली (Pipli)

Evergreen trees. E Nepal: 1300-2300 m. Nepal, E Himalaya, Tibetan Plateau, Assam-Burma, E Asia and SE Asia.

Use(s): Fodder, Ritual, Construction materials.

DAPHNIPHYLLACEAE

Daphniphyllum Blume

Daphniphyllum chartaceum K.Rosenthal

Daphniphyllum himalense var. *chartaceum* (K.Rosenthal) Huang (3, 4, 7)

Evergreen trees. W, C, E Nepal: 1400-2300 m. Nepal, W Himalaya, E Himalaya, Tibetan Plateau, Assam-Burma, E Asia and SE Asia.

Daphniphyllum himalense (Benth.) Müll. Arg. [3, 4, 7]

Goughia himalensis Benth.

रक्त चन्दन, रक्चन (Rakta chandan, Rakchan)

Evergreen trees. W, C, E Nepal: 2000-2800 m. Nepal, E Himalaya, Tibetan Plateau, Assam-Burma and E Asia.

Use(s): Medicine (wood), Carving, Turnery.

GROSSULARIACEAE

Ribes L.

Ribes acuminatum Wall. ex G.Don [6, 10]

Ribes acuminatum var. *desmocarpum* (Hook.f. & Thomson) Jancz.; *Ribes desmocarpum* Hook.f. & Thomson; *Ribes takare* D.Don (2, 4, 7); *Ribes takare* forma *desmocarpum* (Hook.f. & Thomson) Hara (2, 4); *Ribes takare* var. *desmocarpum* (Hook.f. & Thomson) L.T.Lu

टाँफु (Tanphoo)

Deciduous shrubs. W, C, E Nepal: 1800-3600 m. Nepal, W Himalaya, E Himalaya, Tibetan Plateau, Assam-Burma and E Asia.

Use: Food (fruit).

Ribes alpestre Wall. ex Decne. [2, 4, 6, 10]

मसिनो किम्बु (Masino kimbu)

Deciduous shrubs. W, C Nepal: 3100-3800 m. Nepal, W Himalaya, E Himalaya, Tibetan Plateau, E Asia and SW Asia.

Use: Food (fruit).

Ribes glaciale Wall. ex Roxb. [2, 4, 6, 10]

Ribes glaciale var. *laciniatum* (Hook.f. & Thomson) C.B.Clarke; *Ribes laciniatum* Hook.f. & Thomson (2, 4, 6)

टाँफु, केम्बु (Taanfu, Kembu)

Deciduous arching or pendent shrubs. W, C, E Nepal: 2700-4400 m. Nepal, W Himalaya, E Himalaya, Tibetan Plateau, Assam-Burma and E Asia.

Use(s): Food (fruit), Medicine (leaves).

Ribes griffithii Hook.f. & Thomson [2, 4, 6, 10]

Deciduous shrubs. W, C, E Nepal: 2800-4000 m. Nepal, E Himalaya, Tibetan Plateau, Assam-Burma and E Asia.

Ribes himalense Royle ex Decne. [2, 4, 6, 10]

Ribes himalense var. *appendiculatum* Jancz.; *Ribes himalense* var. *urceolatum* Jancz.; *Ribes melanocarpum* K.Koch

काँटा दाख, खमार (Kaanta daakh, Khamaar)

Deciduous shrubs. W, C, E Nepal: 1800-3400 m. Nepal, W Himalaya, E Himalaya, Tibetan Plateau, Assam-Burma and E Asia.

Use(s): Medicine (leaves), Food.

Ribes luridum Hook.f. & Thomson [2, 4, 6, 10]

ग्याक्सिन (Gyaakshin: Sherpa)

Deciduous arching shrubs. W, C, E Nepal: 3400-4100 m. Nepal, E Himalaya, Assam-Burma and E Asia.

Ribes orientale Desf. [2, 4, 6, 10]

Ribes leptostachyum Decne.; *Ribes villosum* Wall.

Deciduous shrubs. W, C, E Nepal: 3600-4200 m. Nepal, W Himalaya, E Himalaya, Tibetan Plateau, E Asia, C Asia, SW Asia and Europe.

Use: Medicine (fruit).

Doubtful/Uncertain Taxa

Ribes vilmorinii Jancz. [2, 4]

SAXIFRAGACEAE

Astilbe Buch.-Ham. ex D.Don

Astilbe rivularis Buch.-Ham. ex D.Don [2, 4, 6, 10]

बुढो ओखती, ठुलो औषधि (Budho okhati, Thulo ausadhi)

Perennial rhizomatous herbs. W, C, E Nepal: 1400-3600 m. Nepal, W Himalaya, E Himalaya, Tibetan Plateau, Assam-Burma and SE Asia.

Use(s): Medicine (rhizome), Ritual.

Bergenia Moench

Bergenia ciliata (Haw.) Sternb. [2, 4, 10]

Megasea ciliata Haw.; *Saxifraga ciliata* Royle; *Saxifraga pacumbis* Buch.-Ham. ex D.Don, nom.inval. (6)

forma **ciliata**

पाखानवेद; गधुर (Paakhanved; Gadhur: Amchi); Rock foil

Perennial herbs. W Nepal: 900-2500 m. Nepal and W Himalaya.

Use: Medicine (root stock).

forma **ligulata** (Wall.) Yeo [2, 4, 10]

Saxifraga ligulata Wall.; *Bergenia ligulata* (Wall.) Engl.

Perennial herbs. W, C Nepal: 1300-3200 m. Nepal, W Himalaya, E Himalaya, Tibetan Plateau, Assam-Burma and SW Asia.

Use(s): Medicine (root stock), Food (flowers), Tannin (root), Commercial.

Bergenia purpurascens (Hook.f. & Thomson) Engl. [2, 4, 6, 10]

Saxifraga purpurascens Hook.f. & Thomson

लि-गा-दुर (Li-gaa-dur: Tibetan)

Perennial herbs. C, E Nepal: 3200-4700 m. Nepal, E Himalaya, Tibetan Plateau and Assam-Burma.

Use: Medicine (rhizome).

Chrysosplenium L.

Chrysosplenium carnosum Hook.f. & Thomson [2, 4, 6, 10]

Chrysosplenium carnosulum Hook.f. & Thomson ex Maxim.

याकिमा (Yaakimaa: Tibetan)

Perennial herbs. W, C, E Nepal: 3700-5500 m. Nepal, W Himalaya, E Himalaya and Tibetan Plateau.

Use: Medicine (whole plant).

Chrysosplenium forrestii Diels [2, 4, 6, 10]

Perennial herbs. C, E Nepal: 3600-4800 m. Nepal, E Himalaya, Tibetan Plateau and Assam-Burma.

Chrysosplenium griffithii Hook.f. & Thomson [2, 4, 10]

Perennial herbs. E Nepal: *ca.* 3600 m. Nepal, E Himalaya and Tibetan Plateau.

Chrysosplenium lanuginosum Hook.f. & Thomson [2, 4, 6, 10]

Chrysosplenium adoxoides Hook.f. & Thomson ex Maxim.; *Saxifraga adoxoides* Griff.

Perennial herbs. C, E Nepal: 2700-3500 m. Nepal, E Himalaya, Tibetan Plateau and Assam-Burma.

Chrysosplenium nepalense D.Don [2, 4, 6, 10]

Perennial herbs. W, C, E Nepal: 1900-4000 m. Nepal, E Himalaya, Tibetan Plateau and Assam-Burma.

Chrysosplenium nudicaule Bunge

Chrysosplenium griffithii var. *intermedium* (H.Hara) J.T.Pan

var. **intermedium** H.Hara [2, 4, 6, 10]

याकि-मा (Yaki-ma: Tibetan)

Perennial herbs. W, C, E Nepal: 3700-4600 m. Nepal, E Himalaya and Tibetan Plateau.

Chrysosplenium singalilense H.Hara [10]

Perennial herbs. E Nepal: *ca.* 3900 m. Nepal and E Himalaya.

Chrysosplenium tenellum Hook.f. & Thomson [2, 4, 10]

Perennial herbs. W Nepal: 2700-4300 m. Nepal, W Himalaya and E Himalaya.

Chrysosplenium trichospermum Edgew. ex Hook.f. & Thomson [2, 4, 10]

Perennial herbs. W Nepal: 2100-3000 m. Nepal and W Himalaya.

Chrysosplenium uniflorum Maxim. [2, 4, 10]

Perennial herbs. C Nepal: *ca.* 4000 m. Nepal, Tibetan Plateau and E Asia.

Micranthes Haw.

Micranthes gageana (W.W.Sm.) Gornall & H.Ohba [10]

Saxifraga gageana W.W.Sm. (2, 4)

Perennial herbs. C, E Nepal: 3600-5400 m. Nepal, E Himalaya and Tibetan Plateau.

Micranthes melanocentra (Franch.) Losinsk. [10]

Saxifraga melanocentra Franch. (2, 4, 6); *Saxifraga atrata* var. *subcorymbosa* Engl.

Perennial herbs. C, E Nepal: 3900-5300 m. Nepal, W Himalaya, E Himalaya, Tibetan Plateau and E Asia.

Micranthes pallida (Wall. ex Ser.) Losinsk. [10]

Saxifraga pallida Wall. ex Ser. (2, 4, 6); *Saxifraga himalaica* M.S.Balakr.; *Saxifraga micrantha* Edgew.

Perennial herbs. W, C, E Nepal: 3200-4900 m. Nepal, E Himalaya and Tibetan Plateau.

Micranthes pseudopallida (Engl. & Irmsch.) Losinsk. [10]

Saxifraga pseudopallida Engl. & Irmsch. (2, 4)

Perennial herbs. W, C, E Nepal: 3400-4800 m. Nepal, W Himalaya, E Himalaya and Tibetan Plateau.

Rodgersia A.Gray

Rodgersia nepalensis Cope ex Cullen [2, 4, 6, 10]

Perennial herbs. E Nepal: 2500-3400 m. Nepal and E Himalaya.

Saxifraga L.

Saxifraga afghanica Aitch. & Hemsl. [2, 4, 10]

Perennial herbs. W, C Nepal: 4600-4800 m. Nepal, W Himalaya, Tibetan Plateau and SW Asia.

Saxifraga alpigena Harry Sm., *Bull. Brit. Mus. (Nat. Hist.) Bot.* **2**: 97. 1958. [2, 4, 6, 10, 12, 28, 29]

Perennial herbs. W, C Nepal: 3400-4200 m. **Endemic to Nepal**.

Central: *Manang*, Marsyangdi, 11500 ft. (3500 m), DG Lowndes 935 (Holotype: BM).

Saxifraga amabilis H.Ohba & Wakab., *J. Jap. Bot.* **62**(6): 162. 1987. [4, 10, 12, 28, 29]

Perennial caespitose herbs. C, E Nepal: 4000-4600 m. **Endemic to Nepal**.

Central: *Dolakha*, Na-Sangma-Khabun, Rolwaling Khola, *ca.* 4300 m, H. Ohba, M. Wakabayashi & S. Akiyama 8320640 (Holotype: TI).

Saxifraga andersonii Engl. [2, 4, 6, 10]

Perennial herbs. W, C, E Nepal: 3400-5500 m. Nepal, E Himalaya and Tibetan Plateau.

Saxifraga aristulata Hook.f. & Thomson [2, 4, 6, 10]

Hirculus aristulata (Hook.f. & Thomson) Losinsk.

Perennial herbs. W, C, E Nepal: 3500-5200 m. Nepal, W Himalaya, E Himalaya and Tibetan Plateau.

Saxifraga asarifolia Sternb. [2, 4, 6, 10]

Saxifraga odontophylla Wall. ex Hook.f. & Thomson

Perennial herbs. W, C Nepal: 3600-4600 m. Nepal, W Himalaya and E Himalaya.

Saxifraga brachypoda D.Don [2, 6, 10]

Hirculus brachypodus (D.Don) Losinsk.; *Saxifraga glandulosa* Wall. ex DC.

Perennial herbs. W, C, E Nepal: (2500-)3300-5000 m. Nepal, W Himalaya, E Himalaya, Tibetan Plateau, Assam-Burma and E Asia.

Saxifraga brunonis Wall. ex Ser. [2, 4, 6, 10]

Hirculus brunonianus Losinsk.; *Saxifraga brunoniana* Wall. ex Sternb., nom.illegit.

Perennial herb. W, C, E Nepal: (1400-)2400-5600 m. Nepal, W Himalaya, E Himalaya and Tibetan Plateau.

Saxifraga caveana W.W.Sm. [2, 4, 6, 10]

Saxifraga diapensia Harry Sm.

Perennial herbs. C, E Nepal: 4200-5000 m. Nepal and E Himalaya.

Saxifraga ciliatopetala (Engl. & Irmsch.) J.T.Pan [11]

Saxifraga hirculus forma *ciliatopetala* Engl. & Irmsch.

Perennial herbs. C Nepal: *ca.* 4800 m. Nepal and Tibetan Plateau.

Saxifraga cinerea Harry Sm., *Bull. Brit. Mus. (Nat. Hist.) Bot.* **2**: 128. 1958. [2, 4, 6, 10, 12, 28, 29]

Perennial caespitose herbs. W, C Nepal: 2700-3300 m. **Endemic to Nepal**.

Central: *Manang*, Marsyangdi, 9000 ft. (2740 m), DG Lowndes 916 (Holotype: BM).

Saxifraga clivorum Harry Sm. [6]

Perennial herbs. E Nepal: *ca.* 4600 m. Nepal, E Himalaya and Tibetan Plateau.

Saxifraga coarctata W.W.Sm. [10]

Saxifraga humilis Engl. & Irmsch. (2, 4, 7)

Perennial herbs. W, E Nepal: 4400-4900 m. Nepal, W Himalaya and Tibetan Plateau.

Saxifraga consanguinea W.W.Sm. [2, 4, 6, 10]

Perennial herbs. W, C, E Nepal: 4400-5400 m. Nepal and Tibetan Plateau.

Saxifraga contraria Harry Sm. [2, 4, 10]

Perennial herbs. E Nepal: 4500-5300 m. Nepal, E Himalaya and Tibetan Plateau.

Saxifraga cordigera Hook.f. & Thomson [2, 4, 6, 10]

Perennial herbs. C, E Nepal: 3900-5100 m. Nepal, E Himalaya and Tibetan Plateau.

Saxifraga dhwojii (Harry Sm.) S.Akiyama & H.Ohba [6, 10]

Saxifraga sphaeradena subsp. *dhwojii* Harry Sm. (2, 4)

Perennial herbs. C, E Nepal: 3600-4500 m. Nepal, E Himalaya and Tibetan Plateau.

Saxifraga diversifolia Wall. ex Ser. [2, 4, 6, 10]

Perennial herbs. W, C, E Nepal: 2400-4800 m. Nepal, W Himalaya, E Himalaya and Tibetan Plateau.

Saxifraga elliptica Engl. & Irmsch. [2, 4, 10]

Perennial herbs. E Nepal: 4200-4300 m. Nepal, E Himalaya and Tibetan Plateau.

Saxifraga engleriana Harry Sm. [2, 4, 6, 10]

Perennial herbs. C, E Nepal: 4400-5000 m. Nepal, E Himalaya and Tibetan Plateau.

Saxifraga excellens Harry Sm., *Bull. Brit. Mus. (Nat. Hist.) Bot.* **2**: 230. 1960. [2, 4, 7, 10, 12, 28, 29]

Perennial herbs. C, E Nepal: 3600-4700 m. **Endemic to Nepal**.

Central: *Kaski*, Mardi Khola, Annapurna Himal, 12500 ft. (3810 m), JDA Stainton, WR Sykes & LHJ Williams 8460 (Holotype: BM; Isotype: E).

Saxifraga filicaulis Wall. ex Ser. [2, 4, 6, 10]

Hirculus filicaulis (Wall. ex Ser.) Losinsk.

Perennial herbs. W, C, E Nepal: (2300-)3000-4500 m. Nepal, W Himalaya, E Himalaya, Tibetan Plateau, Assam-Burma and E Asia.

Saxifraga ganeshii H.Ohba & S.Akiyama, *J. Jap. Bot.* **74**: 223. 1999. [4, 10, 28, 29]

Perennial caespitose herbs. C Nepal: 4200-4300 m. **Endemic to Nepal**.

Central: *Rasuwa*, Jaisuli Kund – Paldol Base Camp, 4250 m, F. Miyamoto *et al* 9420122 (Holotype: TI).

Saxifraga georgei J.Anthony [2, 4, 10]

Perennial herbs. W, C Nepal: 3900-5400 m. Nepal, E Himalaya, Tibetan Plateau and E Asia.

Saxifraga glabricaulis Harry Sm. [2, 4, 6, 10]

Perennial herbs. C, E Nepal: 4100-4500 m. Nepal, E Himalaya and Tibetan Plateau.

Saxifraga granulifera Harry Sm. [2, 4, 6, 10]

Saxifraga sibirica var. *bulbifera* Harry Sm.

Perennial herbs. W, C, E Nepal: (2700-)3100-4400(-5000) m. Nepal, W Himalaya, E Himalaya Tibetan Plateau and E Asia.

Saxifraga harae H.Ohba & Wakab., *J. Jap. Bot.* **62**(6): 163. 1987. [4, 7, 10, 12, 28, 29]

Perennial caespitose herbs. E Nepal: 3900-5000 m. **Endemic to Nepal**.

East: *Solukhumbu*, Beni Kharka – Tschokarma, 4550 m, H. Ohba & M. Wakabayashi 8520377 (Holotype: TI).

Saxifraga hemisphaerica Hook.f. & Thomson [2, 4, 10]

Perennial herbs. C Nepal: 4500-5100 m. Nepal, W Himalaya, E Himalaya and Tibetan Plateau.

Saxifraga hirculoides Decne. [2, 4, 7, 10]

Saxifraga hirculus var. *hirculoides* (Decne.) C.B.Clarke

Perennial herbs. W, C Nepal: 4200-5100 m. Nepal, W Himalaya and Tibetan Plateau.

Saxifraga hispidula D.Don [2, 4, 6, 10]

Hirculus hispidulus (D.Don) Losinsk.

Perennial herbs. W, C, E Nepal: 3000-4500 m. Nepal, W Himalaya, E Himalaya, Tibetan Plateau, Assam-Burma and E Asia.

Saxifraga hookeri Engl. & Irmsch. [2, 4, 6, 10]

Saxifraga corymbosa Hook.f. & Thomson

Perennial herbs. W, C, E Nepal: 3000-4600 m. Nepal, E Himalaya and Tibetan Plateau.

Saxifraga hypostoma Harry Sm., *Bull. Brit. Mus. (Nat. Hist.) Bot.* **2**: 103. 1958. [2, 4, 6, 10, 12, 28, 29]

Perennial stemless herbs. W, C, E Nepal: 3900-5300 m. **Endemic to Nepal**.

Central: *Gorkha*, Jargeng Khola, 16000 ft. (4880 m), DG Lowndes 1108 (Holotype: BM).

Saxifraga jacquemontiana Decne. [2, 4, 7, 10]

Perennial herbs. C, E Nepal: 4600-5300 m. Nepal, W Himalaya, E Himalaya and Tibetan Plateau.

Saxifraga jaljalensis H.Ohba & S.Akiyama, *Alpine Fl. Jaljale Himal.* 31. 1992. [4, 10, 12, 28, 29]

Perennial dwarf herbs. E Nepal: *ca.* 4300 m. **Endemic to Nepal**.

East: *Sankhuwasabha*, Shuwan Kharka – Topke Gola, *ca.* 4300 m, H. Ohba *et al* 91103080 (Holotype: TI).

Saxifraga kingdonii C.Marquand [6, 10]

Saxifraga nayari Wadhwa

Perennial herbs. E Nepal: *ca.* 4000 m. Nepal, Tibetan Plateau and Assam-Burma.

Saxifraga kingiana Engl. & Irmsch. [2, 4, 6, 10]

Saxifraga gageana Engl. & Irmsch., non W.W.Sm. (2, 4, 7)

Perennial herbs. C, E Nepal: 2900-4400 m. Nepal, E Himalaya and Tibetan Plateau.

Saxifraga kumaunensis Engl. [2, 4, 10]

Perennial herbs. W, E Nepal: 3300-4800 m. Nepal and W Himalaya.

Saxifraga lamninamensis H.Ohba [4, 6, 10]

Perennial herbs. E Nepal: 3200-4200 m. Nepal and E Himalaya.

Saxifraga latiflora Hook.f. & Thomson [6, 10]

Perennial herbs. E Nepal: 4100-4600 m. Nepal and E Himalaya.

Saxifraga lepida Harry Sm. [2, 4, 6, 10]

Perennial herbs. C Nepal: 3100-4300 m. Nepal and E Himalaya.

Saxifraga llonakhensis W.W.Sm. [2, 4, 6, 10]

Perennial herbs. W, C Nepal: 4200-5200 m. Nepal, E Himalaya, Tibetan Plateau, Assam-Burma and E Asia.

Saxifraga lowndesii Harry Sm., *Bull. Brit. Mus. (Nat. Hist.) Bot.* **2**: 106. 1958. [2, 4, 10, 12, 28, 29]

Perennial caespitose herbs. C Nepal: 3800-4200 m. **Endemic to Nepal.**

Central: *Manang*, Sabze Khola, 13500 ft. (4110 m), DG Lowndes 958 (Holotype: BM).

Saxifraga lychnitis Hook.f. & Thomson [2, 4, 6, 10]

Perennial herbs. C, E Nepal: 3900-5300 m. Nepal, W Himalaya, E Himalaya and Tibetan Plateau.

Saxifraga mallae H.Ohba & Wakab., *J. Jap. Bot.* **62**(6): 165. 1987. [4, 7, 10, 12, 28, 29]

Perennial caespitose herbs. C, E Nepal: 3900-5000 m. **Endemic to Nepal.**

Central: *Ramechhap*, Neju – Luk Kharka, *ca.* 4500 m, H. Ohba 8530649 (Holotype: TI).

Saxifraga matta-viridis Harry Sm. [2, 4, 10]

Perennial herbs. C Nepal: *ca.* 3500 m. Nepal and E Himalaya.

Saxifraga micans Harry Sm., *Bull. Brit. Mus. (Nat. Hist.) Bot.* **2**: 126. 1958. [2, 4, 10, 12, 28, 29]

Perennial caespitose herbs. C Nepal: *ca.* 3800 m. **Endemic to Nepal.**

Central: *Myagdi*, S. Gurjakhani, 12500 ft. (3810 m), JDA Stainton, WR Sykes & LHJ Williams 3074 (Holotype: BM; Isotype: E).

Saxifraga microphylla Royle ex Hook.f. & Thomson [2, 4, 6, 10]

Saxifraga microviridis H.Hara

Perennial herbs. W, C, E Nepal: 3900-5600 m. Nepal and W Himalaya.

Saxifraga mira Harry Sm., *Bull. Brit. Mus. (Nat. Hist.) Bot.* **2**: 114. 1958. [2, 4, 10, 12, 28, 29]

Perennial caespitose herbs. W Nepal: *ca.* 4400 m. **Endemic to Nepal.**

West: *Dolpa*, Barbung Khola – Kaya Khola,14500 ft. (4420 m), O. Polunin, WR Sykes & LHJ Williams 1094 (Holotype: BM; Isotype: E).

Saxifraga montanella Harry Sm. [2, 4, 6, 10]

Perennial herbs. C, E Nepal: 4100-4800 m. Nepal, E Himalaya and Tibetan Plateau.

Saxifraga moorcroftiana (Ser.) Wall. ex Sternb. [2, 4, 6, 10]

Saxifraga diversifolia var. *moorcroftiana* Ser.

Perennial herbs. W, C, E Nepal: 3500-4600(-4900) m. Nepal, W Himalaya, E Himalaya, Tibetan Plateau and E Asia.

Saxifraga mucronulata Royle [2, 4, 6, 10]

Saxifraga flagellaris subsp. *mucronulata* (Royle) Engl. & Irmsch.; *Saxifraga flagellaris* var. *mucronulata* (Royle) C.B.Clarke

Perennial herbs. W, C, E Nepal: 3700-4800 m. Nepal, W Himalaya, E Himalaya and Tibetan Plateau.

Saxifraga mucronulatoides J.T.Pan [6, 10]

Saxifraga flagellaris subsp. *sikkimensis* Hulten; *Saxifraga mucronulata* subsp. *sikkimensis* (Hulten) H.Hara (2, 4)

Perennial herbs. W, E Nepal: 4500-5100 m. Nepal, E Himalaya and Tibetan Plateau.

Saxifraga nakaoi Kitam. [2, 4, 6, 10]

Perennial herbs. W, C, E Nepal: 3500-5200 m. Nepal, W Himalaya and E Himalaya.

Saxifraga namdoensis Harry Sm., *Bull. Brit. Mus. (Nat. Hist.) Bot.* **2**: 237. 1960. [2, 4, 10, 12, 28, 29]

Perennial herbs. C Nepal: 4500-4700 m. **Endemic to Nepal.**

Central: *Mustang*, Namdo, N. Mustang, 15000 ft. (4570 m), JDA Stainton, WR Sykes & LHJ Williams 2266 (Holotype: BM).

Saxifraga nanella Engl. & Irmsch. [2, 4, 10]

Perennial herbs. C Nepal: 4500-5000 m. Nepal, E Himalaya, Tibetan Plateau and E Asia.

Saxifraga neopropagulifera H.Hara, *J. Jap. Bot.* **51**(1): 7. 1976. [2, 4, 10, 29]

Perennial herbs. W, C Nepal: 4500-5600 m. **Endemic to Nepal.**

East: Basta Bhanjyang, *ca.* 4500 m, S. Einarsson, L. Skarby & B. Wetterhall 2826 (Holotype: BM).

Saxifraga nigroglandulifera N.P.Balakr. [6, 10, 11]

Saxifraga nutans Hook.f. & Thomson (2, 4); *Hirculus nutans* (Hook.f. & Thomson) Losinsk.

Perennial herbs. C, E Nepal: 4000-4800(-5400) m. Nepal, E Himalaya and Tibetan Plateau.

Saxifraga palpebrata Hook.f. & Thomson [2, 4, 6, 10]

Perennial herbs. W, E Nepal: 4100-4200 m. Nepal, W Himalaya and E Himalaya.

Saxifraga parnassifolia D.Don [2, 4, 6, 10]

Saxifraga diversifolia var. *parnassifolia* (D.Don) Ser.

Perennial herbs. W, C, E Nepal: 1400-4900 m. Nepal, W Himalaya, E Himalaya and Tibetan Plateau.

Saxifraga parva Hemsl. [2, 4, 10]

Perennial herbs. C Nepal: 4700- 5000 m. Nepal, E Himalaya and Tibetan Plateau.

Saxifraga perpusilla Hook.f. & Thomson [2, 4, 6, 10]

Perennial herbs. E Nepal: 4700-5300(-5600) m. Nepal, E Himalaya and Tibetan Plateau.

Saxifraga pilifera Hook.f. & Thomson [2, 4, 6, 10]

Perennial herbs. C, E Nepal: 4500-4800 m. Nepal and E Himalaya.

Saxifraga poluniniana Harry Sm., *Bull. Brit. Mus. (Nat. Hist.) Bot.* **2**: 114. 1958. [2, 4, 10, 12, 28, 29]

Perennial herbs. W, C Nepal: 3200-3500 m. **Endemic to Nepal.**

West: *Jumla*, Kabhre, Padmara – Bumra, 11500 ft. (3500 m), O. Polunin, WR Sykes & LHJ Williams 4070 (Holotype: BM).

Saxifraga pulvinaria Harry Sm. [2, 4, 6, 10]

Perennial herbs. W, C, E Nepal: 3800-5900 m. Nepal and W Himalaya.

Saxifraga punctulata Engl. [2, 4, 10]

Perennial herbs. E Nepal: 4500-4900(-5800) m. Nepal, E Himalaya and Tibetan Plateau.

Saxifraga quadrifaria Engl. & Irmsch. [10]

Perennial herbs. C Nepal: *ca.* 4100 m. Nepal, E Himalaya and Tibetan Plateau.

Saxifraga rhodopetala Harry Sm., *Bull. Brit. Mus. (Nat. Hist.) Bot.* **2**: 124. 1958. [2, 4, 10, 12, 28, 29]

Perennial caespitose herbs. W Nepal: 3900-4600 m. **Endemic to Nepal**.

Central: *Dolpa*, Phoksundo Taal, 13000 ft. (3960 m), O. Polunin, WR Sykes & LHJ Williams 2196 (Holotype: BM).

Saxifraga rolwalingensis H.Ohba, *J. Jap. Bot.* **59**: 12, 360. 1984. [4, 10, 12, 28, 29]

Perennial herbs. C, E Nepal: 4000-4600 m. **Endemic to Nepal**.

Central: *Dolakha*, Na – Rinaldung, Rolwaling Khola, 4200 m, H. Ohba, M. Wakabayashi & S. Akiyama 8331589 (Holotype: TI).

Saxifraga roylei Harry Sm., *Bull. Brit. Mus. (Nat. Hist.) Bot.* **2**: 95. 1958. [2, 4, 6, 10, 12, 28, 29]

Perennial herbs. W, C Nepal: 3200-3800 m. **Endemic to Nepal**.

Central: *Myagdi*, S. Gurjakhani, 12500 ft. (3810 m), JDA Stainton, WR Sykes & LHJ Williams 3072A (Holotype: BM).

Saxifraga saginoides Hook.f. & Thomson [2, 4, 6, 10]

Hirculus saginoides (Hook.f. & Thomson) Losinsk.

Perennial herbs. C, E Nepal: 4100-5300 m. Nepal, W Himalaya, E Himalaya and Tibetan Plateau.

Saxifraga serrula Harry Sm. [2, 4, 6, 10]

Perennial herbs. E Nepal: 2800-3500 m. Nepal, E Himalaya and E Asia.

Saxifraga sibirica L. [2, 4, 6, 10]

Saxifraga odontophylla Wall. ex Sternb.

Perennial herbs. W, C, E Nepal: 3200-5000(-5800) m. Nepal, W Himalaya, E Himalaya, Tibetan Plateau, E Asia, N Asia and Europe.

Saxifraga sikkimensis Engl. [2, 4, 6, 10]

Perennial herbs. E Nepal: 3200-4600(-5200) m. Nepal, E Himalaya and Assam-Burma.

Saxifraga sinomontana J.T.Pan & Gornall [10, 11]

Hirculus montanus (Harry Sm.) Losinsk.; *Saxifraga montana* Harry Sm. (2, 4); *Saxifraga montana* forma *rubra* Harry Sm. (2, 4)

Perennial herbs. C Nepal: 4100-5000 m. Nepal, W Himalaya, E Himalaya and Tibetan Plateau.

Saxifraga sphaeradena Harry Sm. [2, 4, 6, 10]

Perennial herbs. C, E Nepal: 3300-4800 m. Nepal, E Himalaya and Tibetan Plateau.

Saxifraga staintonii Harry Sm., *Bull. Brit. Mus. (Nat. Hist.) Bot.* **2**: 118. 1958. [2, 4, 10, 12, 28, 29]

Perennial caespitose herbs. C Nepal: *ca.* 4800 m. **Endemic to Nepal**.

Central: *Mustang*, Samargaon, N. Tukucha, 16000 ft. (4880 m), JDA Stainton, WR Sykes & LHJ Williams 7276 (Holotype: BM).

Saxifraga stella-aurea Hook.f. & Thomson [2, 4, 10]

Saxifraga stella-aurea var. *polyadena* Harry Sm. (2, 4); *Saxifraga jacquemontiana* var. *stella-aurea* (Hook.f. & Thomson) C.B.Clarke

Perennial herbs. C, E Nepal: 4000-4700(-5300) m. Nepal, E Himalaya and Tibetan Plateau.

Saxifraga stenophylla Royle [2, 4, 10]

Saxifraga flagellaris subsp. *hoffmeisteri* (Klotzsch) Hultén; *Saxifraga hoffmeisteri* Klotzsch; *Saxifraga stenophylla* subsp. *hoffmeisteri* (Klotzsch) H.Hara (2, 4)

Perennial herbs. W, C, E Nepal: 4300-5400(-5700) m. Nepal, W Himalaya, Tibetan Plateau and SW Asia.

Saxifraga stolitzkae Duthie ex Engl. & Irmsch. [2, 4, 10]

Perennial herbs. W, C Nepal: 2000-4000(-4300) m. Nepal and W Himalaya.

Saxifraga strigosa Wall. ex Ser. [2, 4, 6, 10]

Hirculus strigosus (Wall. ex Ser.) Losinsk.

Perennial herbs. W, C, E Nepal: 2100-3700(-4200) m. Nepal, W Himalaya, E Himalaya, Tibetan Plateau, Assam-Burma and E Asia.

Saxifraga subsessiliflora Engl. & Irmsch. [10]

Perennial herbs. E Nepal: 4400-4600 m. Nepal and E Himalaya.

Saxifraga substrigosa J.T.Pan [10]

Perennial herbs. C Nepal: 2700-4200 m. Nepal and E Asia.

Saxifraga tangutica Engl. [2, 4, 6, 10]

Hirculus tangutica (Engl.) Losinsk.

Perennial herbs. W, C Nepal: 4300-5400 m. Nepal, W Himalaya, E Himalaya and Tibetan Plateau.

Saxifraga tentaculata C.E.C.Fisch. [2, 4, 6, 10]

Perennial herbs. C Nepal: 3500-4800 m. Nepal, E Himalaya and Tibetan Plateau.

Saxifraga umbellulata Hook.f. & Thomson [2, 4, 6, 10]

सुन-चु-तेग (Sun-chu-teg: Tibetan)

Perennial herbs. E Nepal: *ca.* 3700 m. Nepal, E Himalaya and Tibetan Plateau.

Use: Medicine (whole plant).

Saxifraga viscidula Hook.f. & Thomson [6, 10]

Perennial herbs. E Nepal: 4000-4900 m. Nepal, W Himalaya and E Himalaya.

Saxifraga wallichiana Sternb. [6, 10]

Saxifraga brachypoda var. *fimbriata* Engl. & Irmsch. (2, 4); *Saxifraga fimbriata* Wall. ex Ser., non D.Don

Perennial herbs. W, C, E Nepal: 3300-4600(-5000) m. Nepal, W Himalaya, E Himalaya, Tibetan Plateau, Assam-Burma and E Asia.

Saxifraga williamsii Harry Sm., *Bull. Brit. Mus. (Nat. Hist.) Bot.* **2**: 100. 1958. [2, 4, 6, 10, 12, 28, 29]

Perennial caespitose herbs. W, C, E Nepal: 4000-5100 m. **Endemic to Nepal**.

Central: *Mustang*, Muktinath, 13500 ft. (4110 m), JDA Stainton, WR Sykes & LHJ Williams 2049 (Holotype: BM; Isotype: E).

Saxifraga zimmermannii Baehni, *Candollea* **16**: 226. 1958. [2, 4, 10, 12, 28, 29]

Perennial caespitose herbs. C Nepal: (3600-)4000-4500 m. **Endemic to Nepal**.

Central: *Dolakha*, Beding, 4130 m, A. Zimmermann 1436 (Holotype: G; Isotype: BM).

Tiarella L.

Tiarella polyphylla D.Don [2, 4, 6, 10]

सिस्ने झार (Sisne jhaar); Foam flower

Perennial herbs. C, E Nepal: 2000-3300(-4000) m. Nepal, E Himalaya, Tibetan Plateau and E Asia.

Use: Medicine (root, leaves).

Doubtful/Uncertain Taxa

Bergenia stracheyi (Hook.f. & Thomson) Engl. [102]

Saxifraga duthiei Gand. [98]

Saxifraga gouldii var. **eglandulosa** Harry Sm. [102]

Saxifraga ramulosa Wall. ex Ser. [2, 4]

CRASSULACEAE

Bryophyllum Salisb.

Bryophyllum pinnatum (Lam.) Oken [2, 4, 10]

Cotyledon pinnata Lam.; *Kalanchoe pinnata* (Lam.) Pers. (6)

अजम्मरी (Ajammaree); Cathedral bells

Perennial herbs. C Nepal: 800-1400 m. Asia, Africa, N America, S America and Australasia.

Use(s): Medicine (leaves), Poisonous to cattle.

Kalanchoe Adans.

Kalanchoe spathulata DC.

var. **staintonii** H.Ohba [6, 10]

अजम्मरी झार, हाथी काने (Ajamari jhaar, Haatthi kaane)

Biennial herbs. W, C, E Nepal: 900-1700 m. Nepal, W Himalaya, E Himalaya and Assam-Burma.

Use(s): Medicine (stem, leaves), Poisonous to cattle.

Phedimus Raf.

Phedimus odontophyllus (Fröd.) 't Hart [10]

Sedum odontophyllum Fröd. (2, 4); *Aizopsis odontophyllum* (Fröd.) Grulich.

Perennial herbs. E Nepal: *ca.* 3900 m. Nepal and E Asia.

Rhodiola L.

Rhodiola amabilis (H.Ohba) H.Ohba [2, 4, 6, 10]

Sedum amabile H.Ohba

Perennial herbs. C, E Nepal: 2300-5200 m. Nepal and E Himalaya.

Rhodiola bupleuroides (Wall. ex Hook.f. & Thomson) S.H.Fu [2, 4, 6, 10]

Sedum bupleuroides Wall. ex Hook.f. & Thomson

var. **bupleuroides**

सोहलो-मार्पो (Sohlo-marpo: Tibetan)

Perennial herbs. W, C, E Nepal: 3200-4500 m. Nepal, W Himalaya, E Himalaya, Tibetan Plateau, Assam-Burma and E Asia.

Use: Food.

var. **parva** (Fröd.) H.Ohba [4, 6, 10]

Sedum bupleuroides var. *parva* Fröd.

Perennial herbs. C, E Nepal: 3300-4500 m. Nepal and Tibetan Plateau.

Rhodiola calliantha (H.Ohba) H.Ohba [2, 4, 10]

Sedum callianthum H.Ohba

Perennial herbs. C, E Nepal: 3700-4300 m. Nepal and Tibetan Plateau.

Rhodiola chrysanthemifolia (H.Lev.) S.H.Fu [6, 10]

Sedum chrysanthemifolium H.Lev.; *Rhodiola ovatisepala* (Raym.-Hamet) S.H.Fu (2,4); *Sedum ovatisepalum* (Raym.-Hamet) H.Ohba

subsp. **chrysanthemifolia**

Perennial herbs. W, C, E Nepal: 2700-4600 m. Nepal, E Himalaya, Tibetan Plateau, Assam-Burma and E Asia.

subsp. **sacra** (Prain ex Raym.-Hamet) H.Ohba [4, 6, 10]

Sedum linearifolium var. *sacrum* Prain ex Raym.-Hamet; *Rhodiola sacra* (Prain ex Raym.-Hamet) S.H.Fu (2); *Sedum sacrum* (Prain ex Raym.-Hamet) H.Ohba

Perennial herbs. W, C Nepal: 2200-3300 m. Nepal, E Himalaya and Tibetan Plateau.

Rhodiola coccinea (Royle) Borris. [6, 10]

Sedum coccineum Royle; *Rhodiola asiatica* D.Don; *Sedum asiaticum* (D.Don) DC.

Perennial herbs. W, C, E Nepal: 3200-5800 m. Nepal, W Himalaya, E Himalaya, Tibetan Plateau and SW Asia.

Rhodiola crenulata (Hook.f. & Thomson) H.Ohba [2, 4, 6, 10]

Sedum crenulatum Hook.f. & Thomson; *Rhodiola rotundata* (Hemsl.) S.H.Fu; *Sedum bupleuroides* var. *rotundatum* (Hemsl.) Fröd.; *Sedum rotundatum* Hemsl.

सोहलो-मार्पो (Sohlo-marpo: Tibetan)

Perennial herbs. W, C, E Nepal: 3700-5800 m. Nepal, E Himalaya, Tibetan Plateau and E Asia.

Use: Medicine (root stock).

Rhodiola cretinii (Raym.-Hamet) H.Ohba [2, 4, 6, 10]

Sedum cretinii Raym.-Hamet; *Rhodiola crassipes* var. *cretinii* (Raym.-Hamet) H.Jacobsen; *Sedum crassipes* var. *cretinii* (Raym.-Hamet) Fröd.; *Sedum wallichianum* var. *cretinii* (Raym.-Hamet) H.Hara

Perennial herbs. W, C, E Nepal: 3700-4800 m. Nepal, E Himalaya and Tibetan Plateau.

Rhodiola discolor (Franch.) S.H.Fu [2, 4, 6, 10]

Sedum discolor Franch.; *Sedum bupleuroides* var. *discolor* (Franch.) Fröd.

Perennial herbs. C, E Nepal: 3600-4600 m. Nepal, E Himalaya and E Asia.

Rhodiola fastigiata (Hook.f. & Thomson) S.H.Fu [2, 4, 6, 10]

Sedum fastigiatum Hook.f. & Thomson; *Chamaerhodiola fastigiata* (Hook.f. & Thomson) Nakai; *Sedum quadrifidum* var. *fastigiatum* (Hook.f. & Thomson) Fröd.

Perennial herbs. W, C, E Nepal: 3600-5700 m. Nepal, W Himalaya, E Himalaya, Tibetan Plateau and E Asia.

Rhodiola heterodonta (Hook.f. & Thomson) Boriss. [2, 4, 6, 10]

Sedum heterodontum Hook.f. & Thomson; *Sedum roseum* var. *heterodontum* (Hook.f. & Thomson) Fröd.

Perennial herbs. W, C Nepal: 3900-5100 m. Nepal, W Himalaya, Tibetan Plateau and SW Asia.

Rhodiola himalensis (D.Don) S.H.Fu [2, 4, 6, 10]

Sedum himalense D.Don; *Chamaerhodiola himalensis* (D.Don) Nakai; *Rhodiola quadrifida* (Pall.) Fisch. & C.A.Mey. (2, 4); *Sedum quadrifidum* var. *himalense* (D.Don) Fröd.

subsp. **himalensis**

सोलो मुक्पो (Solo mukpo: Amchi); Himalayan Rhodiola

Perennial herbs. W, C, E Nepal: 3300-4800 m. Nepal, E Himalaya, Tibetan Plateau and E Asia.

Use: Medicine (whole plant).

subsp. **bouvieri** (Raym.-Hamet) Jacobsen [4, 10]

Sedum bouvieri Raym.-Hamet; *Rhodiola bouvieri* (Raym.-Hamet) H.Ohba (2, 4); *Sedum quadrifidum* var. *bouvieri* (Raym.-Hamet) Fröd.

Perennial herbs. W Nepal: *ca.* 3900 m. Nepal and W Himalaya.

Rhodiola hookeri S.H.Fu [10]

Rhodiola bhutanica (Praeger) S.H.Fu; *Sedum bhutanense* Praeger; *Sedum bhutanicum* Praeger; *Sedum hookeri* (S.H.Fu) N.P.Balakr.

Perennial herbs. C, E Nepal: 2900-4200 m. Nepal, E Himalaya and Tibetan Plateau.

Rhodiola humilis (Hook.f. & Thomson) S.H.Fu [2, 4, 7, 10]

Sedum humile Hook.f. & Thomson; *Chamaerhodiola humilis* (Hook.f. & Thomson) Nakai

Perennial herbs. C, E Nepal: 2100-5000 m. Nepal, E Himalaya and Tibetan Plateau.

Rhodiola imbricata Edgew. [2, 4, 6, 10]

Sedum imbricatum (Edgew.) Walp.

Perennial herbs. W, C Nepal: 4100-5700 m. Nepal, W Himalaya and Tibetan Plateau.

Rhodiola lobulata (N.B.Singh & U.C.Bhattach.) H.Ohba [4, 10]

Rhodiola imbricata var. *lobulata* N.B.Singh & U.C.Bhattach.

Perennial herbs. W Nepal: *ca.* 4100 m. Nepal and W Himalaya.

Rhodiola nepalica (H.Ohba) H.Ohba [2, 4, 10, 12, 26, 29]

Sedum nepalicum H.Ohba, *J. Jap. Bot.* **49**: 322. 1974. (29)

Perennial herbs. W, C Nepal: 3500-4600 m. **Endemic to Nepal**.

Central: *Mustang*, Tegar, 15000 ft. (4570 m), JDA Stainton, WR Sykes & LHJ Williams 2222 (Holotype: BM; Isotypes: E, TI).

Rhodiola prainii (Raym.-Hamet) H.Ohba [2, 4, 6, 10]

Sedum prainii Raym.-Hamet; *Sedum stewartii* Craib ex Raym.-Hamet

Perennial herbs. W, C, E Nepal: 3300-5300 m. Nepal, E Himalaya and Tibetan Plateau.

Rhodiola purpureoviridis (Praeger) S.H.Fu

Sedum purpureoviride Praeger

subsp. **phariensis** (H.Ohba) H.Ohba [4, 6, 10]

Sedum phariense H.Ohba; *Rhodiola phariensis* (H.Ohba) S.H.Fu

Perennial herbs. C Nepal: 3100-4800 m. Nepal and Tibetan Plateau.

Rhodiola sinuata (Royle ex Edgew.) S.H.Fu [2, 4, 6, 10]

Sedum sinuatum Royle ex Edgew.; *Sedum linearifolium* Royle; *Sedum pauciflorum* Edgew.; *Sedum trifidum* Wall. ex C.B.Clarke

Perennial herbs. W, C, E Nepal: 1200-4400 m. Nepal, W Himalaya and Tibetan Plateau.

Rhodiola smithii (Raym.-Hamet) S.H.Fu [2, 4, 6, 10]

Sedum smithi Raym.-Hamet

Perennial herbs. C Nepal: 3500-4600 m. Nepal, E Himalaya and Tibetan Plateau.

Rhodiola tibetica (Hook.f. & Thomson) S.H.Fu [6, 10, 11]

Sedum tibeticum Hook.f. & Thomson; *Sedum quadrifidum* var. *tibeticum* (Hook.f. & Thomson) Fröd.

Perennial herbs. W, C Nepal: 4000-4500 m. Nepal, W Himalaya and Tibetan Plateau.

Rhodiola wallichiana (Hook.) S.H.Fu [2, 4, 6, 10]

Sedum wallichianum Hook.; *Rhodiola crassipes* (Wall. ex Hook.f. & Thomson) Boriss.; *Sedum crassipes* Wall. ex Hook.f. & Thomson

अम्बासीं (Ambaasin)

Perennial herbs. W, C, E Nepal: 3000-5500 m. Nepal, W Himalaya, East Himalaya and Tibetan Plateau.

Use: Medicine (whole plant).

Rosularia (DC.) Stapf

Rosularia adenotricha (Wall. ex Edgew.) C.-A.Jansson [2, 4, 10]

Sedum adenotrichum Wall. ex Edgew.

Perennial herbs. W Nepal: 1300-2300 m. Nepal, W Himalaya and SW Asia.

Rosularia marnieri (Raym.-Hamet ex H.Ohba) H.Ohba [2, 4, 6, 10, 12, 26]

Sedum marnieri Raym.-Hamet ex H.Ohba (29); *Rosularia alpestris* subsp. *marnieri* (Raym.-Hamet ex H.Ohba) Eggli

Perennial herbs. W, C Nepal: 3000-4400 m. **Endemic to Nepal**.

Central: *Manang*, Marsyangdi, 12000-13000 ft. (3650-3950 m), DG Lowndes 1506 (Holotype: BM; Isotype: TI).

Rosularia rosulata (Edgew.) H.Ohba [2, 4, 10]

Sedum rosulatum Edgew.; *Sedum pyriforme* Royle ex Hook.f. & Thomson, nom. inval.

Perennial herbs. W Nepal: 2000-3500 m. Nepal, W Himalaya and SW Asia.

Sedum L.

Sedum chauveaudii Raym.-Hamet [2, 4, 10]

Sedum triphyllum Praeger

Perennial herbs. E Nepal: *ca.* 2900 m. Nepal and E Asia.

Sedum filipes Hemsl. [2, 4, 6, 10]

Sedum filipes var. *major* Hemsl.; *Sedum filipes* var. *pseudo-staphii* (Praeger) Fröd; *Sedum pseudo-staphii* Praeger.

Annual or biennial herbs. W, C Nepal: 2200-2700 m. Nepal, E Himalaya, Assam-Burma and E Asia.

Sedum forrestii Raym.-Hamet [6, 10]

Sedum holei Raym.-Hamet (2, 4)

Annual herbs. W, C Nepal: 2800-4600 m. Nepal, Tibetan Plateau and E Asia.

Sedum gagei Raym.-Hamet [2, 4, 6, 10]

Sedum trullipetalum var. *gagei* (Raym.-Hamet) H.Ohba

Perennial herbs. W, C, E Nepal: 3900-4600 m. Nepal, W Himalaya, E Himalaya, Tibetan Plateau and E Asia.

Sedum henrici-roberti Raym.-Hamet [2, 4, 6, 10]

Annual herbs. C, E Nepal: 4200-4700 m. Nepal and Tibetan Plateau.

Sedum magae Raym.-Hamet [2, 4, 10]

Sedum tilaeoides Duthie ex C.E.C.Fisch.

Annual herbs. W, C Nepal: 3000-4600 m. Nepal and W Himalaya.

Sedum multicaule Wall. ex Lindl. [2, 4, 6, 10]

Perennial herbs. W, C, E Nepal: 1300-3200 m. Nepal, W Himalaya, E Himalaya, Tibetan Plateau, Assam-Burma and E Asia.

Use: Medicine (whole plant).

Sedum obtusipetalum Franch.

subsp. **dandyanum** Raym.-Hamet ex H.Ohba, *J. Jap. Bot.* **61**: 274.1986. [4, 10]

Sedum costantinii Raym.-Hamet

Annual herbs. C Nepal: *ca.* 3300 m. **Endemic to Nepal**.

Central: *Baglung*, Phagune Dhuri, *ca.* 3300 m, JDA Stainton, WR Sykes & LHJ Williams 9011 (Holotype: BM; Isotype: TI).

Sedum oreades (Decne.) Raym.-Hamet [2, 4, 6, 10]

Umbilicus oreades Decne.; *Cotyledon oreades* (Decne.) C.B.Clarke

Perennial herbs. W, C, E Nepal: 3200-5200 m. Nepal, W Himalaya, E Himalaya, Tibetan Plateau, Assam-Burma and E Asia.

Sedum perpusillum Hook.f. & Thomson [2, 4, 6, 10]

Annual herbs. E Nepal: 3800-4500 m. Nepal, E Himalaya and Tibetan Plateau.

Sedum przewalskii Maxim. [2, 4, 6, 10]

Annual herbs. W, E Nepal: 4200-4500 m. Nepal, E Himalaya, Tibetan Plateau and E Asia.

Sedum pseudomulticaule H.Ohba, *J. Jap. Bot.* **53**(11): 328. 1978. [2, 4, 7, 10, 12, 26, 29]

Perennial herbs. C Nepal: *ca.* 2100 m. **Endemic to Nepal**.

Central: *Gorkha*, *ca.* 2100 m, T. Namba 1006040-2 (Holotype: KYO).

Sedum roborowskii Maxim. [2, 4, 10]

Annual herbs. W, C Nepal: 2000-4600 m. Nepal, Tibetan Plateau and E Asia.

Sedum triactina A.Berger [2, 4, 6, 10]

Sedum verticillatum (Hook.f. & Thomson) Raym.-Hamet; *Triactina verticillata* Hook.f. & Thomson

Perennial herbs. W, C, E Nepal: 2700-4300 m. Nepal, E Himalaya, Tibetan Plateau, Assam-Burma and E Asia.

Sedum trullipetalum Hook.f. & Thomson [2, 4, 6, 10]

Perennial herbs. W, C, E Nepal: 2400-5500 m. Nepal, W Himalaya, E Himalaya, Tibetan Plateau, Assam-Burma and E Asia.

Sinocrassula A.Berger

Sinocrassula indica (Decne.) A.Berger [2, 4, 6, 10]

Crassula indica Decne.; *Sedum indicum* (Decne.) Raym.-Hamet; *Sedum indicum* var. *obtusifolium* Fröd.

Biennial herbs. W, C Nepal: 2300-4100 m. Nepal, W Himalaya, E Himalaya, Tibetan Plateau and E Asia.

Tillaea L.

Tillaea schimperi (Fisch. & C.A.Mey.) M.G.Gilbert, H.Ohba & K.T.Fu [6, 10, 11]

Crassula schimperi Fisch. & C.A.Mey.; *Crassula pentandra* (Royle ex Edgew.) Schonland; *Tillaea pentandra* Royle ex Edgew. (2, 4)

Perennial herbs. W, C Nepal: 1900-2500 m. Nepal, W Himalaya, E Himalaya, Tibetan Plateau, S Asia, SW Asia and Africa.

Exotic/Cultivated Taxa

Kalanchoe delagoensis Eckl. & Zeyh. (*Kalanchoe verticillata* Elliot); (Devil's backbone). Native to Madagascar. [14]

Sedum sarmentosum Bunge [14]

Sedum spectabile Boreau [14]

Doubtful/Uncertain Taxa

Rhodiola pamiroalaica A.Boriss. [4]

Sedum majus (Hemsl.) Migo [102]

Sedum obtusipetalum Franch. [2, 4]

HALORAGACEAE

Myriophyllum L.

Myriophyllum spicatum L. [2, 4, 6]

Perennial submerged aquatic herbs. W Nepal: *ca.* 3000 m. Nepal, Asia, Europe, Africa, N America and S America.

Use(s): Medicine (whole plant), Dye (leaves).

Exotic/Cultivated Taxa

Myriophyllum aquaticum (Vell.) Verdc., (Parrot's feather). Invasive species in Nepal. Native to S America (Amazon River).

ROSIDS

VITALES

VITACEAE (incl. LEEACEAE)

Ampelocissus Planch.

Ampelocissus barbata (Wall.) Planch. [6, 125]

Vitis barbata Wall.

Woody climbers. E Nepal: 200-300 m. Nepal, E Himalaya, Assam-Burma, S Asia and SE Asia.

Ampelocissus divaricata (Lawson) Planch. [2, 4, 6]

Vitis divaricata Lawson

पुरेनी, कोराखोली (Purenii, Korakholi)

Woody climbers. W, C, E Nepal: 300-1700 m. Nepal, W Himalaya, E Himalaya and SE Asia.

Use(s): Medicine (root), Food (fruit).

Ampelocissus latifolia (Roxb.) Planch. [2, 4, 6]

Vitis latifolia Roxb.

पुरेनी (Purenii); Wild grape

Woody climbers. W, C, E Nepal: 150-1600 m. Nepal and W Himalaya.

Use: Medicine (whole plant).

Ampelocissus nervosa (P.Lawson) Planch. [2, 4]

Vitis nervosa P.Lawson

Woody climbers. C, E Nepal: 1000-1800 m. Nepal and E Himalaya.

Ampelocissus rugosa (Wall.) Planch. [2, 4, 6]

Vitis rugosa Wall.

अईरि लहरा (Aairi laharaa)

Woody climbers. W, C, E Nepal: 1000-2400 m. Nepal, W Himalaya, E Himalaya and Assam-Burma.

Use: Medicine (stem).

Ampelocissus sikkimensis (M.A.Lawson) Planch. [2, 4, 6]

Vitis sikkimensis M.A.Lawson

पुरेनी (Purenii)

Woody climbers. W, C, E Nepal: 200-2000 m. Nepal, E Himalaya and E Asia.

Use: Food (fruit).

Ampelopsis Michx.

Ampelopsis glandulosa (Wall.) Momiy. [2, 4, 6]

Vitis glandulosa Wall.

Woody climbers. C Nepal: 1300-1900 m. Nepal, Assam-Burma, S Asia and E Asia.

Cayratia Juss.

Cayratia japonica (Thunb.) Gagnep. [2, 6]

Vitis japonica Thunb.

var. **japonica**

Perennial herbaceous climbers. E Nepal: 1500-1700 m. Nepal, Asia and Australasia.

var. **mollis** (Wall.) Momiy. [2, 4]

Vitis mollis Wall.

Perennial herbaceous climbers. E Nepal: 1600-1700 m. Nepal, E Himalaya, S Asia and E Asia.

Cayratia pedata (Lam.) Gagnep.

Cissus pedata Lam.

var. **glabra** Gamble [6]

Woody climbers. C Nepal: *ca.* 1500 m. Nepal and S Asia (S India).

Note: **Cayratia pedata** (Lam.) Gagnep. [IUCN Category: Vulnerable A2cd, ver. 3.1].

Cayratia tenuifolia (Wight & Arn.) Gagnep. [2, 4]

Vitis tenuifolia Wight & Arn.

Woody climbers. E Nepal: *ca.* 600 m. Nepal, S Asia and SE Asia.

Cayratia trifolia (L.) Domin [2, 4]

Vitis trifolia L.

करौंज, करौंजा (Karaunj, Karaunja)

Woody climbers. W, C, E Nepal: 100-300 m. Nepal, Assam-Burma, S Asia, E Asia, SE Asia and Australasia.

Use: Medicine (whole plant).

Cissus L.

Cissus adnata Roxb. [2, 4, 6]

Cissus simplex Blanco

जरिल लहरा, चारपाते (Jaril laharaa, Chaarpaate)

Climbing shrubs. C, E Nepal: 300-1100 m. Nepal, S Asia, SE Asia and Australasia.

Use(s): Food (fruit), Medicine (root, leaves, fruit).

Cissus assamica (M.A.Lawson) Craib [2, 4, 6]

Vitis assamica M.A.Lawson

Climbing shrubs. C Nepal: 300-800 m. Nepal, E Himalaya, Tibetan Plateau, Assam-Burma, S Asia, E Asia and SE Asia.

Cissus javana DC. [2, 4, 6]

Cissus discolor Blume

जोगी लहरा (Jogi laharaa)

Herbaceous climbers. C, E Nepal: 300-1200 m. Nepal, E Himalaya, Assam-Burma, S Asia, E Asia and SE Asia.

Use(s): Medicine (whole plant), Fodder.

Cissus repanda Vahl [6]

Vitis repanda (Vahl) Wight & Arn. (4, 98)

Woody climbers. C, E Nepal: 200-300 m. Nepal, E Himalaya, Tibetan Plateau, Assam-Burma, S Asia, E Asia and SE Asia.

Use: Food (fruit).

Cissus repens Lam. [2, 4, 6]

पुरेनी, जोगे लहरो, चरचर लहरा (Purenii, Joge laharo, Charchare laharaa)

Herbaceous climbers. C, E Nepal: 300-1300 m. Nepal, E Himalaya, E Asia, SE Asia and Australasia.

Use(s): Medicine (leaves), Ritual, Ornamental, Construction materials.

Leea D.Royen ex L.

Leea aequata L. [2, 4, 6]

Leea hirta Roxb. ex Hornem.

काक नसिका, काग चुच्चे (Kaak nasika, Kaag chuche)

Shrubs or small trees. E Nepal: *ca.* 200 m. Nepal, E Himalaya, Assam-Burma, S Asia, E Asia and SE Asia.

Use: Medicine (tuber, stem).

Leea alata Edgew. [2, 4, 6]

Shrubs. C Nepal: 100-1200 m. Nepal, E Himalaya, Assam-Burma, S Asia and Australasia.

Leea asiatica (L.) Ridsdale [2, 6]

Phytolacca asiatica L.; *Leea aspera* Edgew.; *Leea crispa* L. (2)

गलेनी (Galeni)

Shrubs or small trees. W, C, E Nepal: 400-900 m. Nepal, W Himalaya, E Himalaya, Assam-Burma, S Asia, E Asia and SE Asia.

Use(s): Food (fruit), Medicine (leaves).

Leea compactiflora Kurz [2, 4]

Leea trifoliata M.A.Lawson

Straggling shrubs. E Nepal: *ca.* 300 m. Nepal, E Himalaya, Tibetan Plateau, Assam-Burma, S Asia and SE Asia.

Use: Medicine (whole plant).

Leea guineensis G.Don [2, 4]

Leea acuminata Wall. ex C.B.Clarke

Herbaceous climbers. C, E Nepal: 200-600 m. Nepal, E Himalaya, Assam-Burma, SE Asia and Africa.

Leea indica (Burm.f.) Merr. [2, 4, 8]

Staphylea indica Burm.f.; *Leea umbraculifera* C.B.Clarke

कुकुर जिब्रे (Kukur jibre)

Shrubs or small trees. W, C, E Nepal: 100-600 m. Nepal, E Himalaya, Assam-Burma, S Asia, E Asia, SE Asia and Australasia.

Use: Medicine (root, leaves).

Leea macrophylla Roxb. ex Hornem. [2, 4]

Leea robusta Roxb.

गलेनी, बुलेभेत्रा (Galeni, Bulebhetra)

Shrubs or perennial herbs. C Nepal: 1500-1700 m. Nepal, E Himalaya, Assam-Burma, S Asia, E Asia and SE Asia.

Use(s): Medicine (root, leaves), Food (fruit).

Parthenocissus Planch.

Parthenocissus semicordata (Wall.) Planch. [2, 4, 6]

Vitis semicordata Wall.; *Parthenocissus himalayana* (Royle) Planch. (2, 4)

var. **semicordata**

चरचरे (Charchare)

Deciduous climbing shrubs. W, C, E Nepal: 1800-2800 m. Nepal, E Himalaya, Tibetan Plateau, Assam-Burma, S Asia, E Asia and SE Asia.

Use(s): Medicine (root), Fodder.

var. **roylei** (King ex R.Parker) Nazim. & Qaiser

Vitis semicordata var. *roylei* King ex R.Parker; *Parthenocissus semicordata* var. *roylei* (King.) Richard & Saxena (6)

Deciduous climbing shrubs. W, C Nepal: 1800-2200 m. Nepal and W Himalaya.

Tetrastigma (Miq.) Planch.

Tetrastigma aplinianum (Collett & Hemsl.) Momiy. [2, 4]

Vitis apliniana Collett & Hemsl.

Woody climbers. C, E Nepal: 1200-1800 m. Nepal, E Himalaya and Assam-Burma.

Tetrastigma bracteolatum (Wall.) Planch. [2, 4, 6]

Vitis bracteolata Wall.

Woody climbers. C, E Nepal: 100-1000 m. Nepal, E Himalaya, Assam-Burma and S Asia.

Tetrastigma dubium (Lawson) Planch. [2, 4, 6]

Vitis dubia Lawson

चर च्यारे लहरो (Char chyaare laharo)

Woody climbers. C, E Nepal: 100-1800 m. Nepal, E Himalaya, S Asia and SE Asia.

Use(s): Medicine (stem), Ritual, Construction materials, Utensils and handicrafts.

Tetrastigma hookeri (Lawson) Planch. [2, 4, 6]

Vitis hookeri Lawson

Woody climbers. W, C, E Nepal: 200-1500 m. Nepal, E Himalaya, Assam-Burma, S Asia and SE Asia.

Tetrastigma leucostaphylum (Dennst.) Alston [6]

Cissus leucostaphyla Dennst.

Indian chesnut vine

Woody scandent shrubs. E Nepal: *ca.* 100 m. Nepal, E Himalaya, Assam-Burma, S Asia and SE Asia.

Tetrastigma obtectum (Wall. ex M.A.Lawson) Planch. ex Franch. [2, 4, 6]

Vitis obtecta Wall. ex M.A.Lawson

var. **obtectum**

Woody climbers. W, C Nepal: 300-2000 m. Nepal, E Himalaya, E Asia and SE Asia.

var. **glabrum** (H.Lév.) Gagnep. [6]

Vitis potentilla var. *glabra* H.Lév.

Woody climbers. W, C Nepal: 1400-1800 m. Nepal and E Asia.

Tetrastigma rumicispermum (M.A.Lawson) Planch. [2, 4, 6]

Vitis rumicisperma M.A.Lawson

Woody climbers. C, E Nepal: 1200-1900 m. Nepal, E Himalaya, Tibetan Plateau, Assam-Burma and SE Asia.

Tetrastigma serrulatum (Roxb.) Planch. [2, 4, 6]

Cissus serrulata Roxb.; *Tetrastigma napaulense* (DC.) C.L.Li (6)

पानी लहरा (Paani laharaa)

Woody climbers. W, C, E Nepal: 500-2400 m. Nepal, W Himalaya, E Himalaya, Assam-Burma, E Asia and SE Asia.

Use(s): Medicine (whole plant), Food (fruit), Fodder.

Vitis L.

Vitis flexuosa Thunb. [2, 6]

Vitis parvifolia Roxb. (2, 4); *Vitis purani* Buch.-Ham. ex D.Don

Creeping grape

Climbing shrubs. W, C Nepal: 100-2200 m. Nepal, W Himalaya, Assam-Burma, S Asia, E Asia and SE Asia.

Use: Medicine (stem sap).

Vitis heyneana Roem. & Schult. [6]

Vitis lanata Roxb. (2, 4)

पुरेनी, महुर, बन अंगूर (Purenii, Mahur, Ban angoor)

Climbing shrubs. W Nepal: *ca.* 2100 m. Nepal, E Himalaya, Tibetan Plateau, S Asia and E Asia.

Use(s): Medicine (stem), Food (fruit).

Vitis jacquemontii R.Parker [2, 4, 6]

Woody climbing shrubs. W, C, E Nepal: 900-2200 m. Nepal, W Himalaya and Asia.

Vitis pedicellata M.A.Lawson [2, 4]

Climbing shrubs. W, C Nepal: 2100-2400 m. Nepal and W Himalaya.

Yua C.L.Li

Yua thomsonii (M.A. Lawson) C.L.Li [2]

Vitis thomsonii M.A.Lawson; *Cayratia thomsonii* (Laws.) Suess. (2, 4)

Woody climbers. C Nepal: 1600-2500 m. Nepal, E Himalaya and E Asia.

Exotic/Cultivated Taxa

Vitis vinifera L. (**दाख,** Daakh; Common grape). Native to SW Asia and SE Europe. *Use*: Food (fruit).

Doubtful/Uncertain Taxa

Vitis oxyphylla Wall. [2, 4, 98]

ZYGOPHYLLALES

ZYGOPHYLLACEAE

Tribulus L.

Tribulus terrestris L. [2, 4, 6, 11]

गोखुर (Gokhur); Cat's head

Annual or biennial herbs. W, C, E Nepal: 100-300 m. Pantropical. Native to Old World in southern Europe, southern Asia, Africa and Australia.

Use: Medicine (whole plant).

FABALES

FABACEAE (LEGUMINOSAE)

Abrus L.

Abrus precatorius L. [2, 4, 6]

रतिगेडी, आँखीगेडी (Ratigedi, Aankhigedi); Crab's eye

Perennial deciduous climbers. W, C, E Nepal: 100-1100 m. Nepal, Asia, Africa and Australasia.

Use(s): Medicine (root, leaves, seeds), Food.

Abrus pulchellus Wall. ex Thwaites [2, 4, 6]

गुंज्जा (Gunjja)

Climbing shrubs. W, C, E Nepal: 200-1300 m. Nepal, E Himalaya, S Asia, E Asia and SE Asia.

Use: Medicine (root, leaves).

Acacia Mill.

Acacia auriculiformis Benth. [6]

Racosperma auriculiforme (A.Cunn. ex Benth.) Pedley

Papuan wattle

Evergeen trees. C Nepal: 300-400 m. Planted as ornamental plant. Native to Australasia.

Acacia caesia (L.) Willd. [6]

Mimosa caesia L.; *Acacia intsia* var. *caesia* (L.) Baker; *Senegalia caesia* (L.) Maslin, Seigler & Ebinger

अर्खु (Arkhu)

Climbing shrubs. C Nepal: 100-200 m. Nepal, Assam-Burma, S Asia, E Asia and SE Asia.

Use(s): Fodder, Fish posion (bark).

Acacia catechu (L.f.) Willd. [2, 4, 6]

Mimosa catechu L.f.; *Acacia wallichiana* DC.; *Mimosa catechoides* Roxb.

खयर (Khayer); Black cutch

Deciduous trees. W, C, E Nepal: 100-1400 m. Nepal, E Himalaya, Assam-Burma, S Asia, E Asia and SE Asia.

Use(s): Medicine (root, bark, heartwood, leaves), Timber, Dye, Charcoal, Masticatory.

Acacia farnesiana (L.) Willd. [2, 4, 6]

Mimosa farnesiana L.; *Vachellia farnesiana* (L.) Wight & Arn.

गन्हाउने खयर, जैत (Ganhaune khayar, Jait); Sweet Acacia

Shrubs or small trees. W, C, E Nepal: 100-1100 m. Pantropical.

Use(s): Medicine (root, bark, leaves, fruit), Fodder, Tannin, Perfume.

Acacia intsia (L.) Willd. [2, 4]

Mimosa intsia L.; *Senegalia intsia* (L.) Maslin, Seigler & Ebinger

अर्खु (Arkhu)

Climbing shrubs. C, E Nepal: 900-1100 m. Nepal, Assam-Burma and S Asia.

Use(s): Medicine (bark, flowers, seeds), Fodder, Fish poison.

Acacia lenticularis Buch.-Ham. ex Benth. [2, 4]

Senegalia lenticularis (Buch.-Ham. ex Benth.) Ragupathy, Seigler, Ebinger & Maslin

Trees. E Nepal: *ca.* 800 m. Nepal and E Himalaya.

Use: Medicine (seeds).

Acacia nilotica (L.) Willd. ex Delile

Mimosa nilotica L.; *Vachellia nilotica* (L.) P.J.H. Hurter & Mabb.

subsp. **indica** (Benth.) Brenan [2, 4, 6]

Acacia arabica var. *indica* Benth.

बबुल, बबुर (Babul, Babur); Gum Arabic-tree

Trees. C Nepal: 100-200 m. Nepal and S Asia.

Use(s): Medicine (bark, flowers, seeds), Agricultural implements, Tannin.

Acrocarpus Wight & Arn.

Acrocarpus fraxinifolius Arn. ex Wight [2, 4, 6]

Pink Cedar

Large deciduous trees. E Nepal: *ca.* 300 m. Nepal, E Himalaya, Assam-Burma, S Asia, E Asia and SE Asia.

Aeschynomene L.

Aeschynomene aspera L. [2, 4, 6]

ढोंध्या, सक्ला (Dhondhya, Saklaa)

Shrubs. W, C, E Nepal: 200-1300 m. Nepal, W Himalaya, E Himalaya, Assam-Burma, S Asia and SE Asia.

Use(s): Fish poison (stem), Decoration materials.

Aeschynomene indica L. [2, 4, 6]

ताल खुकुरी (Taal khukuri); Indian joint vetch

Undershrubs. W, C, E Nepal: 200-1500 m. Nepal, W Himalaya, E Himalaya, Assam-Burma, S Asia, E Asia, SE Asia, Europe, Africa, S America and Australasia.

Use: Medicine (whole plant).

Albizia Durazz.

Albizia chinensis (Osbeck) Merr. [2, 4, 6]

Mimosa chinensis Osbeck; *Albizia stipulata* (DC.) Boiv.; *Mimosa stipulata* Roxb.

कालो शिरिस (Kaalo siris); Chinese Albizia

Deciduous trees. W, C, E Nepal: 200-1500 m. Nepal, S Asia, E Asia and SE Asia.

Use(s): Medicine (bark, flowers, seeds, resin), Fodder.

Albizia julibrissin Durazz. [2, 4, 6]

Acacia julibrissin (Durazz.) Willd.

var. **julibrissin**

पड्के शिरिस, रातो शिरिस (Padke siris, Raato siris); Persian silk tree

Deciduous trees. W, C, E Nepal: 1300-2200 m. Nepal, E Asia, C Asia and SW Asia.

Use(s): Medicine (root bark, leaves, flowers, seeds), Fuel, Furniture.

var. **mollis** (Wall.) Benth. [2, 4]

Acacia mollis Wall.; *Albizia mollis* (Wall.) Boivin

रातो शिरिस (Raato siris)

Deciduous trees. W, C, E Nepal: 1000-3000 m. Nepal, W Himalaya, E Himalaya, Tibetan Plateau and E Asia.

Use: Medicine (root, bark, leaves, seeds).

Albizia lebbeck (L.) Benth. [2, 4, 6]

Mimosa lebbeck L.; *Acacia lebbeck* (L.) Willd.

कालो सिरिष (Kaalo siris); Lebbek tree

Deciduous trees. W, C, E Nepal: 200-800 m. Nepal, native to tropical Africa, naturalized in E Himalaya and S Asia.

Use(s): Medicine (bark, flowers, seeds), Timber, Shade tree, Green manure.

Albizia lucidior (Steud.) I.C.Nielsen [2, 4, 6]

Inga lucidior Steud.; *Albizia gamblei* Prain (2, 4); *Albizia lucida* Benth.

टपरा सिरिष, पड्के (Tapara siris, Padke); Burmese siris

Deciduous trees. C, E Nepal: 200-1000 m. Nepal, W Himalaya, E Himalaya, E Asia and SE Asia.

Use(s): Medicine (root, bark, leaves), Ornamental, Fish poison.

Albizia odoratissima (L.f.) Benth. [2, 4, 6]

Mimosa odoratissima L.f.; *Acacia odoratissima* (L.f.) Willd.

कार्कुर सिरिष, कालो सिरिष (Karkur siris, Kaalo siris)

Deciduous trees. C, E Nepal: 100-500 m. Nepal, W Himalaya, E Himalaya, Assam-Burma, S Asia, E Asia and SE Asia.

Use(s): Medicine (bark, leaves, flowers), Fodder, Construction materials.

Albizia procera (Roxb.) Benth. [2, 4, 6]

Mimosa procera Roxb.; *Acacia procera* (Roxb.) Willd.

सेतो सिरिष, कराही रुख (Seto siris, Karaahi rukh)

Deciduous trees. C, E Nepal: 60-1100 m. Nepal, Assam-Burma, S Asia, E Asia and SE Asia.

Use(s): Medicine (bark, leaves, seeds), Fuel, Charcoal.

Alhagi Adans.

Alhagi nepalensis (D.Don) Shap. [2, 4]

Manna nepalensis D.Don; *Alhagi maurorum* Desv. ex DC.

दुर्लव (Durlav); Camel thorn

Shrubs. C Nepal: *ca.* 1300 m. Nepal and W Himalaya.

Alysicarpus Desv.

Alysicarpus ovalifolius (Schum.) Leonard [8]

Hedysarum ovalifolium Schum.

Annual herbs. C Nepal: 900-1000 m. Widespread in Asian and African tropics.

Use: Medicine (root).

Alysicarpus rugosus (Willd.) DC. [2, 4, 8]

Hedysarum rugosum Willd.; *Alysicarpus violaceus* Schindl.

मधु मूल (Madhu mool)

Perennial herbs. W, C, E Nepal: 60-1200 m. Nepal, Assam-Burma, S Asia, E Asia, SE Asia, Africa and Australasia.

Alysicarpus vaginalis (L.) DC. [2, 4, 6]

Hedysarum vaginale L.; *Alysicarpus nummularifolius* DC.

मधु मूल (Madhu mool)

Perennial herbs. W, C, E Nepal: 200-1200 m. Nepal, Tropics of the Old World, introduced to tropical America.

Use: Medicine (root).

Amphicarpaea Elliott ex Nutt.

Amphicarpaea bracteata (L.) Fernald

Glycine bracteata L.

subsp. **edgeworthii** (Benth.) H.Ohashi [2, 4, 6]

Amphicarpaea edgeworthii Benth.

Annual or perennial climbers. C Nepal: 1500-2700 m. Nepal, W Himalaya, Tibetan Plateau, S Asia, E Asia, SE Asia and N Asia.

Amphicarpaea ferruginea Benth. [2, 4]

Shuteria ferruginea (Benth.) Baker (6)

Perennial climbing herbs. C, E Nepal: 2100-3000 m. Nepal and E Asia.

Apios Fabr.

Apios carnea (Wall.) Benth. ex Baker [2, 4, 6]

Cyrtotropis carnea Wall.

Perennial climbing herbs. C, E Nepal: 1700-2800 m. Nepal, E Himalaya, Tibetan Plateau, E Asia and SE Asia.

Arachis L.

Arachis hypogaea L. [2, 4, 6]

बदाम, मुंगफली (Badaam, Mungphali); Ground nut, Peanut

Annual herbs. W, C Nepal: 500-1500 m. Cultivated through out the world.

Use(s): Medicine (nut), Food (nut).

Archidendron F.Muell.

Archidendron bubalinum (Jack) I.C.Nielson [4]

Inga bubalina Jack

Small trees. C, E Nepal: 1000-1200 m. Nepal, E Himalaya and S Asia.

Archidendron clypearia (Jack) I.C.Nielson [4]

Inga clypearia Jack; *Pithecellobium heterophyllum* (Roxb.) J.F.Macbr. (2)

Evergreen shrubs or small trees. E Nepal: 300-500 m. Nepal, E Himalaya, Assam-Burma, E Asia and SE Asia.

Argyrolobium Eckl. & Zeyh.

Argyrolobium flaccidum (Royle) Jaub. & Spach [2, 4, 6]

Cytisus flaccidus Royle

Prostrate herbs. W Nepal: 2100-2500 m. Nepal and W Himalaya.

Argyrolobium roseum (Cambess.) Jaub. & Spach [2, 4, 6, 11]

Cytisus roseus Cambess.

Procumbent perennial herbs. W, C Nepal: 1900-3400 m. Nepal, W Himalaya and SW Asia.

Use: Medicine (whole plant).

Astragalus L.

Astragalus amherstianus Royle ex Benth. [2, 4, 30]

Herbs. W Nepal: 2100-3700 m. Nepal, W Himalaya and SW Asia.

Astragalus barclayanus Podlech, *Novon* **14**(2): 217. 2004. [27, 29, 30, 110]

Prostrate or ascending herbs. C Nepal: 4800-4900 m. **Endemic to Nepal**.

Central: *Manang*, Tilicho pass, 4820 m, C. Barclay & PM Synge 2620 (Holotype: K).

Astragalus chateri Vassilcz., *Nov. Syst. Pl. Vasc.* **16**: 140. 1979. [4, 12, 27, 29, 30]

Perennial stemless prostrate herbs. W, C Nepal: 3600-3800 m. **Endemic to Nepal**.

West: *Dolpa*, Barbung Khola, Kakkotgaon, 12000 ft. (3660 m); O. Polunin, WR Sykes & LHJ Williams 1084 (Holotype: BM).

Astragalus chlorostachys Lindl. [2, 4, 6, 30]

Astragalus microdontus Baker

Perennial herbs. W, C, E Nepal: 2100-3800 m. Nepal, W Himalaya, E Himalaya, Tibetan Plateau and SW Asia.

Astragalus cobresiiphilus Podlech & L.R.Xu [11]

Perennial herbs. C Nepal: 3600-4600 m. Nepal, Tibetan Plateau and NAsia.

Astragalus concretus Benth. [2, 4, 6, 30]

Astragalus vicioides Graham ex Baker

Perennial herbs. W, C, E Nepal: 2300-4000 m. Nepal, E Himalaya and Tibetan Plateau.

Astragalus confertus Benth. ex Bunge [2, 4, 30]

Herbs. C Nepal: 4200-4800 m. Nepal, W Himalaya and Tibetan Plateau.

Astragalus donianus DC. [2, 4, 6, 30]

Astragalus macrorhizus D.Don; *Astragalus pycnorhizus* Wall. ex Benth.

से-कर (Se-kar: Tibetan)

Perennial prostrate herbs. W, C, E Nepal: 2700-4500 m. Nepal, E Himalaya and Tibetan Plateau.

Astragalus emodi Steud. [2, 4, 8, 30]

Astragalus podocarpus Graham ex Benth.

Shrubs. W, E Nepal: 3200-3600 m. Nepal and W Himalaya.

Astragalus floridulus Podlech [30]

Astragalus floridus Benth. ex Bunge (2, 4)

Herbs. C Nepal: 4200-4700 m. Nepal, E Himalaya, Tibetan Plateau and E Asia.

Use: Medicine (whole plant).

Astragalus frigidus (L.) A.Gray [2, 4, 8]

Phaca frigida L.

Subshrubs. W, C Nepal: 2500-4100 m. Nepal, W Himalaya, N Asia, C Asia and SW Asia.

Astragalus grahamianus Benth ex Royle [8, 30]

Spiny shrubs. C Nepal: 3700-4900 m. Nepal and S Asia.

Astragalus graveolens Buch.-Ham. ex Benth. [2, 4, 30]

Astragalus rotundifolius Royle ex Benth.

Undershrubs. W Nepal: *ca.* 2100 m. Nepal, W Himalaya, E Asia and SW Asia.

Use: Medicine (whole plant).

Astragalus himalayanus Klotzsch [2, 4, 6, 30]

Astragalus carinalis Benth. ex Bunge

Perennial herbs. W, C, E Nepal: 3500-4500 m. Nepal and W Himalaya.

Use: Medicine (root).

Astragalus hoffmeisteri (Klotzsch) Ali [11]

Phaca hoffmeisteri Klotzsch

Perennial herbs. C Nepal: *ca.* 3300 m. Nepal, W Himalaya, Tibetan Plateau and SW Asia.

Astragalus jumlaensis Podlech, *Feddes Repert.* **117**(3-4): 227. 2006. [29, 30, 111]

Herbs. W Nepal: 3800-3900 m. **Endemic to Nepal.**

West: *Jumla*, Bhurchula Lekh, 3820 m, O. Polunin, WR Sykes & LHJ Williams 4526 (Holotype: BM).

Astragalus kongrensis Benth. ex Baker [8, 30, 132]

Perennial rhizomatous herbs. W, C Nepal: 3900-4900 m. Nepal, E Himalaya, Tibetan Plateau and E Asia.

Astragalus lachungensis Chaudhary [30, 112]

Herbs. C Nepal: *ca.* 4900 m. Nepal and S Asia.

Astragalus langtangensis Podlech [8, 30, 113]

Shrubs. C Nepal: 3100-4000 m. Nepal and S Asia (India).

Astragalus lessertioides Benth. ex Bunge [2, 4, 30]

Herbs. C, E Nepal: 4200-4900 m. Nepal, E Himalaya, Tibetan Plateau and E Asia.

Astragalus leucocephalus Graham ex Benth. [2, 4, 6, 30]

पटाका (Pataka: Tibetan)

Perennial prostrate herbs. W, C Nepal: 1500-3700 m. Nepal, W Himalaya and SW Asia.

Use(s): Medicine (whole plant), Plant is toxic to animal.

Astragalus lobbichleri Podlech, *Fedders Repert.* **120**(1-2): 50.2009. [29, 30, 114]

Herbs. C Nepal: *ca.* 3700 m. **Endemic to Nepal.**

Central: *Manang*, Raum Manangbhot, *ca.* 3700 m, F. Lobbichler s.n. (Holotype: M).

Astragalus longiscapus C.C.Ni & P.C.Li [11, 30]

Perennial herbs. C Nepal: 4000-4800 m. Nepal and Tibetan Plateau.

Astragalus melanostachys Benth. ex Bunge [2, 4, 30]

Astragalus bracteosus Klotzsch

Herbs. W Nepal: 3300-5100 m. Nepal, W Himalaya, Tibetan Plateau, C Asia and SW Asia.

Astragalus minutidentatus Y.C.Ho [8, 11, 30]

Perennial herbs. C Nepal: 2100-4900 m. Nepal, E Himalaya and Tibetan Plateau.

Astragalus multiceps Wall. ex Benth. [2, 4, 30]

Astragalus bicuspis Fischer (2, 4); *Astragalus jacquemontii* Bunge

Shrubs. C Nepal: 2300-3500 m. Nepal, W Himalaya and Tibetan Plateau.

Use: Medicine (seeds).

Astragalus nakaoi Kitam., in Kihara, *Peoples Nep. Him.*: 422. 1957. [2, 4, 12, 27, 29, 30]

शिशा (Shisha: Tibetan)

Perennial rhizomatous herbs. C Nepal: *ca.* 3800 m. **Endemic to Nepal**.

Central: *Gorkha*, Manaslu, *ca.* 3800 m, S. Nakao, s.n. (Holotype: KYO).

Astragalus nepalensis Podlech, *Sendtnera* **6**:137. 1999. [29, 30, 113]

Herbs. C Nepal: 2500-2700 m. **Endemic to Nepal**.

Central: *Mustang*, Kali Gandaki, trail above Tukuche, 8500 ft. (2590 m), A. Nicholson 82a (Holotype: US).

Astragalus notabilis Podlech, *Feddes Repert.* **117**: 230. 2006. [29, 30, 111]

Herbs. C Nepal: 3400-3500 m. **Endemic to Nepal**.

Central: *Mustang*, Lower Puyun Khla between Yara and Dhi, 3450 m, S. Miehe, G. Miehe & K. Koch 01-0-45-02 (Holotype: MSB).

Astragalus oplites Benth. ex R.Parker [2, 4, 30]

Astragalus zanskarensis subsp. *oplites* (Benth. ex R.Parker) I.Deml (8, 11)

Dwarf spiny shrubs. W, C Nepal: 2800-4300 m. Nepal, W Himalaya and Tibetan Plateau.

Use: Medicine (whole plant).

Astragalus pindreensis (Benth. ex Baker) Ali [30]

Astragalus candolleanus var. *pindreensis* Benth. ex Baker

Herbs. C Nepal: 2800-4000 m. Nepal, W Himalaya, Tibetan Plateau, N Asia and SW Asia.

Astragalus poluninii Podlech, *Sendtnera* **7**: 178. 2001. [29, 30, 115]

Herbs. W Nepal: 2100-2200 m. **Endemic to Nepal**.

West: *Mugu*, Gum, near Rara, 7000 ft. (2120 m), O. Polunin, WR Sykes & LHJ Williams 3049 (Holotype: E).

Astragalus polyacanthus Royle ex Benth. [2, 4, 30]

Astragalus tenuispinus Bunge

Shrubs. C Nepal: 2000-2600 m. Nepal and W Himalaya.

Use: Medicine (aerial part).

Astragalus pseudorigidulus Podlech, *Feddes Repert.* **120**(1-2): 55. 2009. [29, 30, 114]

Herbs. W Nepal: 4500-4600 m. **Endemic to Nepal**.

West: *Rukum*, Dogadi Khola, 4550 m, JDA Stainton, WR Sykes & LHJ Williams 3194 (Holotype: M; Isotype: BM).

Astragalus rhizanthus Royle ex Benth. [8, 30]

subsp. **rhizanthus**

Astragalus pindreensis (Benth. ex Baker) Ali

Perennial herbs or subshrubs. C Nepal: 4100-4800 m. Nepal, W Himalaya, N Asia and SW Asia.

subsp. **candolleanus** (Royle ex Benth) Podlech [8, 11]

Astragalus candolleanus Royle ex Benth. (2, 4, 30)

जोमोसिंग छुन्गवा (Zomoshing chhungwa: Tibetan)

Dwarf shrubs. W, C Nepal: 2700-4500 m. Nepal and S Asia.

Use: Medicine (root).

Astragalus rigidulus Benth. ex Bunge [2, 4, 30]

Perennial herbs. C Nepal: 3800-4900 m. Nepal, E Himalaya and Tibetan Plateau.

Astragalus sanjappae L.B.Chaudhary & Z.H.Khan [8, 30, 116]

Perennial herbs. C Nepal: 3900-4000 m. Nepal, W Himalaya and E Himalaya.

Astragalus sikkimensis Benth. ex Bunge [2, 4, 6, 30]

Perennial herbs. E Nepal: 2700-3100 m. Nepal, E Himalaya and Tibetan Plateau.

Astragalus stipulatus D.Don ex Sims [2, 4, 6, 30]

var. **stipulatus**

Perennial herbs. C, E Nepal: 2100-3400 m. Nepal, E Himalaya, Tibetan Plateau and E Asia.

var. **phulchokiensis** H.Ohashi, *Fl. E. Himalaya* 3: 57. 1975. [2, 4, 30]

Perennial herbs. C Nepal: *ca.* 2200 m. **Endemic to Nepal**.

Central: *Lalitpur*, Phulchowki, *ca.* 2200 m, H. Kanai 722126 (Holotype: TI; Isotypes: BM, KYO, TI).

Astragalus strictus Graham ex Benth. [2, 4, 6, 11, 30]

Perennial herbs. W, C, E Nepal: 2100-5400 m. Nepal, W Himalaya, E Himalaya, Tibetan Plateau, S Asia, E Asia and N Asia.

Use: Medicine (whole plant).

Astragalus yunnanensis Franch. [2, 4, 8, 11, 30]

से-शेर (Se ser: Tibetan)

Perennial herbs. W, C, E Nepal: 3500-5600 m. Nepal, E Himalaya, Tibetan Plateau and E Asia.

Use: Medicine (whole plant).

Bauhinia L.

Bauhinia malabarica Roxb. [2, 4, 6]

Piliostigma malabaricum (Roxb.) Benth.

अमिल टाँकि (Amil taanki); Malabar ebony

Deciduous trees. W, C, E Nepal: 100-700 m. Nepal, W Himalaya, E Himalaya, S Asia and SE Asia.

Use(s): Medicine (bark, leaves, flowers), Food, Fodder, Tannin (bark).

Bauhinia purpurea L. [2, 4, 6]

Bauhinia triandra Roxb.

टाँकि, रातो कोइरालो, कचनार (Taanki, Raato koiralo, Kachnar); Butterfly tree

Deciduous trees. W, C, E Nepal: 300-1600 m. Nepal, W Himalaya, E Himalaya, Assam-Burma, S Asia, E Asia and SE Asia.

Use(s): Medicine (root, bark, flowers), Food (flower bud), Fodder, Cordage, Tannin.

Bauhinia roxburghiana Voigt [6]

Deciduous trees. C Nepal: 1200-1400 m. Nepal and W Himalaya (N India).

Bauhinia scandens L.

Bauhinia anguina Roxb.; *Bauhinia scandens* var. *anguina* (Roxb.) H.Ohashi; *Bauhinia scandens* var. *horsfieldii* (Watt ex Prain) K.Larsen & S.S.Larsen (2, 4)

Monkey ladders

Climbing shrubs. E Nepal: *ca.* 300 m. Nepal, E Himalaya, Assam-Burma, S Asia, E Asia and SE Asia.

Use: Fodder.

Bauhinia semla Wunderlin [2, 4]

Bauhinia retusa Buch.-Ham. ex Roxb.

Deciduous trees. W, C Nepal: 1200-1400 m. Nepal and W Himalaya.

Use: Fodder.

Bauhinia vahlii Wight & Arn. [2, 4, 6]

Bauhinia racemosa Vahl

भोर्ला, मालु (Bhorlaa, Maalu); Camel's foot climber

Evergreen woody climbers. W, C, E Nepal: 200-1500 m. Nepal, W Himalaya and E Himalaya.

Use(s): Medicine (root, bark, seeds), Food, Fodder, Fiber, Tannin.

Bauhinia variegata L. [2, 4]

Bauhinia candida Ait.

कोइरालो, सेतो कोइरालो (Koiraalo, Seto koiraalo); Orchid-tree, Mountain Ebony

Deciduous trees. W, C, E Nepal: 100-1900 m. Nepal, W Himalaya, E Himalaya, Assam-Burma, E Asia and SE Asia.

Use(s): Medicine (root, bark, leaves, flowers), Food (flowers), Fodder, Fiber, Ornamental.

Butea Roxb. ex Willd.

Butea buteiformis (Voigt) Grierson [4]

Meizotropis buteiformis Voigt (6); *Butea minor* Buch.-Ham. ex Baker (2)

भुजेत्रो, भुलेत्रो (Bhujetro, Bhuletro)

Shrubs or erect perennial herbs. W, C, E Nepal: 300-2000 m. Nepal, E Himalaya, Tibetan Plateau, Assam-Burma and S Asia.

Use(s): Fodder, Medicine (seeds), Cordage (bark), Plate (leaves).

Butea monosperma (Lam.) Kuntze [2, 4, 6]

Erythrina monosperma Lam.; *Butea frondosa* Koen. ex Roxb.

पलाँस, धाँक (Palaans, Dhaank); Flame of the forest

Deciduous trees. W, C, E Nepal: 100-1200 m (Rare). Nepal, E Himalaya, Assam-Burma, S Asia, E Asia and SE Asia.

Use(s): Medicine (root, bark, leaves, flowers, seeds), Fodder, Timber, Cordage (bark), Dye (flowers).

Caesalpinia L.

Caesalpinia bonduc (L.) Roxb. [2, 4, 6]

Guilandina bonduc L.; *Caesalpinia crista* L.

करौंजी, बोक्सी काँडा (Karaunjee, Boksi kaanda); Fever nut

Scandent prickly shrubs. C, E Nepal: 200-1100 m. Nepal and Pantropical.

Use(s): Medicine (tender shoot, seeds), Dye (bark).

Caesalpinia cucullata Roxb. [2, 4, 6]

Mezoneuron cucullatum (Roxb.) Wight & Arn.

Climbing shrubs. W, C, E Nepal: 100-1800 m. Nepal, E Himalaya, Assam-Burma, S Asia, E Asia and SE Asia.

Caesalpinia decapetala (Roth) Alston [2, 4, 6]

Reichardia decapetala Roth; *Caesalpinia sepiaria* Roxb.

उल्टो काँडा, अरिले काँडा (Ulto Kaandaa, Arile kaandaa)

Scandent prickly shrubs. W, C, E Nepal: 100-2200 m. Nepal, E Himalaya, Assam-Burma, S Asia, E Asia and SE Asia.

Use(s): Medicine (root, leaves, pods), Fodder, Fence, Tannin (bark).

Caesalpinia digyna Rottler [2, 4, 6]

Caesalpinia oleosperma Roxb.

बाख्रा लहरा (Baakhraa laharaa)

Scandent shrubs. W, E Nepal: 300-1000 m. Nepal, E Himalaya, S Asia, E Asia and SE Asia.

Use: Medicine (root, leaves).

Caesalpinia pulcherrima (L.) Sw. [2, 4, 6]

Poinciana pulcherrima L.

बास फूल (Baas phool)

Shrubs or small trees. C, E Nepal: 300-1000 m. Widely cultivated in tropics, probably of American origin.

Use: Medicine (root, bark, leaves, flowers).

Cajanus DC.

Cajanus cajan (L.) Huth [2, 4, 8]

Cytisus cajan L.; *Cajanus indicus* Spreng.

रहर (Rahar); Pigeon pea

Shrubs. C, E Nepal: 100-4500 m. Commonly cultivated in tropical regions, probably Native to Africa.

Use(s): Medicine (leaves, seeds), Food, Fodder, Thatching, Brooms.

Cajanus crassus (Prain ex King) Maesen

Atylosia crassa Prain ex King; *Atylosia volubilis* (Blanco) Gamble (2, 4)

Herbaceous climbers. W, C Nepal: 200-1000 m. Nepal, Assam-Burma, S Asia and SE Asia.

Cajanus elongatus (Benth.) Maesen [4, 6]

Atylosia elongata Benth. (2)

जंगली रहर (Jungali rahar); Wild Pigen Pea

Climbers. W, C, E Nepal: 1200-1900 m. Nepal, E Himalaya, Assam-Burma and SE Asia.

Cajanus mollis (Benth.) Maesen [4, 6]

Atylosia mollis Benth. (2)

Twining perennial herbs. W, C, E Nepal: 600-2000 m. Nepal, W Himalaya, E Himalaya and E Asia.

Use: Medicine (whole plant).

Cajanus scarabaeoides (L.) Thouars [6]

Dolichos scarabaeoides L.; *Atylosia scarabaeoides* (L.) Benth. (2, 4)

बन गहत (Ban gahat)

Twining perennial herbs. W, C, E Nepal: 400-1200 m. Nepal, E Himalaya, Assam-Burma, S Asia, E Asia, SE Asia and Africa.

Use: Medicine (whole plant).

Callerya Endl.

Callerya cinerea (Benth.) Schot

Millettia cinerea Benth. (2, 4)

Woody climbing shrubs. E Nepal: 400-500 m. Nepal, E Himalaya, Tibetan Plateau, Assam-Burma, S Asia, E Asia and SE Asia.

Campylotropis Bunge

Campylotropis macrostyla (D.Don) Miq. [2, 4, 6]

Crotalaria macrostyla D.Don; *Lespedeza eriocarpa* DC.; *Oxyramphis macrostyla* (D.Don) Wall.

Undershrubs. W, C Nepal: 500-2300 m. Nepal and W Himalaya.

Campylotropis speciosa (Schindl.) Schindl. [2, 4, 6, 11]

Lespedeza speciosa Schindl.

subsp. **speciosa**

साकिनो, साखिनो (Saakino, Saakhino)

Deciduous shrubs. W, C, E Nepal: 2000-3100 m. Nepal, W Himalaya, E Asia and Assam-Burma.

Use: Food (flowers).

subsp. **eriocarpa** (Schindl.) Iokawa & Ohashi [8]

Campylotropis eriocarpa Schindl.

Deciduous shrubs. C, E Nepal: 2200-2600 m. Nepal, E Himalaya and Tibetan Plateau.

Campylotropis stenocarpa (Klotzsch) Schindl. [8]

Oxyramphis stenocarpa Klotzsch; *Campylotropis macrostyla* var. *stenocarpa* (Klotzsch) H.Ohashi (2, 4); *Lespedeza stenocarpa* (Klotzsch) Maxim.

Woody shrubs. W Nepal: 700-2300 m. Nepal and W Himalaya.

Caragana Lam.

Caragana brevifolia Kom. [11, 98]

Spiny shrubs. C Nepal: 3500-4000 m. Nepal, W Himalaya, Tibetan Plateau and E Asia.

Caragana brevispina Royle [2, 4, 6, 11]

Caragana williamsii Vass.

subsp. **brevispina**

बेबली काँडा, भालु काँडा (Bebali kaanda, Bhaalu kaanda)

Spiny shrubs. W, C Nepal: 2400-3200 m. Nepal and W Himalaya.

Use(s): Medicine (leaves, fruit), Food (young bud), Fodder, Fuel.

subsp. **tenzingii** Vassilcz. [2, 4]

Spiny shrubs. W Nepal. Nepal and W Himalaya.

Caragana campanulata Vass., *Novosti Sist. Vyss. Rast.* **13**: 195. 1976. [2, 4]

Perennial herbs. C Nepal: *ca.* 3200 m. Endemic to Nepal (?)

Central: *Mustang*, Kagbeni, Upper Kali Gandaki, 10500 ft. (3200 m), JDA Stainton 5594 (Holotype: BM).

Caragana gerardiana Royle [2, 4, 6, 11, 29]

Caragana spinosissima Benth.

जोमोसिङ्ग (Zomoshing: Amchi)

Spiny shrubs. W, C Nepal: 3200-4200 m. Nepal, W Himalaya, Tibetan Plateau and SW Asia.

Use(s): Medicine (stem), Food, Fuel.

Caragana jubata (Pall.) Poir. [2, 4, 8, 11]

Robinia jubata Pall.; *Caragana chumbica* Prain (2, 4)

जोमोसिङ्ग (Zomoshing: Tibetan)

Spiny shrubs. C Nepal: 2400-4000 m. Nepal, E Himalaya, Tibetan Plateau, E Asia and N Asia.

Use(s): Medicine (stem), Fodder.

Caragana sukiensis C.K.Schneid. [2, 4, 6]

Caragana hoplites Dunn

पांकचर, फुलसी सेर्मा (Paankchaar, Phulsisherma)

Spiny shrubs. W, C Nepal: 3000-3900 m. Nepal, W Himalaya and E Himalaya.

Use: Medicine (root).

Caragana tragacanthoides (Pall.) Poir.

Robinia tragacanthoides Pall.

var. **himalaica** Kom. [2, 4]

Spiny shrubs. C Nepal: *ca.* 4200 m. Nepal and W Himalaya.

Caragana versicolor (Wall.) Benth. [2, 4, 6]

Genista versicolor Wall.

तापसिंग (Taapsing: Tibetan)

Spiny shrubs. W Nepal: 3300-4600 m. Nepal, W Himalaya, Tibetan Plateau and SW Asia.

Use: Fuel.

Cassia L.

Cassia fistula L. [2, 4, 6]

Cassia rhombifolia Roxb.

राज वृक्ष, अमलतास (Raaj brichhya, Amaltaas); Indian Laburnum

Deciduous trees. W, C, E Nepal: 100-1400 m. Nepal, Native to S Asia (India).

Use(s): Medicine (root, leaves, fruit), Food, Fodder, Charcoal, Tannin.

Chamaecrista Moench.

Chamaecrista leschenaultiana (DC.) Degener

Cassia lechenaultiana DC.; *Cassia mimosoides* var. *auricoma* Benth. (2, 4); *Cassia mimosoides* subsp. *lechenaultiana* (DC.) H.Ohashi (2, 4)

Perennial herbs. W, C Nepal: 500-2000 m. Nepal, S Asia, E Asia and SE Asia.

Chamaecrista mimosoides (L.) Greene [6]

Cassia mimosoides L. (2, 4); *Cassia angustissima* Lam.

अमला झार (Amalaa jhaar); Feather-leaved Cassia

Annual or perennial herbs. W, C, E Nepal: 700-2500 m. Widely cultivated in tropics and subtropics.

Use(s): Medicine (root, leaves), Food.

Chesneya Lindl. ex Endl.

Chesneya nubigena (D.Don) Ali [2, 4]

Astragalus nubigenus D.Don; *Astragalus larkyaensis* Kitam. (8); *Spongiocarpella nubigena* (D.Don) Yakovlev (6, 11); *Spongiocarpella nubigena* subsp. *kumaoensis* Yakovlev (4)

च्याली (Chyaali)

Perennial herbs. W, C, E Nepal: 3600-5200 m. Nepal, W Himalaya, E Himalaya, Tibetan Plateau, Assam-Burma and E Asia.

Cicer L.

Cicer arietinum L. [2, 4, 6]

चना (Chana); Chickpea

Annual herbs. W, C, E Nepal: 100-1300 m. Widely cultivated. Native to SW Asia.

Use: Food (leaves, seeds).

Cicer microphyllum Benth. [2, 4]

Himalayan chickpea

Perennial herbs. W Nepal: 4100-4800 m. Nepal, W Himalaya, Tibetan Plateau and SW Asia.

Clitoria L.

Clitoria ternatea L. [2, 4, 6]

अपराजिता, शंख पुष्पी (Aparaajita, Shankha puspi); Butterfly pea

Herbaceous climbers. W, E Nepal: *ca.* 200 m. Widely cultivated in Tropics; probably Native to S America.

Use: Medicine (root, bark, seeds, leaves).

Cochlianthus Benth.

Cochlianthus gracilis Benth. [2, 4, 6]

खस्रे लहरा, भट्मासे लहरा (Khasre laharaa, Bhatmaase laharaa)

Climbing herbs. C Nepal: 1800-2200 m. Nepal, E Himalaya, Tibetan Plateau and E Asia.

Use: Medicine (bark).

Codariocalyx Hassk.

Codariocalyx gyroides (Roxb.) Hassk. [2, 4, 8]

Hedysarum gyroides Roxb.

Shrubs. C, E Nepal: 100-2000 m. Nepal, E Himalaya, S Asia, E Asia and SE Asia.

Codariocalyx motorius (Houtt.) H.Ohashi [2, 4, 6]

Hedysarum motorium Houtt.; *Desmodium gyrans* (L.f.) DC.; *Hedysarum gyrans* L.f.

Telegraph plant

Shrubs. W, C, E Nepal: 400-2300 m. Nepal, E Himalaya, Assam-Burma, S Asia and SE Asia.

Colutea L.

Colutea multiflora Shap. ex Ali, *Bot. Not.* **112**: 491. 1959. [2, 4, 6, 12, 27, 29]

Deciduous shrubs. C, E Nepal: 2200-4500 m. **Endemic to Nepal**.

Central: *Manang*, Pongsing, 15000 ft. (4570 m), Lal Dhwoj 76 (Holotype: BM).

Colutea nepalensis Sims [2, 4]

Colutea arborescens var. *nepalensis* L.

Deciduous shrubs. Nepal, W Himalaya, Tibetan Plateau and SW Asia.

Use: Medicine (leaves, seeds).

Note: Described from a plant grown from seeds, collected from Nepal. No specimen in BM, K, TI (2).

Crotalaria L.

Crotalaria acicularis Buch.-Ham. ex Benth. [2, 4, 6]

Subshrubs or perennial herbs. E Nepal: 200-600 m. Nepal, Assam-Burma, S Asia, E Asia and SE Asia.

Use: Fodder.

Crotalaria alata Buch.-Ham. ex D.Don [2, 4, 6]

छिन् छिने, ठुलो बोक्सीबाजा (Chhin chhine, Thulo boksibaaja); Winged-stem rattlepod

Herbs or shrublets. W, C, E Nepal: 300-1500 m. Nepal, W Himalaya, E Himalaya, Assam-Burma, S Asia, E Asia and SE Asia.

Use: Medicine (root, tender shoot).

Crotalaria albida Heyne ex Roth [2, 4, 6]

भेडी फूल, पुतलि फूल (Bhedi phool, Putali phool); Mountain rattlepod

Perennial herbs. W, C, E Nepal: 400-2200 m. Nepal, W Himalaya, E Himalaya, Tibetan Plateau, Assam-Burma, S Asia, E Asia, SE Asia and Australasia.

Use(s): Medicine (root, leaves), Food.

Crotalaria bialata Schrank [8]

Perennial herbs or subshrubs. W Nepal: *ca.* 2000 m. Nepal, S Asia and E Asia.

Crotalaria calycina Schrank [2, 4, 6]

Crotalaria nepalensis Link

Silky haired rattlepod

Annual or short-lived perennial herbs. C, E Nepal: 200-1200 m. Nepal, W Himalaya, E Himalaya, Tibetan Plateau, S Asia, E Asia, SE Asia and Australasia.

Crotalaria cytisoides DC. [2, 4, 6]

Crotalaria psoralioides D.Don; *Priotropis cytisoides* (DC.) Wight & Arn.

चेम्गी फूल (Chemgi phool); Broom rattlepod

Deciduous shrubs. W, C, E Nepal: 1200-2000 m. Nepal, E Himalaya, Tibetan Plateau, S Asia, E Asia and SE Asia.

Use: Fodder.

Crotalaria ferruginea Graham ex Benth. [2, 4, 8]

Crotalaria ferruginea var. *pilosissima* Miq.

जुनग्राबन (Jungraaban: Chepang)

Herbs. C, E Nepal: 1400-1900 m. Nepal, E Himalaya, Tibetan Plateau, Assam-Burma, S Asia, E Asia and SE Asia.

Use: Medicine (root).

Crotalaria humifusa Graham ex Benth. [2, 4, 6]

Sprawling rattlepod

Annual or short-lived perennial herbs. W, C, E Nepal: 800-1900 m. Nepal, W Himalaya, E Himalaya, Assam-Burma, E Asia, SE Asia and Australasia.

Crotalaria kanaii H.Ohashi, *J. Jap. Bot.* **51**: 301. 1976. [2, 4, 6, 12, 27, 29]

Perennial prostrate herbs. W, C Nepal: 2100-2900 m. **Endemic to Nepal**.

Central: *Rasuwa*, Dhunche – Singum Gompa, *ca.* 2300 m, H. Kanai, H. Hara & H. Ohba 721801 (Holotype: TI; Isotype: TI).

Crotalaria medicaginea Lam. [2, 4, 6]

Crotalaria procumbens Roxb.

Herbs or undershrubs. W Nepal: 2400-2500 m. Nepal, W Himalaya, S Asia, E Asia, SE Asia, SW Asia and Australasia.

Crotalaria micans Link

Crotalaria anagyroides Kunth (2, 4)

Subshrubs or perennial herbs. E Nepal: 400-500 m. Introduced in Nepal; native to S America and N America (Mexico).

Crotalaria mysorensis Roth [2, 4]

Crotalaria stipulacea Roxb.

Mysore rattlepod

Erect herbs. W Nepal: 1000-1100 m. Nepal, W Himalaya, S Asia, E Asia and SE Asia.

Crotalaria pallida Aiton [2, 4, 6]

Crotalaria mucronata Desv.; *Crotalaria pallida* var. *obovata* Polhill

छिन् छिने (Chhin chhine); Smooth rattlepod

Erect robust herbs. C, E Nepal: 200-1400 m (Naturalized). Native to central and tropical America.

Use: Medicine (whole plant).

Crotalaria prostrata Rottb. ex Willd. [2, 4, 6]

सानो बोक्सीबाजा (Saano boksibaajaa); Prostrate rattlepod

Annual herbs. W, C, E Nepal: 500-2000 m. Nepal, W Himalaya, E Himalaya, Assam-Burma, S Asia, E Asia and SE Asia.

Use: Medicine (whole plant).

Crotalaria quinquefolia L. [2, 4]

Crotalaria heterophylla L.f.

Annual or short-lived perennial herbs. E Nepal: *ca.* 200 m. Nepal and S Asia.

Crotalaria sessiliflora L. [2, 4, 6]

Crotalaria napaulensis Link

सेतो फूल (Seto phool)

Annual herbs. W, C, E Nepal: 200-2800 m. Nepal, W Himalaya, E Himalaya, Tibetan Plateau, Assam-Burma, S Asia, E Asia, SE Asia and Australasia (Pacific Islands).

Use(s): Medicine (whole plant), Food.

Crotalaria spectabilis Roth [2, 4, 6]

Crotalaria sericea Retz.

भुवन झार, छिन् छिन् बिष (Bhuban jhaar, Chhinchin bish); Showy rattlepod

Undershrubs. C, E Nepal: 100-200 m. Nepal, S Asia, E Asia and SE Asia.

Use(s): Medicine (whole plant), Food.

Crotalaria tetragona Andrews [2, 4, 6]

सनई, छिन् छिने (Sanai, Chhin chhine); Eastern rattlepod

Perennial herbs or subshrubs. W, C, E Nepal: 200-1700 m. Nepal, W Himalaya, E Himalaya, Assam-Burma, S Asia, E Asia and SE Asia.

Use: Food.

Dalbergia L.f.

Dalbergia lanceolaria L.f. [2, 4]

Dalbergia frondosa Roxb., *Dalbergia robusta* Wall.

subsp. **lanceolaria**

बाँदरे सिरिष (Baandare siris)

Medium-sized deciduous trees. C Nepal: *ca.* 1400 m. Nepal, Assam-Burma and S Asia.

Use(s): Medicine (bark, seed oil), Timber.

subsp. **paniculata** (Roxb.) Thoth.

Dalbergia paniculata Roxb. (2, 4, 6)

Small deciduous trees. W, C, E Nepal: 200-500 m. Nepal, W Himalaya and Assam-Burma.

Dalbergia latifolia Roxb. [2, 4, 6]

Dalbergia emarginata Roxb.

सति साल (Satisaal); Rose wood

Deciduous large trees. W, C, E Nepal: 200-1000 m. Nepal, W Himalaya, E Himalaya, S Asia and SE Asia.

Use(s): Medicine (bark, heart wood), Furniture, Timber, Ritual.

IUCN Category: **Vulnerable** A1cd (ver 2.3).

Dalbergia pinnata (Lour.) Prain [2, 4, 6]

Derris pinnata Lour.; *Dalbergia tamarindifolia* Roxb.

दामर (Daamar)

Trees or climbing shrubs. C, E Nepal: 200-1100 m. Nepal, E Himalaya, Tibetan Plateau, Assam-Burma, S Asia, E Asia and SE Asia.

Use(s): Medicine (root), Fodder.

Dalbergia sericea G.Don [2, 4, 6]

Dalbergia hircina Wall. ex Benth.

Silky Dalbergia

Medium-sized trees. W, C, E Nepal: 700-2000 m. Nepal, E Himalaya, Tibetan Plateau, Assam-Burma and SE Asia.

Dalbergia sissoo Roxb. ex DC. [2, 4, 6]

सिसौ (Sisau); North Indian rose wood

Deciduous trees. W, C, E Nepal: 200-1400 m. Nepal, W Himalaya, E Himalaya and S Asia.

Use(s): Medicine (root, bark, leaves, heartwood), Timber, Fodder, Fuel, Construction.

Dalbergia stipulacea Roxb. [2, 4, 6]

टाटे बरी, टाटेबीर (Taate bari, Taatebir); East Himalayan Dalbergia

Climbing shrubs. C, E Nepal: 200-1300 m. Nepal, E Himalaya, E Asia and SE Asia.

Use(s): Medicine, Fish poison (root, bark).

Dalbergia volubilis Roxb. [2, 4, 6]

Climbing Dalbergia

Climbing shrubs. W, C Nepal: 300-400 m. Nepal, E Himalaya, Assam-Burma, S Asia and E Asia.

Use: Medicine (root, leaves).

Delonix Raf.

Delonix regia (Hook.) Raf. [2, 4, 6]

गुल मोहर (Gul mohar); Gul mohar tree

Deciduous trees. C, E Nepal: 200-1000 m. Widely planted as shade tree. Native to Africa.

Dendrolobium Benth.

Dendrolobium triangulare (Retz.) Schindl. [2, 4, 6]

Hedysarum triangulare Retz.; *Desmodium triangulare* (Retz.) Merr.

Shrubs. C, E Nepal: *ca.* 200 m. Nepal, Assam-Burma, S Asia, E Asia, SE Asia and Africa.

Derris Lour.

Derris acuminata Benth. [2, 4, 8]

Deguelia acuminata (Benth.) Taub.

डेरी (Deree)

Woody climbing shrubs. C, E Nepal: 1500-2000 m. Nepal and E Himalaya.

Derris cuneifolia Benth. [2, 4]

var. **cuneifolia**

Woody climbers. C Nepal: 1000-1100 m. Nepal, E Himalaya and Assam-Burma.

var. **longipedicellata** Thoth. [2, 4]

Woody climbers. Nepal, E Himalaya and S Asia.

Desmodium Desv.

Desmodium concinnum DC. [2, 4, 6, 11]

Desmodium pendulum Wall.

var. **concinnum**

मसिनो भट्मासे, गहते झार (Masino bhatmaase, Gahate jhaar); Blue Desmodium

Shrubs. W, C, E Nepal: 900-2200 m. Nepal, W Himalaya, E Himalaya, Assam-Burma, E Asia and SE Asia.

Use: Medicine (root).

var. **retusum** (D.Don) H.Ohashi [2, 4]

Hedysarum retusum D.Don; *Desmodium retusum* (D.Don) Sweet

Shrubs. C, E Nepal: 900-2200 m. Nepal, E Himalaya and Assam-Burma.

Desmodium confertum DC. [2, 4, 6]

Hedysarum dioicum Buch.-Ham. ex D.Don

भट्टे, बन भाँट (Bhatte, Ban bhaant)

Shrubs. C, E Nepal: 300-2000 m. Nepal and E Himalaya.

Use: Medicine (whole plant).

Desmodium elegans DC. [2, 4, 6, 11]

Hedysarum tiliaefolium D.Don

var. **elegans**

रातो बाख्रे घाँस, चाम्ले (Raato baakhre ghans, Chaamle)

Shrubs. W, C Nepal: 1200-3000 m. Nepal, W Himalaya, E Himalaya, Tibetan Plateau, E Asia and SW Asia.

Use(s): Medicine (root), Fodder.

var. **nutans** H.Ohashi [2, 4]

Shrubs. W Nepal: 2800-3000 m. Nepal and W Himalaya.

Desmodium gangeticum (L.) DC. [2, 4, 8]

Hedysarum gangeticum L.

बन गहते, शालपार्णी (Ban gahate, Shaalparnee)

Shrubs or under shrubs. W, C, E Nepal: 100-1000 m. Nepal, W Himalaya, E Himalaya, Assam-Burma, S Asia, E Asia, SE Asia, Africa and Australasia.

Use: Medicine (whole plant).

Desmodium heterocarpon (L.) DC. [2, 4, 6]

Hedysarum heterocarpon L.

var. **heterocarpon**

साखिनो झार (Saakhino jhaar)

Shrubs or subshrubs. W, C, E Nepal: 400-1700 m. Nepal, W Himalaya, E Himalaya, Assam-Burma, S Asia, E Asia, SE Asia, Africa and Australasia.

Use: Medicine (root).

var. **strigosum** Meeuwen [2, 4, 8]

Shrubs or subshrubs. C, E Nepal: 1000-2000 m. Nepal, E Himalaya, Assam-Burma, E Asia, SE Asia, Africa and Australasia.

Desmodium heterophyllum (Willd.) DC. [2, 4]

Hedysarum heterophyllum Willd.; *Hedysarum reptans* Roxb.

Perennial prostrate herbs. E Nepal: *ca.* 200 m. Nepal, Assam-Burma, S Asia, E Asia, SE Asia and Australasia.

Use(s): Medicine (whole plant), Fodder.

Desmodium intortum (Mill.) Urb. [8]

Hedysarum intortum Mill.

Perennial herbs. E Nepal: *ca.* 1600 m. Nepal, native to Central and South America.

Desmodium khasianum Prain [2, 4]

Shrubs. E Nepal: *ca.* 1400 m. Nepal, E Himalaya and Assam-Burma.

Use: Fodder.

Desmodium laxiflorum DC. [2, 4, 8]

Desmodium diffusum DC. (8); *Hedysarum recurvatum* Roxb.

कुरो झार (Kuro jhaar); Loose flowered Desmodium

Shrubs. W, C, E Nepal: 100-1600 m. Nepal, Assam-Burma, E Asia and SE Asia.

Use: Fodder.

Desmodium microphyllum (Thunb.) DC. [2, 4, 8]

Hedysarum microphyllum Thunb.; *Codariocalyx microphyllus* (Thunb.) H.Ohashi; *Desmodium microphyllum* forma *glabrum* H.Ohashi & T.T.Chen (4); *Hedysarum tenellum* Buch.-Ham. ex D.Don var. **microphyllum**

बुट्टे कनिके (Butte Kanike)

Perennial herbs. W, C, E Nepal: 800-2500 m. Nepal, Tibetan Plateau, Assam-Burma, S Asia, E Asia, SE Asia and Australasia.

Use: Medicine (whole plant).

var. **macrocarpum** (Baker) H.Ohashi [2, 4]

Desmodium parvifolium var. *macrocarpum* Baker; *Codariocalyx microphyllus* var. *macrocarpus* (Baker) H.Ohashi

बुट्टे कनिके, भमरे फूल (Butte kanike, Bhamare phool)

Perennial herbs. E Nepal. Nepal and Assam-Burma.

Desmodium multiflorum DC. [2, 4, 6]

Hedysarum sambuense D.Don

भटमासे फूल, बाख्रे घाँस (Bhatmase phool, Baakhre ghaans)

Deciduous shrubs. W, C, E Nepal: 1800-2600 m. Nepal, W Himalaya, E Himalaya, Tibetan Plateau, Assam-Burma, E Asia and SE Asia.

Use: Medicine (root).

Desmodium oojeinense (Roxb.) H.Ohashi [2, 4, 6]

Dalbergia oojeinensis Roxb.; *Ougeinia dalbergioides* Benth.; *Ougeinia oojeinensis* (Roxb.) Hochr.

सन्दन, सादन (Sandan, Saadan); Chariot tree, Sandan

Small deciduous trees. W, C Nepal: 400-1300 m. Nepal, W Himalaya and E Asia.

Use(s): Fodder, Medicine (root, bark, heartwood), Fish poison (bark), Agricultural tools.

Desmodium renifolium (L.) Schindl. [2, 4, 6]

Hedysarum renifolium L.; *Desmodium reniforme* (L.) DC.

Subshrubs. W, C, E Nepal: 200-1200 m. Nepal, Assam-Burma, E Asia, SE Asia and Australasia.

Use: Fodder.

Desmodium sequax Wall. [2, 4, 6]

Desmodium sinuatum (Miq.) Baker

Shrubs. W, C, E Nepal: 800-2200 m. Nepal, W Himalaya, E Himalaya, Tibetan Plateau, Assam-Burma, E Asia and SE Asia.

Use: Fodder.

Desmodium triflorum (L.) DC. [2, 4, 6]

Hedysarum triflorum L.

बुट्टे कनिके, मोतोइ (Butte kanike, Motoi)

Perennial herbs or subshrubs. W, C, E Nepal: 600-2300 m. Nepal, Assam-Burma, S Asia, E Asia, SE Asia, SW Asia, Africa, N America and Australasia.

Use(s): Medicine (root, leaves), Forage.

Desmodium velutinum (Willd.) DC. [2, 4, 6]

Hedysarum velutinum Willd.; *Hedysarum latifolium* Roxb.

var. **velutinum**

मोतोइ (Motoi)

Shrubs or subshrubs. C, E Nepal: 100-1500 m. Nepal, Assam-Burma, S Asia, E Asia, SE Asia and Africa.

var. **longibracteatum** (Schindl.) Meeuwen [2, 4, 8]

Desmodium longibracteatum Schindl.; *Desmodium velutinum* subsp. *longibracteatum* (Schindl.) H.Ohashi (8)

Shrubs or subshrubs. C Nepal: 300-400 m. Nepal, Assam-Burma, E Asia and SE Asia.

var. **sikkimense** H.Ohashi [2, 4]

Shrubs or subshrubs. C Nepal and E Himalaya.

Dolichos L.

Dolichos staintonii H.Ohashi & Tateishi [2, 4, 8]

Climbers. C Nepal: 1000-3000 m. Nepal and W Himalaya.

Dolichos tenuicaulis (Baker) Craib [2, 4, 6]

Phaseolus tenuicaulis Baker; *Dolichos appendiculatus* Hand.-Mazz.

Herbaceous climbers. E Nepal: 300-1200 m. Nepal, E Himalaya, Assam-Burma, E Asia and SE Asia.

Dumasia DC.

Dumasia villosa DC. [2, 4, 6]

Dumasia leiocarpa Benth. (6); *Dumasia villosa* var. *leiocarpa* (Benth.) Baker (2, 4)

Himalayan Dumasia

Herbaceous climbers. W, C, E Nepal: 1800-2200 m. Nepal, W Himalaya, E Himalaya, Tibetan Plateau, Assam-Burma, S Asia, E Asia, SE Asia, Africa and Australasia.

Dunbaria Wight & Arn.

Dunbaria rotundifolia (Lour.) Merr. [2, 4, 6]

Indigofera rotundiflium Lour.; *Dunbaria punctata* (Wight & Arn.) Benth.

Perennial climbing herbs. E Nepal: 200-1000 m. Nepal, Assam-Burma, S Asia, E Asia, SE Asia and Australasia.

Entada Adans.

Entada phaseoloides (L.) Merr. [2, 4]

Lens phaseoloides L.; *Entada rheedei* Spreng. (6); *Mimosa scandens* L.

लेक पांग्रा, पांग्रा (Lekh Paangraa, Paangraa)

Woody climbers. E Nepal: 300-1600 m. Nepal, tropical and subtropical Asia and Australasia.

Use(s): Medicine (bark, seeds), Fish poison (seeds), Leech repellent (seeds).

Entada rheedei Spreng. [20]

subsp. **sinohimalensis** (Grierson & D.G.Long) Panigrahi

Entada pursaetha subsp. *sinohimalensis* Grierson & D.G.Long (4); *Entada laotica* Gagnep. (4)

लेक पांग्रा (Lekh paangraa)

Woody climbers. W, E Nepal: 400-1700 m. Nepal, E Himalaya, Assam-Burma and E Asia.

Use: Medicine (stem, bark, seeds).

Eriosema DC.

Eriosema himalaicum H.Ohashi [2, 4, 6]

Crotalaria tuberosa Buch.-Ham. ex D.Don

बाँदर गुली (Baandar guli)

Perennial herbs. W, C, E Nepal: 900-3000 m. Nepal, W Himalaya, E Himalaya and Assam-Burma.

Use: Food (tender shoot).

Erythrina L.

Erythrina arborescens Roxb. [2, 4, 6]

ठेकी काठ, फलेदो (Theki kaath, Phaledo)

Deciduous trees. W, C, E Nepal: 1500-3000 m. Nepal, W Himalaya, E Himalaya, Tibetan Plateau, Assam-Burma, S Asia, E Asia and SE Asia.

Use(s): Medicine (bark, leaves), Fodder.

Erythrina stricta Roxb. [2, 4, 6]

फलेदो (Phaledo)

Spiny deciduous trees. W, C, E Nepal: 1000-1600 m. Nepal, Tibetan Plateau, Assam-Burma, E Asia and SE Asia.

Use(s): Medicine (bark, leaves, flowers), Fodder.

Erythrina suberosa Roxb. [2, 4, 6]

Erythrina suberosa var. *glabrescens* Roxb.

मंदर, फुलिधा (Mander, Phulidha)

Spiny deciduous trees. C Nepal: 900-1200 m. Nepal, W Himalaya, E Himalaya, Assam-Burma and SE Asia.

Use: Medicine (bark).

Flemingia Roxb. ex W.T.Aiton

Flemingia chappar Buch.-Ham. ex Benth. [2, 4]

Moghania chappar (Buch.-Ham. ex Benth.) Kuntze

भट्टे (Bhatte)

Shrubs. W, C Nepal: 100-400 m. Nepal, E Himalaya, Assam-Burma, E Asia and SE Asia.

Use(s): Food, Medicine (root, leaves).

Flemingia macrophylla (Willd.) Merr. [2, 4, 6]

Crotalaria macrophylla Willd.; *Flemingia congesta* Roxb. ex W.T.Aiton

भट्वासी, भट्मासे (Bhatwaasi, Bhatmase)

Shrubs. W, C, E Nepal: 300-1700 m. Nepal, E Himalaya, Assam-Burma, S Asia, E Asia and SE Asia.

Use(s): Medicine (root), Fodder, Toothbrush.

Flemingia paniculata Wall. ex Benth. [2, 4]

Moghania paniculata (Wall. ex Benth.) Li

Shrubs. E Nepal: 100-200 m. Nepal, Assam-Burma, S Asia, E Asia and SE Asia.

Flemingia procumbens Roxb. [2, 4, 6]

Flemingia vestita Benth. ex Baker; *Moghania vestita* (Graham) O.Kuntze

तेंदू (Tendu)

Perennial herbs. W, C Nepal: 2100-2900 m. Nepal, W Himalaya, Assam-Burma, E Asia and SE Asia.

Use: Medicine (whole plant).

Flemingia strobilifera (L.) W.T.Aiton [2, 4, 6]

Hedysarum strobiliferum L.; *Hedysarum bracteatum* Roxb.

भट्वासी, गहते (Bhatwaasi, Gahate)

Shrubs. W, C, E Nepal: 100-2300 m. Nepal, Assam-Burma, S Asia, E Asia and SE Asia.

Use(s): Medicine (root, bark, leaves), Food (fruit), Broom.

Galactia P.Browne

Galactia tenuiflora (Klein ex Willd.) Wight & Arn. [2, 4, 6]

Glycine tenuiflora Klein ex Willd.

Perennial herbaceous climbers. C Nepal: 1000-1100 m. Nepal, S Asia, E Asia, SE Asia and Africa.

Geissaspis Wight & Arn.

Geissaspis cristata Wight & Arn. [2, 4]

Prostrate annual herbs. C Nepal. Nepal, Assam-Burma, S Asia, E Asia and SE Asia.

Glycine L.

Glycine max (L.) Merr. [2, 4, 6]

Phaseolus max L.

भट्मास (Bhatmaas); Soybean

Annual herbs. W, C, E Nepal: 100-1700 m. Widely cultivated as legume crop and oil seeds.

Use: Food (flowers, seeds).

Hedysarum L.

Hedysarum campylocarpon H.Ohashi [2, 4, 6]

Perennial herbs. C Nepal: 2800-4600 m. Nepal and Tibetan Plateau.

Hedysarum kumaonense Benth. ex Baker [2, 4, 6, 11]

Perennial herbs. W, C Nepal: 2400-4600 m. Nepal, W Himalaya and Tibetan Plateau.

Hedysarum manaslense (Kitam.) H.Ohashi [2, 4, 8, 12, 27, 29]

Astragalus manaslensis Kitam., in Kihara, *Peoples Nep. Him*. 422. 1957.

var. **manaslense**

Densely hairy shrubs. C Nepal: 3300-4200 m. **Endemic to Nepal**.

Central: *Gorkha*, Manaslu, *ca.* 3800 m, S. Nakao s.n. (Holotype: KYO; Isotype: KYO).

var. **nepalense** (H.Ohashi) H.Ohashi [2, 4, 8, 11]

Hedysarum nepalense H.Ohashi

Perennial herbs. C Nepal: 3000-4000 m.

Hedysarum sikkimense Benth. ex Baker [2, 4, 6]

Perennial herbs. E Nepal: 3500-4700 m. Nepal, E Himalaya, Tibetan Plateau and E Asia.

Use: Medicine (whole plant).

Hedysarum tibeticum (Benth.) B.H.Choi & H.Ohashi [11]

Stracheya tibetica Benth. (2, 4)

Perennial herbs. W, C, E Nepal: 3700-5700 m. Nepal, W Himalaya, E Himalaya and Tibetan Plateau.

Hylodesmum H.Ohashi & R.R.Mill

Hylodesmum laxum (DC.) H.Ohashi & R.R.Mill.

Desmodium laxum DC. (2, 4)

Perennial herbs. C Nepal. Nepal, E Himalaya, Tibetan Plateau, S Asia, E Asia and SE Asia.

Hylodesmum podocarpum (DC.) H.Ohashi & R.R.Mill.

Desmodium podocarpum DC. (2, 4, 8)

subsp. **podocarpum**

गहते घाँस (Gahate ghaans)

Herbs. W, C, E Nepal: 1500-2000 m. Nepal, W Himalaya, E Himalaya, Tibetan Plateau, E Asia and SE Asia.

subsp. **oxyphyllum** (DC.) H.Ohashi

Desmodium oxyphyllum DC.; *Desmodium podocarpum* var. *mandschuricum* (Maxim.) H.Ohashi & R.R.Mill. (2, 4); *Desmodium podocarpum* subsp. *oxyphyllum* (DC.) H.Ohashi (2, 4, 8)

Herbs. C, E Nepal: 1200-2300 m. Nepal, E Himalaya, Tibetan Plateau, Assam-Burma, E Asia, SE Asia and N Asia.

Hylodesmum williamsii (H.Ohashi) H.Ohashi & R.R.Mill. [11]

Desmodium williamsii H.Ohashi (2, 4, 6)

Perennial herbs. C Nepal: 1800-2600 m. Nepal, E Himalaya, Tibetan Plateau, Assam-Burma and E Asia.

Indigofera L.

Indigofera atropurpurea Buch.-Ham. ex Hornem. [2, 4, 6]

सागिनो, साखिनु (Saagino, Saakhinu)

Shrubs or small trees. W, C, E Nepal: 700-3200 m. Nepal, W Himalaya, E Himalaya, Tibetan Plateau, Assam-Burma, E Asia and SE Asia.

Use(s): Medicine (root, flowers), Food (flowers).

Indigofera bracteata Graham ex Baker [2, 4, 6]

Shrubs. W, C, E Nepal: 1600-3200 m. Nepal, W Himalaya, E Himalaya, Tibetan Plateau and Assam-Burma.

Indigofera cassioides DC. [6]

Indigofera arborea Roxb.; *Indigofera pulchella* Roxb. (2, 4)

फुस्रे घाँस, रातो मिर्मिरे (Phusre ghaans, Raato mirmire); Cassia Indigo

Shrubs. W, C Nepal: 200-1800 m. Nepal, W Himalaya, E Himalaya, Assam-Burma, S Asia, E Asia and SE Asia.

Use(s): Medicine (root, leaves), Food.

Indigofera cylindracea Graham ex Baker [2, 4, 6, 11]

सखिनिया (Sakhiniya)

Small spreading shrubs. W, C, E Nepal: 900-3700 m. Nepal, W Himalaya, E Himalaya and Tibetan Plateau.

Indigofera dosua Buch.-Ham. ex D.Don [2, 4, 6]

Indigofera virgata Roxb.

फुस्रे घाँस, दुसी स्वाँ (Phusre ghaans, Dushi swaan)

Shrubs. W, C, E Nepal: 1000-3000 m. Nepal, W Himalaya, E Himalaya, Assam-Burma, E Asia and SE Asia.

Use(s): Food (flowers), Fodder.

Indigofera exilis Grierson & D.G.Long [4, 6]

Slender Indigo

Shrubs. W, C, E Nepal: 800-3000 m. Nepal, E Himalaya and Assam-Burma.

Indigofera hebepetala Benth. ex Baker [2, 4, 6]

var. **hebepetala**

रोइरो, मसिनो साकिनो (Roiro, Masino saakino); Himalayan Indigo

Shrubs. W, C, E Nepal: 1900-3400 m. Nepal, W Himalaya and E Himalaya.

Use(s): Food (flowers, fruit), Fodder.

var. **glabra** Ali

Indigofera hebepetala forma *glabra* (Ali) H.Ohashi (2, 4, 11)

Shrubs. W, C, E Nepal: 2000-3200 m. Nepal, W Himalaya and E Himalaya.

Indigofera heterantha Wall. ex Brandis [2, 4, 6, 11]

Indigofera gerardiana Wall. ex Baker

साकिनो, सखिनो (Saakino, Saakhino)

Shrubs. W, C, E Nepal: 600-3100 m. Nepal, W Himalaya, E Himalaya, Tibetan Plateau, Assam-Burma, S Asia, SW Asia and Africa.

Use(s): Food (flowers), Fodder.

Indigofera linifolia (L.f.) Retz. [2, 4, 6]

Hedysarum linifolium L.f.

Perennial herbs. W, C, E Nepal: 100-1300 m. Nepal, W Himalaya, E Himalaya, Assam-Burma, S Asia, E Asia, SE Asia, SW Asia, Africa and Australasia.

Use: Medicine (whole plant).

Indigofera pseudoreticulata Grierson & D.G.Long [4, 6]

Shrubs. C Nepal: 1700-3700 m. Nepal and E Himalaya.

Indigofera stachyodes Lindl.

Indigofera dosua var. *tomentosa* Baker (2, 4)

Shrubs. C, E Nepal: 100-2700 m. Nepal, E Himalaya, Assam-Burma, E Asia and SE Asia.

Indigofera trifoliata L. [2, 4, 6]

var. **trifoliata**

Perennial herbs. C Nepal: 1600-1700 m. Nepal, W Himalaya, E Himalaya, Assam-Burma, S Asia, E Asia, SE Asia and Australasia.

Use: Medicine (seeds).

var. **nepalensis** H.Ohashi [2, 4]

Perennial herbs. E Nepal: *ca.* 1200 m. Endemic to Nepal (?)

Indopiptadenia Brenan

Indopiptadenia oudhensis (Brandis) Brenan [2, 4, 6, 128]

Piptadenis oudhensis Brandis

कफाली, हात्ती पाईल्ले (Kaphaali, Haatti paaile)

Small to medium-sized evergreen trees. W Nepal: 200-600 m. Nepal and W Himalaya.

IUCN Category: **Near Threatened** (ver 3.1).

Lablab Adans.

Lablab purpureus (L.) Sweet [2, 4, 11]

Dolichos purpureus L.

हिउँदे सिमि, टाटे सिमि (Hiunde simi, Taate simi); Hyacinth bean

Annual twining herbs. Widely cultivated as vegetable and legume crop. Native to Africa.

Lathyrus L.

Lathyrus aphaca L. [2, 4, 6]

बन खेसरी, जंगली केराउ (Ban khesari, Jangali keraau); Wild Vetch

Annual herbaceous climbers. W, E Nepal: 200-1200 m. Nepal, W Himalaya, SW Asia, Europe and Africa.

Use: Fodder.

Lathyrus laevigatus (Waldst. & Kit.) Gren.

Orobus laevigatus Waldst. & Kit.

subsp. **emodi** H.Ohashi [2, 4]

Lathyrus emodi (Wall. ex Fritsch) Fritsch

Herbaceous climbers. W Nepal: 2100-3100 m. Nepal and W Himalaya.

Lathyrus pratensis L. [2, 4, 6]

Meadow pea

Herbaceous climbers. W, C Nepal: 2300-3000 m. Nepal, Asia and Europe.

Lathyrus sativus L. [2, 4, 6]

खेसरी (Khesari); Grass pea

Annual herbaceous climbers. W, C, E Nepal: 200-1100 m. Widely cultivated and naturalized in the Old World.

Use(s): Medicine (seeds, seed oil), Food (seeds), Fodder.

Lathyrus sphaericus Retz. [2, 4]

Lathyrus angulatus Benth.

Red grass pea

Herbaceous climbers. W Nepal: 1300-1600 m. Nepal, W Himalaya, C Asia, SW Asia, Europe and Africa.

Lens Mill.

Lens culinaris Medik. [2, 4, 6]

मुसुरो (Musuro); Lentil

Annual herbs. W, C Nepal: 200-1000 m. Widely cultivated as legume crop.

Use: Food (seeds).

Lespedeza Michx.

Lespedeza cuneata (Dum.Cours.) G.Don

Anthyllis cuneata Dum.Cours.; *Lespedeza juncea* var. *sericea* F.B.Forbes & Hemsl. (2, 4)

Chinese bush clover

Subshrubs or perennial herbs. W, C Nepal: 1900-2000 m. Nepal, E Himalaya, E Asia, SE Asia and Australasia.

Lespedeza gerardiana Graham [2, 4, 6, 11]

Subshrubs or perennial herbs. W, C Nepal: 1600-2600 m. Nepal, W Himalaya, E Himalaya and Tibetan Plateau.

Lespedeza juncea (L.f.) Pers. [2, 4, 8]

Hedysarum junceum L.f.

var. **juncea**

भैंसानो (Bhainsaano)

Subshrubs or perennial herbs. W, C Nepal: 1500-3000 m. Nepal, E Asia and N Asia.

Use: Medicine (root).

var. **variegata** (Cambess.) Ali

Lespedeza variegata Cambess.; *Lespedeza juncea* forma *variegata* (Cambess.) H.Ohashi (2, 4, 11)

Subshrubs or perennial herbs. W, C Nepal: 2300-3000 m. Nepal, W Himalaya and E Himalaya.

Lespedeza tomentosa (Thunb.) Siebold ex Maxim. [2, 4]

Hedysarum tomentosum Thunb.

Subshrubs or perennial herbs. W Nepal: 2200-2300 m. Nepal, W Himalaya, Tibetan Plateau, E Asia and N Asia.

Leucaena Benth.

Leucaena leucocephala (Lam.) deWit [2, 4]

Mimosa leucocephala Lam.; *Leucaena latisiliqua* (L.) Gillis (6)

सेतो बबुल (Seto babul); Ipil-ipil, White lead tree

Evergreen small trees. W, C Nepal: 100-1500 m. Widely planted as popular fodder plant. Native to N America (Mexico and C America).

Use: Fodder.

Lotus L.

Lotus corniculatus L. [2, 4, 6, 11]

var. **corniculatus**

Lotus corniculatus var. *minor* Baker

नखर सिम्बी (Nakhar simbi); Bird's foot-trefoil

Perennial herbs. W, C Nepal: 1400-3900 m. Nepal, W Himalaya, Tibetan Plateau, SW Asia, E Asia, Europe, Africa and introduced in Australasia.

Use: Forage.

var. **japonicus** Regel [2, 4, 11]

Lotus corniculatus subsp. *japonicus* (Regel) H.Ohashi (11); *Lotus japonicus* (Regel) K.Larsen

Perennial herbs. W, C Nepal: 1500-3900 m. Nepal, Tibetan Plateau, S Asia and E Asia.

Macrotyloma Verdc.

Macrotyloma uniflorum (Lam.) Verdc. [2, 4, 6]

Dolichos uniflorus Lam.

गहत (Gahat); Horse gram

Perennial or annual climbing herbs. W, C, E Nepal: 100-2800 m. Widely cultivated. Nepal, E Himalaya, Assam-Burma, S Asia, E Asia, Africa and Australasia.

Use(s): Food, Medicine (seeds), Forage.

Medicago L.

Medicago edgeworthii Sirj.

Trigonella pubescens Edgew. ex Baker (2, 4, 6)

Bur clover

Perennial herbs. W, C Nepal: 2200-3600 m. Nepal, W Himalaya, Tibetan Plateau, E Asia and SW Asia.

Medicago falcata L. [2, 4, 11]

Medicago sativa subsp. *falcata* (L.) Arcang. (6, 11)

भिरिन साग, कोटे मान्द्रो (Bhirin saag, Kote maandro)

Perennial herbs. C Nepal: 2600-4000 m. Nepal, Asia and Europe.

Medicago lupulina L. [2, 4, 6, 11]

पहेंलो प्याउली, मेथ्या (Pahenlo pyaauli, Methya); Hop clover

Annual or short-lived perennial herbs. W, C Nepal: 1000-4000 m. Nepal, Asia, Europe, Africa, N America, S America and Australasia.

Use(s): Medicine (stem, leaves, flowers, fruit), Forage.

Medicago sativa L. [2, 4, 6, 11]

Perennial herbs. C Nepal: *ca.* 2900 m (Naturalized). Native to the N Asia and SW Asia.

Medicago × varia Martyn [2, 4, 8, 11]

Perennial herbs. C Nepal: 2700-3500 m. Widely cultivated elsewhere.

Melilotus Mill.

Melilotus albus Desr.

Melilotus albus Medik. (2, 4), *Melilotus officinalis* subsp. *alba* (Medik.) H.Ohashi & Tateishi (6, 11)

मेथी घाँस, मेथी झार (Methi ghaans, Methi jhaar); Yellow sweet clover

Annual herbs. W, C, E Nepal: 200-3800 m. Nepal, E Himalaya, E Asia, N Asia and Africa.

Use: Forage.

Melilotus indicus (L.) All. [2, 4, 6, 11]

Trifolium indicum L.; *Melilotus parviflora* Desf.

मेथी घाँस, मेथी झार (Methi ghaans, Methi jhaar); Sour clover

Annual herbs. C Nepal: 2700-3600 m. Nepal, S Asia, E Asia, C Asia and Europe.

Use(s): Medicine (seeds), Forage.

Melilotus suaveolens Ledeb. [21]

Annual or biennial herbs. C Nepal: *ca.* 2900 m. Nepal and E Asia.

Use(s): Medicine, Food (leaves).

Millettia Wight & Arn.

Millettia extensa (Benth.) Baker [2, 4, 6]

Otosema extensa Benth.; *Millettia auriculata* Brandis

गाउजो (Gaaujo)

Woody climbers. W, C, E Nepal: 100-1000 m. Nepal, W Himalaya, E Himalaya and SE Asia.

Use(s): Fodder, Medicine (root, bark), Cordage, Fish poison (root, bark).

Millettia fruticosa Benth. ex Baker f. [2, 4, 6]

Phaseolodes fruticosum (Benth. ex Baker f.) Kuntze; *Robinia fruticosa* (DC.) Roxb.

टाँटरी (Taantari)

Woody climbers. C, E Nepal: 300-1000 m. Nepal and S Asia (India).

Use(s): Medicine (whole plant), Fish poison.

Millettia glaucescens Kurz [2, 4]

Pongamia angustifolia Buch.-Ham. ex Baker

Woody climbers. C, E Nepal: 300-600 m. Nepal, E Himalaya and Assam-Burma.

Millettia nepalensis R.Parker, *Kew Bull.* **1931**: 42. 1931. [2, 4, 27, 29]

Scandent shrubs. W Nepal: 1500-2500 m. **Endemic to Nepal**.

West: *Doti*, Chusiakana, Bis Ram 48 (Holotype: K).

Millettia pachycarpa Benth. [2, 4]

कुखुरे बिष, काकुष बिष (Kukhure bish, Kaakush bish)

Woody climbers. E Nepal: *ca.* 300 m. Nepal, E Himalaya, Tibetan Plateau, Assam-Burma, S Asia, E Asia and SE Asia.

Use: Medicine (root).

Mimosa L.

Mimosa himalayana Gamble [6]

Mimosa rubicaulis subsp. *himalayana* (Gamble) H.Ohashi (2, 4)

बोक्सी घाँस, अरेली काँडा (Boksi ghaans, Areli kaanda); Himalayan Mimosa

Deciduous shrubs. W, C, E Nepal: 100-1900 m. Nepal, W Himalaya, E Himalaya and SW Asia.

Use(s): Medicine (whole plant), Fence, Charcoal.

Mimosa pudica L. [2, 4, 6]

लज्जावती, बुहारी झार (Lajjawati, buhaari jhaar); Sensitive-plant

Spinous undershrubs. C, E Nepal: 200-1200 m. Invasive species in Nepal. Native to S America (Brazil); naturalized in tropical regions of the world.

Use(s): Medicine (root, leaves), Ornamental.

Mucuna Adans.

Mucuna macrocarpa Wall. [2, 4, 6]

बलदेंग्रा (Baldengra)

Woody climbers. C, E Nepal: 300-1800 m. Nepal, E Himalaya, Assam-Burma, E Asia and SE Asia.

Use(s): Medicine (seeds), Food (seeds), Fodder.

Mucuna nigricans (Lour.) Steud. [2, 4, 6]

Citta nigricans Lour.; *Mucuna imbricata* DC.

काउसो (Kaauso)

Woody climbers. W, C, E Nepal: 600-1200 m. Nepal, W Himalaya, E Himalaya, Assam-Burma, E Asia and SE Asia.

Use(s): Medicine (sap), Fodder.

Mucuna pruriens (L.) DC. [2, 4, 6]

Dolichos pruriens L.; *Stizolobium pruriens* (L.) Medik.

काउसे सिमि, काउसो (Kaause simi, Kaauso); Velvet bean, Cowhage

Herbaceous climbers. W, C, E Nepal: 100-1200 m. Nepal; widely distributed in the tropics.

Use: Medicine (root, seeds).

Ormosia Jacks.

Ormosia glauca Wall. [2, 4, 29]

Fedorovia glauca (Wall.) Yakovlev

Evergreen trees. W, C Nepal: 600-1000 m. Nepal and E Himalaya.

Oxytropis DC.

Oxytropis arenae-ripariae Vassilcz., *Nov. Syst. Pl. Vasc.* **14**: 170. 1977. [2, 4, 12, 27, 29]

Perennial stemless herbs. W Nepal: 4500-4700 m. **Endemic to Nepal**.

West: *Mugu*, Sisne Himal, 15000 ft. (4570 m), O. Polunin, WR Sykes & LHJ Williams 261 (Holotype: BM).

Oxytropis duthieana Ali [2, 4, 8]

Perennial herbs. W, C Nepal: 3500-5000 m. Nepal and W Himalaya.

Oxytropis fasciculiflorum Vassilcz., *Nov. Syst. Pl. Vasc.* **14**: 176. 1977. [2, 4, 12, 27, 29]

Perennial stemless herbs. W, C Nepal: *ca.* 5600 m. **Endemic to Nepal**.

West: *Dolpa*, Chharka bhot, Mukdem Khola, 18500 ft. (5640 m), O. Polunin, WR Sykes & LHJ Williams 1176 (Holotype: BM).

Oxytropis graminetorum Vassilcz., *Nov. Syst. Pl. Vasc.* **14**: 174. 1977. [2, 4, 11, 12, 27, 29]

Perennial stemless herbs. W, C Nepal: 3800-4300 m. **Endemic to Nepal**.

Central: *Mustang*, Taglung, S. Tukucha, 14000 ft. (4270 m), JDA Stainton, WR Sykes & LHJ Williams 1833 (Holotype: BM).

Oxytropis humifusa Kar. & Kir. [2, 4, 8]

Oxytropis lapponica var. *jacquemontiana* Benth. ex Baker

Perennial herbs. C, E Nepal: 3600-4800 m. Nepal, W Himalaya, Tibetan Plateau, N Asia, C Asia and SW Asia.

Oxytropis kansuensis Bunge [2, 4]

Oxytropis leucocephala Ulbr.

Perennial herbs. W, E Nepal: 4200-4500 m. Nepal and Tibetan Plateau.

Oxytropis lapponica (Wahlenb.) Gay [2, 4, 8]

Phaca lapponica Wahlenb.

Perennial herbs. C, E Nepal: 3300-5100 m. Nepal, W Himalaya, Tibetan Plateau, E Asia, N Asia, C Asia and Europe.

Oxytropis microphylla (Pall.) DC. [2, 4, 6, 11]

Phaca microphylla Pall.; *Astragalus microphyllus* (Pall.) Pall.

Perennial herbs. C Nepal: 2100-4100 m. Nepal, W Himalaya, Tibetan Plateau, N Asia, C Asia and SW Asia.

Oxytropis mollis Royle [2, 4, 6, 11]

Oxytropis mollis var. *nepalensis* Kitam.

Perennial herbs. C Nepal: 2500-3900 m. Nepal, W Himalaya and Tibetan Plateau.

Oxytropis morenarum Vassilcz., *Nov. Syst. Pl. Vasc.* **14**: 172. 1977. [2, 4, 12, 27, 29]

Perennial herbs. W, C Nepal: 4000-4600 m. **Endemic to Nepal.**

West: *Rukum*, Chalikhe Pahad, 13000 ft. (3960 m), JDA Stainton, WR Sykes & LHJ Williams 3139 (Holotype: BM).

Oxytropis nepalensis Vassilcz., *Nov. Syst. Pl. Vasc.* **14**: 175. 1977. [2, 4, 8, 11, 12, 27, 29]

Perennial stemless herbs. C, E Nepal: 3500-4500 m. **Endemic to Nepal.**

Central: *Mustang*, Tukucha, Kali Gandaki, 13500 ft. (4110 m), JDA Stainton, WR Sykes & LHJ Williams 1135 (Holotype: BM).

Oxytropis tatarica Cambess. ex Bunge [8, 11]

Perennial herbs. C Nepal: 3500-5200 m. Nepal and W Himalaya.

Oxytropis torrentium Vassilcz., *Nov. Syst. Pl. Vasc.* **14**: 174. 1977. [2, 4, 12, 27, 29]

Perennial stemless herbs. W Nepal: *ca.* 4200 m. **Endemic to Nepal.**

West: *Dolpa*, Phoksumdo Khola, 14000 ft. (4270 m), JDA Stainton 4323 (Holotype: BM).

Oxytropis williamsii Vassilcz., *Nov. Syst. Pl. Vasc.* **14**: 168. 1977. [2, 4, 6, 11, 12, 27, 29]

सेकार (Se-kar: Tibetan)

Perennial herbs. W, C Nepal: 2500-4400 m. **Endemic to Nepal.**

West: *Dolpa*, Barbung Khola, Bandar Phatka, 11500 ft. (3500 m); O. Polunin, WR Sykes & LHJ Williams 1058 (Holotype: BM).

Use(s): Medicine (root, seeds), Fodder.

Paracalyx Ali

Paracalyx scariosus (Roxb.) Ali [2, 4]

Cylista scariosa Roxb.

Woody climbers. E Nepal: 400-500 m. Nepal, Assam-Burma and SE Asia.

Use: Medicine (root).

Parochetus Buch.-Ham. ex D.Don

Parochetus communis Buch.-Ham. ex D.Don [2, 4, 6]

Parochetus major D.Don

चेम्गीफुल, जंगली बदामे झार (Chemgi phul, Jangali Badaame jhaar); Blue Oxalis

Perennial creeping herbs. W, C, E Nepal: 900-4000 m. Nepal, W Himalaya, E Himalaya, Tibetan Plateau, Assam-Burma, S Asia, E Asia and SE Asia.

Use(s): Medicine (root, leaves), Food, Fodder.

Phaseolus L.

Phaseolus lunatus L. [2, 4]

Herbaceous climbers. Widely cultivated in Nepal. Native to tropical America.

Use(s): Food (seeds), Fodder.

Phaseolus vulgaris L. [2, 4, 6]

असारे सिमि, राजमा (Asare simi, Rajma); Common bean

Herbaceous climbers. C, E Nepal: 500-1300 m. Widely cultivated as legume crop. Native to tropical America.

Use(s): Food (fruit/seeds), Medicine (green pods).

Phyllodium Desv.

Phyllodium pulchellum (L.) Desv. [2, 4, 6]

Hedysarum pulchellum L.; *Desmodium pulchellum* (L.) Benth.

एकता झार, भट्टे (Ekta jhaar; Bhatte: Majhi)

Shrubs. W, C, E Nepal: 100-500 m. Nepal, S Asia, E Asia, SE Asia and Australasia.

Use: Medicine (seeds).

Phyllolobium Fisch.

Phyllolobium tribulifolium (Benth. ex Bunge) M.L.Zhang & Podlech [8]

Astragalus tribulifolius Benth. ex Bunge (11)

Perennial herbs. C Nepal: 2900-4400 m. Nepal, Tibetan Plateau, S Asia and E Asia.

Piptanthus D.Don

Piptanthus nepalensis (Hook.) D.Don [2, 4, 6]

Baptisia nepalensis Hook.; *Thermopsis nepaulensis* DC.

सुगा फूल (Suga phool)

Deciduous shrubs. W, C, E Nepal: 2000-3800 m. Nepal, W Himalaya, E Himalaya, Tibetan Plateau, Assam-Burma and E Asia.

Use(s): Fodder, Beams, Huts, Walking sticks, Fish poison.

Pisum L.

Pisum sativum L. [2, 4, 11]

केराउ, मटर, सानो केराउ (Keraau, Matar, Saano keraau); Garden pea

Annual herbaceous climbers. W, C, E Nepal: 1200-4000 m. Widely cultivated, Cosmopolitan.

Pithecellobium Mart.

Pithecellobium dulce (Roxb.) Benth. [2, 4]

Mimosa dulcis Roxb.; *Inga dulcis* (Roxb.) Willd.

जलेबी (Jalebi); Madras thorn

Semi-evergreen trees. W, C Nepal: 700-800 m (Cultivated). Native to tropical America.

Use(s): Medicine (root, bark, leaves), Food (aril), Timber, Tannin (bark).

Pongamia (L.) Merr.

Pongamia pinnata (L.) Merr. [6]

Cytisus pinnatus L.

करेंगी (Karengi); Pongam tree

Evergreen trees. W, C Nepal: 200-300 m. Nepal, Assam-Burma, S Asia, E Asia, SE Asia, Africa, America and Australasia.

Use(s): Medicine (bark, leaves, flowers, seeds), Timber.

Pterocarpus Jacq.

Pterocarpus marsupium Roxb. [6]

बिजय साल (Bijay Saal); Malabar kino

Large deciduous trees. W Nepal: 100-300 m (Rare). Nepal, native to S Asia.

Use(s): Timber, Medicine (bark, heartwood).

IUCN Category: **Vulnerable** A1cd (ver 2.3).

Pueraria DC.

Pueraria montana (Lour.) Merr.

Dolichos montanus Lour.

var. **lobata** (Willd.) Maesen & S.M.Almeida ex Sanjappa & Predeep [8]

Dolichos lobatus Willd.

Robust climbers. C Nepal: *ca.* 2600 m. Native to SE Asia.

Pueraria peduncularis (Graham ex Benth.) Benth. [2, 4, 6]

Neustanthus peduncularis Graham ex Benth.

कदुजा घाँस (Kadujaa ghaans)

Herbaceous climbers. W, C, E Nepal: 2000-3000 m. Nepal, W Himalaya, E Himalaya, Tibetan Plateau and Assam-Burma.

Use(s): Medicine (root), Insecticidal (root), Ritual, Fodder.

Pueraria phaseoloides (Roxb.) Benth. [2, 4, 6]

Dolichos phaseoloides Roxb.; *Pueraria subspicata* Benth.

बिराली लहरो, बिडारी लहरो (Biraali laharo, Bidaari laharo)

Woody climbers. C, E Nepal: 100-1900 m. Nepal, E Himalaya, Assam-Burma, E Asia and SE Asia.

Use(s): Forage, Food (tubers), Fiber (stem).

Pueraria tuberosa (Willd.) DC. [2, 4]

Hedysarum tuberosum Willd.

बिराली कंद (Biraalee kand); Indian kudzu

Deciduous woody climbers. W, C Nepal: 300-1500 m. Nepal and W Himalaya.

Use(s): Medicine (root/tuber), Fodder.

Pueraria wallichii DC. [2, 4, 8]

Dolichos frutescens Buch.-Ham. ex D.Don

Climbing shrubs. C, E Nepal: 900-2000 m. Nepal, E Himalaya, Tibetan Plateau, E Asia and SE Asia.

Rhynchosia Lour.

Rhynchosia falconeri Baker [2, 4, 8]

Twining perennial herbs. W, C Nepal: 1900-2400 m. Nepal and W Himalaya.

Rhynchosia harae H.Ohashi & Tateishi [2, 4]

Twining perennial herbs. E Nepal: 600-700 m. Nepal and E Himalaya.

Rhynchosia himalensis Benth. ex Baker [2, 4, 6]

बन भाटा (Ban Bhaata)

Twining perennial herbs. W Nepal: 2500-2800 m. Nepal, W Himalaya, Tibetan Plateau and E Asia.

Use(s): Medicine (root), Food (fruit).

Rhynchosia minima (L.) DC. [2, 4, 6]

Dolichos minimus L.; *Rhynchosia minima* var. *laxiflora* (Cambess.) Baker

Annual twining herbs. W Nepal: 1800-1900 m. Pantropical.

Rhynchosia nepalensis H.Ohashi & Tateishi, *Bot. Mag.* (Tokyo) **90**: 228. 1977. [2, 4, 12, 27, 29]

Twining perennial herbs. W Nepal: 1000-1100 m. **Endemic to Nepal**.

West: *Salyan*, Sitalpati, 3500 ft. (1070 m), O. Polunin, WR Sykes & LHJ Williams 1235 (Holotype: BM).

Rhynchosia pseudo-cajan Cambess. [2, 4]

Erect shrubs. W Nepal: 1300-2300 m. Nepal and W Himalaya.

Saraca L.

Saraca asoca (Roxb.) deWilde [2, 4, 6]

Jonesia asoca Roxb.

अशोक, असाउ (Ashoka, Ashaau); Ashok tree

Evergreen trees. C, E Nepal: 100-1400 m. Nepal, W Himalaya, Assam-Burma and S Asia.

Use(s): Medicine (bark, flowers), Sacred tree.

IUCN Category: **Vulnerable** B1+2c (ver 2.3).

Senegalia Raf.

Senegalia gageana (Craib) Maslin, Seigler & Ebinger [68]

Acacia gageana Craib

Scrambling shrubs. Nepal, W Himalaya, E Himalaya, Assam-Burma and S Asia.

Senegalia megaladena (Desv.) Maslin, Seigler & Ebinger [68]

Acacia megaladena Desv.

Climbing shrubs. Nepal, Assam-Burma, S Asia, E Asia and SE Asia.

Senegalia pennata (L.) Maslin [2, 4, 6]

Mimosa pennata L.; *Acacia pennata* (L.) Willd.

subsp. **pennata**

अर्खु, अर्फु, अरारे (Arkhu, Arphu, Araare); Climbing wattle

Woody climbers. W, C, E Nepal: 200-1100 m. Nepal, S Asia and SE Asia.

Use(s): Medicine (root, bark, leaves, pods), Food, Fodder, Tannin, Washing clothes, Fish poison.

subsp. **kerri** (I.C.Nielsen) Maslin [67]

Acacia pennata subsp. *kerri* I.C.Nielsen

Climbing shrubs. Nepal, E Himalaya, Assam-Burma, S Asia, E Asia, SE Asia and Australasia.

Senegalia rugata (Lam.) Britton & Pose

Mimosa rugata Lam.; *Acacia rugata* (Lam.) Voigt (2, 4); *Mimosa concinna* Willd.; *Acacia sinuata* (Lour.) Merr.

सिकाकाई, अरारे (Sikakaee, Araare)

Small trees or climbing shrubs. W, C Nepal: 100-800 m. Nepal, E Himalaya, Assam-Burma, S Asia, E Asia and SE Asia.

Use(s): Medicine (leaves, pods), Washing hair and clothes (fruit).

Senna Mill.

Senna floribunda (Cav.) H.S.Irwin & Barneby [6]

Cassia floribunda Cav. (2, 4); *Cassia laevigata* Willd.

छिन् छिने बिष (Chhin chhine bish)

Evergreen large shrubs or small trees. W, C, E Nepal: 700-2200 m. Naturalized in tropical Himalaya. Native to tropical America.

Use(s): Food (fruit), Medicine (whole plant).

Senna occidentalis (L.) Link [6]

Cassia occidentalis L. (2, 4)

कसौडी, ताप्रे (Kasaudi, Taapre)

Shrubs or subshrubs. W, C, E Nepal: 200-1400 m. Invasive species in Nepal. Probably native to tropical America.

Use: Medicine (whole plant).

Senna sophera (L.) Roxb. [6]

Cassia sophera L. (2, 4)

ताप्रे (Taapre)

Shrubs or subshrubs. C, E Nepal: 100-1100 m. Cultivated or naturalized in W Himalaya, E Himalaya, E Asia, S Asia, SE Asia, Africa and Australasia.

Use: Medicine (whole plant).

Senna tora (L.) Roxb. [6]

Cassia tora L. (2, 4)

ताप्रे (Taapre); Fever weed

Annual suffrutescent herbs. W, C, E Nepal: 400-1300 m. Invasive species in Nepal. Native to tropical America.

Use: Medicine (root, leaves, seeds).

Sesbania Scop.

Sesbania bispinosa (Jacq.) W.Wight [2, 4, 6]

Aeschynomene bispinosa Jacq.

धैंचा, काँडे धैंचा (Dhaincha, Kaande dhaincha)

Perennial suffrutescent herbs. Widely planted as shade plant and green manure.

Sesbania grandiflora (L.) Poir. [2, 4, 6]

Robinia grandiflora L.; *Aeschynomene grandiflora* (L.) L.

अगस्ती, धैंचा (Agasti, Dhaincha); Swamp pea

Small deciduous trees. E Nepal: 200-300 m (Cultivated). Probably native to SE Asia.

Use(s): Ornamental, Medicine (root, flowers, leaves), Food (young leaves), Tannin (bark).

Sesbania sesban (L.) Merr. [2, 4, 6]

Aeschynomene sesban L.; *Sesbania aegyptiacus* Poir.

जईता, धनदैयाँ जईता (Jaitaa, Dhandaian jaitaa); Common sesban

Shrubs or small trees. W, E Nepal: 200-400 m (Cultivated). Probably native to tropical Asia.

Use(s): Medicine (leaves), Fodder, Green manure, Fiber.

Shuteria Wight & Arn.

Shuteria hirsuta Baker [2, 4]

Herbaceous climbers. E Nepal: *ca.* 1200 m. Nepal, E Himalaya, Assam-Burma and SE Asia.

Shuteria involucrata (Wall.) Wight & Arn. [2, 4]

Glycine involucrata Wall.; *Shuteria vestita* var. *involucrata* (Wall.) Baker

var. **involucrata**

दोंगसिंग नोबा (Dongsing nobaa: Tamang)

Herbaceous climbers. C, E Nepal: 1000-2800 m. Nepal, E Himalaya, Assam-Burma, E Asia and SE Asia.

Use: Medicine (whole plant).

var. **glabrata** (Wight & Arn.) H.Ohashi [2, 4]

Shuteria glabrata Wight & Arn.; *Shutera vestita* var. *glabrata* (Wight & Arn.) Baker

Herbaceous climbers. E Nepal: 1000-2800 m. Nepal, W Himalaya, E Himalaya, S Asia and SE Asia.

var. **vestita** (Wight & Arn.) H.Ohashi

Shuteria vestita Wight & Arn. (6)

Herbaceous climbers. C Nepal: *ca.* 2000 m. Nepal and E Asia (?)

Use: Green manure.

Smithia Aiton

Smithia ciliata Royle [2, 4, 6]

नरीमुरागो (Narimuraago)

Annual herbs. W, C, E Nepal: 1200-2900 m. Nepal, W Himalaya, E Himalaya, E Asia and SE Asia.

Use: Medicine (whole plant).

Smithia conferta Sm. [2, 4]

Smithia geminiflora Roth

लक्ष्मण बुटी (Laxaman buti)

Annual herbs. C Nepal: *ca.* 1500 m. Nepal, S Asia, E Asia, SE Asia and Australasia.

Use(s): Medicine (whole plant), Food (leaves).

Smithia sensitiva Aiton [2, 4, 6]

अम्पी (Ampi: Danuwar)

Annual herbs. W, C, E Nepal: 200-2000 m. Nepal, W Himalaya, E Himalaya, S Asia, E Asia, SE Asia, Africa and Australasia.

Use(s): Medicine (whole plant), Forage.

Sophora L.

Sophora mollis (Royle) Baker [2, 4]

Edwardsia mollis Royle

Deciduous shrubs. W, C Nepal: *ca.* 1700 m. Nepal, W Himalaya, E Asia and SW Asia.

Use: Medicine (root, flowers, seeds).

Sophora moorcroftiana Benth. ex Baker [8]

Caragana moorcroftiana Benth.

var. **moorcroftiana**

Spiny shrubs. C Nepal: 2800-3000 m. Nepal, E Himalaya, Tibetan Plateau, Assam-Burma and E Asia.

var. **nepalensis** Kitam. [2, 4, 6, 11]

प्रवल सिम्बी (Prawal simbi)

Spiny shrubs. C Nepal: 2800-3900 m. Endemic to Nepal (?)

Use(s): Medicine (seeds), Firewood.

Sophora velutina Lindl.

Sophora glauca Lesch. ex DC. (2, 4)

Shrubs. C Nepal: 2400-2500 m. Nepal, Assam-Burma, S Asia and E Asia.

Sophora wightii Baker

subsp. **bhutanica** (H.Ohashi) Grierson & D.G.Long [4]

Sophora bhutanica H.Ohashi

Shrubs or small trees. Nepal, E Himalaya and S Asia.

IUCN Category: **Endangered** B1+2c (ver 2.3).

Note: *Sophora wightii* Baker is treated as the Endangered species (www.iucnredlist.org).

Spatholobus Hassk.

Spatholobus parviflorus (Roxb.) Kuntze [2, 4, 6]

Butea parviflora Roxb.; *Spatholobus roxburghii* Benth.

देब्रे लहरा (Debre laharaa)

Perennial woody climbers. W, C, E Nepal: 100-2000 m. Nepal, S Asia, E Asia and SE Asia.

Use(s): Fodder, Fiber (bark).

Spongiocarpella Yakovlev & N.Ulziykh.

Spongiocarpella purpurea (P.C.Li) Yakovlev

Chesneya purpurea P.C.Li; *Chesneya polystichoides* subsp. *bhutanica* H.Ohashi (2, 4)

Perennial herbs. C Nepal: *ca.* 4000 m. Nepal and E Himalaya.

Tadehagi H.Ohashi

Tadehagi pseudotriquetrum (DC.) Y.C.Yang & P.H.Huang

Desmodium pseudotriquetrum DC.; *Pteroloma triquetrum* Benth.; *Tadehagi triquetrum* subsp. *pseudotriquetrum* (DC.) H.Ohashi (2, 4, 6)

सानो भट्टे (Saano bhatte)

Shrubs or subshrubs. W, C, E Nepal: 400-1100 m. Nepal, W Himalaya, E Himalaya, E Asia and SE Asia.

Use: Medicine (leaves).

Tamarindus L.

Tamarindus indica L. [2, 4, 6]

Tamarindus officinalis Hook.

इमली, तित्रि (Imli, Titri); Tamarind

Evergreen trees. W, C, E Nepal: 200-900 m. Pantropical.

Use(s): Medicine (bark, leaves, fruit), Food (fruit), Fodder, Tannin.

Tephrosia Pers.

Tephrosia candida DC. [2, 4, 6]

काँडे साखिनु (Kaande saakhinu)

Perennial suffrutescent herbs. C, E Nepal: 300-1300 m. Nepal, W Himalaya, Assam-Burma, E Asia and SE Asia.

Use: Medicine (whole plant).

Tephrosia purpurea (L.) Pers. [2, 4, 6]

Cracca purpurea L.; *Galega lanceaefolia* Roxb.

काँडे साखिनु (Kaande saakhinu)

Perennial herbs. C Nepal: 100-200 m. Nepal, S Asia, E Asia and SE Asia.

Use(s): Medicine (whole plant), Green manure, Fish poison.

Thermopsis R.Br.

Thermopsis barbata Royle [2, 4, 6]

पेल्चे मेंदो (Pelche mendo: Sherpa)

Perennial herbs. W, C Nepal: 2700-4600 m. Nepal, W Himalaya, Tibetan Plateau and E Asia.

Use(s): Medicine (young root stocks), Food.

Thermopsis lanceolata R.Br. ex W.T.Aiton [2, 4, 6, 11]

Perennial herbs. W, C Nepal: 3600-4300 m. Nepal, Tibetan Plateau, N Asia and C Asia.

Use: Medicine (young root stocks).

Tibetia (Ali) H.B.Cui

Tibetia himalaica (Baker) H.P.Tsui [6, 11]

Gueldenstaedtia himalaica Baker (2, 4)

से-मा (Se-maa: Tibetan)

Perennial herbs. W, C Nepal: 3300-4600 m. Nepal, W Himalaya, E Himalaya, Tibetan Plateau and E Asia.

Use: Medicine (whole plant).

Trifolium L.

Trifolium pratense L. [6, 98]

बेउली (Beulee); Red clover

Perennial herbs. C Nepal: 1300-1500 m (Naturalized). Native to SW Asia, Europe and N Africa.

Use: Forage.

Trifolium repens L. [2, 4]

बेउली, सेतो बेउली (Beulee, Seto beulee); White clover

Perennial herbs. C, E Nepal: 1300-2800 m (Naturalized). Native to C Asia, Europe and N Africa.

Use(s): Fodder, Medicine (leaves, flowers).

Trifolium resupinatum L. [22]

Perennial herbs. W Nepal: *ca.* 2400 m. Nepal, W Himalaya, SW Asia and Europe.

Trigonella L.

Trigonella corniculata (L.) L. [2, 4, 6]

Trifolium corniculatum L.

Annual herbs. W, C Nepal: 3700-3800 m. Nepal, W Himalaya, E Himalaya, C Asia, SW Asia and Europe.

Use: Medicine (fruit).

Trigonella emodi Benth. [2, 4, 6, 11]

Trigonella emodi var. *himalaica* Sirj.

बुशुहांग (Busuhaang: Tibetan)

Perennial herbs. W, C, E Nepal: 1300-4900 m. Nepal, W Himalaya, C Asia and SW Asia.

Use: Medicine (root).

Trigonella fimbriata Royle ex Benth.

Trigonella emodi subsp. *fimbriata* (Royle ex Benth.) H.Ohashi (2, 4)

Perennial herbs. C Nepal: 1800-1900 m. Nepal and W Himalaya.

Trigonella foenum-graceum L. [15]

मेथी (Methi); Fenugreek

Annual herbs. Widely cultivated as vegetable crop and condiments.

Trigonella gracilis Benth. [2, 4, 6, 11]

तिन पाते झार (Tinpaate jhaar)

Perennial herbs. W, C Nepal: 2200-3400 m. Nepal and W Himalaya.

Use: Medicine (whole plant).

Uraria Desv.

Uraria lagopodioides (L.) Desv. [2, 4, 6]

Hedysarum lagopodioides L.; *Uraria lagopoides* DC.

प्रिश्नी पार्णी, सेतो भट्टे (Prishnee parnee, Seto bhatte)

Trailing herbs or undershrubs. C, E Nepal: 100-1400 m. Nepal, W Himalaya, E Himalaya, Assam-Burma, E Asia, SE Asia and Australasia.

Use(s): Medicine (root, leaves), Tannin.

Uraria lagopus DC. [2, 4, 6]

निलो टाँकि (Nilo taanki)

Shrubs. W, C, E Nepal: 700-2400 m. Nepal, W Himalaya and E Himalaya.

Uraria neglecta Prain

Uraria lagopus var. *neglecta* (Prain) H.Ohashi (2, 4)

Perennial herbs or subshrubs. C Nepal: 700-1500 m. Nepal, W Himalaya, S Asia and E Asia.

Uraria picta (Jacq.) Desv. ex DC. [2, 4, 6]

Hedysarum pictum Jacq.

प्रिश्नी पार्णी (Prishni paarnee)

Subshrubs or shrubs. W Nepal: 100-1100 m. Nepal, E Himalaya, Assam-Burma, S Asia, E Asia, SE Asia, Africa and Australasia.

Use: Medicine (root, leaves, fruit).

Uraria rufescens (DC.) Schindl. [2, 4, 6]

Desmodium rufescens DC.

Subshrubs. C Nepal: 100-1000 m. Nepal, Tibetan Plateau, Assam-Burma, S Asia, E Asia and SE Asia.

Vicia L.

Vicia bakeri Ali [2, 4, 6]

बन कुलो (Ban kulo); Wood vetch

Annual herbaceous climbers. W, C Nepal: 2700-4200 m. Nepal, W Himalaya, Tibetan Plateau and E Asia.

Use: Food (tender fruit).

Vicia faba L. [6, 98]

बकुला (Bakula); Broad bean

Annual herbs. W, C Nepal: 100-1500 m. Widely cultivated as legume crop. Native to SW Asia (Middle East).

Vicia hirsuta (L.) Gray [2, 4, 6]

Ervum hirsutum L.

झुसे कुटुली कोसा, मुनमुन (Jhuse kutulikosa, Munmun); Tiny vetch

Annual herbaceous climbers. W, C, E Nepal: 200-2700 m. Nepal, Asia (except for SE Asia), Europe and Africa.

Use(s): Food (tender shoot, leaves, fruit), Medicine (seeds).

Vicia sativa L. [98]

subsp. **sativa**

Common vetch

Annual herbaceous climbers. Widely cultivated throughout the world.

subsp. **nigra** (L.) Ehrh.

Vicia sativa var. *nigra* L.; *Vicia angustifolia* L. (2); *Vicia sativa* var. *angustifolia* L. (6, 11)

कुटुली कोसा (Kutuli kosa)

Annual climbing herbs. W, C, E Nepal: 200-4000 m. Nepal, Asia, Europe and Africa.

Use(s): Food, Fodder.

Vicia tenuifolia Roth [2, 4]

कुटुली कोसा (Kutuli kosa); Vetch

Annual herbaceous climbers. W, C Nepal: 1600-2200 m. Nepal, W Himalaya, N Asia, C Asia, SW Asia, Europe and Africa.

Vicia tetrasperma (L.) Schreb. [6]

Ervum tetraspermum L.

सानो कुटुली कोसा (Sano kutulikosa); Lentil vetch

Annual herbs. C Nepal: 1500-1600 m (Naturalized). Native to Asia, Europe and N Africa.

Vigna Savi

Vigna angularis (Willd.) Ohwi & H.Ohashi

Dolichos angularis Willd.; *Vigna angularis* var. *nipponensis* (Ohwi) Ohwi & H.Ohashi (2, 4, 6)

मास लहरी (Maas lahari); Red mung bean, Azuki bean

Annual herbaceous climbers. W, C Nepal: 1500-2600 m. Nepal and E Asia.

Vigna mungo (L.) Hepper [2, 4, 6]

Phaseolus mungo L.

मस, उरद (Maas, Urad); Black gram

Annual climbing herbs. C, E Nepal: 400-2100 m. Widely cultivated as legume crop. Native to S Asia (India).

Use(s): Medicine/Food (seeds), Green manure.

Vigna nepalensis Tateishi & Maxted [88]

Annual herbs. E Nepal: 1300-1400 m. Nepal, E Himalaya and Assam-Burma.

Use(s): Food, Fodder.

Vigna umbellata (Thunb.) Ohwi & H.Ohashi [2, 4, 8]

Dolichos umbellatus Thunb.; *Phaseolus pubescens* Blume

मस्याङ्ंग (Masyaang); Rice bean

Annual climbers. C, E Nepal: 400-2100 m. Widely cultivated as legume crop. Native to Cultivated in tropics to warm temperate regions in Asia and Africa.

Vigna unguiculata (L.) Walp. [2, 4]

Dolichos unguiculatus L.; *Vigna sinensis* var. *elongata* Miq.

subsp. **unguiculata**

कात्तिके बोडी, बोडी (Kaattike bodi, Bodee); Cow pea

Annual herbaceous climbers. E Nepal: 800-1300 m. Cultivated in Tropical to warm Temperate regions in Asia and Africa.

Use(s): Food, Medicine (seeds).

subsp. **cylindrica** (L.) Verdc.

Phaseolus cylindricus L.; *Vigna unguiculata* var. *catjang* (Burm.f.) Bertoni (2, 4)

Annual herbaceous climbers. Widely cultivated in Nepal. Native to Africa.

subsp. **sesquipedalis** (L.) Verdc. [4]

Dolichos sesquipedalis L.; *Vigna unguiculata* var. *sesquipedalis* (L.) H.Ohashi (2, 4)

ताने बोडी (Taane bodee); Asparagus bean

Annual herbaceous climbers. E Nepal: 800-1300 m. Widely cultivated.

Vigna vexillata (L.) A.Rich. [8]

Phaseolus vexillatus L.

var. **vexillata**

Wild cowpea

Annual climbers. W, C Nepal: 100-1400 m. Nepal, widely distributed in tropical and subtropical regions.

var. **angustifolia** (Schumach. & Thonn.) Baker [2, 4]

Phaseolus angustifolius Schumach. & Thonn.

Annual herbaceous climbers. W, C Nepal: 900-1900 m. Nepal, Asia, Africa and Australasia.

Zornia J.F.Gmel.

Zornia gibbosa Span. [2, 4, 6]

Annual prostrate herbs. W, C, E Nepal: 400-2200 m. Nepal, E Himalaya, Assam-Burma, S Asia, E Asia, SE Asia and Australasia.

Use(s): Medicine (root), Forage.

Exotic/Cultivated Taxa

Canavalia ensiformis (L.) DC., **खुँडे सिमि, तरवारे सिमि** (Khunde simi, Tarbaare simi); Sword bean

Crotalaria juncea L. (Sunn hemp). Cultivated. Native to S Asia (India). [98]

Cyamopsis tetragonoloba (L.) Taub. (**चारपाते सिमि, ग्वार सिमि**; Chaar-paate simi, Gwaar simi; Cluster bean). Probably native to W Himalaya (NW India).

Erythrina blakei R.Parker (**पंगार,** Pangaar; Blake's coral tree). Native to America. [14]

Erythrina crista-galli L. (Cockspur coral tree). Native to S America. [14]

Glycyrrhiza glabra L. (**जेठी मधु;** Jethi madhu; Liquorice, Licorice). Native to Southern Europe, part of Asia (India). *Use*: Medicine (root).

Lathyrus odoratus L. (**केराउ फूल,** Kerau phul; Sweet pea). Planted in garden as ornamental plant. Native to Italy. [4, 14, 54, 98]

Pachyrhizus erosus (L.) Urb. (**मिश्री कन्द, केशौर;** Mishrikand, Keshaur); Yam bean.

Pisum sativum var. **arvense** (L.) Poir., **सानो केराउ** (Sano kerau).

Prosopis juliflera (Sw.) DC., (Prosopis, Mesquite). Native to N America (Mexico) and S America.

Senna alata (L.) Roxb. Native to Tropical America. [66]

Senna didymobotrya (Fresen.) H.S.Inwin & Barneby (*Cassia nairobi-ensis* L.H.Bailey), (**गैंडे फूल;** Gainde phool; African Senna). Native to Africa. [14]

Sesbania cannabina (Retz.) Pers. (**धैंचा, रातो दांठे धैंचा;** Dhaincha, Rato danthe dhaincha). Native to Australasia.

Sesbania rostrata Bremek. & Oberm. (**धैंचा, गिर्खे धैंचा;** Dhaincha, Girkhe dhaincha). Native to Africa.

Stylosanthes fruticosa (Retz.) Alston (Pencil flower). Native to Asia and tropical Africa.

Vigna radiata (L.) R.Wilczek (**मूँगी,** Mungi; Mung bean, Green gram). Native to Indian subcontinent (?)

Wisteria sinensis (Sims) Sweet (**निलहार फूल,** Nilhaar phool; Chinese wisteria). Woody climbers. Planted in the garden as ornamental plant. Native to E Asia (China). [6, 14]

Doubtful/Uncertain Taxa

Adenanthera pavoniana L. (Rakta chandan); (Coral wood tree); E Nepal: *ca.* 300 m. [123]

Alysicarpus glumaceus (Vahl) DC. W Nepal: *ca.* 200 m (?) [20]

Astragalus arnoldii Hemsl. & H.Pearson [30, 105]

Astragalus heydei Baker [30]

Astragalus oxyodon Baker [30]

Astragalus tenuicaulis Benth. ex Bunge [30]

Astragalus thomsonii Podlech [30, 105]

Astragalus tribulifolius Bunge [30]

Astragalus tumbatsicus C.Marquand & Airy Shaw [30]

Astragalus webbianus Graham ex Benth. [30]

Caragana crassispina C.Marquand [2, 4]

Caragana limprichtii Harms [2, 4]

Caragana polyacantha Royle [2, 4]

Derris hainesiana Thoth. [2, 4]

Indigofera astragalina DC. [120]

Indigofera constricta (Thwaites) Trimen [98]

Indigofera linnaei Ali [120, 121]

Medicago denticulata Willd. "Sankhe Pyauli' [98]

Pithecellobium bigeminum (L.) Benth. [2]

Rhynchosia viscosa DC. [98]

Vachellia eburnea (L.f.) P.J.H.Hurter & Mabb. (*Acacia eburnea* Willd.) [98]

POLYGALACEAE

Polygala L.

Polygala abyssinica R.Br. ex Fresen. [2, 4, 8]

Polygala scabra Edgew.

Perennial herbs. W, C Nepal: 1200-2700 m. Nepal, W Himalaya, S Asia, SW Asia and Africa.

Use: Medicine (root).

Polygala arillata Buch.-Ham. ex D.Don [2, 4, 6]

Chamaebuxus arillata (Buch.-Ham. ex D.Don) Hassk.

var. **arillata**

लुइँचे फूल (Luinche phool)

Shrubs or small trees. C, E Nepal: 1500-2700 m. Nepal, E Himalaya, Tibetan Plateau, Assam-Burma, S Asia, E Asia and SE Asia.

Use: Medicine (root, leaves).

var. **laevicarpa** R.N. Banerjee & G.S.Giri [4]

Shrubs. E Nepal: 1500-1600 m. Nepal and E Himalaya.

Polygala crotalarioides Buch.-Ham. ex DC. [2, 4, 6]

Perennial herbs. W, C Nepal: 100-3000 m. Nepal, W Himalaya, E Himalaya, Tibetan Plateau, Assam-Burma, S Asia, E Asia and SE Asia.

Use: Medicine (whole plant).

Polygala elegans Wall. ex Royle

Polygala sibirica var. *elegans* (Wall. ex Royle) H.Hara (2, 4); *Polygala sibirica* subsp. *elegans* (Wall. ex Royle) Kit Tan (8)

Perennial herbs. W, C, E Nepal: 1600-3600 m. Nepal, W Himalaya, E Himalaya, Tibetan Plateau and Assam-Burma.

Use: Medicine (root).

Polygala furcata Royle [2, 4, 6]

Polygala triphylla Buch.-Ham. ex D.Don

Annual herbs. W, C, E Nepal: 600-2200 m. Nepal, W Himalaya, E Himalaya, Assam-Burma, S Asia and E Asia.

Polygala linariifolia Willd. [2, 4, 6]

Polygala brachystachya Blume; *Polygala chinensis* var. *linariifolia* (Willd.) Chodat

Perennial herbs. W, C, E Nepal: 200-2900 m. Nepal, W Himalaya, E Himalaya, Assam-Burma, S Asia, E Asia, SE Asia and Australasia.

Polygala longifolia Poir. [2, 4, 6]

Polygala leptalea DC.; *Polygala oligophylla* DC.

Annual herbs. W, C Nepal: 900-1500 m. Nepal, W Himalaya, E Himalaya, Assam-Burma, S Asia, E Asia, SE Asia and Australasia.

Use: Medicine (root).

Polygala persicariifolia DC. [2, 4, 6]

Polygala buchanani D.Don

बिजकन्ने (Bijkanne: Chepang)

Annual herbs. W, C, E Nepal: 700-2200 m. Nepal, W Himalaya, E Himalaya, Assam-Burma, S Asia, E Asia, SE Asia, SW Asia, Africa and Australasia.

Use: Fermenter.

Polygala sibirica L. [2, 4, 6, 11]

Polygala monopetala Cambess.

Perennial herbs. W, C Nepal: 1800-3600 m. Nepal, W Himalaya, E Himalaya, Tibetan Plateau, S Asia, E Asia, SE Asia, N Asia and Europe.

Use: Medicine (root, leaves).

Polygala tatarinowii Regel [2, 4, 6, 11]

कुरकुरे झार, फरबरे घाँस (Kurkure jhaar, Pharware ghaans)

Annual herbs. W, C Nepal: 1800-2900 m. Nepal, W Himalaya, E Himalaya, Tibetan Plateau, Assam-Burma, S Asia, E Asia, SE Asia and N Asia.

Use: Medicine (whole plant).

Salomonia Lour.

Salomonia cantoniensis Lour. [2, 4, 6]

Salomonia edentula DC.

मेथी घाँस (Methi ghaans)

Annual herbs. C, E Nepal: 700-1600 m. Nepal, E Himalaya, Assam-Burma, S Asia, E Asia and SE Asia.

Use: Medicine (leaves).

Salomonia ciliata (L.) DC. [2, 4]

Polygala ciliata L.; *Salomonia oblongifolia* DC.

Annual herbs. C Nepal: 800-1000 m. Nepal, E Himalaya, Assam-Burma, S Asia, E Asia, SE Asia and Australasia.

Securidaca L.

Securidaca inappendiculata Hassk. [2, 4]

Securidaca scandens Buch.-Ham. ex Hassk.

Climbing shrubs. C Nepal: *ca.* 500 m. Nepal, E Himalaya, Assam-Burma, S Asia, E Asia and SE Asia.

ROSALES

ROSACEAE

Agrimonia L.

Agrimonia pilosa Ledeb.

var. **nepalensis** (D.Don) Nakai [2, 4, 10]

Agrimonia nepalensis D.Don; *Agrimonia eupatoria* var. *nepalensis* (D.Don) Kuntze; *Agrimonia pilosa* forma *nepalensis* (D.Don) Murata (6)

कठलंगे, शिला झार (Kathlange, Shilaa jhaar)

Perennial herbs. W, C, E Nepal: 1000-3900 m. Nepal, W Himalaya, E Himalaya, Tibetan Plateau, Assam-Burma and E Asia.

Use: Medicine (root).

Alchemilla L.

Alchemilla vulgaris L. [10]

Perennial herbs. C Nepal: 4000-4300 m. Nepal, W Himalaya, SW Asia and Europe.

Argentina Hill

Argentina stenophylla (Franch.) Soják [74]

Potentilla peduncularis var. *stenophylla* Franch.; *Potentilla stenophylla* var. *emergens* Cardot (6)

Perennial herbs. E Nepal: *ca.* 4400 m. Nepal, E Himalaya and E Asia.

Argentina turfosa (Hand.-Mazz.) Soják [74]

Potentilla turfosa Hand.-Mazz. (6); *Potentilla turfosa* var. *gracilescens* (T.T.Yu & C.L.Li) H.Ikeda & H.Ohba

Perennial herbs. E Nepal: 3800-4200 m. Nepal, Tibetan Plateau, Assam-Burma and E Asia.

Aruncus L.

Aruncus dioicus (Walter) Fernald

Actaea dioica Walter; *Spiraea aruncus* L.

var. **triternatus** (Maxim.) H.Hara [2, 4, 6, 10]

Aruncus sylvester var. *triternausa* Maxim.; *Spirarea triternata* Wall., nom.nud.

Perennial herbs. W, C, E Nepal: 3000-4200 m. Nepal, W Himalaya, E Himalaya, Tibetan Plateau and E Asia.

Chaenomeles Lindl.

Chaenomeles japonica (Thunb.) Lindl. ex Spach. [6, 14]

Pyrus japonica Thunb.

Japanese Quince

Deciduous spiny shrubs. C Nepal: *ca.* 1300 m. Planted in the garden as ornamental plant. Native to E Asia.

Cotoneaster Medik.

Cotoneaster acuminatus Wall. ex Lindl. [2, 4, 6, 10, 29]

Cotoneaster bisramianus G.Klotz (2, 4); *Cotoneaster paradoxus* G.Klotz (2, 4); *Cotoneaster sanguineus* T.T.Yu (2, 4); *Cotoneaster staintonii* G.Klotz (2, 4)

ढल्के फूल (Dhalke phool)

Deciduous shrubs. W, C, E Nepal: (1500-)2500-3700(-4200) m. Nepal, W Himalaya, E Himalaya, Tibetan Plateau, Assam-Burma and E Asia.

Use: Walking sticks.

Cotoneaster adpressus Bois [2, 4, 6, 10]

Cotoneaster distichus var. *duthieanus* C.K.Schneid.; *Cotoneaster duthieanus* (C.K.Schneid.) G.Klotz (2, 4, 8)

Deciduous shrubs. W, C, E Nepal: (1900-)3000-4500 m. Nepal, W Himalaya, E Himalaya, Assam-Burma and E Asia.

Cotoneaster affinis Lindl. [2, 4, 6, 10, 29]

Cotoneaster bacillaris Wall. ex Lindl. (2, 4); *Cotoneaster bacillaris* var. *affinis* Hook.f.; *Cotoneaster confusus* G.Klotz ex Kumar & Panigrahi (2, 4); *Cotoneaster frigidus* var. *affinis* Wenz.; *Cotoneaster obtusus* Wall. ex Lindl. (4, 65); *Cotoneaster virgatus* G.Klotz (2, 4); *Cotoneaster wattii* G.Klotz (2, 4); *Mespilus affinis* (Lindl.) D.Don

काउसे फूल, राईन्स (Kaause phool, Raaeens)

Deciduous shrubs. W, C, E Nepal: 1100-(2200-3000)-3900 m. Nepal, W Himalaya, E Himalaya, Tibetan Plateau, Assam-Burma, S Asia, E Asia and C Asia.

Use(s): Medicine (root), Walking sticks.

Cotoneaster buxifolius Wall. ex Lindl. [10]

Evergreen shrubs. W Nepal: 2200-2900 m. Nepal, W Himalaya, E Himalaya, Tibetan Plateau, Assam-Burma and E Asia.

Cotoneaster frigidus Wall. ex Lindl. [2, 4, 6, 10]

छार्वा (Chharwaa: Tibetan)

Deciduous shrubs or small trees. W, C, E Nepal: (900-)2200-3400 m. Nepal, W Himalaya, E Himalaya and Tibetan Plateau.

Use: Medicine (fruit).

Cotoneaster hebephyllus Diels [6, 10]

Cotoneaster ludlowii G.Klotz (2, 4); *Cotoneaster rotundifolius* Wall. ex Lindl. (4, 6, 65); *Cotoneaster schlechtendalii* G.Klotz (2, 4); *Cotoneaster sherriffii* G.Klotz (2, 4).

Semi-evergreen shrubs. W, C Nepal: 2000-4100 m. Nepal, E Himalaya, Tibetan Plateau, Assam-Burma and E Asia.

Use: Medicine (fruit).

Cotoneaster integrifolius (Roxb.) G.Klotz [2, 4, 10, 29]

Crataegus integrifolius Roxb.; *Cotoneaster congestus* Baker (2, 4); *Cotoneaster linearifolius* (G.Klotz) G.Klotz (2, 4); *Cotoneaster marginatus* (Lindl. ex Loudon) Schltdl. (2, 4); *Cotoneaster meuselii* G.Klotz (2, 4); *Cotoneaster prostratus* Baker (4, 65); *Cotoneaster poluninii* G.Klotz (2, 4); *Cotoneaster rotundifolius* var. *lanatus* (Dippel) C.K.Schneid. (2)

Semi-evergreen shrubs. W, C, E Nepal: 1100-3500(-3900) m. Nepal, W Himalaya, E Himalaya, Tibetan Plateau, Assam-Burma and E Asia.

Cotoneaster microphyllus Wall. ex Lindl. [2, 4, 6, 10]

Cotoneaster microphyllus var. *cochleatus* (Franch.) Rehder & E.H.Wilson; *Cotoneaster microphyllus* var. *glacialis* (Baker) Koehne; *Cotoneaster nivalis* (G.Klotz) Panigrahi & A.Kumar (4, 65)

खरेटो, घंगारू (Khareto, Ghangaaru)

Evergreen shrubs. W, C, E Nepal: 900-5400 m. Nepal, W Himalaya, E Himalaya, Tibetan Plateau, Assam-Burma and E Asia.

Use(s): Medicine (fruit), Food (fruit), Firewood, Fence.

Cotoneaster nitidus Jacq. [2, 4, 6, 10]

Cotoneaster cavei G.Klotz (2, 4); *Cotoneaster distichus* Lange; *Cotoneaster milkedandaensis* J.Fryer & B.Hylmo; *Cotoneaster nitidus* subsp. *taylorii* (T.T.Wu) H.Ohashi; *Cotoneaster taylori* T.T.Yu (4, 65)

Semi-evergreen shrubs. C, E Nepal: 1600-2600(-4000) m. Nepal, E Himalaya, Tibetan Plateau, Assam-Burma and E Asia.

Cotoneaster racemiflorus (Desf.) K.Koch [6, 10, 29]

Mespilus racemiflora Desf.; *Cotoneaster inexspectus* G.Klotz (2, 4); *Cotoneaster tibeticus* G.Klotz (2, 4, 8); *Cotoneaster racemiflorus* var. *royleanus* Dippel (2, 4); *Cotoneaster zayulensis* G.Klotz (2, 4)

Deciduous shrubs. W, C Nepal: (1400-)2700-4000 m. Nepal, W Himalaya, E Himalaya, Tibetan Plateau, Assam-Burma, E Asia, N Asia, C Asia, SW Asia and Africa.

Cotoneaster symondsii T.Moore [6, 10]

Cotoneaster simonsii Baker (2, 4)

Semi-evergreen shrubs. W, C, E Nepal: 3000-3800 m. Nepal, E Himalaya and Assam-Burma.

Cotoneaster verruculosus Diels [2, 4, 6, 10]

Cotoneaster distichus var. *verruculosus* (Diels) T.T.Yu; *Cotoneaster sandakphuensis* G.Klotz (2, 4)

Deciduous shrubs. C, E Nepal: 2800-3800 m. Nepal, E Himalaya, Tibetan Plateau, Assam-Burma and E Asia.

Docynia Decne.

Docynia indica (Colebr.) Decne. [2, 4, 6, 10]

Pyrus indica Colebr.; *Docynia hookeriana* Decne.

पासी, मेहुल (Paasi, Mehul)

Semi-evergreen trees. E Nepal: (700-)1200-2000 m. Nepal, E Himalaya, Assam-Burma, E Asia and SE Asia.

Use: Food (fruit).

Eriobotrya Lindl.

Eriobotrya dubia (Lindl.) Decne. [2, 4, 6, 10]

Photinia dubia Lindl.

जुरे काफल (Jure kaaphal)

Evergreen trees. C Nepal: 1300-2800 m. Nepal, W Himalaya and E Himalaya.

Use(s): Food (fruit), Fodder.

Eriobotrya elliptica Lindl. [2, 4, 6, 10]

माया (Maaya)

Evergreen trees. W, C, E Nepal: 500-2800 m. Nepal and Tibetan Plateau.

Eriobotrya hookeriana Decne. [2, 4, 6, 10]

Evergreen trees. E Nepal: 1400-2400 m. Nepal and E Himalaya.

Eriobotrya japonica (Thunb.) Lindl. [2, 4, 6]

Mespilus japonica Thunb.

लोकट, सुवर्ण आलु (Lokat, Suwarna alu)

Small evergreen trees. C Nepal: 2000-2100 m (Cultivated). Native to E Asia.

Use: Medicine (leaves, flowers, fruit).

Filipendula Mill.

Filipendula vestita (Wall. ex G.Don) Maxim. [2, 4, 10]

Spiraea vestita Wall. ex G.Don; *Spiraea camtschatica* var. *himalensis* Lindl.

Perennial herbs. W Nepal: 2200-3200 m. Nepal, W Himalaya, E Asia and SW Asia.

Fragaria L.

Fragaria daltoniana J.Gay [2, 4, 10]

Fragaria rubiginosa Lacaita (2, 4); *Fragaria sikkimensis* Kurz; *Potentilla daltoniana* (J.Gay) Mabb. (6)

कर्ण फूल झार (Karna phool jhaar)

Perennial herbs. W, C, E Nepal: 2000-4000 m. Nepal, W Himalaya, E Himalaya, Tibetan Plateau and Assam-Burma.

Use(s): Food (fruit), Medicine (root, fruit).

Fragaria nilgerrensis Schltdl. ex J.Gay [2, 4, 10]

Potentilla nilgerrensis (J.Gay) Mabb. (2, 4, 6)

Perennial herbs. C, E Nepal: 2800-4200 m. Nepal, E Himalaya, Assam-Burma and E Asia.

Use(s): Medicine (whole plant), Food (fruit).

Fragaria nubicola (Hook.f.) Lindl. ex Lacaita [2, 4, 6, 10]

Fragaria vesca var. *nubicola* Hook.f.; *Potentilla nubicola* (Hook.f.) Mabb.

भुँइ ऐंसेलु; डीठा-सजिन (Bhuin aainselu; Ditha-sazin: Amchi)

Prostrate perennial herbs. W, C, E Nepal: 1600-4000 m. Nepal, W Himalaya, E Himalaya and Tibetan Plateau.

Use(s): Food (fruit), Medicine (leaves, flowers, fruit).

Geum L.

Geum elatum Wall. ex G.Don [2, 4, 6, 10]

Acomastylis elata (Wall. ex G.Don) F.Bolle; *Acomastylis elata* var. *humilis* (Royle) F.Bolle; *Geum elatum* var. *humile* (Royle) Hook.f. (2, 4)

forma **elatum**

बेलोचन (Belochan)

Perennial herbs. W, C, E Nepal: 2900-4500 m. Nepal, W Himalaya, E Himalaya, Tibetan Plateau and SW Asia.

Use: Medicine (root).

forma **rubrum** Ludlow [2, 4, 6, 10]

Perennial herbs. W, C Nepal: 3000-4400 m. Nepal and W Himalaya.

Geum roylei Wall. ex F.Bolle [2, 4, 6, 10]

Perennial herbs. W Nepal: 1900-3500 m. Nepal, W Himalaya and SW Asia.

Use: Medicine (root).

Geum sikkimense Prain [2, 4, 6, 10]

Perennial herbs. C, E Nepal: 3000-4400 m. Nepal and E Himalaya.

Maddenia Hook.f. & Thomson

Maddenia himalaica Hook.f. & Thomson [2, 4, 8, 10]

Maddenia himalaica var. *glabrifolia* H.Hara (2, 4)

Deciduous shrubs or small trees. E Nepal: 2500-3400 m. Nepal, E Himalaya, Tibetan Plateau and Assam-Burma.

Malus L.

Malus baccata (L.) Borkh. [2, 4, 6, 10]

Pyrus baccata L.; *Malus baccata* subsp. *himalaica* (Maxim.) Likhonos; *Malus baccata* var. *himalaica* (Maxim.) C.K.Schneid.; *Pyrus baccata* var. *himalaica* Maxim.

जंगली स्याउ, चिनिया नासपाती (Jangali syau, Chiniya naaspaati); Siberian Crab Apple

Deciduous shrubs or small trees. W, C, E Nepal: 1800-4100 m. Nepal, W Himalaya, E Himalaya, Tibetan Plateau, Assam-Burma, E Asia and N Asia.

Use(s): Food (fruit), Medicine (fruit), Ornamental.

Malus pumila Mill. [2, 4, 6, 10]

स्याउ (Syaau); Apple

Deciduous shrubs or small trees. W, C Nepal: 2000-3300 m. Widely cultivated for edible fruit. Native to SW Asia and Euope.

Use(s): Food (Fruit), Medicine (root-bark), Agricultural implements.

Malus sikkimensis (Wenz.) Koehne [2, 4, 10]

Pyrus pashia var. *sikkimensis* Wenz.; *Malus baccata* subsp. *sikkimensis* (Wenz.) Likhonos; *Pyrus sikkimensis* (Wenz.) Hook.f.

जंगली स्याउ (Jangali syau); Sikkim Crab Apple

Deciduous trees. W, C Nepal: 2400-3000 m. Nepal, E Himalaya, Tibetan Plateau, Assam-Burma and E Asia.

Use: Food (fruit).

Neillia D.Don

Neillia rubiflora D.Don [2, 4, 6, 10]

हात्ती लहरो (Haatti laharo)

Deciduous shrubs. C, E Nepal: 1500-3500 m. Nepal, E Himalaya, Tibetan Plateau and E Asia.

Neillia thrysiflora D.Don [2, 4, 6, 10]

हात्ती लहरो, भोगला (Haatti laharo, Bhogla: Gurung)

Deciduous shrubs. C, E Nepal: 1600-2400 m. Nepal, E Himalaya, Assam-Burma and E Asia.

Use: Food (fruit).

Neillia velutina Pendry [10]

Deciduous shrubs. E Nepal: 2000-2600 m. Nepal and E Himalaya.

Photinia Lindl.

Photinia integrifolia Lindl. [2, 4, 6, 10]

Eriobotrya integrifolia (Lindl.) Kurz; *Pyrus integerrima* D.Don

गज फूल, लहरे बाकलपाते (Gaj phool, Lahare baakalpate)

Evergreen climbing shrubs or small trees. C, E Nepal: 1300-3000 m. Nepal, W Himalaya, E Himalaya, Assam-Burma, E Asia and SE Asia.

Use: Fodder.

Photinia nussia (Buch.-Ham. ex D.Don) Kalkman [6, 10]

Pyrus nussia Buch.-Ham. ex D.Don; *Stranvaesia nussia* (D.Don) Decne. (2, 4); *Stranvaesia glaucescens* Lindl.

जुरे मयल (Jure mayal)

Evergreen shrubs or small trees. W, C Nepal: 900-2500 m. Nepal, W Himalaya, Tibetan Plateau, Assam-Burma, E Asia and SE Asia.

Use(s): Food (fruit), Fodder, Medicine (leaves).

Potentilla L.

Potentilla anserina L. [2, 4, 6, 10]

मसिनो झार, येचुरुक (Masino jhaar, Yechuruk: Tibetan)

Perennial stoloniferous herbs. C, E Nepal: 3100-4600 m. Asia, Europe, N America and Australasia.

Use(s): Medicine (root, leaves, flowers), Food (root).

Potentilla argyrophylla Wall. ex Lehm. [2, 4, 6, 10]

Potentilla insignis Royle ex Lindl.; *Potentilla jacquemontiana* Cambess.; *Potentilla nivea* var. *himalaica* Kitam.; *Potentilla venusta* Soják var. **argyrophylla**

आटे (Aate)

Perennial herbs. W, C, E Nepal: 2800-4800 m. Nepal, W Himalaya, E Himalaya, Tibetan Plateau and SW Asia.

Use(s): Medicine (root), Forage.

var. **atrosanguinea** (Lodd.) Hook.f. [2, 4, 6, 10]

Potentilla atrosanguinea Lodd.

Perennial herbs. W, C, E Nepal: 2900-4600 m. Nepal, W Himalaya and Tibetan Plateau.

Potentilla aristata Soják [4, 6, 10]

Argentina aristata (Soják) Soják (74); *Potentilla microphylla* var. *achilleifolia* Hook.f. (2)

Perennial herbs. C, E Nepal: 3600-4700 m. Nepal, E Himalaya and Tibetan Plateau.

Potentilla biflora Willd. ex Schltdl. [2, 4, 6, 10]

Perennial herbs. W, C, E Nepal: 4000-5000 m. Nepal, W Himalaya, Tibetan Plateau, N Asia and N America.

Potentilla bifurca L. [2, 4, 6, 10]

Sibbaldianthe bifurca (L.) Kurtto & T.Erikss.

Perennial herbs. W, C Nepal: 3100-5200 m. Nepal, Asia and Europe.

Potentilla cardotiana Hand.-Mazz.

var. **nepalensis** H.Ikeda & H.Ohba, *Bull. Univ. Mus. Univ. Tokyo* **39**: 72. 1999. [4, 10, 129]

Perennial herbs. E Nepal: *ca.* 3800 m. **Endemic to Nepal**.

East: *Sankhuwasabha*, Cha Ding Kharka – Khongma, 3800 m, Suzuki *et al* 888085 (Holotype: TI; Isotypes: A, E, BM, KATH).

Potentilla commutata Lehm. [4, 6, 10]

var. **commutata**

Argentina commutata (Lehm.) Soják (74); *Potentilla microphylla* var. *commutata* (Lehm.) Hook.f. (2)

Perennial herbs. C, E Nepal: 3500-4500 m. Nepal, W Himalaya and E Himalaya.

var. **polyandra** Soják [4, 6, 10]

Potentilla mieheorum Soják

Perennial herbs. C, E Nepal: 4000-4500 m. Nepal, W Himalaya, E Himalaya and Tibetan Plateau.

Potentilla contigua Soják [4, 6, 10]

Potentilla peduncularis var. *clarkei* Hook.f.

Perennial rhizomatous herbs. C, E Nepal: 3500-4700 m. Nepal, E Himalaya and Tibetan Plateau.

Potentilla coriandrifolia D.Don [2, 4, 6, 10]

Potentilla meifolia Wall. ex Lehm.

Perennial herbs. C, E Nepal: 3700-5600 m. Nepal, W Himalaya, E Himalaya and Tibetan Plateau.

Potentilla cuneata Wall. ex Lehm. [2, 4, 6, 10]

Sibbaldia parviflora Willd.

Subshrubs. W, C, E Nepal: 1900-4900 m. Nepal, W Himalaya, E Himalaya, Tibetan Plateau and E Asia.

Potentilla eriocarpa Wall. ex Lehm. [2, 4, 6, 10]

var. **eriocarpa**

Subshrubs. W, C, E Nepal: 3500-5100 m. Nepal, W Himalaya, E Himalaya, Tibetan Plateau and E Asia.

var. **major** Kitam., *Acta Phytotax. Geobot.* **15**(5): 130. 1954. [2, 4, 10, 130]

Subshrubs. C Nepal: 3400-3900 m. **Endemic to Nepal**.

Central:*Gorkha*, Bangu Khola, *ca.* 3500 m, S. Nakao s.n. (Holotype: KYO?).

Potentilla festiva Soják [4, 6, 10]

Argentina festiva (Soják) Soják (74)

Perennial herbs. W, C, E Nepal: 2200-3800 m. Nepal, W Himalaya, E Himalaya, Tibetan Plateau, Assam-Burma and E Asia.

Potentilla fruticosa L. [2, 4, 6, 10]

Dasiphora fruticosa (L.) Rydb.

चिनिया फल, भैरुंग पाते; पेम्मा-नाक्पो (Chiniyaa phal, Bhairung paate; Pemma-nakpo: Amchi)

Use(s): Medicine (stem, leaves, flowers), Incense, Forage.

var. **arbuscula** (D.Don) Maxim. [10]

Potentilla arbuscula D.Don; *Dasiphora arbuscula* Soják; *Potentilla fruticosa* var. *rigida* (Wall. ex Lehm.) Wolf (2, 4); *Potentilla rigida* Wall. ex Lehm.

Subshrubs. W, C, E Nepal: 2700-4500(-6000) m. Nepal, W Himalaya, E Himalaya and Tibetan Plateau.

Use(s): Food (leaves), Medicine (root), Incense.

var. **ochreata** Lindl. ex Lehm. [2, 4, 6, 10]

Potentilla ochreata Lindl. ex Wall., nom. nud.

Subshrubs. W, C Nepal: 3300-4600 m. Nepal and Tibetan Plateau.

var. **pumila** Hook.f. [2, 4, 6, 10]

Potentilla arbuscula var. *pumila* (Hook.f.) Hand.-Mazz.

चिनिया फल (Chiniyaa Phal)

Prostrate shrubs. W, C, E Nepal: 3800-6000 m. Nepal, W Himalaya, E Himalaya and Tibetan Plateau.

Use: Medicine.

Potentilla glabriuscula (T.T.Yu & C.L.Li) Soják [4, 6, 10]

Sibbaldia glabriuscula T.T.Yu & C.L.Li; *Argentina glabriuscula* (T.T.Yu & C.L.Li) Soják (74)

Perennial herbs. E Nepal: 3500-5000 m. Nepal, E Himalaya, Tibetan Plateau, Assam-Burma and E Asia.

Potentilla griffithii Hook.f. [2, 4, 6, 10]

Potentilla griffithii var. *metallica* Soják (4); *Potentilla leschenaultiana* Ser. (2, 4); *Potentilla spodiochlora* Soják (2, 4)

हयुमी (Hyumi)

Perennial herbs. W, C, E Nepal: 1600-4000 m. Nepal, E Himalaya, Tibetan Plateau and E Asia.

Potentilla indica (Andrews) Wolf [6, 10]

Fragaria indica Andrews; *Duchesnea indica* (Andrews) Focke (2, 4); *Potentilla wallichiana* Ser.

सर्पे काफल, भुँइ ऐंसेलु (Sarpe Kaaphal, Bhuin aainselu)

Perennial herbs. W, C, E Nepal: 900-2500 m. Nepal, W Himalaya, E Himalaya, E Asia, SE Asia and SW Asia.

Use(s): Food (fruit), Medicine (fruit).

Potentilla josephiana H.Ikeda & H.Ohba [4, 6, 10]

Potentilla fulgens var. *intermedia* Hook.f.; *Potentilla lineata* var. *intermedia* (Hook.f.) R.D.Dixit & Panigrahi

Perennial herbs. W, C, E Nepal: 3200-4100 m. Nepal, W Himalaya and E Himalaya.

Potentilla leuconota D.Don [2, 4, 6, 10]

Argentina leuconota (D.Don) Soják (74)

सक्कली जिब्रे झार (Sakkali jibre jhaar)

Perennial herbs. W, C, E Nepal: 2700-4500 m. Nepal, W Himalaya, E Himalaya, Tibetan Plateau, Assam-Burma and E Asia.

Use: Medicine (root).

Potentilla lineata Trevir. [4, 6, 10]

Argentina lineata (Trevir.) Soják (74); *Potentilla fulgens* Wall. ex Hook. (2); *Potentilla martinii* H.Lev.

बज्रदन्ति (Bajradanti); Himalayan cinquefoil

Perennial herbs. W, C, E Nepal: 1700-3700 m. Nepal, W Himalaya, E Himalaya, Tibetan Plateau, Assam-Burma, E Asia and SE Asia.

Use: Medicine (root).

Potentilla makaluensis H.Ikeda & H.Ohba; *J. Jap. Bot.* **67**: 149. 1992. [4, 10, 12, 28, 29]

Perennial rhizomatous herbs. E Nepal: *ca.* 4100 m. **Endemic to Nepal**.

East: *Sankhuwasabha*, Shipton Pass, 4120 m, M. Suzuki *et al* 8880826 (Holotype: TI; Isotypes: A, BM, E, KATH).

Potentilla microphylla D.Don [2, 4, 6, 10]

Argentina microphylla (D.Don) Soják (74); *Potentilla microphylla* var. *depressa* Wall. ex Lehm.; *Potentilla microphylla* var. *glabriuscula* Wall. ex Lehm.; *Potentilla microphylla* var. *latifolia* Wall. ex Lehm.

ताछेर (Tachher: Tibetan)

Perennial herbs. W, C, E Nepal: 3400-5200 m. Nepal, W Himalaya, E Himalaya, Tibetan Plateau and Assam-Burma.

Use: Medicine (root).

Potentilla monanthes Wall. ex Lehm. [2, 4, 6, 10]

Potentilla monanthes var. *alata* Soják (4)

सपलंग, सल्यांगसाइ (Saplang, Salyangsai: Chepang)

var. **monanthes**

Perennial herbs. E Nepal: 3000-4800 m. Nepal, W Himalaya and E Himalaya.

Use: Food (fruit).

var. **sibthorpioides** Hook.f. [10]

Perennial herbs. C, E Nepal: 4100-4700 m. Nepal and E Himalaya.

Potentilla multifida L. [2, 4, 8, 10]

Potentilla exigua Soják (2, 4); *Potentilla plurijuga* Hand.-Mazz. (2, 4, 11)

Perennial herbs. C, E Nepal: 2900-4500 m. Nepal, Asia and Europe.

Potentilla nepalensis Hook. [2, 4, 6, 10]

Perennial herbs. W Nepal: 2000-2600 m. Nepal and W Himalaya.

Use: Medicine (root).

Potentilla peduncularis D.Don [2, 4, 6, 10]

Argentina peduncularis (D.Don) Soják (74); *Potentilla peduncularis* var. *subcontigua* Soják

मुला झार, काली ऐंसुलु (Mula jhaar, Kali ainselu)

var. **peduncularis**

Perennial herbs. W, C, E Nepal: 3000-4700 m. Nepal, E Himalaya and Tibetan Plateau.

Use(s): Food, Medicine (root, leaves bud).

var. **ganeshii** H.Ikeda & H.Ohba, *J. Jpn. Bot.* **71**: 252. 1996. [8, 10]

Perennial herbs. C Nepal: 3400-4300 m. **Endemic to Nepal**.

Central: *Rasuwa*, Thulo Bhera Kharka-Jaisuli Kund, 4250 m, Miyamoto *et al* 9430046 (Holotype: TI; Isotypes: KATH, BM, GH-iso).

Potentilla polyphylla Wall. ex Lehm. [2, 4, 6, 10]

Argentina polyphylla (Wall. ex Lehm.) Soják (74)

var. **polyphylla**

Perennial herbs. C, E Nepal: 1400-4000 m. Nepal, W Himalaya, E Himalaya, Assam-Burma, S Asia and SE Asia.

var. **himalaica** H.Ikeda & H.Ohba [4, 10]

Perennial herbs. E Nepal: 3400-4000 m. Nepal and E Himalaya.

var. **interrupta** (T.T.Yu & C.L.Li) H.Ikeda & H.Ohba [4, 6, 10]

Potentilla interrupta T.T.Yu & C.L.Li; *Argentina interrupta* (T.T.Yu & C.L.Li) Soják (74)

Perennial herbs. C, E Nepal: 3500-4200 m. Nepal, E Himalaya, Tibetan Plateau and E Asia.

Potentilla saundersiana Royle [2, 4, 6, 10, 29]

Potentilla caespitosa Lehm.; *Potentilla caliginosa* Soják (2, 4); *Potentilla forrestii* W.W.Sm. (2, 4); *Potentilla forrestii* var. *caespitosa* (Wolf) Soják (2, 4, 6); *Potentilla illudens* Soják (4); *Potentilla jacquemontii* (Franch.) Soják; *Potentilla thibetica* Cardot; *Potentilla williamsii* Soják (4)

Perennial herbs. C, E Nepal: 3100-4900 m. Nepal, E Himalaya, Tibetan Plateau and E Asia.

Potentilla sundaica (Blume) Kuntze [10]

Fragaria sundaica Blume; *Potentilla kleiniana* Wight (2, 4, 6)

कौवा काफल (Kauwaa kaaphal)

Perennial herbs. W, E Nepal: 1000-2400 m. Nepal, W Himalaya, E Himalaya, Tibetan Plateau, Assam-Burma, S Asia, E Asia and SE Asia.

Use(s): Medicine (root, leaves), Food (fruit).

Potentilla supina L. [2, 4, 6, 10]

Comarum flavum Buch.-Ham. ex Roxb.

बज्रदन्ति (Bajradanti)

Annual or biennial herbs. W, E Nepal: 1000-2000 m. Nepal, Asia and Europe.

Use: Medicine (root).

Potentilla tristris Soják [4, 8, 10]

Argentina tristis (Soják) Soják (74); *Potentilla peduncularis* var. *obscura* Hook.f.; *Potentilla tristis* forma *ciliata* H.Ikeda & H.Ohba (2, 4)

Perennial herbs. W, C, E Nepal: 3300-5000 m. Nepal and W Himalaya.

Potentilla turfosoides H.Ikeda & H.Ohba; *Himal. Pl.* **3**: 62. 1999. [4, 10, 28, 29]

Potentilla turfosa var. *caudiculata* Soják

Perennial herbs. E Nepal: 3500-4100 m. **Endemic to Nepal**.

East: *Sankhuwasabha*, Khongma, *ca.* 3500 m, M. Suzuki *et al* 8880570 (Holotype: TI).

Prinsepia Royle

Prinsepia utilis Royle [2, 4, 6, 10]

धतेलो (Dhatelo)

Deciduous shrubs. W, C, E Nepal: 1100-3400 m. Nepal, W Himalaya, E Himalaya, Tibetan Plateau and E Asia.

Use(s): Food (fruit, seed oil), Fence, Medicine (seed oil, oil cake), Fodder, Washing clothes, Painting (fruit), Lightining.

Prunus L.

Prunus armeniaca L. [2, 4, 10]

Armeniaca vulgaris Lam.

खुर्पानी, जंगली खुर्पानी (Khurpani, Jangalee khurpanee); Apricot

Deciduous trees. C Nepal: 2900-3500 m (Naturalized). Native to C Asia.

Use(s): Food (fruit, seed oil), Medicine (seeds).

Prunus cerasifera Ehrh. [2, 4, 6, 10]

आरुबखडा (Arubakhadaa); Plum

Deciduous shrubs or trees. C, E Nepal: 1300-2200 m. Widely cultivated for fruit. Native to Europe and W Asia.

Use(s): Food/Medicine (fruit), Fodder.

Prunus cerasoides D.Don [2, 4, 6, 10, 14]

Cerasus cerasoides (D.Don) S.Ya.Sokolov; *Cerasus phoshia* Buch.-Ham. ex D.Don; *Cerasus puddum* Roxb. ex Ser.; *Prunus carmesina* H.Hara (2, 4); *Prunus puddum* (Roxb. ex Ser.) Brandis

पैंयु (Painyu); Himalayan cherry

Deciduous trees. W, C, E Nepal: 1300-2700 m. Nepal, W Himalaya, E Himalaya, Tibetan Plateau, Assam-Burma, E Asia and SE Asia.

Use(s): Food (fruit), Ornamental, Medicine (bark, heartwood, kernel), Fodder, Walking stick (stem).

Prunus cornuta (Wall. ex Royle) Steud. [2, 4, 6, 10]

Cerasus cornuta Wall. ex Royle; *Padus cornuta* (Wall. ex Royle) Carrière; *Prunus cornuta* forma *villosa* (H.Hara) H.Hara (2, 4, 8); *Padus cornuta* var. *villosa* G.Singh; *Prunus cornuta* var. *villosa* H.Hara

लेक आरु, आरुपाते (Lekh aaru, Aarupaate)

Deciduous trees. W, C, E Nepal: 2100-3500 m. Nepal, W Himalaya, E Himalaya, Tibetan Plateau and SW Asia.

Use(s): Food (fruit), Fodder, Medicine (fruit, seeds, oil).

Prunus davidiana (Carrière) Franch. [2, 4, 10]

Persica davidiana Carrière

झुसे आरु, खमु (Jhuse aaru, Khamu)

Deciduous trees. W, C Nepal: 2100-2700 m. Nepal, Tibetan Plateau and E Asia.

Use(s): Food (fruit), Medicine (fruit, seed oil), Essential oil (seeds).

Prunus himalaica Kitam., *Acta Phytotax. Geobot.* **15**: 131. 1954. [2, 4, 10, 12, 28, 29]

Deciduous trees. C Nepal: *ca.* 3900 m. **Endemic to Nepal**.

Central: *Rasuwa*, Chum Gompa, *ca.* 3900 m, S. Nakao s.n. (Holotype:KYO).

Prunus jajarkotensis H.Hara, *J. Jap. Bot.* **52**: 355. 1977. [2, 4, 10, 12, 28, 29]

Evergreen trees. W Nepal: 900-1000 m. **Endemic to Nepal**.

West: *Jajarkot*, Jajarkot, 3000 ft. (915 m), O. Polunin, WR Sykes & LHJ Williams 5784 (Holotype: BM; Isotype: BM).

Prunus mira Koehne

Persica mira (Koehne) Kovalev & Kostina

subsp. **nepalensis** (Seraf.) H.Hara [2, 4, 10]

Persica mira subsp. *nepalensis* Seraf.

Deciduous trees. W, C Nepal: 2700-4000 m. Nepal, Tibetan Plateau and E Asia.

Prunus napaulensis (Ser.) Steud. [2, 4, 6, 10]

Cerasus napaulensis Ser.; *Padus napaulensis* (Ser.) C.K.Schneid.

जंगली पैंयु (Jangali painyu)

Deciduous trees. W, C, E Nepal: 1600-3000 m. Nepal, W Himalaya, E Himalaya, Tibetan Plateau, Assam-Burma and E Asia.

Use(s): Food (fruit), Timber, Firewood, Poisonous to cattle.

Prunus persica (L.) Batsch [2, 4, 6, 10]

Amygdalus persica L.

आरु (Aaru); Peach

Deciduous trees. W, C Nepal: 1300-3600 m. Widely planted for fruit. Probably native to W China.

Use(s): Medicine (leaves, seeds), Food (fruit).

Prunus rufa Hook.f. [2, 4, 6, 10]

Cerasus rufa Wall., nom. nud.; *Cerasus rufa* T.T.Yu & C.L.Li; *Prunus rufa* var. *trichantha* (Koehne) H.Hara (2, 4, 8)

Deciduous trees. W, C, E Nepal: 3200-3800 m. Nepal, E Himalaya and Tibetan Plateau.

Prunus singalilaensis H.Ohba & S.Akiyama [35]

Shrubs. E Nepal: 2600-3500 m. Nepal, E Himalaya and S Asia.

Prunus taplejungnica H.Ohba & S.Akiyama, *Bull. Natl. Mus. Nat. Sci., Tokyo, B.* **36**(4): 135, 2010. [8, 10, 29, 35]

Deciduous shrubs. E Nepal: 2400-3600 m. **Endemic to Nepal**.

East: *Taplejung*, Chairam – Dorongden, 3250 m, S. Noshiro, S. Akiyama & N. Acharya 9261245 (Holotype: TI).

Prunus topkegolensis H.Ohba & S.Akiyama, *Bull. Natl. Sci. Mus. Tokyo, B*, **36**(4): 134. 2010. [10, 29, 35]

Deciduous shrubs. E Nepal: *ca.* 3700 m. **Endemic to Nepal**.

East: *Taplejung*, Topke Gola, *ca.* 3700 m, H. Kanai *et al* 720610 (Holotype: TI; Isotype: TNS).

Prunus undulata Buch.-Ham. ex D.Don [2, 4, 6, 10]

Cerasus acuminata Wall., nom. nud.; *Cerasus undulata* (Buch.-Ham. ex D.Don) Ser.; *Laurocerasus undulata* (Buch.-Ham. ex D.Don) M.Roem.; *Prunus wallichii* Steud.

खोसिनी (Khoshini)

Evergreen shrubs or small trees. C, E Nepal: 1500-2600 m. Nepal, E Himalaya, Assam-Burma, E Asia and SE Asia.

Prunus venosa Koehne [2, 4, 6, 10]

Prunus undulata forma *venosa* (Koehne) Koehne

Deciduous trees. W, C, E Nepal: 1600-3200 m. Nepal, W Himalaya, E Himalaya, Tibetan Plateau, Assam-Burma and E Asia.

Use: Young foliage poisonous to cattle.

Pygeum Gaertn.

Pygeum zeylanicum Gaertn. [2, 4, 10]

Polydontia ceylanica Wight; *Prunus ceylanica* (Wight) Miq.

Evergreen trees. C, E Nepal: 300-500 m. Nepal and S Asia.

Pyracantha M.Roem.

Pyracantha crenulata (D.Don) M.Roem. [2, 4, 6, 10]

Mespilus crenulata D.Don; *Crataegus crenulata* (D.Don) Roxb.

घंगारू, काठ गेडी (Ghangaaru, Kaath gedi)

Evergreen shrubs or small trees. W, C, E Nepal: 800-2800 m. Nepal, W Himalaya, E Himalaya, Tibetan Plateau, Assam-Burma and E Asia.

Use(s): Medicine (fruit), Food (fruit), Hedge, Agricultural implements, Walking sticks.

Pyracantha fortuneana (Maxim.) H.L.Li [6]

Photinia fortuneana Maxim.

Evergreen shrubs. W Nepal: *ca.* 1000 m. Nepal, Tibetan Plateau and E Asia.

Pyrus L.

Pyrus communis L. [6]

नासपाती (Naaspati), Common pear

Deciduous trees. Widely planted for edible fruit. Native to CE Europe and SW Asia.

Use: Food (fruit).

Pyrus pashia Buch.-Ham. ex D.Don [2, 4, 6, 10]

Pyrus nepalensis Buch.-Ham. ex Hook.f., nom. inval.

मयल (Mayal); Himalayan pear

Deciduous trees. W, C, E Nepal: 700-3100 m. Nepal, W Himalaya, E Himalaya, Tibetan Plateau, Assam-Burma, E Asia and SE Asia.

Use(s): Food (fruit), Fodder, Medicine (fruit), Firewood, Walking stick (stem).

Rosa L.

Rosa brunonii Lindl. [2, 4, 6, 10]

Rosa moschata var. *nepalensis* Lindl.; *Rosa pubescens* Roxb.

भैंसी काँडा (Bhainsi kaanda)

Climbing shrubs. W, C, E Nepal: 1300-3000 m. Nepal, W Himalaya, E Himalaya, Tibetan Plateau, Assam-Burma and E Asia.

Use(s): Ornamental, Food (fruit).

Rosa clinophylla Thory [2, 4, 10]

Rosa involucrata Roxb. ex Lindl.; *Rosa lyellii* Lindl.; *Rosa palustris* Buch.-Ham. ex Lindl.

Erect shrubs. C Nepal: 100-1000 m. Nepal, W Himalaya, Assam-Burma, S Asia and SE Asia.

Rosa macrophylla Lindl. [2, 4, 6, 10]

Rosa hookeriana Wall., nom. nud.; *Rosa macrophylla* var. *hookeriana* Hook.f.

जंगली गुलाब; सेगोए-फो (Jangali gulaab; Segoe-fo: Amchi)

Shrubs. W, C, E Nepal: 2100-4400 m. Nepal, W Himalaya, E Himalaya and Tibetan Plateau.

Use(s): Ornamental, Medicine (fruit), Food (fruit), Hedge, Fodder.

Rosa multiflora Thunb. [2, 4, 6]

Rosa multiflora var. *carnea* Thory

Climbing shrubs. W, C, E Nepal: 800-1500 m. Nepal and E Asia. Often Cultivated in S Asia.

Rosa sericea Lindl. [2, 4, 6, 10]

भोटे गुलाफ, दारिम पाते; सेवा (Bhote gulaaph, Daarim paate; Sewa: Amchi)

Erect shrubs. W, C, E Nepal: 2100-4600 m. Nepal, W Himalaya, E Himalaya, Tibetan Plateau, Assam-Burma and E Asia.

Use(s): Ornamental, Medicine (bark, flowers, fruit), Food (fruit), Fuel, Hedge.

Rosa webbiana Wall. ex Royle [2, 4, 6, 10]

भोटे गुलाफ (Bhote gulaaf)

Deciduous shrubs. W Nepal: 2200-2900 m. Nepal, W Himalaya, Tibetan Plateau and N Asia.

Use(s): Food/Medicine (fruit).

Rubus L.

Rubus acuminatus Sm. [2, 4, 6, 10]

रातो ऐंसेलु (Raato aainselu)

Scandent shrubs. W, C, E Nepal: 1000-2500 m. Nepal, E Himalaya, Assam-Burma, E Asia and SE Asia.

Use(s): Food (fruit), Beverage, Medicine (bark).

Rubus alexeterius Focke [2, 4, 6, 10]

Rubus acaenocalyx H.Hara; *Rubus alexeterius* var. *acaenocalyx* (H.Hara) T.T.Yu & L.T.Lu

Scandent shrubs. C Nepal: 2000-3700 m. Nepal, E Himalaya, Tibetan Plateau and E Asia.

Rubus amabilis Focke [2, 4, 10]

Shrubs. W Nepal: (1000-)3300-3800 m. Nepal, Tibetan Plateau and E Asia.

Use: Food (fruit).

Rubus biflorus Buch.-Ham. ex Sm. [2, 4, 6, 10]

सानो गुलाफ (Saano gulaf)

Scandent shrubs. W, C, E Nepal: 1500-3500 m. Nepal, W Himalaya, E Himalaya, Tibetan Plateau, Assam-Burma and E Asia.

Use: Food (fruit).

Rubus calycinoides Kuntze [2, 4, 6, 10]

Rubus himalaicus Kuntze

Erect or arching shrubs. C, E Nepal: 1200-2200 m. Nepal, E Himalaya and Assam-Burma.

Rubus calycinus Wall. ex D.Don [2, 4, 6, 10]

Dalibarda calycinus (Wall. ex D.Don) Ser.; *Rubus lobatus* Wall., nom. nud.

भुईं ऐंसेलु (Bhuin ainselu)

Perennial creeping herbs. C, E Nepal: 1200-3000 m. Nepal, W Himalaya, E Himalaya, Assam-Burma, E Asia and SE Asia.

Use: Food (fruit).

Rubus efferatus Craib [10]

Rubus ferox Wall. ex Kurz, non Vest ex Tratt. (2, 4)

Scandent shrubs. C Nepal: *ca.* 1200 m. Nepal, E Himalaya, Assam-Burma and SE Asia.

Rubus ellipticus Sm. [2, 4, 6, 10]

Rubus fasciculatus Duthie; *Rubus flavus* Buch.-Ham. ex D.Don; *Rubus wallichianus* Wight & Arn.

ऐंसेलु, पहेंलो ऐंसेलु (Ainselu, Pahenlo ainselu); Golden Himalayan raspberry

Scandent shrubs. W, C, E Nepal: 300-2600 m. Nepal, W Himalaya, E Himalaya, Tibetan Plateau, Assam-Burma, S Asia, E Asia and SE Asia.

Use(s): Food (fruit), Medicine (root, stem, fruit).

Rubus fockeanus Kurz [2, 4, 6, 10]

Rubus nutans var. *fockeanus* (Kurz) Kuntze

भुँई ऐंसेलु (Bhuin ainselu)

Prostrate perennial herbs. C Nepal: 2000-3000(-4000) m. Nepal, E Himalaya, Tibetan Plateau, Assam-Burma and E Asia.

Use: Food (fruit).

Rubus foliolosus D.Don [2, 4, 6, 10]

Rubus pedunculosus D.Don (4); *Rubus hypargyrus* var. *concolor* (Hook.f.) H.Hara (2, 4); *Rubus hypargyrus* Edgew. (2, 4, 6); *Rubus hypargyrus* var. *niveus* (Wall. ex G.Don) H.Hara (2, 4)

कालो ऐंसेलु, सानु ऐंसेलु; कण्ड-करी (Kaalo ainselu, Saanu ainselu; Kanda-kari: Amchi)

Straggling shrubs. W, C, E Nepal: 1600-3200 m. Nepal, W Himalaya, E Himalaya and Tibetan Plateau.

Use(s): Food (fruit), Medicine (leaves, stem pith).

Rubus franchetianus H.Lév. [10]

Rubus fragarioides Bertol. (2, 4, 6)

Prostrate perennial herbs. C, E Nepal: 3000-4200 m. Nepal, E Himalaya, Tibetan Plateau, Assam-Burma and E Asia.

Rubus glandulifer N.P.Balakr. [10, 29]

Rubus lanatus Wall. ex Hook.f., non Focke (2, 4)

Scandent shrubs. C Nepal: 900-2500 m. Nepal and W Himalaya.

Note: Presence of *Rubus glandulifer*, a West Himalayan species in C Nepal requires confirmation (10).

Rubus griffithii Hook.f. [2, 4, 6, 10]

भुँई ऐंसेलु (Bhuin ainselu)

Erect or ascending shrubs. E Nepal: 1400-2400 m. Nepal and E Himalaya.

Use: Food (fruit).

Rubus hamiltonii Hook.f. [2, 4, 10]

Large climbing shrubs. E Nepal: 300-700 m. Nepal, E Himalaya and Assam-Burma.

Rubus hoffmeisterianus Kunth ex Bouché [2, 4, 10]

Rubus niveus var. *aitchisoni* Hook.f.

Erect or reclining shrubs. W Nepal: 1600-2300 m. Nepal, W Himalaya and SW Asia.

Rubus inopertus (Focke) Focke [10]

Rubus niveus subsp. *inopertus* Focke; *Rubus niveus* var. *rosifolius* (Hook.f.) H.Hara (2, 4)

Scandent shrubs. C, E Nepal: 2400-2600 m. Nepal, E Asia and SE Asia.

Rubus kumaonensis Balakr. [6, 10]

Rubus reticulatus Wall. ex Hook.f., non A. Kern. (2, 4)

Scandent shrubs. C, E Nepal: 1100-2800 m. Nepal, W Himalaya, E Himalaya and Tibetan Plateau.

Use(s): Food (fruit), Medicine (fruit).

Rubus lineatus Reinw. [2, 4, 6, 10]

Rubus pulcherrimus Hook.

घ्याम्पे ऐंसेलु (Ghyampe ainselu)

Scrambling shrubs. E Nepal: 1400-3100 m. Nepal, E Himalaya, Tibetan Plateau, Assam-Burma, E Asia and SE Asia.

Use: Food (fruit).

Rubus macilentus Cambess. [2, 4, 6, 10]

Rubus trichopetalus Hand.-Mazz.

जोगी ऐंसेलु (Jogi ainselu)

Scrambling shrubs. W, C, E Nepal: 900-3300 m. Nepal, W Himalaya, E Himalaya, Tibetan Plateau and E Asia.

Use: Food (fruit).

Rubus mesogaeus Focke [2, 4, 10]

Scandent shrubs. C, E Nepal: 900-3600 m. Nepal, E Himalaya, Tibetan Plateau, E Asia and N Asia.

Rubus nepalensis (Hook.f.) Kuntze [2, 4, 6, 10]

Rubus nutans var. *nepalensis* Hook.f.

भुइँ ऐंसेलु, न्हयलांग (Bhuin Ainselu; Nhyalang: Sherpa)

Prostrate herbs. W, C, E Nepal: 1800-3500 m. Nepal, W Himalaya and E Himalaya.

Use: Food (fruit).

Rubus niveus Thunb. [2, 4, 6, 10]

Rubus niveus var. *micranthus* (D.Don) H.Hara (2, 4)

कालों ऐंसेलु (Kaalo ainselu)

Scandent shrubs. W, C, E Nepal: 1000-3100 m. Nepal, E Himalaya, Tibetan Plateau, Assam-Burma, E Asia, SE Asia and SW Asia.

Use(s): Food (fruit), Medicine (leaves).

Rubus paniculatus Sm. [2, 4, 6, 10]

Rubus paniculatus forma *tiliaceus* (Sm.) H.Hara (2, 4)

रुख ऐंसेलु, भालु ऐंसेलु (Rukh ainselu, Bhaalu ainselu)

Scandent shrubs. W, C, E Nepal: 1500-3200 m. Nepal, W Himalaya, E Himalaya, Tibetan Plateau, Assam-Burma and E Asia.

Use(s): Food (fruit), Medicine (bark, leaves).

Rubus pentagonus Wall. ex Focke [2, 4, 6, 10]

Rubus alpestris Blume

लेक ऐंसेलु (Lekh ainselu)

Scrambling shrubs. W, C, E Nepal: 2000-3100 m. Nepal, W Himalaya, E Himalaya, Tibetan Plateau, Assam-Burma, E Asia and SE Asia.

Use: Food (fruit).

Rubus pungens Cambess. [2, 4, 10]

Scrambling shrubs. C Nepal: 2200-3300 m. Nepal, W Himalaya, E Himalaya, Tibetan Plateau, Assam-Burma and E Asia.

Rubus rosifolius Sm. [2, 4, 10]

Rubus rosifolius forma *coronarius* (Sims) Kuntze (2, 4); *Rubus rosifolius* var. *coronarius* Sims

रातों ऐंसेलु (Raato ainselu)

Erect or climbing shrubs. C, E Nepal: 600-2500 m. Nepal, E Himalaya, Assam-Burma, E Asia, Africa and Australasia.

Use: Food (fruit).

Note: *Rubus rosifolius* may not be truly wild in Nepal (10).

Rubus rugosus Sm. [2, 4, 6, 10]

Rubus hamiltonianus Ser.

गोरु ऐंसेलु, बिपेन काँडा (Goru ainselu, Bipen kaanda)

Climbing or scandent shrubs. C, E Nepal: 600-2700 m. Nepal, E Himalaya and Assam-Burma.

Use(s): Food (fruit), Medicine (root, fruit).

Rubus × seminepalensis Naruh., *J. Jpn. Bot.* **65**(6): 187. 1960. [4, 10, 12, 28, 29]

Prostrate unarmed shrubs. W, C, E Nepal: 2300-3100 m. **Endemic to Nepal**.

Central: *Ramechap*, Neju-Choarma, 3010 m, H. Ohba, T. Kikuchi, M. Wakabayashi, M. Suzuki, N. Kurosaki, KR Rajbhandari and SK Wu, 8541181 (Holotype: TI; Isotype: BM).

Note: *Rubus × seminepalensis* Naruh. is an apparently hybrid between the compound-leaved *Rubus nepalensis* (Hook.f.) Kuntze and the simple-leaved *Rubus treutleri* Hook.f. (10).

Rubus splendidissimus H.Hara [2, 4, 6, 10]

चांदे ऐंसेलु (Chaande ainselu)

Erect or scrambling shrubs. C, E Nepal: 2200-3100 m. Nepal and E Himalaya.

Use: Food (fruit).

Rubus sumatranus Miq. [10]

Rubus indotibetanus Koidz. (2, 4); *Rubus rosifolius* subsp. *sumatranus* (Miq.) Focke

Erect or scandent shrubs. C, E Nepal: 700-2500 m. Nepal, E Himalaya, Tibetan Plateau, Assam-Burma, E Asia and SE Asia.

Rubus thomsonii Focke [2, 4, 10]

ऐंसेलु (Ainselu)

Scrambling shrubs. E Nepal: 2600-3700 m. Nepal and E Himalaya.

Use: Food (fruit).

Rubus treutleri Hook.f. [2, 4, 6, 10]

ठुलो ऐंसेलु (Thulo ainselu)

Scandent shrubs. W, C, E Nepal: 2300-3700 m. Nepal, E Himalaya, Tibetan Plateau and E Asia.

Use: Food (fruit).

Sanguisorba L.

Sanguisorba diandra (Hook.f.) Nordborg [2, 4, 6, 10]

Poterium diandrum Hook.f.

Perennial herbs. W, C, E Nepal: 2400-4400 m. Nepal, W Himalaya, E Himalaya and Tibetan Plateau.

Sibbaldia L.

Sibbaldia adpressa Bunge [8, 10, 11, 29]

Sibbaldia minutissima Kitam. (2, 4)

Perennial herbs. C Nepal: 2800-4500 m. Nepal, Tibetan Plateau, E Asia and N Asia.

Sibbaldia cuneata Hornem. ex Kuntze [2, 4, 6, 10]

Potentilla sibbaldi Hook.f.

Perennial herbs. W, C, E Nepal: 3000-4900 m. Nepal, W Himalaya, E Himalaya, Tibetan Plateau, E Asia, N Asia and SW Asia.

Sibbaldia emodi H.Ikeda & H.Ohba, *J. Jap. Bot.* **71**(4): 188.1996. [4, 10, 12, 28, 29]

Argentina emodi (H.Ikeda & H.Ohba) Soják (74)

Perennial rosulate herbs. C, E Nepal: 4000-4100 m. **Endemic to Nepal**.

East: *Solukhumbu*, Tangna – Dik Kharka, 4015 m, F. Miyamoto *et al* 9588122 (Holotype: TI; Isotypes: BM, E, G, GH, KATH, KYO).

Sibbaldia micropetala (D.Don) Hand.-Mazz. [2, 4, 6, 10]

Potentilla micropetala D.Don; *Sibbaldia potentilloides* Cambess.

Perennial herbs. W, C, E Nepal: 2300-4800 m. Nepal, W Himalaya, E Himalaya, Tibetan Plateau and E Asia.

Sibbaldia perpusilloides (W.W.Sm.) Hand.-Mazz. [2, 4, 6, 10]

Potentilla perpusilloides W.W.Sm.

Perennial herbs. C, E Nepal: 4000-4200 m. Nepal, E Himalaya, Tibetan Plateau and E Asia.

Sibbaldia purpurea Royle [2, 4, 6, 10]

Potentilla purpurea (Royle) Hook.f.

Tufted shrublets. W, C, E Nepal: 3600-5700 m. Nepal, W Himalaya, E Himalaya and Tibetan Plateau.

Sibbaldia sikkimensis (Prain) Chatterjee [2, 4, 10, 11]

Potentilla sikkimensis Prain; *Sibbaldia melinotricha* Hand.-Mazz. (6)

Perennial herbs. C, E Nepal: 3300-4800 m. Nepal, E Himalaya, Assam-Burma and E Asia.

Sibbaldia tetrandra Bunge [2, 4, 6, 10]

Potentilla tetrandra (Bunge) Bunge ex Hook.f.

Perennial herbs. W, C Nepal: 4500-6000 m. Nepal, W Himalaya, E Himalaya, Tibetan Plateau, N Asia and C Asia.

Sorbaria (Ser.) A.Braun

Sorbaria tomentosa (Lindl.) Rehder [2, 4, 6, 10]

Schizonotus tomentosus Lindl.; *Spiraea lindleyana* Wall. ex Lindl.; *Spiraea sorbifolia* L.

बाख्रे झार (Baakhre jhaar); Himalayan Sorbaria

Deciduous shrubs. W, C Nepal: 1800-2900 m. Nepal, W Himalaya, C Asia and SW Asia.

Use(s): Fodder, Medicine (seeds).

Sorbus L.

Sorbus arachnoidea Koehne [4, 6, 10]

Deciduous shrubs or trees. W, C, E Nepal: 2800-4200 m. Nepal and E Himalaya.

Sorbus foliolosa (Wall.) Spach [2, 4, 6, 10]

Pyrus foliolosa Wall.; *Photinia foliolosa* (Wall.) Koehne; *Sorbus ursina* (Wall. ex G.Don) S.Schauer (2,4); *Sorbus ursina* var. *wenzigiana* C.K.Schneid. (2,4); *Sorbus wenziana* (C.K.Schneid.) Koehne

Deciduous shrubs or small trees. W, C, E Nepal: 2200-4300 m. Nepal, W Himalaya, E Himalaya, Tibetan Plateau, Assam-Burma and E Asia.

Use(s): Medicine (leaves), Fodder.

Sorbus hedlundii C.K.Schneid. [2, 4, 6, 10]

Pyrus hedlundii (C.K.Schneid.) Lacaita

नाझिल (Naajhil)

Deciduous trees. E Nepal: 2500-3400 m. Nepal and E Himalaya.

Use: Food (fruit).

Sorbus himalaica Gabrieljan [6, 10]

Deciduous shrubs or small trees. W, C, E Nepal: 2500-4100 m. Nepal and E Himalaya.

Sorbus insignis (Hook.f.) Hedl. [2, 4, 6, 10]

Pyrus insignis Hook.f.

Deciduous trees. E Nepal: 2800-3200 m. Nepal, E Himalaya, Tibetan Plateau and Assam-Burma.

Sorbus kurzii (Watt ex Prain) C.K.Schneid. [2, 4, 6, 10]

Pyrus kurzii Watt ex Prain

Deciduous trees. C, E Nepal: 2800-3700 m. Nepal, E Himalaya, Tibetan Plateau and Assam-Burma.

Sorbus lanata (D.Don) Schauer [2, 4, 6, 10]

Pyrus lanata D.Don

पान्घर्मे (Paangmhe)

Deciduous trees. W, C, E Nepal: 2200-3400 m. Nepal and W Himalaya.

Use(s): Food (fruit), Fodder.

Sorbus microphylla Wenz. [2, 4, 6, 10]

Pyrus microphylla (Wenz.) Wall. ex Hook.f.; *Pyrus microphylla* Wall., nom.nud.; *Sorbus khumbuensis* McAll.

बझार (Bajhaar)

Deciduous shrubs or small trees. W, C, E Nepal: 2000-4200 m. Nepal, W Himalaya, E Himalaya, Tibetan Plateau, Assam-Burma and E Asia.

Use(s): Food, Fodder.

Sorbus rhamnoides (Decne.) Rehder [2, 4, 6, 10]

Micromeles rhamnoides Decne.; *Pyrus rhamnoides* (Decne.) Hook.f.

Deciduous shrubs or small trees. E Nepal: 2500-3000(-3500) m. Nepal, E Himalaya, Assam-Burma and E Asia.

Sorbus sharmae M.F. Watson, V. Manandhar & Rushforth, *Int. Dendrological Soc. Yearbook* **2009**: 79. 2010. [6, 10, 28, 29]

Deciduous trees. C, E Nepal: 2500-3200 m. **Endemic to Nepal**.

Central: *Rasuwa*, Langtang, 3170 m, DG Long *et al* 487 (Holotype: E).

Sorbus thomsonii (King ex Hook.f.) Rehder [2, 4, 8, 10]

Pyrus thomsonii King ex Hook.f.

Deciduous trees or large shrubs. E Nepal: 1700-2800 m. Nepal, E Himalaya, Tibetan Plateau, Assam-Burma and E Asia.

Sorbus vestita (Wall. ex G.Don) Lodd. [6, 10]

Pyrus vestita Wall. ex G.Don; *Sorbus cuspidata* (Spach) Hedl. (2, 4)

च्युली, लेक मयल (Chyuli, Lekh mayal)

Deciduous trees. W, C, E Nepal: (1350-)2000-3700 m. Nepal and W Himalaya.

Use: Food (fruit).

Sorbus wallichii (Hook.f.) T.T.Yu [2, 4, 6, 10]

Pyrus wallichii Hook.f.

Deciduous shrubs or trees. W, C, E Nepal: 1900-4100 m. Nepal, E Himalaya, Assam-Burma and E Asia.

Spiraea L.

Spiraea arcuata Hook.f. [2, 4, 6, 10]

पान्डा, पानस (Paanda, Paanas)

Deciduous arched shrubs. W, C, E Nepal: (2100-)3300-4900 m. Nepal, W Himalaya, E Himalaya, Tibetan Plateau, Assam-Burma and E Asia.

Use: Food.

Spiraea bella Sims [2, 4, 6, 10]

Spiraea amoena Spae; *Spiraea expansa* Wall., nom.nud.

सेतो खरेटो (Seto khareto)

Deciduous shrubs. W, C, E Nepal: 1800-4300 m. Nepal, W Himalaya, E Himalaya, Tibetan Plateau, Assam-Burma and E Asia.

Use: Fodder.

Note: Purohit & Panigrahi (1984) described *Spiraea bella* var. *fastigiata* (Wall. ex Schneider) K.M.Purohit & G. Panigrahi from Gossainthan, Nepal (Wallich 708.1, K-W); and also reported from India and Tibet (73).

Spiraea canescens D.Don [2, 4, 6, 10]

Spiraea cuneifolia Wall. ex Cambess.

झिल्लेटी (Jhilleti)

Deciduous shrubs. W, C, E Nepal: 1500-3000 m. Nepal, W Himalaya, E Himalaya, Tibetan Plateau and E Asia.

Use: Fodder.

Spiraea cantoniensis Lour. [2, 4, 10]

Deciduous shrubs. C Nepal: (200-)500-2200 m. Nepal and E Asia; probably cultivated in Nepal.

Spiraea hemicryptophyta A.J.C.Grierson [4, 6, 10]

Deciduous subshrubs. E Nepal: 3500-4300 m. Nepal, E Himalaya, Tibetan Plateau, Assam-Burma and E Asia.

Spiraea hypericifolia L. [2, 4, 6, 10]

Deciduous shrubs. W, C Nepal: *ca.* 1700 m. Nepal, W Himalaya, E Asia, N Asia, SW Asia and Europe.

Spiraea hypoleuca Dunn [2, 4, 6, 10]

Spiraea diversifolia Dunn (2, 4, 6)

Deciduous shrubs. W, C Nepal: 2100-3300(-4400) m. Nepal and W Himalaya.

Spiraea micrantha Hook.f. [2, 4, 6, 10]

Spiraea japonica subsp. *micrantha* (Hook.f.) Kitam.; *Spiraea japonica* var. *himalaica* Ktam.

Deciduous shrubs. W, C, E Nepal: 1500-3000 m. Nepal, E Himalaya, Tibetan Plateau and Assam-Burma.

Spiraea vacciniifolia D.Don [2, 4, 6, 10]

Spiraea laxiflora Lindl.

Deciduous shrubs. W, C Nepal: 1500-2300 m. Nepal and W Himalaya.

Exotic/Cultivated Taxa

Cydonia oblonga Mill. (Quince). *Use(s)*: Ornamental, Medicine (fruit, seeds). [2, 4]

Prunus dulcis (Mill.) D.A.Webb. (**कागजी बदाम,** Kaagaji badaam)

Rosa alba L. (**गुलाफ,** Gulaaf; Rose) [14]

Rosa banksiae Aiton [2, 4]

Rosa chinensis Jacq. (**अत्तर गुलाफ,** Attar gulaaf), Planted for ornamental purpose. [2, 4]

Rosa moschata Herrm. (**कस्तुरी गुलाफ,** Kasturi gulaaf; Musk rose). [2, 4]

Doubtful/Uncertain Taxa

Alchemilla subcrenata Buser [2, 4]

Cotoneaster horizontalis Decne [4, 65]

Cotoneaster lindleyi Steud. [4, 65]

Potentilla × microcontigua H.Ikeda & H.Ohba [4, 12, 28, 29]

Potentilla × micropeduncularis H.Ikeda & H.Ohba [4, 28, 29]

Potentilla × polyjosephiana H.Ikeda & H.Ohba [4, 28, 29]

Pyrus crenata D.Don [2, 4]

Rosa laevigata Michx. (**गुलाफ,** Gulaaf) [2, 4]

Rubus hibiscifolius Focke [2, 4, 28]

Sorbus thibetica (Cardot) Hand.-Mazz. [2, 4, 6]

ELAEAGNACEAE

Elaeagnus L.

Elaeagnus caudata Schltdl. ex Momiy. [3, 4, 7]

var. **caudata**

Shrubs. E Nepal: 2400-3500 m. Nepal and E Himalaya.

var. **aliena** Momiy. [3, 4]

Shrubs. E Nepal: 2500-2700 m. Endemic to Nepal (?)

Elaeagnus infundibularis Momiy. [3, 4, 7]

Elaeagnus armatus Buch.-Ham. ex D.Don

गुएँलो, मदिलो (Guenlo, Madilo)

Shrubs. C, E Nepal: 1500-2500 m. Nepal and E Himalaya.

Use(s): Medicine (flowers, fruit), Food (fruit).

Elaeagnus kanaii Momiy. [3, 4, 8]

कालगेडी (Kaalgedi)

Shrubs or small trees. W, C Nepal: 1400-2700 m. Nepal and W Himalaya.

Use: Food (Fruit).

Elaeagnus parvifolia Wall. ex Royle [3, 4, 11]

Elaeagnus umbellata Thunb. (7); *Elaeagnus umbellata* subsp. *parvifolia* (Wall. ex Royle) Servett.

गुएँली (Gunyeli)

Deciduous spiny shrubs. W, C, E Nepal: 1300-3000 m. Nepal, W Himalaya, E Himalaya, Tibetan Plateau, E Asia and SW Asia.

Use(s): Medicine (flowers, fruit), Food (fruit), Fodder.

Elaeagnus tricholepis Momiy., *J. Jap. Bot.* **48**: 262. 1973. [3, 4, 12, 26, 29]

Rambling shrubs. C Nepal: 1600-2500 m. **Endemic to Nepal**.

Central: *Lalitpur*, Godavari, 5300 ft. (1620 m), AD Schilling 1065 (Holotype: TI; Isotype: K).

Hippophae L.

Hippophae salicifolia D.Don [3, 4, 7, 11]

Elaeagnus salicifolia (D.Don) A.Nelson; *Hippophae rhamnoides* subsp. *salicifolia* (D.Don) Servett.

डाले चुक, चुगो (Daale chuk, Chugo); Willow-leaved sea buckthorn

Deciduous spiny shrubs or small trees. W, C Nepal: 2200-3500 m. Nepal, W Himalaya, E Himalaya and Tibetan Plateau.

Use(s): Medicine (leaves, fruit), Food, Wood, Cosmetic.

Hippophae tibetana Schltdl. [3, 4, 7]

Hippophae rhamnoides subsp. *tibetana* (Schlecht.) Servett.

भुई चुक, सेर्के; तर्बु-सतार (Bhuinchuk, Serke; Tarbu-satar: Amchi); Tibetan sea buckthorn

Deciduous subshrubs. W, C, E Nepal: 3600-4500 m. Nepal, W Himalaya, E Himalaya, Tibetan Plateau and E Asia.

Use(s): Medicine (fruit), Food (fruit), Fence, Fuel, Commercial.

RHAMNACEAE

Berchemia Neck. ex DC.

Berchemia edgeworthii M.A.Lawson [2, 4, 6, 11]

Berchemia nana W.W.Sm.

मुसे लरी (Muse lari)

Small shrubs. W, C Nepal: 2400-4400 m. Nepal, W Himalaya, E Himalaya, E Asia and SW Asia.

Use(s): Medicine (root, leaves), Food.

Berchemia flavescens (Wall.) Brongn. [2, 4, 6, 11]

Ziziphus flavescens Wall.

घुगि (Ghugi: Sherpa)

Climbing shrubs. W, C, E Nepal: 2300-3500 m. Nepal and E Himalaya.

Use: Food.

Berchemia floribunda (Wall.) Brongn. [2, 4]

Ziziphus floribunda Wall.

Climbing shrubs. C, E Nepal: 1600-2300 m. Nepal, W Himalaya, E Himalaya, Tibetan Plateau, Assam-Burma, E Asia and SE Asia.

Gouania Jacq.

Gouania leptostachya DC. [2, 4, 6]

भट्वासी (Bhatwaasi)

Climbing shrubs. C, E Nepal: 200-2000 m. Nepal, W Himalaya, E Himalaya, Assam-Burma, E Asia and SE Asia.

Use: Medicine (root, bark, leaves).

Gouania napalensis Wall. [2, 4]

Climbing shrubs. C, E Nepal: 900-1800 m. Nepal, W Himalaya, E Himalaya, Assam-Burma and S Asia.

Helinus E.Mey. ex Endl.

Helinus lanceolatus Brandis [2, 4, 6]

Climbing shrubs. W Nepal: 400-900 m. Nepal, W Himalaya, Assam-Burma and S Asia.

Hovenia Thunb.

Hovenia acerba Lindl. [2, 4, 6]

बाँगी काठ (Baangi kaath)

Deciduous trees. C, E Nepal: 500-1500 m. Nepal, W Himalaya, E Himalaya, E Asia, Assam-Burma and S Asia.

Use: Medicine (fruit).

Paliurus Mill.

Paliurus spina-christi Mill. [2, 4, 8]

Christ's thorn

Shrubs or rarely small trees. C Nepal: 800-1500 m. Planted in Nepal, Native to Mediterranean region and SW Asia.

Rhamnella Miq.

Rhamnella martinii (H.Lév.) C.K.Schneid. [2, 4, 8]

Rhamnus martini H. Lév.

Shrubs or small trees. C Nepal: 1300-2100 m. Nepal, Tibetan Plateau and E Asia.

Rhamnus L.

Rhamnus bodinieri H.Lév. [2, 4]

Evergreen shrubs. E Nepal: *ca.* 2400 m. Nepal, E Asia and SE Asia.

Rhamnus napalensis (Wall.) M.A.Lawson [2, 4, 6]

Ceanothus nepalensis Wall.

चिल्ले काठ (Chile kaath)

Deciduous shrubs or small trees. C, E Nepal: 600-1700 m. Nepal, E Himalaya, Tibetan Plateau, Assam-Burma, S Asia, E Asia and SE Asia.

Use(s): Medicine (bark, fruit), Fodder.

Rhamnus procumbens Edgew. [2, 4, 6, 11]

Rhamnus rupestris Royle

Climbing subshrubs. W, C, E Nepal: 1800-3000 m. Nepal, W Himalaya, E Himalaya and S Asia.

Rhamnus prostrata Jacq. [8, 11]

Deciduous prostrate subshrubs. C Nepal: 3600-3700 m. Nepal, W Himalaya, Tibetan Plateau and SW Asia.

Rhamnus purpurea Edgew. [2, 4, 6]

Purple buckthorn

Deciduous shrubs or small trees. W, C Nepal: 2100-2900 m. Nepal, W Himalaya and E Himalaya.

Use(s): Medicine (fruit), Agricultural implements.

Rhamnus triquetra (Wall.) Brandis [2, 4, 6]

Ceanothus triquetrus Wall.

Medium sized trees. W Nepal: 1500-2100 m. Nepal and W Himalaya.

Use: Medicine (bark).

Rhamnus virgata Roxb. [2, 4, 6, 11]

var. **virgata**

काँडे पैंयु, भालु काँडा (Kaande paiyun, Bhaalu kaanda)

Shrubs or small trees. W, C, E Nepal: 1000-3000 m. Nepal, W Himalaya, E Himalaya, Tibetan Plateau, E Asia and SE Asia.

Use(s): Medicine (fruit), Fodder, Agricultural implements.

var. **flavidus** (Momiy.) Momiy. [2, 4]

Rhamnus flavidus Momiy.

Small trees or shrubs. C Nepal: *ca.* 1300 m. Endemic to Nepal (?)

Rhamnus wightii Wight & Arn. [2, 4]

Climbing shrubs. W, C Nepal: 700-1400 m. Nepal and S Asia.

Use: Medicine (bark).

Sageretia Brongn.

Sageretia filiformis (Roth ex Schult.) G.Don [2, 4, 6]

Rhamnus filiformis Roth ex Schult.; *Rhamnus parviflorus* Klein ex Willd. ex Schult.

Climbing shrubs. W, C Nepal: 800-1800 m. Nepal, W Himalaya and SE Asia.

Sageretia hamosa (Wall.) Brongn. [2, 4]

Ziziphus hamosa Wall.; *Berchemia hamosa* (Wall.) Wall.

Climbing shrubs. C Nepal. Nepal, Tibetan Plateau, S Asia, E Asia and SE Asia.

Sageretia thea (Osbeck.) M.C.Johnst.

Rhamnus thea Osbeck.

var. **bornmuelleri** (C.K.Schneid.) M.C.Johnst. ex H.Hara [2, 4, 6]

Sageretia brandrethiana var. *bornmuelleri* C.K.Schneid.

Evergreen shrubs. W Nepal: 2500-2900 m. Nepal, W Himalaya, C Asia and SW Asia.

Ventilago Gaertn.

Ventilago denticulata Willd. [2, 4, 6]

Ventilago calyculata Tul.

सनरजोम (Sanarjom)

Climbing shrubs. W, C Nepal: *ca.* 400 m. Nepal, W Himalaya, E Himalaya, Assam-Burma, S Asia, E Asia and SE Asia.

Use: Medicine (root, stem bark).

Ziziphus Mill.

Ziziphus apetala Hook.f. ex M.A.Lawson [2, 4]

Climbing shrubs. C Nepal: *ca.* 900 m. Nepal, E Himalaya and Assam-Burma.

Ziziphus incurva Roxb. [2, 4, 6]

हाडे बयर (Haade bayar); Common Jujube

Medium-sized evergreen trees. C, E Nepal: 900-1900 m. Nepal, W Himalaya, E Himalaya, Tibetan Plateau, Assam-Burma, E Asia and SE Asia.

Use(s): Food (fruit), Fodder.

Ziziphus mauritiana Lam. [2, 4, 6]

Rhamnus jujuba L.; *Ziziphus jujuba* (L.) Lam.

अगुजे बयर, बयर (Aguje bayar, bayar)

Shrubs or small trees. W, C, E Nepal: 200-1200 m. Native to SE Asia, widely cultivated in E Tropical Asia and Australasia.

Use(s): Food (fruit), Fodder, Fence, Medicine (root, bark, leaves, fruit), Agricultural implements, Tannin.

Ziziphus nummularia (Burm.f.) Wight & Arn. [2, 4, 6]

Rhamnus nummularia Burm.f.; *Ziziphus rotundifolia* Lam.

बयर (Bayar); Wild jujube

Straggling shrubs. W, C Nepal: 200-500 m. Nepal, S Asia, SE Asia and C Asia.

Use: Medicine (leaves, fruit).

Ziziphus oenopolia (L.) Mill. [2, 4, 6]

Rhamnus oenopolia L.

बोक्सी बयर, औले बयर (Boksi bayar, Aule bayar); Jackal Jujube

Thorny straggling shrubs. W, C Nepal: 100-500 m. Nepal, Assam-Burma, S Asia, E Asia, SE Asia and Australasia.

Use(s): Food (fruit), Fence, Hedge, Medicine (root, bark, fruit), Tannin.

Ziziphus oxyphylla Edgew. [2, 4, 8]

Shrubs or small trees. W Nepal: 800-1400 m. Nepal, W Himalaya and SW Asia.

Ziziphus rugosa Lam. [2, 4, 6]

Ziziphus glabra Roxb.

काँटे बयर, हर्‍रायाँ बयर (Kaante bayar, Harraya bayar)

Climbing shrubs. C, E Nepal: 100-800 m. Nepal, W Himalaya, E Himalaya, Assam-Burma, S Asia, E Asia and SE Asia.

Use(s): Food (fruit), Fodder, Medicine (bark, flowers), Fuel, Fence.

Ziziphus xiangchengensis Y.L.Chen & P.K.Chou

Ziziphus budhensis Bhattarai & M.L.Pathak (25)

बोधिचित्त (Bodhi chitta)

Small trees. C Nepal: 1000-2000 m. Nepal and E Asia.

Use: Ritual (fruit).

Ziziphus xylopyrus (Retz.) Willd. [2, 4, 6]

Rhamnus xylopyrus Retz.; *Ziziphus caracutta* Roxb.

बन बयर (Ban bayar)

Small trees. C, E Nepal: 100-300 m. Nepal, W Himalaya and S Asia.

Use: Medicine (fruit).

Doubtful/Uncertain Taxa

Rhamnus minuta Grubov [4, 98]

Rhamnus persica Boiss. [98]

ULMACEAE

Holoptelea Planch.

Holoptelea integrifolia (Roxb.) Planch. [3, 4, 7]

Ulmus integrifolia Roxb.

चेप्टे पांग्रो, सानो पांग्रो (Chepte paangro, Saano paangro); Indian elm

Large deciduous trees. W, E Nepal: 100-500 m. Nepal, W Himalaya, Assam-Burma, S Asia and SE Asia.

Use: Medicine (bark, leaves).

Ulmus L.

Ulmus androssowii Litv.

var. **subhirsuta** (C.K.Schneid.) P.H.Huang

Ulmus wilsoniana var. *subhirsuta* C.K.Schneid.; *Ulmus chumlia* Melville & Heybroek (3, 4, 8)

कुम्लिया (Kumliya)

Deciduous trees. W, C, E Nepal: 1500-1900 m. Nepal, W Himalaya, S Asia, E Asia and SW Asia.

Ulmus lanceifolia Roxb. ex Wall. [3, 4, 7]

चमलायो (Chamlaayo); Vietnam elm

Evergreen trees. E Nepal: 100-2500 m. Nepal, W Himalaya, E Himalaya, Assam-Burma, E Asia and SE Asia.

Use(s): Medicine (bark), Timber.

Ulmus wallichiana Planch. [3, 4, 7]

धमिना (Dhamina); Himalayan elm

Deciduous trees. W, C Nepal: 1000-3000 m. Nepal, W Himalaya and SW Asia.

Use(s): Timber, Fodder, Fiber/Tannin (bark), Construction materials, Fish poison (bark, leaves).

IUCN Category: **Vulnerable** A1c (ver 2.3).

CANNABACEAE (incl. ULMACEAE, p.p.)

Cannabis L.

Cannabis sativa L. [7]

Cannabis sativa subsp. *indica* (Lam.) Small & Cronquist (3, 4); *Cannabis sativa* var. *kafiristanica* (Vavilov) Small & Cronquist (3, 4)

भांग, गाँजा (Bhaang, Gaanja); Hemp, Marijuana

Annual herbs. W, C, E Nepal: 200-2800 m (Naturalized). Probably Native to C Asia; widely cultivated and naturalized in most temperate and tropical areas of the world.

Use(s): Medicine (flowering tops of female plant, seeds), Food (seeds), Narcotic, Essential oil (Hemp oil), Fiber.

Celtis L.

Celtis australis L. [3, 4, 7]

खरी (Khari); Honey-berry tree, Southern nettle tree

Deciduous trees. C Nepal: 1300-2200 m. Nepal, W Himalaya and Europe.

Use(s): Medicine (bark, stem, leaves, fruit), Food, Fodder.

Celtis tetrandra Roxb. [3, 4, 7]

खरी (Khari); Eastern nettle tree

Deciduous trees. W, C, E Nepal: 700-2800 m. Nepal, W Himalaya, E Himalaya, Tibetan Plateau, Assam-Burma, E Asia, S Asia and SE Asia.

Use(s): Medicine (seeds), Fodder.

Celtis timorensis Span. [3, 4]

Stink wood

Evergreen trees. E Nepal: 300-1400 m. Nepal, E Himalaya, Tibetan Plateau, Assam-Burma, S Asia, E Asia and SE Asia.

Use: Medicine (wood).

Trema Lour.

Trema cannabina Lour. [3, 4, 7]

खाक्सी (Khaaksi)

Shrubs or small trees. W, C, E Nepal: 100-2000 m. Nepal, Assam-Burma, S Asia, E Asia, SE Asia and Australasia.

Use(s): Fodder, Cordage.

Trema orientalis (L.) Blume [3, 4, 7]

Celtis orientalis L.

मदिसे खरी, कुयेल (Madise Khari, Kuyel); Pigeon wood; Oriental Trema

Evergreen shrubs or small trees. E Nepal: 300-1200 m. Nepal, W Himalaya, E Himalaya, Tibetan Plateau, Assam-Burma, S Asia, E Asia, SE Asia and Australasia.

Use(s): Fodder, Fiber, Charcoal, Gunpowder, Tannin, Medicine (bark, wood).

Trema politoria (Planch.) Blume [3, 4, 8]

Sponia politoria Planch.

Shrubs or small trees. W, C, E Nepal: 200-1500 m. Nepal, W Himalaya, E Himalaya and SE Asia.

Trema tomentosa (Roxb.) H.Hara [3, 4, 7]

Celtis tomentosa Roxb.

Poison peach

Evergreen shrubs or small trees. W, C, E Nepal: 900-1500 m. Nepal, E Himalaya, Assam-Burma, S Asia, E Asia and SE Asia.

MORACEAE

Artocarpus J.R.Forst. & G.Forst.

Artocarpus heterophyllus Lam. [3, 4]

रुख कटहर (Rukh katahar); Jackfruit

Evergreen trees. C Nepal: *ca.* 800 m. Native to S Asia and SE Asia.

Use(s): Food (Fruit), Medicine (root, bark, latex, leaves, seeds), Tannin, Dye.

Artocarpus integer (Thunb.) Merr. [3, 4, 7]

Radermachia integra Thunb.; *Artocarpus integrifolia* L.f.

पानस (Paanas)

Evergreen trees. E Nepal: 200-1700 m. Widely cultivated. Native to SE Asia (Malaysia).

Use(s): Medicine (leaves, fruit, seeds, pith), Fodder.

Artocarpus lakoocha Wall. ex Roxb. [3, 4, 7]

Artocarpus lakucha Buch.-Ham. ex D.Don

बढहर (Badhar); Monkey jack

Evergreen trees. W, C, E Nepal: 100-1300(-2200) m. Nepal, E Himalaya, Assam-Burma and SE Asia.

Use(s): Fodder, Timber, Food, Medicine (bark, leaves, fruit, seeds).

Broussonetia L.Her. ex Vent.

Broussonetia kurzii (Hook.f.) Corner

Allaeanthus kurzii Hook.f. (4, 38)

Scandent shrubs. E Nepal: 200-600 m. Nepal, E Himalaya, Assam-Burma, E Asia and SE Asia.

Ficus L.

Ficus abelii Miq. [3, 4]

Ficus subpyriformis Miq.

Scandent shrubs. C Nepal, E Himalaya, Assam-Burma, S Asia, E Asia and SE Asia.

Ficus arnottiana (Miq.) Miq. [3, 4, 7]

Urostigma arnottianum Miq.

var. **arnottiana**

पारस पिपल (Paaras pipal)

Deciduous small trees. W, C, E Nepal: 800-1500 m. Nepal and S Asia.

Use(s): Medicine (bark), Fodder.

var. **subcostata** Corner [3, 4]

Deciduous small trees. W Nepal: 2300 m. Nepal and W Himalaya.

Ficus assamica Miq.

Ficus heterophylla var. *assamica* (Miq.) Corner ex Chater (3, 4)

Small shrubs. W Nepal: *ca.* 300 m. Nepal, E Himalaya, Assam-Burma, S Asia, E Asia and SE Asia.

Ficus auriculata Lour. [3, 4, 7]

Ficus oligodon Miq. (3, 4, 7); *Ficus roxburghii* Wall. ex Miq.

तिमिला, निमारो (Timilaa, Nimaaro); Elephant-ear fig

Small trees. C Nepal: 200-2000 m. Nepal, W Himalaya, E Himalaya, Assam-Burma, S Asia and SE Asia.

Use(s): Medicine (bark, fruit), Food (fruit), Fodder, Plates (leaves).

Ficus benghalensis L. [3, 4, 7]

Ficus indica L.

बर (Bar); Banyan tree

Evergreen trees. W, C, E Nepal: 500-1500 m. Widely planted in Nepal, S Asia.

Use(s): Medicine (bark, latex, leaves), Food (fruit), Fodder, Shade tree, Ritual.

Ficus benjamina L. [3, 4, 7]

Ficus benjamina var. *nuda* (Miq.) Barrett (3, 4); *Urostigma benjaminum* var. *nudum* Miq.

समी, स्वामी (Sami, Swaami); Weeping fig

Evergreen trees. W, C Nepal: 600-1200 m. Widely planted. Nepal, Assam-Burma, E Asia, S Asia, SE Asia and Australasia.

Use(s): Shade tree, Medicine (latex, leaves), Fodder, Fiber (bark).

Ficus carica L. [7]

अँजिर (Anjeer); Common fig

Deciduous shrubs or small trees. W Nepal: 100-200 m (Cultivated). Native to SW Asia.

Use(s): Food (fruit), Medicine (latex, root, bark, fruit).

Ficus curtipes Corner [3, 4, 7]

Ficus obtusifolia Roxb.

Eastern laurel fig

Small trees (epiphytic when young). E Nepal: 400-600 m. Nepal, E Himalaya, Assam-Burma, E Asia and SE Asia.

Ficus cyrtophylla (Wall. ex Miq.) Miq. [8]

Covellia cyrtophylla Wall. ex Miq.

Shrubs or small trees. C, E Nepal: 700-1600 m. Nepal, E Himalaya, Assam-Burma, S Asia, E Asia and SE Asia.

Ficus drupacea Thunb.

Ficus drupacea var. *pubescens* (Roth) Corner (3, 4, 7)

Mysore fig

Medium-sized trees. E Nepal: 100-1600 m. Nepal, E Himalaya, Assam-Burma, S Asia, E Asia, SE Asia and Australasia.

Ficus gasparriniana Miq. [7]

Shrubs or small trees. W, C, E Nepal: 500-1600 m. Nepal, E Himalaya, Assam-Burma and SE Asia.

Ficus geniculata Kurz [3, 4, 7]

पकर (Pakar)

Large trees. W, E Nepal: 100-700 m. Nepal, E Himalaya, Assam-Burma, S Asia, E Asia and SE Asia.

Ficus glaberrima Blume [3, 4, 7]

Ficus angustifolia Roxb.

पाखुरी (Paakhuri)

Large trees (epiphytic when young). W, C Nepal: 600-1500 m. Nepal, E Himalaya, Tibetan Plateau, Assam-Burma, E Asia and SE Asia.

Use(s): Fodder, Fiber (bark).

Ficus hederacea Roxb. [3, 4, 7]

Ficus scandens Roxb.

दुधे लहरी (Dudhe lahari); Ivy fig

Climbing shrubs. W, C, E Nepal: 500-1400 m. Nepal, W Himalaya, E Himalaya, Assam-Burma, S Asia, E Asia and SE Asia.

Use: Fodder.

Ficus heterophylla L.f. [20]

Large or procumbent shrubs. W Nepal: 100-200 m. Nepal, Assam-Burma, S Asia, E Asia and SE Asia.

Use(s): Food (fruit), Medicine (root, leaves).

Ficus hirta Vahl [3, 4, 8]

Ficus hirsuta Roxb.

खरसेटो, खस्रेटो (Kharseto, Khasreto)

Shrubs or small trees. C, E Nepal: 900-1200 m. Nepal, E Himalaya, Assam-Burma and SE Asia.

Use(s): Medicine (bark, latex), Food (fruit).

Ficus hispida L.f. [3, 4, 7]

Ficus daemonum Koenig ex Vahl

खस्रेटो, कोठाया डुमरी (Khasreto, Kothaya-dumri); Hairy fig

Shrubs or small trees. W, C, E Nepal: 400-1800 m. Nepal, W Himalaya, E Himalaya, S Asia, E Asia, SE Asia and Australasia.

Use(s): Medicine (root, bark, fruit), Food (fruit), Fodder, Fiber (bark), Cordage.

Ficus hookeriana Corner [3, 4, 7]

Ficus hookeri Miq.

Medium-sized trees. E Nepal: *ca.* 1800 m. Nepal, E Himalaya and E Asia.

Use: Fodder.

Ficus lacor Buch.-Ham. [3, 4]

Ficus concinna (Miq.) Miq.

काब्रो (Kaabro); Java fig

Large evergreen trees. W, C Nepal: *ca.* 500 m. Nepal, W Himalaya, E Himalaya, Assam-Burma and SE Asia.

Use(s): Fodder, Cordage, Medicine (bark, latex), Food (young shoot).

Ficus laevis Blume [3, 4, 7]

Ficus vagans Roxb.

Scandent shrubs (often epiphytic). C, E Nepal: *ca.* 300 m. Nepal, E Himalaya, Tibetan Plateau, Assam-Burma, S Asia, E Asia and SE Asia.

Ficus microcarpa L.f. [3, 4, 7]

जामु (Jaamu); Chinese banyan, Laurel fig

Small or medium-sized trees. W, C Nepal: 100-1900 m. Nepal, E Himalaya, Assam-Burma, S Asia, E Asia, SE Asia and Australasia.

Use: Medicine (root, bark, leaves).

Ficus neriifolia Sm. [3, 4, 7]

Ficus nemoralis Wall. ex Miq.; *Ficus neriifolia* var. *fieldingii* (Miq.) Corner (3, 4); *Ficus neriifolia* var. *nemoralis* (Wall. ex Miq.) Corner (3, 4)

दुधिलो (Dudhilo); Willow leaf fig

Medium-sized trees. W, C, E Nepal: 1200-2900 m. Nepal, E Himalaya, Tibetan Plateau, Assam-Burma and E Asia.

Use(s): Fodder, Medicine (latex), Food (fruit).

Ficus nervosa Heyne ex Roth [3, 4, 7]

Evergreen tall trees. E Nepal: 400-1000 m. Nepal, E Himalaya, Assam-Burma, S Asia, E Asia and SE Asia.

Ficus palmata Forssk. [3, 4, 7]

Ficus caricoides Roxb.

बेडु, निमारो (Bedu, Nimaaro); Punjab fig

Shrubs or small trees. W Nepal: 600-2300 m. Nepal, W Himalaya, SW Asia and Africa.

Use(s): Medicine (latex, fruit), Food (young shoot, fruit), Fodder.

Ficus pubigera (Wall. ex Miq.) Miq. [3, 4, 7]

Pogonotrophe pubigera Wall. ex Miq.

Scandent shrubs. C Nepal: *ca.* 2100 m. Nepal, W Himalaya, E Himalaya, Assam-Burma, E Asia and SE Asia.

Ficus pumila L. [3, 4, 7]

Creeping fig

Woody creeping vine. C Nepal: *ca.* 1400 m. Nepal, E Asia and SE Asia.

Ficus racemosa L. [3, 4, 7]

Ficus glomerata Roxb.

डुम्री, गुलर (Dumri, Gular); Cluster fig, Gular fig

Tall trees. W, C Nepal: 300-1000 m. Nepal, Assam-Burma, S Asia, E Asia, SE Asia and Australasia.

Use(s): Food (root, bark, latex, fruit), Fodder, Medicine (bark, latex, fruit), Fuel, Tannin.

Ficus religiosa L. [3, 4, 7]

पीपल (Peepal); Peepal tree, Sacred fig

Large trees (epiphytic when young). W, C, E, Nepal: 100-1500 m. Nepal, Native to S Asia.

Use(s): Shade tree, Ritual, Medicine (bark, latex, leaves, fruit), Fodder, Dye, Tannin.

Ficus rumphii Blume [3, 4, 7]

Ficus cordifolia Roxb.

पहरे पिपल, काठे पिपल (Pahare pipal, Kaathe pipal)

Medium-sized trees (sometime epiphytic). W Nepal: 200-1500 m. Nepal, E Himalaya, Assam-Burma, S Asia, E Asia and SE Asia.

Use(s): Fodder, Medicine (fruit), Food (fruit), Fiber (bark).

Ficus sarmentosa Buch.-Ham. ex Sm. [3, 4, 7]

Ficus nepalensis Spreng. (3, 4).

बन तिमिला, काठे डुम्री (Ban timila, Kaathe dumri)

Climbing shrubs. W, C, E Nepal: 1400-2500 m. Nepal, E Himalaya and Tibetan Plateau.

Use(s): Medicine (bark, latex), Food (fruit), Fodder.

Ficus semicordata Buch.-Ham. ex Sm. [3, 4, 7]

Ficus cunia Buch.-Ham. ex Roxb.

खनायो, खन्यु (Khanaayo, Khanyu); Drooping fig

Medium-sized trees. W, C, E Nepal: 200-1700 m. Nepal, E Himalaya, Tibetan Plateau, Assam-Burma, S Asia, E Asia and SE Asia.

Use(s): Food (fruit), Fodder, Medicine (root, bark, latex, fruit), Fiber.

Ficus squamosa Roxb. [3, 4]

Ficus laminosa Hardw. ex Roxb.

Gregarious shrubs. C, E Nepal: 500-600 m. Nepal, E Himalaya, Assam-Burma, S Asia, E Asia and SE Asia.

Ficus subincisa Buch.-Ham. ex Sm. [3, 4, 7]

Ficus clavata Wall. ex Miq.

var. **subincisa**

बेडुलो, आँखा पकुवा (Bedulo; Aankha pakuwa)

Shrubs or small trees. W, C, E Nepal: 300-1800 m. Nepal, W Himalaya, E Himalaya, Assam-Burma, E Asia and SE Asia.

Use(s): Food (fruit), Fodder.

var. **trachycarpa** (Miq.) Corner ex Chater [3, 4]

Ficus trachycarpa Miq.

Shrubs or small trees. C Nepal: *ca.* 1100 m. Nepal and W Himalaya.

Ficus subulata Blume [3, 4]

Ficus acuminata Roxb.

Climbing shrubs. E Nepal: *ca.* 300 m. Nepal, E Himalaya, Tibetan Plateau, Assam-Burma, E Asia and SE Asia.

Ficus tinctoria G.Forst.

subsp. **gibbosa** (Blume) Corner

Ficus gibbosa Blume; *Ficus tinctoria* subsp. *parasitica* (Willd.) Corner (3, 4)

Epiphytic shrubs. W Nepal. Nepal, E Himalaya, Tibetan Plateau, Assam-Burma, S Asia, E Asia and SE Asia.

Ficus virens Aiton [4]

White fig

Trees (epiphytic when young): E Nepal: 80-200 m. Nepal, E Himalaya, Tibetan Plateau, S Asia, E Asia, SE Asia and Australasia.

Use(s): Medicine (bark), Food (young leaves).

Maclura Nutt.

Maclura cochinchinensis (Lour.) Corner [3, 4, 7]

Vanieria cochinchinensis Lour.; *Cudrania javanensis* Trecul

डमरू, डम्बरे काँडा (Damaru, Dambare kaanda); Cockspur thorn

Scrambling shrubs. W, C, E Nepal: 300-1300 m. Nepal, W Himalaya, E Himalaya, Tibetan Plateau, Assam-Burma, S Asia, E Asia, SE Asia and Australasia.

Use(s): Medicine (bark, latex, stem), Food (fruit).

Maclura pubescens (Trecul) Z.K.Zhou & M.G.Gilbert [8]

Cudrania pubescens Trecul (23)

Woody vines. E Nepal: 1200-2000 m. Nepal, Assam-Burma, E Asia and SE Asia.

Morus L.

Morus alba L. [7]

Morus indica L.

किम्बु (Kimbu); Common mulberry, White mulberry

Small deciduous trees. Widely planted for edible fruit. Native to E Asia.

Use(s): Food (fruit), Fodder.

Morus australis Poir. [3, 4, 7]

किम्बु, सानो किम्बु (Kimbu, Saano kimbu); Chinese mulberry

Small deciduous trees. W, C, E Nepal: 900-2400 m. Nepal, W Himalaya, E Himalaya, Tibetan Plateau, Assam-Burma and E Asia.

Use(s): Medicine (root, bark, fruit), Food (fruit), Avenue tree.

Morus macroura Miq. [3, 4]

Morus alba var. *laevigala* Bureau

किम्बु (Kimbu); Tibetan mulberry

Deciduous small trees. E Nepal: 1200-1700 m. Nepal, W Himalaya, E Himalaya, Tibetan Plateau, Assam-Burma, E Asia and SE Asia.

Use(s): Food (fruit), Food (for silkworms), Medicine (bark, root, leaves, fruit), Furniture.

Morus serrata Roxb. [3, 4, 7]

Morus alba var. *serrata* Bureau

कालो काफल, किम्बु (Kaalo kaaphal, Kimbu); Himalayan mulberry

Deciduous medium-sized trees. W, C Nepal: 1600-2400 m. Nepal, W Himalaya and Tibetan Plateau.

Use(s): Food (fruit), Medicine (root), Furniture, Agricultural implements.

Streblus Lour.

Streblus asper Lour. [3, 4, 7]

खाक्सी (Khaaksi); Toothbrush tree

Evergreen small or medium-sized trees. C, E Nepal: 100-500 m. Nepal, W Himalaya, E Himalaya, S Asia, E Asia and SE Asia.

Use(s): Medicine (bark, root, twigs), Food (fruit), Fiber (bark), Tooth brushes (twigs), Fodder.

Exotic/Cultivated Taxa

Ficus elastica Roxb. ex Hornem. (**रबर,** Rabar; Rubber tree). E Himalaya, Assam-Burma, S Asia, E Asia and SE Asia. [3, 4, 7]

Doubtful/Uncertain taxa

Ficus altissima Blume [3, 4]

Ficus semicordata var. **montana** Amatya [78]

URTICACEAE

Boehmeria Jacq.

Boehmeria clidemioides Miq. [3, 4, 7, 134]

काम्ले (Kaamle)

Boehmeria platyphylla var. *cinerascens* Hook.f.

var. **clidemioides**

Perennial herbs or subshrubs. C, E Nepal: 1300-2200 m. Nepal, E Himalaya, Tibetan Plateau, Assam-Burma, E Asia and SE Asia.

Use(s): Medicine (whole plant), Rough clothes, Sacks, Fishing nets.

var. **diffusa** (Wedd.) Hamd.-Mazz. [134]

Boehmeria diffusa Wedd.

Perennial herbs. E Nepal: *ca.* 900 m. Nepal, E Himalaya, Assam-Burma, SE Asia, E Asia.

Boehmeria conica C.J.Chen [134]

Subshrubs or shrubs. Nepal: 1200-2800 m. Nepal, E Himalaya, Assam-Burma, E Asia.

Boehmeria hamiltoniana Wedd. [3, 4, 7, 134]

Shrubs. C, E Nepal: 1600-2400 m. Nepal, E Himalaya, Assam-Burma, E Asia and SE Asia.

Boehmeria penduliflora Wedd. ex D.G.Long [134]

Boehmeria densiflora var. *intermedia* Acharya & Yonek. (7, 133); *Boehmeria densiflora* var. *penduliflora* (Wedd. ex D.G.Long) Acharya & Yonek. (7, 133); *Boehmeria macrophylla* D.Don (3, 4)

Subshrubs. W, C, E Nepal: 500-1700 m. Nepal, E Himalaya, Tibetan Plateau, Assam-Burma, E Asia and SE Asia.

Boehmeria polystachya Wedd. [3, 4, 7, 11, 134]

Boehmeria tibetica C.J.Chen

भरौं (Bharaun)

Perennial herbs or subshrubs. W, C, E Nepal: 1600-3000 m. Nepal, E Himalaya, Tibetan Plateau and E Asia.

Use: Fishnets (bark).

Boehmeria siamensis Craib [134]

Shrubs. E Nepal (?): 500-2700 m. Nepal, Assam-Burma, SE Asia and E Asia.

IUCN Category: **Near Threatened** (ver. 3.1)

Boehmeria ternifolia D.Don [3, 4, 7, 134]

काम्ले (Kaamle)

Shrubs or small trees. W, C, E Nepal: 900-2300 m. Nepal, W Himalaya and E Himalaya.

Use(s): Medicine (leaves), Food, Fodder.

Boehmeria virgata (G.Forst.) Guill. [134]

subsp. **macrophylla** (Hornem.) Friis & Wilmot-Dear

Boehmeria macrophylla Hornem.

भरुन (Bharun)

Shrubs or subshrubs. W, C, E Nepal: 800-2700 m. Nepal, W Himalaya, E Himalaya, Tibetan Plateau, Assam-Burma, S Asia, E Asia and SE Asia.

Use(s): Medicine (whole plant), Fiber, Fishnets.

var. **canescens** (Wedd.) Friis & Wilmot-Dear [134]

Boehmeria canescens Wedd. (7); *Boehmeria macrophylla* var. *canescens* (Wedd) D.G.Long;

Boehmeria platphylla var. *canescens* Wedd. (3, 4)

Subshrubs. C, E Nepal: 400-1500 m. Nepal and E Himalaya.

var. **kamley** (Acharya & Yonek.) Friis & Wilmot-Dear [134]

Boehmeria kamley Acharya & Yonek. (7, 133)

Sparsely branched shrubs. E Nepal: 1000-2200 m. Nepal and E Himalaya.

var. **macrostachya** (Wight) Friis & Wilmot-Dear [134]

Splitgerbera macrostachya Wight; *Boehmeria macrostachya* (Wight) F.M.Bailey; *Boehmeria platyphylla* D.Don (3, 4, 7)

Perennial herbs or subshrubs. Nepal: 100-2600 m. Nepal, E Himalaya and S Asia.

var. **minuticymosa** Acharya, Friis & Wilmot-Dear [134]

Perennial or subshrubs. C Nepal: *ca.* 1800 m. Nepal, Assam-Burma, S Asia, E Asia and SE Asia.

var. **rotundifolia** (D.Don) Friis & Wilmot-Dear [134]

Boehmeria rotundifolia D.Don (7, 11); *Boehmeria macrophylla* var. *rotundifolia* (D.Don) W.T.Wang

Perennial herbs. W, C, E Nepal: 400-2700 m. Nepal, E Himalaya, Tibetan Plateau, Assam-Burma and E Asia.

var. **scabrella** (Roxb.) Friis & Wilmot-Dear [134]

Urtica scabrella Roxb.; *Boehmeria macrophylla* var. *scabrella* (Roxb.) D.G.Long ; *Boehmeria platyphylla* var. *scabrella* (Roxb.) Wedd.

Perennial herbs. Nepal, W Himalaya, SE Asia and E Asia.

Chamabainia Wight

Chamabainia cuspidata Wight [3, 4, 7]

पाम्ले (Paamle)

Erect or ascending herbs. W, C, E Nepal: 1800-2900 m. Nepal, W Himalaya, E Himalaya, Tibetan Plateau, Assam-Burma, S Asia, E Asia and SE Asia.

Use: Medicine (whole plant).

Debregeasia Gaudich.

Debregeasia longifolia (Burm.f.) Wedd. [3, 4, 7]

Urtica longifolia Burm.f.

तुषारे (Tushaare)

Evergreen shrubs. W, C, E Nepal: 100-2000 m. Nepal, W Himalaya, E Himalaya, Tibetan Plateau, Assam-Burma, S Asia, E Asia and SE Asia.

Use(s): Medicine (bark, leaves), Food (fruit), Fodder, Fuel, Fiber.

Debregeasia saeneb (Forssk.) Hepper & Wood [7, 11]

Rhus saeneb Forssk.; *Boehmeria salicifolia* D.Don; *Debregeasia salicifolia* (D.Don) Rendle (3, 4)

तुषारे, दार (Tushaare, Daar)

Evergreen shrubs. W, C, E Nepal: 1200-2500 m. Nepal, W Himalaya, Tibetan Plateau, SW Asia and Africa.

Use(s): Food (fruit), Fodder, Fuel, Fiber (bark).

Debregeasia wallichiana (Wedd.) Wedd. [3, 4, 7]

Missiessya wallichiana Wedd.

Shrubs or small trees. E Nepal: 400-1200 m. Nepal, E Himalaya, Assam-Burma, S Asia, E Asia and SE Asia.

Dendrocnide Miq.

Dendrocnide sinuata (Blume) Chew [3, 4, 8]

Urtica sinuata Blume

मोरंगे, मरिंगे (Morange, Maringe); Stinging tree

Shrubs or small trees. E Nepal: 200-1000 m. Nepal, E Himalaya, Tibetan Plateau, Assam-Burma, S Asia, E Asia and SE Asia.

Use: Medicine (root, leaves).

Elatostema J.R.Forst. & G.Forst.

Elatostema acuminatum (Poir.) Brongn. [8]

Procris acuminata Poir.

Subshrubs. C, E Nepal: 1800-2100 m. Nepal, E Himalaya, Assam-Burma, E Asia and SE Asia.

Elatostema dissectum Wedd. [3, 4, 8]

Perennial herbs. E Nepal: 200-2300 m. Nepal, E Himalaya, Assam-Burma, E Asia and SE Asia.

Elatostema ficoides Wedd. [3, 4, 7]

Perennial herbs. C Nepal: 800-2500 m. Nepal, E Himalaya, E Asia and SE Asia.

Elatostema grande (Wedd.) P.S.Green

Elatostema sessile var. *grande* Wedd.; *Elatostema punctatum* Wedd. (11)

Perennial herbs. C Nepal: 1700-2500 m. Endemic to Australasia. Introduced elsewhere.

Elatostema hookerianum Wedd. [3, 4, 8]

Perennial herbs. E Nepal: 1900-2500 m. Nepal, E Himalaya, Tibetan Plateau and E Asia.

Elatostema integrifolium (D.Don) Wedd. [3, 4, 7]

Procris integrifolia D.Don

लिपे झार (Lipe jhaar)

Perennial herbs or subshrubs. C Nepal: 900-1800 m. Nepal, E Himalaya, Assam-Burma, S Asia, E Asia and SE Asia.

Use: Medicine (whole plant).

Elatostema lineolatum Wight [3, 4, 7]

Subshrubs. W, C, E Nepal: 200-1800 m. Nepal, E Himalaya, Tibetan Plateau, Assam-Burma, S Asia, E Asia and SE Asia.

Elatostema monandrum (D.Don) H.Hara [3, 4, 7, 11]

Procris monandra D.Don; *Elatostema diversifolium* Wedd.; *Elatostema monandrum* var. *ciliatum* (Hook.f.) S.K.Murti

Perennial herbs. W, C, E Nepal: 800-3000 m. Nepal, W Himalaya, E Himalaya, Tibetan Plateau, Assam-Burma, S Asia, E Asia and SE Asia.

Elatostema obtusum Wedd. [3, 4, 7]

Perennial herbs. W, C, E Nepal: 2600-3200 m. Nepal, W Himalaya, E Himalaya, Tibetan Plateau, Assam-Burma, E Asia and SE Asia.

Elatostema parvum (Blume) Blume ex Miq.

Procris parva Blume; *Elatostema stracheyanum* Wedd.(3, 4)

Perennial herbs. C, E Nepal: 500-2800 m. Nepal, W Himalaya, E Himalaya, E Asia and SE Asia.

Elatostema platyphyllum Wedd. [3, 4, 7]

Elatostema platyphyllum var. *polycephalum* H.Hara (3, 4)

सानो गग्लेटो (Saano gagleto)

Undershrubs or perennial herbs. C, E Nepal: 700-1900 m. Nepal, W Himalaya, E Himalaya, Tibetan Plateau, E Asia and SE Asia.

Use(s): Food (tender shoot and leaves), Fodder.

Elatostema pusillum C.B.Clarke ex Hook.f. [3, 4, 8]

Annual herbs. C, E Nepal: 2300-2900 m. Nepal and E Himalaya.

Elatostema rupestre (Buch.-Ham. ex D.Don) Wedd. [3, 4, 7]

Procris rupestris Buch.-Ham. ex D.Don

Perennial herbs. C, E Nepal: 300-2000 m. Nepal, E Himalaya, E Asia and SE Asia.

Elatostema sessile J.R.Forst. & G.Forst. [3, 4, 7]

var. **sessile**

गग्लेटो (Gagleto)

Gregarious herbs. W, C, E Nepal: 300-3000 m. Nepal, W Himalaya, E Himalaya, E Asia, SE Asia and Africa.

Use(s): Medicine (root), Food (tender shoot, leaves).

var. **polycephalum** Wedd. [3, 4, 8]

Gregarious herbs. C Nepal: 900-2600 m. Nepal, W Himalaya, E Himalaya and E Asia.

Elatostema subincisum Wedd. [3, 4, 7]

Perennial herbs. W, E Nepal: 2500-3000 m. Nepal and E Himalaya.

Girardinia Gaudich.

Girardinia diversifolia (Link) Friis [3, 4, 7, 11]

Urtica diversifolia Link

अल्लो, ठुलो सिस्नु, गैंडा सिस्नु (Allo, Thulo sisnu, Gainda sisnu); Himalayan nettle

Perennial herbs. W, C, E Nepal: 1700-3000 m. Nepal, W Himalaya, E Himalaya, Tibetan Plateau, E Asia, SE Asia and Africa.

Use(s): Fiber (bark), Medicine (root, leaves), Food (young leaves), Construction materials.

Gonostegia Turcz.

Gonostegia hirta (Blume ex Hassk.) Miq. [3, 4, 7, 11]

Pouzolzia hirta Blume ex Hassk.

चिप्ले घाँस, चिप्ले लहरी (Chiple ghaans, Chiple lahari)

Perennial herbs. W, C, E Nepal: 500-2500 m. Nepal, Asia and Australasia.

Use(s): Medicine (whole plant), Food (root), Detergent.

Gonostegia pentandra (Roxb.) Miq. [3, 4, 7]

Urtica pentandra Roxb.

Prostrate or ascending herbs or subshrubs. W, C, E Nepal: 100-900 m. Nepal, W Himalaya, E Himalaya, Assam-Burma, S Asia, E Asia, SE Asia and SW Asia.

Use: Fiber.

Laportea Gaudich.

Laportea bulbifera (Siebold & Zucc.) Wedd. [7]

Urtica bulbifera Siebold & Zucc.; *Laportea terminalis* Wight (3, 4)

पातले सिस्नु (Paatle sisnu)

Perennial herbs. W, C, E Nepal: 700-3500 m. Nepal, E Himalaya, Tibetan Plateau, Assam-Burma, S Asia, E Asia and SE Asia.

Lecanthus Wedd.

Lecanthus peduncularis (Wall. ex Royle) Wedd. [3, 4, 7]

Procris peduncularis Wall. ex Royle

खोले झार (Khole jhaar)

Perennial herbs. W, C, E Nepal: 1200-3200 m. Nepal, E Himalaya, Tibetan Plateau, E Asia, S Asia, SE Asia and Africa.

Use(s): Medicine (whole plant), Food (tender shoot).

Maoutia Wedd.

Maoutia puya (Hook.) Wedd. [3, 4, 7]

Boehmeria puya Hook.

नानी पालो, क्यीनकी (Naani palo, Kyinki)

Shrubs. W, C, E Nepal: 900-1100 m. Nepal, E Himalaya, Tibetan Plateau, E Asia and SE Asia.

Use(s): Medicine (root), Fodder, Fiber (bark).

Oreocnide Miq.

Oreocnide frutescens (Thunb.) Miq. [3, 4, 7]

Urtica frutescens Thunb.; *Villebrunea frutescens* (Thunb.) Blume

subsp. **frutescens**

प्लेथा (Pletha)

Shrubs or small trees. W, C, E Nepal: 500-2600 m. Nepal, W Himalaya, E Himalaya, Tibetan Plateau, Assam-Burma, S Asia, E Asia and SE Asia.

subsp. **occidentalis** C.J.Chen [8]

चिप्ले घाँस (Chiple ghaans)

Small trees. W, C, E Nepal: 1400-2500 m. Nepal, E Himalaya, Tibetan Plateau and E Asia.

Use(s): Medicine (root, leaves), Food, Fishnets, Sacks.

Oreocnide integrifolia (Gaudich.) Miq. [3, 4]

Villebrunea integrifolia Gaudich.

Wild Rhea

Evergreen shrubs or small trees. E Nepal: 300-500 m. Nepal, E Himalaya, Tibetan Plateau, Assam-Burma, E Asia and SE Asia.

Parietaria L.

Parietaria judaica L. [3, 4]

Perennial herbs. E Nepal: 700-800 m. Nepal, Asia, Europe and Africa.

Parietaria micrantha Ledeb. [3, 4, 7, 11]

Annual herbs. W, C Nepal: 2600-4000 m. Nepal, Asia, Europe and Africa.

Pilea Lindl.

Pilea anisophylla (Hook.f.) Wedd. [3, 4, 7, 11]

Pilea anisophylla var. *khasiana* Hook.f.

छाल (Chhaal: Tamang)

Perennial stoloniferous herbs. C, E Nepal: 600-2200 m. Nepal, E Himalaya, Tibetan Plateau, Assam-Burma and E Asia.

Use(s): Medicine (root), Fodder.

Pilea approximata C.B.Clarke [3, 4, 7]

Perennial herbs. E Nepal: 2500-3200 m. Nepal, E Himalaya, Tibetan Plateau and E Asia.

Pilea bracteosa Wedd. [3, 4, 8]

Perennial herbs. C, E Nepal: 500-2800 m. Nepal, W Himalaya, E Himalaya, Tibetan Plateau, Assam-Burma and E Asia.

Pilea clarkei Hook.f. [3, 4]

Herbs. E Nepal: *ca.* 1800 m. Nepal and E Himalaya.

Pilea cordifolia Hook.f. [2, 4, 7]

Perennial stoloniferous herbs. W Nepal: *ca.* 2700 m. Nepal, E Himalaya, Tibetan Plateau and E Asia.

Pilea glaberrima (Blume) Blume [3, 4, 7]

Urtica glaberrima Blume

हरे गग्लेटो (Hare gagleto)

Perennial herbs. C, E Nepal; 500-1300 m. Nepal, E Himalaya, Assam-Burma, S Asia, E Asia and SE Asia.

Use: Food (tender leaves and shoot).

Pilea hookeriana Wedd. [3, 4, 7]

Perennial herbs. E Nepal: 1200-1800 m. Nepal, E Himalaya, E Asia and SE Asia.

Pilea kanaii H.Hara, *J. Jap. Bot.* **49**: 193. 1974. [3, 4, 12, 28, 29]

Annual erect herbs. W, C Nepal: 1500-3600 m. **Endemic to Nepal**.

Central: *Rasuwa*, Sim Chotala – Ramche, *ca.* 1500 m, H. Kanai, H. Hara & H. Ohba 721665 (Holotype: TI; Isotype: TI).

Pilea martinii (H.Lév.) Hand.-Mazz. [3, 4, 8]

Boehmeria martinii H.Lév.

Perennial stoloniferous herbs. W, C, E Nepal: 2000-3200 m. Nepal, E Himalaya, Tibetan Plateau and E Asia.

Pilea oxyodon Wedd.

Pilea bracteosa var. *oxyodon* (Wedd.) H.Hara (3,4,8)

Perennial herbs. C, E Nepal: 1500-2900 m. Nepal, E Himalaya and Tibetan Plateau.

Pilea racemosa (Royle) Tuyama [3, 4, 8, 11]

Procris racemosa Royle

लेंडे साग (Lende saag)

Perennial herbs. W, C, E Nepal: 2000-4000 m. Nepal, W Himalaya, E Asia and Tibetan Plateau.

Use: Food (tender leaves, shoot).

Pilea scripta (Buch.-Ham. ex D.Don) Wedd. [3, 4, 7, 11]

Urtica scripta Buch.-Ham. ex D.Don

Perennial herbs. W, C, E Nepal: 1300-2700 m. Nepal, W Himalaya, E Himalaya, Tibetan Plateau and E Asia.

Use: Fiber.

Pilea symmeria Wedd. [3, 4, 7]3

काम्ले, गग्लेटो (Kaamle, Gagleto)

Perennial herbs. W, C, E Nepal: 1700-3300 m. Nepal, W Himalaya, E Himalaya and Tibetan Plateau.

Use(s): Food (tender shoot, leaves), Fodder.

Pilea ternifolia Wedd. [3, 4]

var. **ternifolia**

Perennial stoloniferous herbs. E Nepal: 2900-3100 m. Nepal, W Himalaya, E Himalaya and Tibetan Plateau.

var. **exmucronata** H.Hara [3, 4]

Perennial stoloniferous herbs. E Nepal: 2400-2600 m. Nepal and E Himalaya.

Pilea umbrosa Wall. ex Blume [3, 4, 7, 11]

var. **umbrosa**

छोल्यांग (Chholyaang)

Perennial stoloniferous herbs. W, C, E Nepal: 1500-2800 m. Nepal, W Himalaya, E Himalaya, Tibetan Plateau and E Asia.

Use: Food (tender shoot, leaves).

var. **obesa** Wall. ex Wedd.

Pilea obesa (Wedd.) Wedd. ex H.Hara (3, 4)

Perennial stoloniferous herbs. C, E Nepal: 2300-2600 m. Nepal, Tibetan Plateau and E Asia.

Poikilospermum Zipp. ex Miq.

Poikilospermum lanceolatum (Trecul) Merr. [3, 4, 7]

Conocephalus lanceolatus Trecul

Woody climbers. E Nepal: *ca.* 700 m. Nepal, E Himalaya, Tibetan Plateau, Assam-Burma, E Asia and SE Asia.

Poikilospermum naucleiflorum (Roxb. ex Lindl.) Chew [3, 4]

Conocephalus naucleiflorus Roxb. ex Lindl.

Mandarin

Woody climbers. C, E Nepal: *ca.* 1600 m. Nepal, E Himalaya, Tibetan Plateau, Assam-Burma and SE Asia.

Pouzolzia Gaudich.

Pouzolzia calophylla W.T.Wang & C.J.Chen

Pouzolzia sanguinea var. *fulgens* (Wedd.) H.Hara (3, 4, 8)

Evergreen shrubs. E Nepal: 400-2800 m. Nepal, E Himalaya, Tibetan Plateau, Assam-Burma and E Asia.

Pouzolzia rugulosa (Wedd.) Acharya & Kravtsova [7, 134, 135]

Boehmeria rugulosa Wedd. (3, 4)

दार, गेथा (Daar, Gethaa)

Small to medium-sized evergreen trees. W, C, E Nepal: 300-2000 m. Nepal, W Himalaya and E Himalaya.

Use(s): Medicine (bark), Food (bark), Fodder, Household utensils (wood).

Pouzolzia sanguinea (Blume) Merr. [3, 4, 7]

Urtica sanguinea Blume; *Boehmeria nepalensis* Wedd.; *Pouzolzia sanguinea* var. *nepalensis* (Wedd.) H.Hara (3, 4, 8)

सेतो नीचे साग (Seto niche saag)

Evergreen shrubs. W, C, E Nepal: 1000-3000 m. Nepal, W Himalaya, E Himalaya, Tibetan Plateau, Assam-Burma, E Asia and SE Asia.

Use(s): Medicine (root), Food.

Pouzolzia zeylanica (L.) Benn. & R.Br. [3, 4, 7]

Parietaria zeylanica L.

चिप्ले झार, बारबेरे (Chiple jhaar, Baarbere)

Perennial herbs. W, C, E Nepal: 200-2400 m. Nepal, W Himalaya, E Himalaya, Assam-Burma, S Asia, E Asia, SE Asia, SW Asia and Africa.

Use(s): Medicine (whole plant), Food.

Procris Comm. ex Juss.

Procris crenata C.B.Rob. [3, 4, 7]

Procris wightiana Wall. ex Wedd.

Perennial herbs or subshrubs. E Nepal: 300-2000 m. Nepal, E Himalaya, Tibetan Plateau, Assam-Burma, S Asia, E Asia, SE Asia and Africa.

Urtica L.

Urtica ardens Link [3, 4, 8]

Himalayan nettle

Annual herbs. W, C, E Nepal: 600-3200 m. Nepal, W Himalaya, E Himalaya, Tibetan Plateau and E Asia.

Urtica dioica L. [3, 4, 7, 11]

सिस्नु, लेक् सिस्नु (Sisnu, Lek sisnu); Nettle, Stinging nettle

Perennial herbs. W, C Nepal: 500-4500 m. Nepal, Asia, Europe, Africa and N America.

Use(s): Medicine (root, aerial part), Food (young shoot, leaves), Fodder, Fiber (bark).

Urtica hyperborea Jacq. ex Wedd. [3, 4, 8, 11]

सिस्नु (Sisnu); Northern nettle

Perennial herbs. W, C Nepal: 4100-5100 m. Nepal, W Himalaya, E Himalaya, Tibetan Plateau, E Asia and N Asia.

Use(s): Medicine (leaves), Food.

Urtica mairei H.Lév. [3, 4]

Perennial herbs. C, E Nepal: 2600-3200 m. Nepal, W Himalaya, E Himalaya, Tibetan Plateau, Assam-Burma and E Asia.

Urtica parviflora Roxb. [7, 11]

Perennial herbs. C, E Nepal: 100-3000 m. Nepal, W Himalaya, E Himalaya, Tibetan Plateau and E Asia.

Use(s): Food (tender leaves), Medicine (root, leaves).

Doubtful/Uncertain Taxa

Boehmeria clidemioides var. **diffusa** (Wedd.) Hand.-Mazz. [100]

Boehmeria depauperata Wedd. (*Boehmeria glomerulifera* Miq.); Shrubs or small trees. C, E Nepal: 500-2200 m. Nepal, E Himalaya, Tibetan Plateau, Assam-Burma, S Asia, E Asia and SE Asia. [3, 4, 7]

Boehmeria nivea (L.) Gaudich. [100]

Debregeasia orientalis C.J.Chen [4]

Elatostema cuspidatum Wight [100]

Elatostema nasutum Hook.f. [100]

Elatostema obtusum var. **polycephalum** Wedd. [8]

Pilea insolens Wedd. [100]

Pilea linearifolia C.J.Chen [100]

Pilea peploides (Gaudich.) Hook. & Arn. [*Bull. B. Sur. Ind.* 20: 44. 1978]

Pilea squamosa C.J.Chen [100]

FAGALES

FAGACEAE

Castanopsis (D.Don) Spach

Castanopsis echinocarpa Miq. [3, 4]

Castanopsis echidnocarpa A.DC.

Small to medium-sized evergreen trees. C Nepal, E Himalaya, Tibetan Plateau, Assam-Burma, S Asia, E Asia and SE Asia.

Castanopsis hystrix Hook.f. & Thomson ex A.DC. [3, 4, 7]

पातले कटुस (Paatale katus)

Evergreen trees. C, E Nepal: 1000-2500 m. Nepal, E Himalaya, Tibetan Plateau, Assam-Burma, E Asia and SE Asia.

Use(s): Food (fruit), Fuel.

Castanopsis indica (J.Roxb. ex Lindl.) A.DC. [3, 4, 7]

Castanea indica J.Roxb. ex Lindl.

ढाले कटुस, ढाल्ने कटुस (Dhaale katus, Dhaalne katus); Indian chestnut

Evergreen trees. W, C, E Nepal: 1000-2900 m. Nepal, E Himalaya, Tibetan Plateau, E Asia and SE Asia.

Use(s): Medicine (bark, resin, young leaf), Food (fruit), Fodder, Fuel, Construction.

Castanopsis lanceifolia (Kurz) Hickel & A.Camus [3, 4, 7]

Castanea lanceifolia Kurz

औले कटुस (Aule katus)

Evergreen trees. E Nepal: *ca.* 300 m. Nepal, E Himalaya, Assam-Burma and SE Asia.

Castanopsis tribuloides (Sm.) A.DC. [3, 4, 7]

Quercus tribuloides Sm.; *Quercus armata* D.Don

मुसुरे कटुस (Musure katus); Hill chestnut

Evergreen trees. W, C, E Nepal: 400-2300 m. Nepal, W Himalaya, E Himalaya, Tibetan Plateau, Assam-Burma, E Asia and SE Asia.

Use(s): Food (seeds), Fodder, Planks.

Cyclobalanopsis Oerst.

Cyclobalanopsis glauca (Thunb.) Oerst. [4]

Quercus glauca Thunb. (3, 7)

फलाँट (Phalaant); Ring cupped Oak

Evergreen trees. W, C, E Nepal: 400-3100 m. Nepal, W Himalaya, E Himalaya, Assam-Burma and E Asia.

Use(s): Timber, Fodder, Tannin (bark).

Cyclobalanopsis lamellosa (Sm.) Oerst. [4, 7]

Quercus lamellosa Sm. (3)

बंगसेट, ठुलो फलाँट (Bangset, Thulo phalaant)

Evergreen trees. C, E Nepal: 1600-2800 m. Nepal, E Himalaya, Tibetan Plateau, Assam-Burma and E Asia.

Use(s): Timber, Fodder, Medicine (bark, acorns), Tannin (bark).

Cyclobalanopsis oxyodon (Miq.) Oerst. [4]

Quercus oxyodon Miq. (3, 7); *Quercus glauca* var. *nakaoi* Kitam. & Horik. (3); *Quercus lineata* Blume

फलाँट (Phalaant)

Evergreen trees. C, E Nepal: 1700-3200 m. Nepal, E Himalaya, Assam-Burma, S Asia, E Asia and SE Asia.

Use: Timber.

Lithocarpus Blume

Lithocarpus elegans (Blume) Hatus. ex Soepadmo [3]

Quercus elegans Blume

Evergreen trees. W, C, E Nepal: 1400-2000 m. Nepal, E Himalaya, Assam-Burma and SE Asia.

Lithocarpus fenestratus (Roxb.) Rehder [3, 4, 7]

Quercus fenestrata Roxb.

अर्खटे (Arkhate)

Evergreen large trees. E Nepal: 1900-2000 m. Nepal, E Himalaya, Tibetan Plateau, Assam-Burma, E Asia and SE Asia.

Use: Fodder.

Lithocarpus grandifolius (D.Don) S.N.Biswas [4, 7]

Quercus grandifolia D.Don; *Lithocarpus spicatus* (Sm.) Rehder & E.H. Wilson; *Quercus spicata* Sm., not Humboldt & Bonpland

अखौंलो (Arkhaulo)

Evergreen medium-sized trees. W, C, E Nepal: 1400-2500 m. Nepal, E Himalaya, Assam-Burma, E Asia and SE Asia.

Use(s): Medicine (seed oil), Food (seeds), Fodder, Firewood.

Lithocarpus pachyphyllus (Kurz) Rehder [3, 4, 7]

Quercus pachyphylla Kurz

बानटे, ठुलो कटुस, सुंगुरे कटुस (Baante, Thulo katus, Sungure katus); Thick-leaved Oak

Evergreen medium-sized trees. E Nepal: 2100-2900 m. Nepal, E Himalaya, Tibetan Plateau, Assam-Burma and E Asia.

Use(s): Food (seeds), Medicine (bark, seeds), Fodder.

Quercus L.

Quercus floribunda Lindl. ex A.Camus [3, 4, 7]

Quercus dilatata Lindl. ex A.DC.

मोरु, बेले-खरमेंदो (Moru, Bele-kharmendo); Moru Oak

Large evergreen trees. W, C Nepal: 2000-2700 m. Nepal, W Himalaya and SW Asia.

Use(s): Fodder, Timber.

Quercus lanata Sm. [3, 4, 7]

Quercus leucotrichophora A.Camus (3)

बाँझ (Baanjh); Woolly leaved Oak

Medium-sized evergreen trees. W, C, E Nepal: 400-2600 m. Nepal, W Himalaya, E Himalaya, Assam-Burma and SE Asia.

Use(s): Medicine (bark, seeds), Food, Fodder, Firewood, Construction, Agricultural implements.

Quercus mespilifolioides A.Camus [3, 4, 7]

Quercus mespilifolia Wall. ex A.DC.

Evergreen trees. E Nepal: 1800-2500 m. Nepal, Assam-Burma and SE Asia.

Quercus semecarpifolia Sm. [3, 4, 7]

Quercus cassura Buch.-Ham. ex D.Don

खस्रु, घेसी (Khasru, Ghesi)

Medium-sized evergreen trees. W, C, E Nepal: 1700-3800 m. Nepal, W Himalaya, E Himalaya, Tibetan Plateau, Assam-Burma and SW Asia.

Use(s): Medicine (bark), Food (sap), Fodder, Fuel, Tannin.

Exotic/Cultivated Taxa

Castanea sativa Mill. [7]

(Sweet chestnut); Deciduous trees, native to Asia Minor and Europe.

Use(s): Food (nuts), Timber, Medicine

Doubtful/Uncertain Taxa

Cyclobalanopsis annulata (Sm.) Oerst. [4, 99]

Lithocarpus arcaula (Buch.-Ham. ex D.Don) C.C.Huang & Y.T.Chang [4, 99]

Lithocarpus listeri (King) Grierson & D.G.Long [4, 99]

Quercus acutissima Carruth. [3, 4]

Quercus griffithii Hook.f. & Thomson ex Miq. [3, 4]

Quercus incana Roxb.

MYRICACEAE

Myrica L.

Myrica esculenta Buch.-Ham. ex D.Don [3, 4, 7]

काफल, कोबशी (Kaafal, Kobashi-Newari); Bayberry, Box myrtle

Medium-sized evergreen trees. W, C, E Nepal: 1200-2300 m. Nepal, W Himalaya, E Himalaya, Assam-Burma, S Asia, E Asia and SE Asia.

Use(s): Food (fruit), Firewood, Medicine/Dye/Fish poison/Tannin (bark).

JUGLANDACEAE

Engelhardia Lesch. ex Blume

Engelhardia spicata Lesch. ex Blume [3, 4]

Engelhardia roxburghiana Lindl. ex Wall.

var. **spicata**

मौवा (Mauwa)

Medium-sized deciduous trees. W, C, E Nepal: 400-1900 m. Nepal, E Himalaya, Tibetan Plateau, S Asia, E Asia and SE Asia.

Use(s): Medicine (bark), Tannin (bark), Fiber, Fish poison (young twig).

var. **acerifolia** (Reinw.) Koord. & Valeton [4, 7]

Pterilema aceriflorum Reinw.

Deciduous trees. E Nepal: 1300-1400 m. Nepal, Assam-Burma, E Asia and SE Asia.

var. **colebrookeana** (Lindl. ex Wall.) Koord. & Valeton [3, 4, 7]

Engelhardia colebrookeana Lindl. ex Wall.

Deciduous trees. W, C, E Nepal: 500-1700 m. Nepal, W Himalaya, E Himalaya, Tibetan Plateau, Assam-Burma, E Asia and SE Asia.

Juglans L.

Juglans regia L.

Juglans kamaonia (C.DC.) Dode; *Juglans regia* var. *kamaonia* C.DC. (3,4,7)

ओखर, तार-का (Okhar; Tar-ka: Amchi); Common walnut

Medium-sized deciduous trees. W, C, E Nepal: 1200-2700 m. Nepal, W Himalaya, E Himalaya and Tibetan Plateau.

Use(s): Food (seeds), Fodder, Furniture, Medicine (bark, leaves, seeds), Dye, Tannin, Fish poison.

IUCN Category: **Near Threatened** (ver 3.1).

Exotic/Cultivated Taxa

Carya illinoinensis (Wangenh.) K.Koch., Native to N America.

CASUARINACEAE

Casuarina equisetifolia L. (Australian pine tree). Planted as avenue tree. Native to SE Asia and Australasia.

BETULACEAE (incl. CORYLACEAE)

Alnus Mill.

Alnus nepalensis D.Don [3, 4, 7]

उतिस (Utis); Nepalese Alder

Medium-sized deciduous trees. W, C, E Nepal: 500-2600 m. Nepal, W Himalaya, E Himalaya, Tibetan Plateau, Assam-Burma, E Asia and SE Asia.

Use(s): Medicine (bark), Fodder, Timber, Dye, Tannin.

Alnus nitida (Spach) Endl. [3, 4, 7]

Clethropis nitida Spach

उतिस (Utis); West Himalayan black Cedar

Medium-sized deciduous trees. W, C Nepal: 2100-2400 m. Nepal and W Himalaya.

Use(s): Medicine (bark), Construction, Furniture, Dye, Tannin.

Betula L.

Betula alnoides Buch.-Ham. ex D.Don [3, 4, 7, 11]

Betula nitida D.Don

लेक पैंयु, सैार (Lekh painyu, Saur); Birch

Medium-sized deciduous trees. W, C, E Nepal: 1200-2600 m. Nepal, W Himalaya, E Himalaya, Assam-Burma, E Asia and SE Asia.

Use(s): Medicine (bark, fruit), Fodder, Timber.

Betula utilis D.Don [3, 4, 7, 11]

Betula bhojpattra Lindl.

भोज पत्र (Bhoj patra); Himalayan birch

Large deciduous trees. W, C, E Nepal: 2700-4300 m. Nepal, E Himalaya, Tibetan Plateau, E Asia and SW Asia.

Use(s): Medicine (bark, resin), Fodder, Timber, Incense.

Carpinus L.

Carpinus faginea Lindl. [3, 4, 7]

Medium-sized deciduous trees. W Nepal: 1500-2100 m. Nepal and W Himalaya.

Carpinus viminea Lindl. [3, 4, 7]

खादिक (Khaadik); Himalayan horn beam

Medium-sized deciduous trees. W, C, E Nepal: 1200-2600 m. Nepal, W Himalaya, E Himalaya, Tibetan Plateau, Assam-Burma, E Asia and SE Asia.

Corylus L.

Corylus ferox Wall. [3, 4, 7]

लेक कटुस (Lek katus); Himalayan hazel nut

Medium-sized deciduous trees. C, E Nepal: 2500-3600 m. Nepal, E Himalaya, Assam-Burma and E Asia.

Corylus jacquemontii Decne. [3, 4, 7]

दाँते ओखर (Daante okhar); Indian tree hazel

Medium-sized deciduous trees. W Nepal: 2400-3000 m. Nepal and W Himalaya.

REFERENCES

1. Hara H, Stearn WT, and Williams LHJ 1978 *An Enumeration of the Flowering Plants of Nepal*, Vol. I, London: The British Museum (Natural History)
2. Hara H and Williams LHJ 1979 *An Enumeration of the Flowering Plants of Nepal*, Vol. II, London: The British Museum (Natural History)
3. Hara H, Chater AO, and Williams LHJ 1982 *An Enumeration of the Flowering Plants of Nepal*, Vol. III, London: The British Museum (Natural History)
4. Press JR, Shrestha KK, and Sutton DA 2000 *Annotated Checklist of the Flowering Plants of Nepal*, London: The Natural History Museum
5. Rajbhandari KR and Baral SR (eds) 2010 *Catalogue of Nepalese Flowering Plants I*, Lalitpur: National Herbarium & Plant Laboratories, Department of Plant Resources, Ministry of Forest and Soil Conservation
6. Rajbhandari KR, Bhattarai KR, and Baral SR (eds) 2011 *Catalogue of Nepalese Flowering Plants Part II*, Lalitpur: National Herbarium & Plant Laboratories, Department of Plant Resources, Ministry of Forest and Soil Conservation
7. Rajbhandari KR, Bhattarai KR, and Baral SR (eds) 2012 *Catalogue of Nepalese Flowering Plants III*, Lalitpur: National Herbarium & Plant Laboratories, Department of Plant Resources, Ministry of Forest and Soil Conservation
8. Rajbhandari KR, Bhatt GD, and Chhetri R (eds) 2015 *Catalogue of Nepalese Flowering Plants Supplement 1*, Lalitpur: National Herbarium & Plant Laboratories, Department of Plant Resources, Ministry of Forest and Soil Conservation

9. Sharma BK 2014 *Bioresources of Nepal*, Kathmandu: Subidhya Sharma

10. Watson MF, Akiyama S, Ikeda H, Pendry C, Rajbhandari KR, and Shrestha KK (eds) 2011 *Flora of Nepal* Vol. 3 (Magnoliaceae – Rosaceae), Edinburgh: Royal Botanic Garden Edinburgh

11. Ohba H, Iokawa Y, and Sharma LR 2008 *Flora of Mustang, Nepal*, Tokyo: Kodansha Scientific

12. Shrestha KK and Press JR 2000 *Catalogue of Type Specimens from Nepal*: A catalogue of type specimens of flowering plants described from Nepal, London: The Natural History Museum

13. Rajbhandari KR 2015 *A Handbook of the Orchids of Nepal*, Kathmandu: Department of Plant Resources, Ministry of Forest and Soil Conservation

14. Bajracharya D, Shrestha KK, and Chaudhary RP 1997 *Garden Flowers: An illustrated guide to indoor and outdoor garden plants in Nepal*, Lalitpur: The King Mahendra Trust for Nature Conservation

15. Gurung K 2009 Essential Oils in Nepal: A Practical Guide to Essential Oils and Aromatherapy, Kathmandu: Himalayan Bio Trade Pvt. Ltd.

16. Sharma I and Adhikari MK (eds) 2003 *Flora of Royal Botanical Garden* (Phanerogams), Kathmandu: Department of Plant Resources

17. Stapleton C 1994 Bamboos of Nepal: *An Illustrated Guide*, Kew: The Royal Botanic Gardens

18. Adhikari B, Pendry CA, Pennington RT, and Milne RI 2012 A Revision of *Berberis* s.s. (Berberidaceae) in Nepal, *Edinburgh Journal of Botany* **69** 447–522

19. Poudel RC, Möller M, Gao L-M, Ahrends A, Baral SR, Liu J, Thomas P, and Li D-Z 2012 Using morphological, molecular and climatic data to delimitate Yews along the Hindu Kush Himalaya and Adjacent Regions, *PLoS One* **7** DOI: 10.1371/journal.pone.0046873

20. Rajbhandari KR, Thapa Magar MS, Kandel DR, and Khanal C 2016 *Plant Resources of Kailali, West Nepal*, Kailali: District Plant Resources Office

21. Sukhorukov AP 2010 Additions to the flora of Nepal, *Newsletter of Himalayan Botany* **43** 14

22. Sukhorukov AP 2014 *Trifolium resupinatum* and *Sium medium* – new species for the flora of Nepal, *Newsletter of Himalayan Botany* **48** 14–17

23. Akiyama S, Ibaragi Y, Katsuyama T, Koba H, Mouri C, and Noshiro S 2010 New records of Berberidaceae, Celastraceae, Cyperaceae, Gramineae, Lauraceae, Moraceae, Rutaceae, and Schisandraceae in Nepal, *Newsletter of Himalayan Botany* **43** 15–16

24. Akiyama S, Amano M, Kurosawa T, Noshiro S, Yonekura K, and Ohba H 2009 New records and treatments of Nepalese flowering plants (4), *Newsletter of Himalayan Botany* **42** 28–34

25. Bhattarai KR and Pathak ML 2015 A New species of *Ziziphus* (Rhamnaceae) from Nepal Himalayas, *Indian Journal of Plant Science* **4** 71–77

26. Rajbhandari KR and Adhikari MK 2009 *Endemic Flowering Plants of Nepal* Part 1. (Acanthaceae – Euphorbiaceae), Kathmandu: Department of Plant Resources, Ministry of Forest and Soil Conservation

27. Rajbhandari KR and Dhungana SK 2010 *Endemic Flowering Plants of Nepal* Part 2. (Gentianaceae – Primulaceae), Kathmandu: Department of Plant Resources, Ministry of Forest and Soil Conservation
28. Rajbhandari KR and Dhungana SK 2011 *Endemic Flowering Plants of Nepal* Part 3. (Ranunculaceae – Zingiberaceae), Kathmandu: Department of Plant Resources, Ministry of Forest and Soil Conservation
29. Rajbhandari KR, Rai SK, and Bhatt GD 2016 Endemic Flowering Plants of Nepal: An Update, *Plant Resources (A Scientific Publication), Bulletin of Department of Plant Resources* **38** 106–144
30. Chaudhary LB and Siwakoti M 2015 A complete list of the species of *Astragalus* L. (Fabaceae) of Nepal with three new records, *International Journal of Current Research in Biosciences and Plant Biology*, **2** 33–44
31. Egan, PA 2011 *Meconopsis autumnalis* and *M. manasluensis* (Papaveraceae), two new species of Himalayan poppy endemic to central Nepal with sympatric congeners, *Phytotaxa* **20** 47–56
32. Schilling T and Watson M 2014 Clarifying the Identities of two Nepalese Mahonia, *The Plantsman* June 2014 94–100
33. Erst AS and Sukhorukov AP 2011 Records of *Ranunculus longicaulis* and *Ranunculus pseudohirculus* (Ranunculaceae) from Nepal, *Taiwania* **56** 218–222
34. Grey-Wilson C 2006 The True Identity of *Meconopsis napaulensis* DC., *Curtis's Botanical Magine* **23** 176-209
35. Ohba H and Akiayama S 2010 Four New Species of Himalayan *Prunus* Subgenus *Cerasus* (Rosaceae-Prunoideae), *Bulletin of the National Museum of Nature and Science Series B,* **36** 133–140
36. Raskoti BB, Bhatt GD, and Ale R 2014 *Nymphaea tetragona* (Nymphaeaceae) a new record for Flora of Nepal, *Banko Janakari* **24** 55–56
37. Yoshida T, Hang S, and Grey-Wilson C 2012 A Revision of *Meconopsis lyrata* (Cummins & Prain) Fedde ex Prain and its Allies, *Curtis's Botanical Magazine* **29** 194–216
38. Siwakoti M and Varma SK 1999 *Plant Diversity of Eastern Nepal: Flora of Plains of Eastern Nepal*, Dehradun: Bishen Singh, Mahendra Pal Singh
39. Chaudhary RP, Uprety Y, Joshi SP, Shrestha KK, Basnet KB, Basnet G, Shrestha KR, Bhatta KP, Acharya KP, and Chettri N 2015 *Kangchenjunga Landscape Nepal: from conservation and development prospectives*, Kathmandu: MoFSC/GoN, RECAST/TU and ICIMOD
40. Rokaya MB, Raskoti BB, Timsina B, and Münzbergová Z 2013 An annotated checklist of the orchids of Nepal, *Nordic Journal of Botany* **31** 511–550
41. Raskoti BB 2015 A new species of *Gastrochilus* and new records for the orchids of Nepal, *Phytotaxa* **233** 179–184
42. Bhandari P, Raskoti BB, and Shrestha KK 2015 *Phreatia elegans* Lindley (Orchidaceae) – a new record for the flora of Nepal, *Pleione* **9** 365–368
43. Kolanowska M, Kras M, and Szlachetko DL 2015 A New Spurless *Habenella* species (Orchidaceae) from Nepal, *Polish Botanical Journal* **60** 27–31
44. Raskoti BB and Kurzweil H 2015 *Odontochilus nandae* (Orchidaceae; Cranichideae; Goodyerinae), a new species from Nepal, *Phytotaxa* **233** 293–297

45. Kolanowska M 2016 A new species of *Habenaria* (Orchidaceae) from Nepal, *Phytotaxa* **261** 092–096

46. Bhandari P, Shakya LR, and Shrestha KK 2016 *Zeuxine membranacea* Lindl. (Orchidaceae) – An addition to the orchid flora of Nepal, *The Japanese Journal of Botany* **91** 317–320

47. Parmer G 2016 *Zephyranthes citrina* Baker (Amaryllidaceae), a new record for flora of Nepal, *Bulletin of the Department of Plant Resources* **38** 14–15

48. Rajbhandhari KR, Bhatt GD, Chhetri R, and Gajurel JP 2016 *Eragrostris cilianensis* (Poaceae), a new record for Nepal, *Bulletin of the Department of Plant Resources* **38** 12–13

49. Rajbhandari KR, Bhatt GD, and Chhetri R 2015 *Paspalidium geminatum* (Forssk.) Stapf (Poaceae), a new record for Nepal, *Bulletin of the Department of Plant Resources* **37** 19–20

50. Jin X, Ikeda H, Yano O, Chen W, and Zhou Y 2015 *Carex staintonii*, a new species of Cyperaceae from Nepal, *The Japanese Journal of Botany* **90** 109–114

51. Global Carex Group 2015 Making *Carex* monophyletic (Cyperaceae, tribe Caricaecae): a new broader circumscription, *Journal of the Linnean Society* **179** 1–42

52. Efimov PR 2016 A Revision of *Platanthera* (Orchidaceae; Orchidoideae; Orchideae) in Asia, *Phytotaxa* **254** 145–147

53. Jha S and Jha PK 2000 Contributions to the flora of Morang district and adjoining areas of Nepal, *Lidia* **5** 25–64

54. Rai SK, Thapa Magar MS, Joshi MD, and Khatri S 2010 *Basics of Gardening*, Kathmandu: Department of Plant Resources

55. Rajbhandari KR, Bhatt GD, and Chhetri R 2015 A new record of *Puccinellia, P. himalaica* (Poaceae) from Nepal, *The Japanese Journal of Botany* **90** 285–287

56. Yano O, Katsuyama T, Ikeda H, Pendry CA, and Rajbhandari KR 2015 A new record of *Eleocharis uniglumis* (Link) Schult. (Cyperaceae) from Nepal, *The Japanese Journal of Botany* **90** 58–60

57. Yano O, Katuyama T, Ikeda H, Watson MF, and Rajbhandari KR 2013 Nomenclature and lectotypification of *Carex henryi* (Cyperaceae) with a new record from Nepal, *Acta Phytotaxonomica et Geobotanica* **63** 143–148

58. Lin T-P, Liu H-Y, Hsieh C-F, and Wang K-H 2016 Complete list of the native orchids of Taiwan and their type information, *Taiwania* **61** 78–126

59. Gajurel JP, Rai SK, Shrestha KK, Scheidegger C, and Shakya LR 2013 *Bulbophyllum griffithii* (Lindley) Reichenbach f. and *Platanthera cumminsiana* (King & Pantling) J. Renz – two new records of orchids from Nepal, *Pleione* **7** 253–257

60. Rai, SK, Gajurel JP, Shrestha KK, Scheidegger C, and Shakya LR 2013 *Peristylus manii* (H.G. Reichenbach) Mukerjee [Orchidaceae] – a new record for Nepal, *Pleione* **7** 250–252

61. Rajbhandhari KR 2005 *Calanthe densiflora* Lindl. (Orchidaceae), a new addition to the flora of Nepal, *The Japanese Journal of Botany* **80** 106–108

62. Raskoti BB and Ale R 2013 Four new additions to the orchid flora of Nepal, *The Japanese Journal of Botany* **88** 378–382

63. Raskoti BB and Ale R 2012 *Liparis ferruginea* Lindley (Orchidaceae) – a new record for flora of Nepal, *Taiwania* **57** 308–311

64. Raskoti BB and Ale R 2009 *Nervilia makinnoni* Duthie and *Nervilia plicata* (Andrews) Schltr. (Orchidaceae): new records for flora of Nepal, *Botanica Orientalis–Journal of Plant Science* **6** 109–110

65. Panigrahi G and Kumar A 1986 A Conspectus of the 74 taxa of *Cotoneaster* Ehrhart ex Medic. (Rosaceae) in India, *Bulletin of the Botanical Survey of India* **28** 63–80

66. Pathak ML and Chhetri DB 2015 *Senna alata* (Linnaeus) Roxburgh (Leguminosae), a new addition to the Flora of Nepal, *Bulletin of the Department of Plant Resources* **37** 21–22

67. Maslin BR 2012 New Combinations in *Senegalia* (Fabaceae: Mimosoideae) for Australia, *Nuytsia* **22** 465–468

68. Maslin BR, Seigler DS, and Ebinger J 2013 New combinations in *Senegalia* and *Vachellia* (Leguminosae: Mimosoideae) for Southeast Asia and China, *Blumea* **58** 39–44

69. Rajbhandari KR, Siwakoti M, Rajbhandary S, and Ghimire SK 2016 *Catalogue of Herbarium specimens deposited at TUCH*, Kirtipur: Central Department of Botany, TU

70. Rajbhandari KR 2016 A New Species of *Poa* (Poaceae) from Mustang, Central Nepal, *The Japanese Journal of Botany* **91** suppl. 175–177

71. Polunin O and Stainton A 1984 *Flowers of the Himalaya*, Delhi: Oxford University Press

72. Stainton A 1988 *Flowers of the Himalaya: A supplement*, Delhi: Oxford University Press

73. Purohit KM and Panigrahi G 1984 Notes on Nomenclature and Taxonomy of the *Spiraea bella* Sims complex (Rosaceae), *The Japanese Journal of Botany* **59** 10

74. Sojak J 2010 Argentina Hill, a genus distinct from Potentilla (Rosaceae), *Thaiszia* **20** 91–97

75. Rajbhandari KR 2016 A new species of Poa (Poaceae) from Mustang, Central Nepal, *The Japanese Journal of Botany* **91** suppl. 175–177

76. Gudkova PD, Pendry CA, Nobis M, and Bayahmetov E 2017 *Stipa krylovii* Roshev. (Poaceae), a new record for the flora of Nepal, *Check list* **13** 1–6

77. Subedi S, Chaudhary RP, Vermeulen JJ, and Gravendeed B 2011 *Panisea panchaseensis* sp. nov. (Orchidaceae) from central Nepal, *Nordic Journal of Botany* **29** 361–365

78. Amatya SM 1996 A New Variety of *Ficus semicordata* (Moraceae) from Nepal, *Novon* **6** 323–324

79. Noltie HJ 1994 Notes relating to the flora of Bhutan: XXVII Eriocaulaceae (*Eriocaulon*), Musaceae (*Musa*), Cyperaceae (*Actinoscirpus*), *Edinburgh Journal of Botany* **51** 169–174

80. Bhatt GD, Poudel RC, Pandey TR, and Basnet R 2017 *Yews of Nepal*, Lalitpur: National Herbarium and Plant Laboratories (NHPL)

81. Raskoti BB, Schuiteman A, Jin W-T, and Jin X-H 2017 A taxonomic revision of *Herminium* L. (Orchidoideae, Orchidaceae), *PhytoKeys* **79** 1–74

82. Raskoti BB, Amatya G, and Shakya LR 2008 *Zeuxine affnis* (Lindl.) Bentham ex Hook.f. (Orchidaceae) – a new record for Nepal, *Pleione* **2** 252–254

83. Raskoti BB, Ale R, and Bhatt GD 2011 *Zeuxine grandis* Seidenf. (Orchidaceae) – a new addition to the orchid flora of Nepal, *Bulletin of the Department of Plant Resources* **33** 5–6

84. Rai SK, Shrestha MR, Maden K, and Shakya LR 2007 *Didymoplexis pallens* Griffith (Orchidaceae): A new record for Nepal, *Pleione* **1** 55–56

85. Subedi A and Shakya LR 2008 *Taeniophyllum scaberulum* Hook.f. (Orchidaceae) – a new record for Nepal, *Pleione* **2** 27–28

86. Shrestha MR, Shakya LR, and Ghimire SK 2007 A new species of *Malaxis* Solander ex Swartz (Orchidaceae) from Dolpa, Nepal, *Pleione* **1** 6–7

87. Wen-Tsai W 2000 Notes on the genus *Clematis* (Ranunculaceae) (I), *Acta Phytotaxonomica Sinica* **38** 305–336

88. Tateishi Y and Maxted N 2002 New Species and Combinations in *Vigna* Subgenus *Ceratropis* (Piper) Verdc. (Leguminosae, Phaseoleae), *Kew Bulletin* **57** 625–633

89. Raskoti BB, Bhatt GD, and Rajbhandari KR 2013 *Acrochaene punctata* L. and *Bhutanthera alpina* (Hand.-Mazz.) Renz (Orchidaceae), new records for flora of Nepal, *Journal of Plant Resources, Department of Plant Resources* **35** 67–69

90. Bajracharya DM 2001 *Eria apertiflora* Summerh. and *E. bipunctata* Lindl. (Orchidaceae), new records for the Nepal Himalaya, *The Japanese Journal of Botany* **76** 297–302

91. Bajracharya DM and Shrestha KK 2006 *Eria clausa* King & Prantl. (Orchidaceae), a new record from Nepal, *The Japanese Journal of Botany* **81** 304–307

92. Raskoti BB and Ale R 2010 *Malaxis biaurita* (Lindley) Kuntze (Orchidaceae): a new record for Nepal, *Himalayan Journal of Sciences* **6** 41–42

93. Raskoti BB, Ale R, and Shrestha K 2012 *Peristylus intrudens* (Ames) Ormerod (Orchidaceae) – a new record for flora of Nepal, *The Japanese Journal of Botany* **87** 137–139

94. Rai SJ, Gajurel JP, Shrestha KK, Scheidegger C, and Shakya LR 2014 *Risleya* (Orchidaceae), a new record for flora of Nepal, *The Japanese Journal of Botany* **89** 409–411

95. Shakya LR, Jin, X-H, Rajbhandari KR, and Ikeda H 2013 *Robinquetia succisa* (Orchidaceae) – a new record for Nepal, *The Japanese Journal of Botany* **88** 191–193

96. Acharya KP, Wood JJ, Berwian R, and Sharma A 2010 *Spiranthes spiralis* (Orchidaceae), a new record for the Nepal Himalaya, *Harvard Papers in Botany* **15** 71–72

97. Bajracharya DM and Shrestha KK 2004 Two new records of *Eria* Lindl. (Orchidaceae) for Nepal, *Himalayan Journal of Sciences* **2** 46–50

98. Malla SB, Rajbhandari AB, Rajbhandari SB, Shrestha TB, Adhikari PM, and Adhikari SR (eds) 1976 *Catalogue of Nepalese Vascular Plants*, Kathmandu: Bulletin of the Department of Medicinal Plants Nepal **7** HMG

99. Wu ZY and Raven PH (eds) 1999 *Flora of China. Vol. 4 (Cycadaceae through Fagaceae)*, Beijing and St. Louis: Science Press and Missouri Botanical Garden Press

100. Wu Z Y, Raven PH, and Hong DY (eds) 2003 *Flora of China. Vol. 5 (Ulmaceae through Basellaceae)*, Beijing and St. Louis: Science Press and Missouri Botanical Garden Press

101. Wu Z Y, Raven PH, and Hong DY (eds) 2001 *Flora of China. Vol. 6 (Caryophyllaceae through Lardizabalaceae)*, Beijing and St. Louis: Science Press and Missouri Botanical Garden Press

102. Wu ZY and Raven PH (eds) 2001 *Flora of China. Vol. 8 (Brassicaceae through Saxifragaceae)*, Beijing and St. Louis: Science Press and Missouri Botanical Garden Press

103. Wu ZY, Raven PH, and Hong DY (eds) 2006 *Flora of China. Vol. 22 (Poaceae)*, Beijing and St. Louis: Science Press and Missouri Botanical Garden Press

104. Wu ZY and Raven PH (eds) 2000 *Flora of China. Vol. 24 (Flagellariaceae through Marantaceae)*, Beijing and St. Louis: Science Press and Missouri Botanical Garden Press

105. Wu ZY, Raven PH, and Hong DY (eds) 2010 *Flora of China. Vol. 10 (Fabaceae)*, Beijing and St. Louis: Science Press and Missouri Botanical Garden Press

106. Tiwari NN and Joshi MP 1990 Medicinal Plants of Nepal, *Journal of Nepal Medical Association* **28** 181–190

107. Nathaniel Wallich's *A numerical list of dried specimens of plants in the East India Company's Museum, collected under the superintendence of Dr. Wallich of the Company's Botanic Garden at Calcutta* (1829 [as 1828] –1849), London

108. Raskoti BB 2012 A new species of *Bhutanthera* (Orchidaceae; Orchidoideae) from Nepal, *Phytotaxa* **62** 57–60

109. Wen-Tsai W 2001 Notes on the genus *Clematis* (Ranunculaceae) (IV), *Acta Phytotaxonomica Sinica* **39** 1–19

110. Podlech D and Xu LR 2004 New species and combinations in *Astragalus* (Leguminosae) from China and the Himalayas, *Novon* **14** 216–226

111. Xu LR and Podlech D 2006 Some new taxa of *Astragalus* L. (Leguminosae) from China and Nepal, *Feddes Repert* **117** 225–231

112. Chaudhary LB 2006 *Astragalus kongrensis* Benth. ex Baker (Fabaceae), a new record for Central and North-West Himalayas, *Taiwania* **51** 237–239

113. Podlech D 1999 New Astragali and Oxytropis from North Africa and Asia, including some new combinations and remarks on some species, *Sendtnera* **6** 135–174

114. Podlech D 2009 Some new species and combination in *Astragalus* and *Phyllolobium* (Fabaceae), *Feddes Repert* **120** 48–58

115. Podlech D 2001 Contribution to the knowledge of the genus *Astragalus* L. (Leguminosae) VII X, *Sendtnera* **7** 163–201

116. Chaudhary LB and Khan ZH 2005 A new species of *Astragalus* L. (Fabaceae) from Indo-Nepal region, *Rheedea* **15** 129–131

117. Miyamoto F, Ohba H, and Subedi MN 2003 Studies of the Flora of Mustang, Central Nepal 1. The Genus *Juncus* (Juncaceae), *The Japanese Journal of Botany* **78** 152–157

118. Stapleton CMA 1994 The Bamboo of Nepal and Bhutan Part II. *Arundinaria*, *Thamnocalamus*, *Borinda*, and *Yushania* (Gramineae: Poaceae, Bambusoideae), *Edinburgh Journal of Botany* **51** 275–295

119. Manandhar R and Bajracharya DM 2012 *Sinarundinaria langtangensis* sp. nov. (Gramineae: Bambuseae) from Nepal Himalaya, *Pleione* **6** 381–383

120. Thapa B and Shrestha KK 1995 Indigofera L. In *Nepal Himalaya* S Rajbhandary and SB Karmacharya (eds), Proceedings of Second National Botanical Conference 1994, Nepal Botanical Society, Kathmandu, Nepal, 26–37 pages

121. Sanjappa M 1992 Notes on identity and nomenclature of some species of *Indigofera* of China, *Taiwania* **37** 1–4

122. Stapleton CMA, Chonghaile GN, and Hodkinson TR 2004 *Sarocalamus*, a new Sino-Himalayan bamboo genus (Poaceae: Bambusoideae), *Novon* **14** 345–349

123. Shrestha K and Tiwari NN 2002 New Report of *Adenanthera pavoniana* Linn. in and its Ethnobotany in Nepal, *Journal of the Natural History Museum* **21** 61–66

124. Sah JP, Singh RL, and Bhatta N 2002 Floristic diversity and use of plants in Ghodaghodi Lake area, Nepal, *Journal of the Natural History Museum* **21** 243–266

125. Bhatt GD, Baral SR, and Rajbhandari KR 2010 New addition of *Ampelocissus barbata* (Wall.) Planchon (Vitaceae) to the Flora of Nepal, *Journal of the Natural History Museum* **25** 339–341

126. Yano O, Ikeda H, Watson MF, Rajbhandari KR, and Ohba H 2010 Cytological studies on Cyperaceae in the Nepal Himalaya I. Chromosome counts of 14 species collected from the Manaslu Himalaya, Central Nepal, *The Japanese Journal of Botany* **85** 157–165

127. Rhodes DG 1975 A Revision of the Genus *Cissampelos*, *Phytologia* **30** 415–484

128. Bajpai O, Srivastava AK, Kushwaha AK, and Chaudhary LB 2014 Taxonomy of a monotypic genus *Indopiptadenia* (Leguminosae-Mimosoideae), *Phytotaxa* **164** 061–078

129. Ikeda H and Ohba H 1999 A systematic revision of Potentilla L. section Leptostylae (Rosaceae) in the Himalaya and adjacent regions, *Bulletin, University Museum, University of Tokyo* **39** 31–118

130. Kitamura S 1954 New species from Nepal collected by Sasuke Nakao, a member of the Japan Himalayan Expedition III, *Acta Phytotaxonomica et Geobotanica* **15** 129–134

131. Maden K 2008 *Alphonsea ventricosa* (Roxb.) Hook.f. & Thomson (Annonaceae): new record for the flora of Nepal, *Pleione* **2** 244–245

132. Chaudhary LB (2006) *Astragalus kongrensis* Benth. ex Baker (Fabaceae), a new record for central and north-west Himalayas, *Taiwania* **51** 237–239

133. Acharya N, Yonekura K, and Suzuki M 2002 A new species and a new variety of *Boehmeria* (Urticaceae) from the Himalaya with special reference to the status of *B. pendulifera* Wedd. ex D.G. Long, *Acta Phytotaxonomica et Geobotanica* **53** 1–9

134. Wilmot-Dear CM and Friis I 2013 The old world species of *Boehmeria* (Urticaceae, tribus Boehmerieae) – A taxonomic revision, *Blumea* **58** 85–216

135. Wilmot-Dear CM, Acharya N, Kravtsova TI, and Friis I 2009 *Pouzolzia rugulosa* transferred from *Boehmeria*, and the distinction between *Boehmeria* and *Pouzolzia* (Urticaceae), *Edinburgh Journal of Botany* **66** 51–64

136. Christenhusz MJM, Reveal JL, Farjon A, Gardner MF, Mill RR, and Chase MW 2011 A new classification and linear sequence of extant gymnosperms, *Phytotaxa* **19** 55–70

137. APG IV 2016 An Update of the Angiosperm Phylogeny Group Classification for the Orders and Families of Flowering Plants: APG IV, *Botanical Journal of the Linnean Society* **181** 1–20

APPENDICES

APPENDIX - 1

Synopsis of Families (Number of Genera and Species)

(Handbook of Flowering plants of Nepal Vol. 1)

Family	Genera	Species	Infraspecific
Acoraceae	1	1	
Alismataceae	4	5	
Amaryllidaceae	5	16	
Annonaceae	7	12	
Aponogetonaceae	1	1	
Araceae	17	42	4
Araucariaceae	1	1	
Arecaceae	10	16	
Aristolochiaceae	2	6	
Asparagaceae	12	38	
Asphodelaceae	3	3	
Berberidaceae	3	24	5
Betulaceae	4	8	
Bromeliaceae	1	1	
Burmanniaceae	1	3	
Butomaceae	1	1	
Buxaceae	2	6	
Cannabaceae	3	8	
Cannaceae	1	1	
Chloranthaceae	1	1	
Circaeasteraceae	1	1	
Colchicaceae	3	4	
Commelinaceae	9	26	
Costaceae	2	2	

Family	Genera	Species	Infraspecific
Crassulaceae	8	45	3
Cupressaceae	3	9	
Cycadaceae	1	2	
Cyperaceae	21	220	14
Daphiphyllaceae	1	2	
Dilleniaceae	1	4	
Dioscoreaceae	1	13	
Elaeagnaceae	2	7	1
Ephedraceae	1	4	
Eriocaulaceae	1	13	
Fabaceae	93	322	28
Fagaceae	4	16	
Ginkgoaceae	1	1	
Gnetaceae	1	1	
Grossulariaceae	1	7	
Haloragaceae	1	1	
Hamamelidaceae	1	1	
Hydrocharitaceae	6	10	
Hypoxidaceae	2	5	
Iridaceae	2	7	
Juglandaceae	2	2	2
Juncaceae	2	48	
Juncaginaceae	1	2	
Lardizabalaceae	2	3	
Lauraceae	11	47	1
Liliaceae	10	22	3
Magnoliaceae	1	9	
Marantaceae	1	1	
Melanthiaceae	3	6	
Menispermaceae	7	15	1
Moraceae	6	48	4
Musaceae	2	2	
Myricaceae	1	1	
Myristicaceae	3	3	

Family	Genera	Species	Infraspecific
Nartheciaceae	1	4	
Nelumbonaceae	1	1	
Nymphaeaceae	2	6	1
Orchidaceae	112	480	7
Paeoniaceae	1	1	
Pandanaceae	1	3	
Papaveraceae	9	83	5
Pinaceae	6	11	3
Piperaceae	2	10	
Poaceae	134	446	9
Podocarpaceae	1	1	
Polygalaceae	3	13	1
Pontederiaceae	2	3	
Potamogetonaceae	3	11	
Proteaceae	2	2	
Ranunculaceae	20	167	24
Rhamnaceae	10	31	1
Rosaceae	28	162	10
Sabiaceae	2	9	
Saururaceae	1	1	
Saxifragaceae	7	108	
Schisandraceae	2	4	
Smilacaceae	2	17	2
Taxaceae	1	3	
Tofieldiaceae	1	1	
Trochodendraceae	1	1	
Typhaceae	1	2	
Ulmaceae	2	4	
Urticaceae	17	65	10
Vitaceae	9	37	3
Xyridaceae	1	2	
Zingiberaceae	11	41	4
Zygophyllaceae	1	1	

APPENDIX - 2

Index of Endemic plants of Nepal

(Handbook of Flowering plants of Nepal Vol. 1)

Family	Latin name	Distribution
Amaryllidaceae (1/1)	**Allium hypsistum** Stearn	W,C
Asparagaceae (1/1)	**Asparagus penicillatus** H.Hara	W,C
Berberidaceae (1/2)	**Berberis mucrifolia** Ahrendt	W,C,E
	Berberis pendryi Bh.Adhikari	C
Crassulaceae (2/4)	**Rhodiola nepalica** (H.Ohba) H.Ohba [*Sedum nepalicum* H.Ohba]	W,C
	Rosularia marnieri (Raym.-Hamet ex H.Ohba) H.Ohba [*Sedum marnieri* Raym.-Hamet ex H.Ohba]	W,C
	Sedum obtusipetalum subsp. **dandyanum** Raym.-Hamet ex H.Ohba	C
	Sedum pseudomulticaule H.Ohba	C
Cyperaceae (1/7)	**Carex esbirajbhandarii** (Rajbh. & H.Ohba) O.Yano	C,E
	Carex gandakiensis Katsuy.	W,C,E
	Carex himalaica T.Koyama	E
	Carex mallae (Rajbh. & H.Ohba) O.Yano	W,C
	Carex rhombifructus Ohwi	W,C
	Carex rufulistolon T.Koyama	W,C
	Carex staintonii X.F.Jin, H.Ikeda & O.Yano	C
Elaeagnaceae (1/1)	**Elaeagnus tricholepis** Momiy.	C
Eriocaulaceae (1/5)	**Eriocaulon exsertum** Satake	E
	Eriocaulon kathmanduense Satake	C

Family	Latin name	Distribution
	Eriocaulon obclavatum Satake	C
	Eriocaulon staintonii Satake	C,E
	Eriocaulon trisectoides Satake	C,E
Fabaceae (7/22)	**Astragalus barclayanus** Podlech	C
	Astragalus chateri Vassilcz.	W,C
	Astragalus jumlaensis Podlech	W
	Astragalus lobbichleri Podlech	C
	Astragalus nakaoi Kitam.	C
	Astragalus nepalensis Podlech	C
	Astragalus notabilis Podlech	C
	Astragalus poluninii Podlech	W
	Astragalus pseudorigidulus Podlech	W
	Astragalus stipulatus var. **phulchokiensis** H.Ohashi	C
	Colutea multiflora Shap. ex Ali	C,E
	Crotalaria kanaii H.Ohashi	W,C
	Hedysarum manaslense (Kitam.) H.Ohashi var. **manaslense** [*Astragalus manaslensis* Kitam.]	C
	Millettia nepalensis R.Parker	W
	Oxytropis arenae-ripariae Vassilcz.	W
	Oxytropis fasciculiflorum Vassilcz.	W,C
	Oxytropis graminetorum Vassilcz.	W,C
	Oxytropis morenarum Vassilcz.	W,C
	Oxytropis nepalensis Vassilcz.	C,E
	Oxytropis torrentium Vassilcz.	W
	Oxytropis williamsii Vassilcz.	W,C
	Rhynchosia nepalensis H.Ohashi & Tateishi	W
Iridaceae (1/1)	**Iris staintonii** H.Hara	C
Juncaceae (1/2)	**Juncus ganeshii** Miyam. & H.Ohba	C,E
	Juncus mustagensis Miyam. & H.Ohba	C
Lauraceae (2/2)	**Litsea doshia** (D.Don) Kosterm. [*Tetranthera doshia* D.Don]	W,C,E
	Machilus pubescens Blume	C
Orchidaceae (15/22)	**Bhutanthera fimbriata** Raskoti	C
	Bulbophyllum nepalense Raskoti & Ale	C

Family	Latin name	Distribution
	Eria annapurnensis L.R.Shakya & M.R.Shrestha	C
	Eria baniae Bajrach., L.R.Shakya & Chettri	C
	Eria nepalensis Bajrach. & K.K.Shrestha	C
	Eria pokharensis Bajrach., Subedi & K.K.Shrestha	C
	Gastrochilus nepalensis Raskoti	C
	Habenaria nepalensis Kolan.	C
	Habenella wallichii Kolan.	C
	Herminium hongdeyuanii Raskoti	C
	Liparis langtangensis Raskoti & Ale	C
	Liparis olivacea Lindl.	C
	Malaxis dolpensis M.R.Shrestha, L.R.Shrestha & Ghimire	W
	Malaxis monophyllos var. **obtusa** Tsukaya & H.Okada	E
	Malaxis tamurensis Tuyama	E
	Neottia chandrae Raskoti	C
	Neottia nepalensis (N.P.Balakr.) Szlach.	C
	Oberonia nepalensis L.R.Shakya & R.P.Chaudhary	C
	Odontochilus nandae Raskoti & H.Kurzweil	C
	Oreorchis porphyranthes Tuyama	C
	Pleione coronaria P.J.Cribb & C.Z.Tang	C
	Sunipia nepalensis Raskoti & Ale	C
Papaveraceae (2/24)	**Corydalis calycina** Lidén	W,C
	Corydalis clavibracteata Ludlow	W
	Corydalis elegans subsp. **robusta** Lidén	W
	Corydalis megacalyx Ludlow	W,C,E
	Corydalis simplex Lidén	W
	Corydalis spicata Lidén	W
	Corydalis stipulata Lidén	C
	Corydalis terracina Lidén	W,C
	Corydalis uncinata Lidén	W,C
	Corydalis uncinatella Lidén	W,E
	Meconopsis autumnalis P.Egan	C
	Meconopsis chankheliensis Grey-Wilson	W
	Meconopsis dhwojii G.Tayl. ex Hay	C,E

Family	Latin name	Distribution
	Meconopsis ganeshensis Grey-Wilson	C
	Meconopsis gracilipes G.Tayl.	C
	Meconopsis grandis subsp. **jumlaensis** Grey-Wilson	W
	Meconopsis lamjungensis T.Yoshida	C
	Meconopsis manasluensis P.Egan	C
	Meconopsis napaulensis DC.	C
	Meconopsis regia G.Tayl.	W,C
	Meconopsis simikotensis Grey-Wilson	W
	Meconopsis staintonii Grey-Wilson	C
	Meconopsis taylorii L.H.J.Williams	C
Poaceae (10/18)	**Elymus nepalensis** (Melderis) Melderis	W,C
	Eulaliopsis sykesii Bor	W
	Festuca eriobasis H.Scholz	C
	Festuca nepalica E.B.Alexeev	W,C
	Festuca poluninii E.B.Alexeev	W
	Himalayacalamus asper Stapleton	C
	Himalayacalamus cupreus Stapleton	C
	Himalayacalamus fimbriatus Stapleton	W,C
	Himalayacalamus planatus Stapleton	C
	Himalayacalamus porcatus Stapleton	C
	Poa hideaki-ohbae Rajbh.	C
	Poa imperialis Bor	E
	Poa langtangensis Melderis	C
	Poa muktinathensis Rajbh.	C
	Saccharum williamsii (Bor) Bor	C
	Sinarundinaria langtangensis R.Manandhar & Bajracharya	C
	Stipa staintonii Bor	W,C
	Thamnocalamus chigar (Stapleton) Stapleton	W,C
	Yushania emeryi (Stapleton) Demoly	E
Ranunculaceae (5/20)	**Aconitum amplexicaule** Lauener	W,C
	Aconitum angulatum Tamura	C,E
	Aconitum bhedingense Lauener	C,E
	Aconitum deltoideum Lauener	E

Family	Latin name	Distribution
	Aconitum dhwojii Lauener	C
	Aconitum poluninii Lauener	W,E
	Aconitum staintonii Lauener	E
	Aconitum tabatae Tamura	W
	Aconitum tamuranum Lauener	W
	Aconitum williamsii Lauener	W,C
	Anemone fuscopurpurea H.Hara	E
	Clematis bracteolata Tamura	W
	Clematis phlebantha L.H.J.Williams	W,C
	Clematis staintonii W.T.Wang	C
	Delphinium himalayae Munz	W,C
	Delphinium unifolium Tamura	C
	Delphinium williamsii Munz	W
	Oxygraphis nepalensis Tamura	C
	Ranunculus himalaicus Tamura	W
	Ranunculus makaluensis Kadota	E
Rosaceae (6/13)	**Geum elatum** forma **rubrum** Ludlow	W,C
	Potentilla cardotiana subsp. **nepalensis** H.Ikeda & H.Ohba	E
	Potentilla eriocarpa var. **major** Kitam.	C
	Potentilla makaluensis H. Ikeda & H.Ohba	E
	Potentilla peduncularis var. **ganeshii** H.Ikeda & H.Ohba	C
	Potentilla turfosoides H.Ikeda & H.Ohba	E
	Prunus himalaica Kitam.	C
	Prunus jajarkotensis H.Hara	W
	Prunus taplejungnica H.Ohba & S.Akiyama	E
	Prunus topkegolensis H.Ohba & S.Akiyama	E
	Rubus × seminepalensis Naruh.	W,C,E
	Sibbaldia emodi H.Ikeda & H.Ohba	C,E
	Sorbus sharmae M.F.Watson, V.Manandhar & Rushforth	C,E
Saxifragaceae (1/21)	**Saxifraga alpigena** Harry Sm.	W,C
	Saxifraga amabilis H.Ohba & Wakab.	C,E

Family	Latin name	Distribution
	Saxifraga cinerea Harry Sm.	W,C
	Saxifraga excellens Harry Sm.	C,E
	Saxifraga ganeshii H.Ohba & S.Akiyama	C
	Saxifraga harae H.Ohba & Wakab.	E
	Saxifraga hypostoma Harry Sm.	W,C,E
	Saxifraga jaljalensis H.Ohba & S.Akiyama	E
	Saxifraga lowndesii Harry Sm.	C
	Saxifraga mallae H.Ohba & Wakab.	C,E
	Saxifraga micans Harry Sm.	C
	Saxifraga mira Harry Sm.	W
	Saxifraga namdoensis Harry Sm.	C
	Saxifraga neopropagulifera H.Hara	W,C
	Saxifraga poluniniana Harry Sm.	W,C
	Saxifraga rhodopetala Harry Sm.	W
	Saxifraga rolwalingensis H.Ohba	C,E
	Saxifraga roylei Harry Sm.	W,C
	Saxifraga staintonii Harry Sm.	C
	Saxifraga williamsii Harry Sm.	W,C,E
	Saxifraga zimmermannii Baehni	C
Urticaceae (1/1)	**Pilea kanaii** H.Hara	W,C
Zingiberaceae (1/3)	**Roscoea ganeshensis** Cowley & W.J.Baker	C
	Roscoea nepalensis Cowley	W,C
	Roscoea tumjensis Cowley	C,E

Doubtful endemic flowering plants of Nepal
(Handbook of Flowering plants of Nepal Vol. 1)

Family	Latin name	Distribution
Cyperaceae	**Cyperus trisulcus** D.Don	C
Elaeagnaceae	**Elaeagnus caudata** var. **aliena** Momiy.	E
Fabaceae	**Caragana campanulata** Vass.	C
	Indigofera trifoliata var. **nepalensis** H.Ohashi	E
	Shuteria involucrata var. **vestita** (Wight & Arn.) H.Ohashi	C

Family	Latin name	Distribution
Menispermaceae	**Cissampelos nepalensis** D.G.Rhodes	W,C,E
Pinaceae	**Abies spectabilis** var. **langtangensis** Silba	C
	Pinus wallichiana var. **manangensis** H.Ohba & M.Suzuki	C
	Picea smithiana var. **nepalensis** Franco	W
Ranunculaceae	**Aconitum balangrense** Lauener	W,C,E
Rhamnaceae	**Rhamnus virgata** var. **flavidus** (Momiy.) Momiy.	C
Zingiberaceae	**Hedychium spicatum** var. **trilobum** (Wall. ex Roscoe) Wall.	C

APPENDIX - 3

Index of Endangered and Threatened Plants of Nepal

(Handbook of Flowering plants of Nepal Vol. 1)

S.N.	Family	Latin name	*CITES Appendices	**IUCN Red List category	Distri-bution
1	Berberidaceae	*Podophyllum hexandrum*	II		W,C,E
2	Cycadaceae	*Cycas pectinata*	II	Vulnerable	E
3	Dioscoreaceae	*Dioscorea deltoidea*	II		W,C,E
4	Fabaceae	*Dalbergia latifolia*		Vulnerable	W,C,E
5		*Indopiptadenia oudhensis*		Near Threatened	W
6		*Pterocarpus marsupium*		Vulnerable	W
7		*Saraca asoca*		Vulnerable	C,E
8		*Sophora wightii* (?)		Endangered	
9	Gnetaceae	*Gnetum montanum*	III	LC	CE
10	Juglandaceae	*Juglans regia*		Near Threatened	W,C,E
11	Orchidaceae	*Cypripedium cordigerum*	II	Vulnerable	W,C
12		*Cypripedium elegans*	II	Endangered	C,E
13		*Cypripedium himalaicum*	II	Endangered	W,C,E
14		*Gastrochilus calceolaris*	II	Critically Endangered	W,C,E
15		*Gastrodia elata*	II	Vulnerable	W,C,E
16		*Malaxis muscifera*	II	Vulnerable	W,C,E

S.N.	Family	Latin name	*CITES Appendices	**IUCN Red List category	Distribution
17		*Paphiopedilum fairrieanum*	I	Critically Endangered	E
18		*Paphiopedilum venustum*	I	Endangered	E
19	Papaveraceae	*Meconopsis regia*	III		W,C
20	Podocarpaceae	*Podocarpus neriifolius*	III	LC	C,E
21	Ranunculaceae	*Aconitum heterophyllum*		Endangered	W,C
22		*Aconitum violaceum*		Vulnerable	W,C
23	Rosaceae	*Sorbus wallichii*		Lower Risk	W,C,E
24	Taxaceae	*Taxus contorta*	II	Endangered	W,C
25		*Taxus wallichiana*	II	Endangered	W,C,E
26	Trochodendraceae	*Tetracentron sinense*	III		E
27	Ulmaceae	*Ulmus wallichiana*		Vulnerable	W,C
28	Urticaceae	*Boehmeria siamensis*		Near Threatened	E

Notes: All orchids of Nepal are protected under CITES II, except *Paphiopedilum fairrieanum* and *Paphiopedilum venustum.*

*CITES Appendices (https://www.cites.org/eng/app/appendices.php)

** IUCN Red List Category (http://www.iucnredlist.org/search)

APPENDIX - 4

Index to Vernacular Names (Transliteration)

■■■■

APPENDIX - 5

Index to Vernacular Names (Devanagari)

■■■■

APPENDIX - 6

Index to Common English Names

APPENDIX 7

Index to Scientific Names

G

L

P

Y

Z